Active Array Antennas for High Resolution Microwave Imaging Radar

Jiaguo Lu · Wei Wang · Xiaolu Wang ·
Yongxin Guo

Active Array Antennas for High Resolution Microwave Imaging Radar

国防工業出版社
National Defense Industry Press

Springer

Jiaguo Lu
Department of Science and Technology
East China Research Institute of Electronic Engineering
Hefei, China

Xiaolu Wang
Research Center for Microwave and Antenna Technology
East China Research Institute of Electronic Engineering
Hefei, China

Wei Wang
Research Center for Microwave and Antenna Technology
East China Research Institute of Electronic Engineering
Hefei, China

Yongxin Guo
Department of Electrical and Computer Engineering
National University of Singapore
Singapore, Singapore

ISBN 978-981-99-1474-6 ISBN 978-981-99-1475-3 (eBook)
https://doi.org/10.1007/978-981-99-1475-3

Jointly published with National Defense Industry Press
The print edition is not for sale in China (Mainland). Customers from China (Mainland) please order the print book from: National Defense Industry Press.

This Springer imprint is published by the registered company Springer Nature Singapore Pte Ltd.
The registered company address is: 152 Beach Road, #21-01/04 Gateway East, Singapore 189721, Singapore

Preface

Human beings and the earth they live on are facing severe challenges. The development of advanced earth observation radar technology is of great significance to national security, ecological monitoring, and disaster monitoring. For a long time, active phased array antennas have been hotspots and challenging in the research of the earth observation radar systems. The purpose of writing this book is to enable readers to systematically, comprehensively, and deeply understand and master the basic concepts, working principles, and types of active array antennas, understand and master the working modes, architecture compositions, and analysis methods of active array antennas and the differences with conventional active phased array antennas, as well as the ideas, methods and special considerations in the design and research of active array antennas.

On the basis of discussing and introducing the basic principles of active array antennas, array synthesis and analysis methods, and modeling techniques for antenna radiation characteristics, this book systematically illustrates the architectures, analysis methods, and engineering practices of active array antennas with the goal of being low profile, high efficiency, and light weight to realize broadband, multi-band, multi-polarization, and shared-aperture properties, and studies digital array antennas, microwave photonic array antennas, active package antennas, and other hot technologies.

This book is divided into 9 chapters. Chapter 1 introduces the characteristics of high-resolution microwave imaging radar and active array antennas, puts forward the new concept of "antenna array microsystems," and discusses the outlooks of the new technologies and development directions of active array antennas; Chap. 2 analyzes linear arrays, planar arrays, sparse arrays, and beamforming optimization techniques based on the application scenarios by using microwave imaging; Chap. 3 analyzes the influence of active array antenna error factors, introduces antenna measurement techniques, and provides rapid measurement of microwave imaging radar two-dimensional phased array antennas and precise modeling technology; Chap. 4 starts from the mechanism of antenna instantaneous broad bandwidth, analyzes the configuration methods of real-time delay lines, and introduces the design methods and experimental results of common microwave delay components in detail; Chap. 5 focuses on the research of "tile" array modules, miniaturization of chip transceiver components, and three-dimensional heterogeneous integration methods on the basis of discussing the low profile, high efficiency and light weight of active array antennas; Chap. 6 introduces the requirements and implementation methods of broadband multi-band multi-polarization shared-aperture antennas, focuses on the multi-polarization/multi-band shared-aperture technologies of microstrip patches and waveguide slots antennas, and introduces the triple-band dual-polarization shared-aperture antennas; Chap. 7 analyzes the coupling and mutual interference mechanism of multiple physical quantities at a small scale, studies the parasitic effects between multiple parameters on the basis of introducing the classification of packaged antennas and broadband packaged antenna elements, and explores embedded devices to realize the miniaturization, light weight and high integration of microwave passive devices. Chapter 8 introduces the basic principle of digital array antennas, DDS spectrum characteristic analysis methods, digital signal generation based on DDS spurious suppression, digital sampling, digital down-conversion, and other digital receiving technologies, researches on reducing the noise figure of digital array antenna systems and improving the technical approach to the dynamic range of the system, and proposes a distributed frequency source design idea beyond the traditional frequency synthesizer; Chap. 9 studies and analyzes microwave photonic digital array antennas and optically controlled phased array antennas, and elaborates on the solution to broadband active array antennas. Basic principles and implementation methods of optical real-time delay, microwave signal modulation and demodulation, optical analog-to-digital conversion, and microwave photon filtering.

The authors of this book have published more than 100 academic papers on active array antennas. These papers are the outcomes and findings of the author's research work on active array antennas for more than 30 years. The authors are closely aiming at the frontier of high-resolution earth observation microwave imaging radar technology, have been engaged in the research of synthetic aperture radar systems and active array antennas, and have rich professional knowledge and practical experience. Therefore, this book can be used as a reference for engineering and technical design and scientific researchers engaged in high-resolution synthetic aperture radar, phased array radar, and other new system phased array antenna technologies, and also has reference value for teachers and students of related majors in colleges and universities.

Chapters 1, 7, 8, and 9 of this book were written by Jiaguo Lu, Chaps. 2, 3, and 6 were written by Wei Wang, and Chaps. 4 and 5 were written by Xiaolu Wang. Yongxin Guo revised the full text and translated it into English. The whole book is planned and drafted by Jiaguo Luo and Yongxin Guo. Thanks to National Defense Industry Press and Springer Nature for their support, as well as the responsible editor for their hard work.

Hefei, China
December 2022

Jiaguo Lu

About This Book

This book addresses the "invisible and indistinguishable" technical challenges encountered in the high-resolution microwave imaging radar's earth observation by focusing on the two elements of "frequency and polarization," and researching and discussing the analysis, optimization, and design methods of active array antennas for high-resolution microwave imaging Radar. On the basis of discussing and introducing the basic principles, analysis methods, and performance parameters of active array antennas, aiming at low profile, high efficiency, and light weight, it systematically illustrates the realization of broadband, multi-band, multi-polarization, and shared-aperture architectures, analysis methods and engineering practices for active array antennas, studies hot technologies such as digital array antennas, microwave photonic array antennas, and active package antennas, proposes that the advanced stage of "active array antennas" is the new concept of "antenna array microsystems," and discusses the outlook of the new technologies and development directions of active array antennas.

This book can be used as a teaching reference book for senior undergraduates and postgraduates majoring in radar, communications, microwave, and antennas, and can also be used as a reference for relevant professional scientific research and engineering technicians.

Contents

Acronyms

ADC	Analog to Digital Converter
AGC	Automatic Gain Control
AIP	Antenna in Package
AOC	Antenna on Chip
APD	Avalanche Photo Diode
ASK	Amplitude-shift key
AUT	Antenna under Test
BGA	Ball Grid Array
CDR	Compression Dynamic Range
CGA	Cylinder Grid Array
CPW	Coplanar Waveguide
CWDM	Coarse Wavelength Division Multiplexer
D	Directivity
DAC	Digital to Analog Converter
DAM	Digital Array Module
DBC	Direct Bond Copper
DBF-SAR	Digital Beam Forming Synthetic Aperture Radar
DDC	Digital Down Conversion
DDFS	Direct Digital Frequency Synthesis
DDS	Direct Digital Synthesis
DDWS	Direct Digital Waveform Synthesis
DFB	Distributed Feedback Laser
DNL	Differential Nonlinearity
DTR	Digital Transceiver
DWDM	Dense Wavelength Division Multiplexer
EDFA	Erbium Doped Fiber Amplifier
EIRP	Equivalent Isotropic Radiation Power
FET	Field Effect Transistor
FM	Frequency Modulated
FNBW	First-Null Beam Width
FO-WLCSP	Fan Out-Wafer Level Chip Scale Packaging

FRA	Fiber Raman Amplifier
FSK	Frequency-shift Keying
GMTI	Ground moving target indication
HBT	Heterojunction Bipolar Transistor
HEMT	High Electron Mobility Transistor
HIC	Hybrid Integrated Circuit
HMSIW	Half Model Substrate Integrated Waveguide
HPA	High Power Amplifier
HTCC	High Temperature Co-fired Ceramic
IDS	Interface Data Sheet
INL	Integral Nonlinearity
INSAR	Interferometric Synthetic Aperture Radar
IPD	Integrated Passive Device
ISAR	Inverse-Synthetic-Aperture-Radar
LFM	Linear Frequency Modulation
LGA	Land Grid Array
LNA	Low Noise Amplifier
LTCC	Low Temperature Co-fired Ceramic
MCM	Multi Chip Module
MEMS	Micro-Electromechanical Systems
MESFET	Metal Semiconductor Field Effect Transistor
MIC	Monolithic Integrated Circuit
MIM	Mental Insulator-Mental
MLPCB	Multi-Layer Printed Circuit Board
MMIC	Monolithic Microwave Integrated Circuit
MPL	Microwave Photonic Link
NCO	Numerical Control Oscillator
NEσ0	Noise Equivalent Sigma Zero
NLFM	Nonlinear frequency modulation signal
OEST	Ohmic Electrode Sharing Technology
OTTD	Optical True Time Delay
PA	Power Amplifier
PCB	Printed Circuit Board
PGA	Pin Grid Array
PIN	Positive Intrinsic Negative
PLF	Polarization Loss Factor
PM	Phase Modulation
PSK	Phase-shift Keying
PSO	Particle Swarm Optimization
QFN	Quad Flat No-lead Package
RDL	Redistribution Layer
RF	Radio Frequency
RF-MEMS	Radio Frequency Microelectro Mechanical Systems
RIN	Relative Intensity Noise
RMS	Root Mean Square

SAM	Scalable Array Module
SAR	Synthetic Aperture Radar
ScanSAR	Scanning Synthetic Aperture Radar
SFDR	Spurious Free Dynamic Rang
SIC	Semiconductor Integrated Circuit
SIP	System In Package
SIW	Substrate Integrated Waveguide
SLL	Side Lobe Level
SNR	Signal-Noise Ratio
SOA	Semiconductor Optical Amplifier
SOC	System on Chip
SOI	Silicon-On-Insulator
SPDT	Single Pole Double Throw
SpotlightSAR	Spotlight Synthetic Aperture Radar
SRF	Self Resonance Frequency
STC	Sensitivity Time Control
StripSAR	Strip Synthetic Aperture Radar
T/R	Transmit / Receive
TGV	Through Glass Via
T-SAM	Tile Scalable Array Module
TSV	Through Silicon Via
TTD	True Time Delay
TTDL	True Time Delay Line
UWB	Ultra-Wide Band
VCSEL	Vertical Cavity Surface Emitting Laser
VIC	Vertical Interdigital Capacitor
VSWR	Voltage Standing Wave Ratio
WDM	Wavelength Division Multiplexing
WLP	Wafer level packaging

Chapter 1
Introduction

1.1 High-Resolution Microwave Imaging Radar

Microwave imaging radars can be categorized as either active or passive, with Synthetic Aperture Radar (SAR) being the most common active type. SAR is a typical active sensor and one of the important means for high-resolution earth observation. As shown in Fig. 1.1, the operating wavelengths are usually located in meter wave, microwave, millimeter wave, and submillimeter wave, and the side-view operation mode is generally adopted. Compared with other passive high-resolution sensors, such as visible imagers or infrared sensors, SAR has demonstrated obvious advantages in reconnaissance and surveillance performance. Its main features include (1) imaging is not limited by natural conditions, such as day and night, weather, etc., and can work 24/7 and in all-weather; (2) selecting the appropriate radar working wavelength can penetrate certain shelters to find the target; and (3) radar image resolution has nothing to do with wavelength, flight height, and radar range. Therefore, SAR is an essential milestone in the development of radar. Furthermore, SAR imaging technology enables the radar to determine the position and motion parameters of the observed objects and obtain images of targets and scenes. In addition, the SAR's capability to effectively differentiate moving targets from stationary backgrounds has contributed to its widespread adoption and popularity [1].

Because of the combination of high resolution, penetrability, all-weather, and all-day operation capabilities of SAR, these characteristics have established the vital position of SAR in the fields of intelligence reconnaissance, target reconnaissance, surveillance, and remote sensing technology, making it an indispensable and irreplaceable means of obtaining information on the earth's surface today.

SAR is an advanced surveying and mapping sensor. The surveying and mapping tasks include rapidly drawing and repairing basic surveying and mapping maps (topographic map scales of 1:10,000 and 1:50,000) for overseas areas and "hot spots." Spaceborne SAR is also an accurate and fast surveying and mapping method that processes SAR images to provide surveying and mapping for various purposes.

J. Lu et al., *Active Array Antennas for High Resolution Microwave Imaging Radar*,
https://doi.org/10.1007/978-981-99-1475-3_1

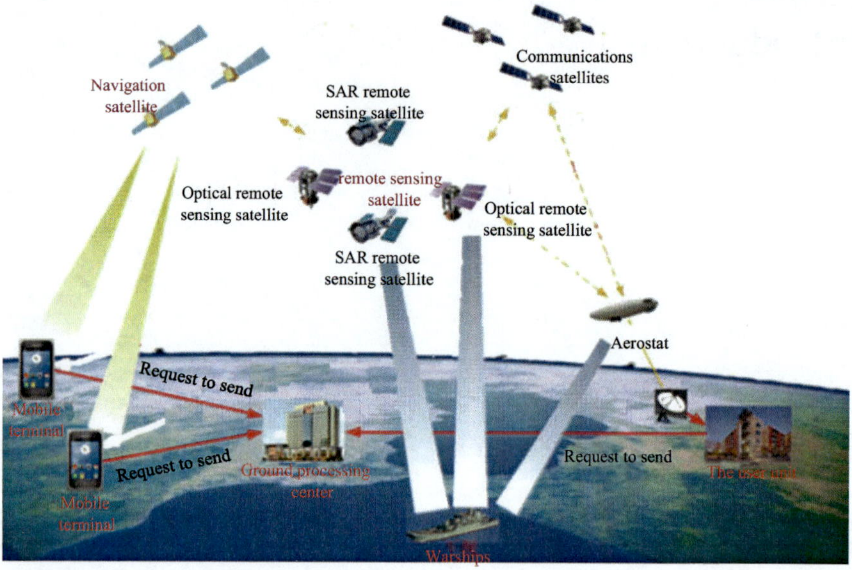

Fig. 1.1 Schematic diagram of high-resolution Earth observation system

As one of the most critical remote sensing and imaging methods, SAR has been widely used in resource and environmental surveys, disasters (flood, drought, storm surge, etc.) and ocean monitoring, agricultural yield estimation, geological, hydrological, and engineering surveys.

SAR is an advanced scientific device that integrates platforms, antennas, RF modules, signal processing, data transmission, image processing, and other modules. In fact, with the development of electronic technology and its application in imaging radar, today's SAR is entirely different from the original. In terms of resolution, SAR has developed from tens of meters to meters and decimeters. From the polarization perspective, it has grown from single to multi-polarization and full polarization. From the frequency point of view, it has developed from meter wave and microwave bands to millimeter wave and submillimeter-wave bands. In terms of applications, it has developed from military use to civilian use and scientific research. From the platform perspective, SAR technology has evolved from being mounted on unmanned aerial vehicles, helicopters, and manned fixed-wing aircraft, to being integrated into satellites and missiles.

How to obtain better images from imaging radar and how to turn images into more useful intelligence [2] require solving the following theory problems:

(1) Improve the individual quality of SAR as soon as possible. First, improve the existing SAR technology, with a particular focus on the realization of high-resolution imaging of small targets in strong interference, scattering, and high-density electromagnetic signals. Given that current SAR technology is a "zero IQ" system, it faces significant challenges in effectively tracking and competing with "smart targets".Therefore, the intellectualization of SAR is imperative.

(2) Maximize the group advantage of SAR combined with other types of sensors. These groups include SAR of different platforms, passive detectors, and other sensors such as infrared, optical, and acoustic. To this end, solving the fusion for the sensors of the same type and multi-type sensors data is necessary.
(3) Strengthen the image intelligence processing capability of SAR. SAR image intelligence processing is to transform massive SAR image data into usable and effective information, and it is the only way to realize the application of SAR technology. However, it is still difficult to quickly and automatically detect and identify the target of interest from SAR images, which seriously affects the application value of SAR technology and has become an urgent problem and bottleneck technology in the application of SAR technology.
(4) In-depth study of the new system and new technology of SAR so that SAR truly becomes the "jewel in the crown." New SAR systems and technologies are fundamental to improving performance, such as digital array imaging radar and optically controlled phased array imaging radar. Effectively solve the core problems of ultra-large instantaneous bandwidth signal generation, amplification, radiation, processing, and image application, especially the high efficiency, low profile, and light weight of large active array antennas, and welcome the new era of imaging/sensing in the network information system.

1.2 Development of Antenna Technology

Antennas and signal processing are the two core units of an electronic information system. The system of the high-resolution SAR antenna determines the radar system. Usually, the antenna accounts for about 90% of the cost, weight, and power consumption in an active phased array radar. The rapid progress of the antenna technology has promoted the development of radar system technology. Some say that if the painter Vincent van Gogh used his hands to create a masterpiece that amazed the world, then the antenna engineer used his wisdom to draw a pleasing radiation pattern in cyberspace [3], as shown in Fig. 1.2. It allows human beings to perceive and utilize information.

An antenna is a converter that converts the guided waves on the transmission structure into space free waves. The essence of antenna radiation is the problem of the macroscopic electromagnetic field. Electromagnetic waves are transmitted from the feed port of the antenna to the antenna radiation unit through the feeder. The prerequisite for an antenna to convert electromagnetic guided waves into space electromagnetic waves is to satisfy the boundary conditions of Maxwell's equations. The differential form of Maxwell's equations is usually composed of the full current law (Eq. 1.1), Faraday's law of electromagnetic induction (Eq. 1.2), the principle of magnetic flux continuity (Eq. 1.3), and Gauss's law (Eq. 1.4), which describe the changes in the electromagnetic field in space.

$$\nabla \times H = \frac{\partial D}{\partial t} + J \tag{1.1}$$

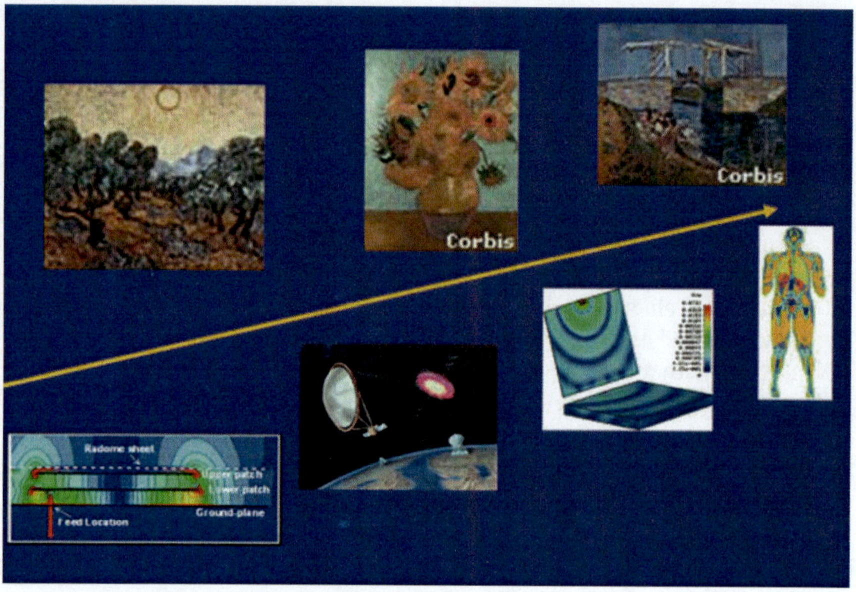

Fig. 1.2 The radiated electromagnetic field of the antenna

$$\nabla \times E = \frac{\partial B}{\partial t} \tag{1.2}$$

$$\nabla \cdot B = 0 \tag{1.3}$$

$$\nabla \cdot D = \rho \tag{1.4}$$

Equation 1.1 indicates that the magnetic field strength H's curl equals the full current density at this point (the sum of the conduction current density J and the displacement current density $\frac{\partial D}{\partial t}$). That is, the vortex source of the magnetic field is the full current density, and the displacement current can generate the magnetic field as well as the conduction current. Equation 1.2 indicates that the electric field strength E's curl equals the negative value of the time rate of change of the magnetic flux density B at this point. That is, the vortex source of the electric field is the time rate of change of the magnetic flux density. Equation 1.3 indicates that the magnetic flux density B's divergence always equals zero. That is, the B line has no beginning and no end. Finally, Eq. 1.4 is a generalization of Gauss's law of electrostatic field. That is, under time-varying conditions, the divergence of electric displacement D is still equal to the density of free charge bodies at this point.

When analyzing antenna radiation, because the time-varying charge and current are difficult to determine, and the electromagnetic field excited by the radiation source also affects the boundary conditions of the radiation source, solving Maxwell's equations will lead to mathematically complex problems. Therefore, an approximate solution method is often used in practice. The antenna radiation problem can be decomposed into two relatively independent problems, namely, determining the current distribution on the antenna and solving the spatial radiation field characteristics, that is, solving the problem of the external field of the antenna.

As a transmitting and receiving equipment of electromagnetic wave signal, an antenna directly affects the quality of the electromagnetic wave signal. Therefore, an antenna occupies a significant position in the electronic information system. An antenna system with a reasonable structure and excellent performance can not only minimize the requirements of an electronic information system for other parts of the system and save the overall cost of the system, but also improve the performance of the entire electronic information system. In modern electronic information systems, by processing the received signal, a phased array antenna can respond quickly to the environment, control its beam to point in the desired direction, and simultaneously aim its beam null direction to the unwanted interfering signal so as to maximize the signal-to-noise ratio of the desired signal. For example, with the development of modern cities and the increasing number of high-rise buildings, the electromagnetic environment in which an antenna is located is increasingly complex. Therefore, in order to improve the signal quality of electronic information systems, especially mobile communication systems, the research and development of active array antennas have attracted much attention.

As early as 1887, to verify Maxwell's electromagnetic wave theory, Hertz designed a transmitting antenna composed of a metal rod, a metal plate, a metal ball, and an induction coil, which was the first antenna for humanity. Since this antenna's generation, the antenna's development can be roughly divided into four historical periods.

1.2.1 Wire Antennas

In the early twentieth century, most research was on conducted wire antennas. Since the longer the wavelength, the smaller the attenuation in the propagation signal, the wavelengths used to achieve long-distance communication were above 1000 m. In this period, various asymmetric antennas appeared, such as inverted L-shaped, T-shaped, and umbrella antennas. Since the height of an antenna is limited by its structure, the size of an antenna is much smaller than its operation wavelength; this type of wire antenna belongs to the category of electrically small antennas. Later, due to the discovery of the existence of the ionosphere and its reflection on short waves, the research field of the short-wave band and medium-wave band wire antennas was opened up. At this time, an antenna size can be compared with its wavelength, which promotes the rapid development of antennas. During this period, tower broadcasting antennas, and other forms of antennas and antenna arrays, such as dipole antennas, loop antennas, long wire antennas, in-phase horizontal antennas,

Yagi antennas, diamond antennas, and fishbone antennas, appeared. These antennas have higher gain, stronger directivity, and a wider frequency band than the initial long-wave antennas. During this period, theoretical work on antennas was also developed. Pocklington established the integral equation for wire antennas, proving that thin wire antennas' currents are approximately sinusoidally distributed.

1.2.2 Planar Antennas

During World War II, Radar greatly promoted microwave antenna technology development. Several theoretical frameworks for planar antennas were established, such as geometric optics, physical optics, and aperture field theory. At that time, due to the urgent need for warfare, antenna theory was not yet complete. Therefore, experimental research became an essential means of developing new antennas. Concepts such as testing equipment and error analysis were established, and methods such as field and model measurements were proposed. A large number of reflector antennas were used in radar systems. The primary beam electric scanning antenna appeared in this period to quickly capture targets.

From the end of World War II to the 1970s, antennas for microwave relay communication, troposcatter communication, radio astronomy, and television broadcasting have been extensively developed. Especially, the successful development of satellites and intercontinental missiles requires antennas to have high gain, high resolution, circular polarization, and broadband performance and to have fast beam scanning and precise tracking capabilities. During this period, the development of antennas was unprecedentedly rapid. On one hand, large-scale ground station antennas were constructed and improved, including the emergence of Cassegrain antennas, the correction of main and sub-reflectors, and the application of beams waveguide technology and high-efficiency antenna feeds such as corrugated horns. On the other hand, due to the advent of new phase shifters and computers and the need for simultaneous search and tracking of multiple targets, electronically scanned antennas have received renewed attention and have been widely used and developed. Furthermore, during this period, breakthroughs were also made in the research of broadband antennas, and broadband or ultra-broadband antennas such as equiangular helical antennas and log-periodic antennas appeared. At the same time, the statistical theory for analyzing antenna tolerances and the comprehensive theory for antenna arrays were developed.

1.2.3 Planar Array Antennas

In the 1980s, with the development of radar and communication technology, antenna frequency reuse, orthogonal polarization, fast beam scanning, and multi-beam antennas began to receive attention. Ground-based large-scale phased array radar and microwave synthetic aperture imaging radar and other technologies have entered the stage of practical research, and planar array antenna technology has developed rapidly. It began to be applied to ground intelligence radar and airborne radar. On the other hand, it began to study large-scale active phased array radar on the ground. Each antenna element of an active phased array radar antenna array contains active circuit modules. A transmit/receive (T/R) module is the critical component of an active phased array radar, which largely determines its performance. A T/R module with an integrated transceiver includes a transmitting branch, a receiving branch, a radio frequency switch, and a phase shifter. Each T/R module has a high power amplifier (HPA), filters, a limiter, a low noise amplifier (LNA), attenuators and phase shifters, a beam control circuit, and so on. It can be seen that the amount of equipment and cost of an active phased array radar that uses two-dimensional phase variation to control beam scanning are considerable. Among the phased array radars, passive phased array radars, used for satellite measurement and control and long-range detection of strategic targets such as ballistic missiles, came out the earliest. In contrast, active phased array radars appeared relatively late.

With the development of electronic information systems toward millimeter and submillimeter waves with shorter and shorter wavelengths, new millimeter wave antennas such as dielectric waveguides, surface waves, and leaky wave antennas have appeared. In addition, antenna arrays were developed from linear arrays to circular arrays and from planar arrays to conformal arrays. At the same time, due to the need for anti-interference, the ultra-low sidelobe antenna has been significantly developed. Furthermore, due to the emergence of high-speed and large-capacity computers, the method of moments and geometric diffraction theory have been applied in the simulation calculation and design of antennas to solve many simulation analysis problems that could not be solved or were difficult to be solved in the past.

The antenna structure and process technology also made significant progress during this period. In terms of antenna measurement technology, microwave anechoic chamber, near-field measurement technology, the use of celestial radio sources to measure antenna technology, and the establishment of an automatic measurement system controlled by a computer, etc., appeared in this period. The application of these technologies solves the measurement problem of large antennas and improves the accuracy and speed of antenna measurement.

1.2.4 Active Array Antennas

In the past 20 years, with the development of interdisciplinary fields, such as integrated circuits, microsystems, new materials, and other fields, the active phased array antenna technology has been developed in terms of integration, digitization, multiple function, high frequency, and ultra-wide bandwidth. Different from a traditional active phased array antenna, a distributed miniature transceiver unit is directly connected to the back of each antenna element. A miniature transceiver unit includes

Power Amplifier (PA), LNA, attenuator, phase shifter, and duplexer. Sometimes traditional frequency source, signal generator, RF signal receiving/transmitting channel, and digital-analog/analog-digital converter are also integrated into an active antenna module to realize a unified integration of antenna, RF and beam control in the electronic information system. We call this active phased array system as an active array antenna. Compared with traditional active phased array antennas, radio frequency systems, and modules, a large number of distributed, miniaturized, highly integrated, low-power active and passive chips have been adopted, and heterogeneous structures and heterogeneous 3D integration have been implemented in their internal structure. The active array antenna of this structure can reduce the feeder connection's power loss and give the system a higher signal-to-noise ratio, better impedance matching, and a wider frequency band.

A traditional radar system comprises antennas, transmitting, receiving, signal processing, data processing, system monitoring, and other subsystems. The emergence and development of active array antennas have changed the form of electronic information systems. For example, a radar system comprises only three parts: an active array antenna, connecting cable, and a general digital processor [4]. High-density, high-efficiency, high-power, and multi-functional active array antennas are closely related to develop new semiconductor devices, materials, and advanced integration and packaging technologies.

1.3 Active Array Antennas

Phased array antennas are divided into passive array antennas and active array antennas. Active phased array radars have become a mainstream radar development, including high-resolution SARs. In high-resolution SAR, an active array antenna is the only choice to effectively alleviate the contradiction between high resolution and wide swath. The large aperture, low profile, high efficiency, and light weight of the antenna are eternal pursuits of antenna engineers. Figure 1.3 shows a schematic diagram of a large active array antenna for the spaceborne SAR. The large aperture of the antenna is the most direct way to obtain the high-power aperture product of the radar. Due to the rocket's launch envelope limitation, a larger antenna aperture can be obtained with a lower antenna section thickness. The high efficiency of the antenna enables the antenna to achieve two-way effects of transmission and reception. Therefore, antenna efficiency is an essential parameter that spaceborne SAR prioritizes. The phased array radar technology can meet the requirements of various advanced radars and has excellent potential to improve the "four countermeasures" capability of the radar in modern warfare. In addition, the development of computers, integrated circuits, and hybrid integration technology has laid a solid foundation for developing and applying active array antenna technology.

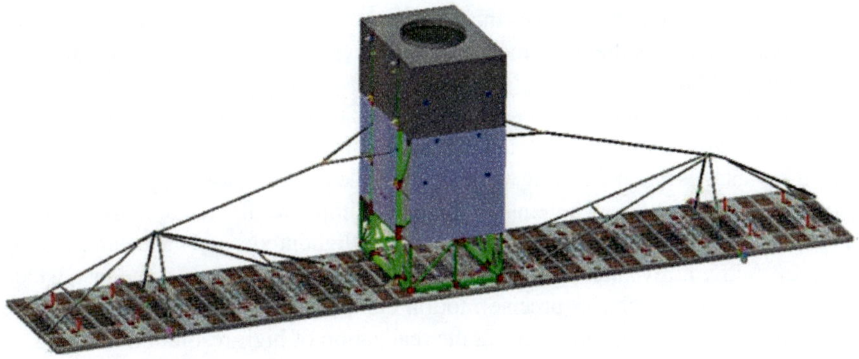

Fig. 1.3 Large aperture spaceborne active array antenna

1.3.1 Characteristics of Active Array Antennas

Active array antenna technology is applied to SAR, which is especially suitable for multimode fast switching to realize multiple functions of SAR and enable a radar to have the ability of rapid response, self-adaptation, and fault weakening. According to the characteristics of SAR and the development of technology, the following characteristics of active array antenna technology are most worthy of attention:

(1) **An active array antenna is a critical way to improve the performance of SAR**

High resolution, multimode, multi-polarization, and multi-band are essential development directions of synthetic aperture imaging radar. Active array antennas have significant advantages in high-resolution imaging and multimode realization.

SARs with different loading platforms have different requirements for effective radiated power. Reasonable realization of effective radiated power is the basis of radar work. Therefore, we must pay attention to the product of antenna aperture and average transmit power, that is, the antenna size is as large as possible, and the transmit power is as large as possible. As we all know, the azimuth resolution of SAR is half of the azimuth size of the antenna. Therefore, a smaller antenna size is better. On the other hand, a large antenna aperture is an important way to reduce the cost of SAR. Therefore, achieving high resolution and using a large antenna aperture is a contradiction. An active array antenna effectively alleviates this contradiction. When the resolution is low, the large aperture of the antenna can be effectively used. On the other hand, when the resolution is high, the antenna beam can be broadened by phase weighting, equivalent to shortening the antenna aperture.

Compared with a conventional vacuum tube or solid-state transmitter-based SAR, an active array antenna-based SAR system can reduce the loss of transmitting and receiving feeder, realize signal power synthesis in space, and improve the effective radiation power of radar. At the same time, for the active array antenna, the low noise amplifier is generally placed behind the antenna element, which reduces the noise of the receiving system and improves the sensitivity of the radar receiving system. The active array antenna is the main measure to achieve a high signal-to-noise ratio and sensitivity of SAR.

Compared with a mechanical scanning antenna, an active array antenna has the characteristics of flexible beam scanning, no inertia, and fast speed. It shows unparalleled advantages when the SAR works in multiple modes. Especially for the spaceborne SAR, an active array antenna dramatically improves the range imaging width of the ScanSAR mode, the implementation of spotlight mode, and the beam pointing accuracy. Active array antenna beam scanning is flexible, inertia-free, and fast, which enables the SAR to achieve precise motion compensation, thereby improving the quality of radar imaging and ensuring the realization of high-resolution imaging.

(2) **An active array antenna is beneficial to improving the anti-jamming capability of SAR**

The purpose of SAR is to obtain the intelligence information of the selected area, and the purpose of jamming the radar is to prevent, confuse or delay obtaining the information of the selected area. For intelligence or tracking radars, the effectiveness of jammers is generally measured by the reduction in the radar's search or tracking range. For SAR, a jammer is to prevent the reconnaissance of image information in an area, and the effectiveness of a jammer is generally measured by the sensitivity reduction of the SAR.

In order to improve the anti-jamming capability, the commonly used technical means include high effective radiated power, low or ultra-low sidelobe antennas, large time-bandwidth product signals, and dual/multistatic radar systems. These techniques are crucial to improve the radar's ability to anti-interference. In addition, active array antennas are beneficial to improve the total antenna radiation power and form high-gain, low-sidelobe antennas. Spatial filtering technology can also achieve adaptive antenna zeroing and suppress interference and clutter. At the same time, it is also conducive to realizing signal energy management, rational use of signal energy, and improving radar anti-jamming self-defense distance.

(3) **An active array antenna is conducive to the standardization and modularization of SAR, thereby reducing cost and improving reliability**

The improvement of radar performance requirements and the deterioration of the working environment make the composition of the radar system more and more complex, the development cycle prolonged, the development cost increased, and the technical risk increased. To adapt to the evolving needs of modern radar systems, researchers have focused on strengthening the research on basic radar technologies, such as simulation technology, special testing equipment, new processes, structures, and materials. However, another promising solution to address these needs is the use of active array antennas. An active array radar can use many consistent standard components (such as T/R modules), which is conducive to standardization, modularization, and lower production costs.

Microelectronics technology is one of the effective measures to reduce the volume and weight of radar, and it is an important condition to ensure that the radar can normally work in harsh environments. Microelectronics technology is also significant in improving system reliability and signal and data processing speed. It is also a key measure to reduce the production cost of an active array radar. In addition, the development of microelectronics has facilitated the rapid miniaturization of a series of components.

Despite the many advantages of active array antennas, the decision to use them in practical applications should be based on the specific needs and requirements of the application. Firstly, we should focus on analyzing radar tasks. Secondly, we should analyze the cost of using active array antennas and consider technical risks, development cycles, and the impact of production costs. Then a reasonable choice can be made.

Active array antenna technology is indeed a technology that will give SAR a "new life." Moreover, with technological progress and large-scale use, its cost is more affordable. However, there are still many challenges in terms of its technology, which mainly lie in:

(1) The use of large-scale active array antennas in space pays more attention to the low profile, high efficiency, and light weight performance of the antenna, and it is necessary to solve the problem of effective integration of electromechanical and thermal integration of the antenna (including radio frequency, analog, digital, power supply, etc.).
(2) In the case of wide-band and wide-angle scanning, it is necessary to solve the problems of mutual coupling between antenna elements and the "dispersion" of the beam in space and time.
(3) Active array antenna multi-channel technology is one of the most effective methods to alleviate the contradiction between high resolution and wide swaths of SAR. In a wide frequency band, it is necessary to solve the consistency and stability of the amplitude and phase of multiple receiving channels.
(4) An active array antenna needs to solve the monitoring and compensation of signal amplitude and phase in the static state, meanwhile solve the real-time measurement of antenna array deformation and real-time compensation of signal amplitude and phase in the dynamic situation (in flight). However, these real-time measurements and real-time compensations under dynamic conditions are extremely difficult.

In addition, technologies such as high thermal conductivity materials, semiconductor integrated circuits, and three-dimensional hetero-heterogeneous hybrid integration need to be developed simultaneously.

1.3.2 Semiconductor Integrated Circuit Technology

The rapid progress of semiconductor integrated circuit (SIC) and packaging technology has extensively promoted the development and advancement of active array

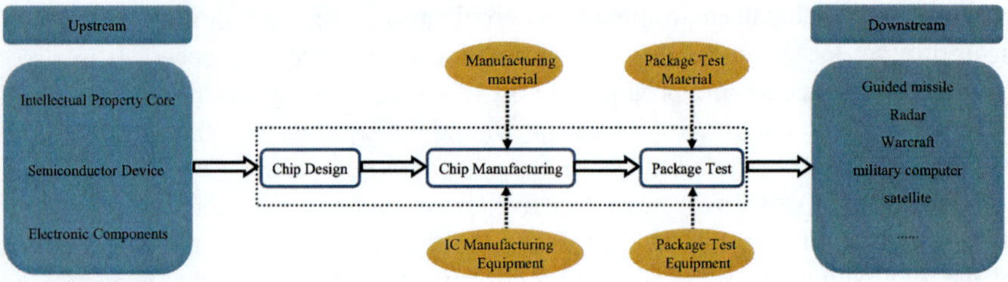

Fig. 1.4 Integrated circuit industry chain

antennas. The relationship between SIC and electronic information-related industries is shown in Fig. 1.4.

The core of an active array antenna is to change the antenna from a passive system to an active system. Therefore, an active antenna contains a large number of active circuits, and most of them are microwave active integrated circuits. At the same time, due to the arraying and expansibility of the antenna system, the scale of the antenna array of this system is huge, and the number of channels of the antenna array has reached thousands. Moreover, it is under development toward digitization. In general, an active phased array antenna system is a planar structure. Therefore, low profile, high efficiency, and light weight are significant for high-resolution SAR active array antennas. The microwave integrated circuit is the key to realizing these characteristics of an active array antenna.

Microwave integrated circuits (MICs) are a type of integrated circuit that utilizes advanced RF complementary metal oxide semiconductor (CMOS), silicon germanium (SiGe), gallium arsenide (GaAs), and other semiconductor processes to process microwave and analog signals. This processing includes amplification, transformation, calibration, comparison, and transmission of signals at microwave frequencies. A microwave system in radar is mainly composed of a microwave transceiver channel, modulation and demodulation circuit, and microwave signal generation. With the development of microwave integrated circuit technology and digital technology, the integration degree of microwave chips is getting higher and higher. An integrated circuit combines multiple components on a chip, improving the chip's performance and reducing cost. The multi-functional microwave single-chip can integrate small-signal downstream (receiving) and upstream (transmitting) circuits. The downstream circuits include low-noise amplifiers, frequency mixing, gain control, and even high-performance Analog to Digital Converter (ADC). The upstream part of the circuit also includes signal generation, frequency mixing, power amplifier, etc. The improvement of the performance and reliability of the multi-functional microwave integrated circuit chip and the cost reduction have greatly promoted the improvement of the sampling frequency and bandwidth of the radar microwave signal.

Developing semiconductor materials and process levels is the basis and prerequisite for the rapid development of integrated circuits. A generation of materials and a generation of processes will produce a generation of devices and a generation of circuits. Silicon (Si) has always been the dominant material in semiconductor technology. In recent years, integrated silicon (CMOS and BiCMOS) RF technology has greatly progressed. However, many applications can only use the compound semiconductors such as indium phosphide (InP) and gallium nitride (GaN).

In the semiconductor industry, silicon (Si) and germanium (Ge) are generally referred to as first-generation semiconductor materials, while Gallium Arsenic (GaAs), Indium Phosphide (InP), and their ternary and quaternary alloys are called second-generation semiconductor materials. Wide-bandgap (E_g > 2.3 eV) semiconductor materials have developed very rapidly and become the third generation of semiconductor materials, mainly including silicon carbide (SiC), diamond, and Gallium Nitride (GaN). Wide-bandgap semiconductor devices, as the third generation of semiconductor devices, will effectively improve the power level, radiation resistance, high junction temperature, and other capabilities of current microwave devices. InP devices are very suitable for the high-end millimeter-wave band and solve the solid-state problem of current millimeter-wave devices. Improving Microelectromechanical Systems (MEMS) devices and RF-MEMS devices technology mainly applies switches, phase shifters, attenuators, and so on. GaAs is a reasonable microwave transmission medium material, which is very suitable for the substrate of monolithic microwave integrated circuits. Because of the general promotion of GaAs technology, the development of monolithic microwave integrated circuit technology is promoted.

The development of multi-chip integration technology can integrate different types of semiconductor devices (silicon and compound semiconductors) and passive components (including filters and antennas) on a single substrate, and passive components are embedded in multiple stacks to achieve high Q value and miniaturization. Short-distance interconnects enable higher performance and denser circuits than traditional printed circuit boards and have been achieved in recent years in multilayer substrate materials and integration and assembly technologies, including laminates, ceramics, and integrated passives. In addition, the tremendous progress of an integrated circuit has led to the continuous improvement of integrated circuit density and system performance. However, as frequencies increase, interconnect losses between multiple integrated circuits increase rapidly, while multi-chip integration typically lacks the geometric and interconnect definition to achieve high-density integration. This requires the study of new technologies to solve the parasitic effects of micro-scale interconnection.

In the wafer-level heterogeneous integration (small chip, wafer bonding, and epitaxy transfer) approach, Si and compound semiconductor devices are integrated after the respective processes of Si and compound semiconductors are independently completed. This poses minimal risk to existing manufacturing processes and provides tight vertical integration between compound semiconductor (InP) and Si (CMOS and BiCMOS) devices. Among them, small chip integration can integrate various semiconductor chips on a complete CMOS wafer, such as gallium nitride High Electron Mobility Transistor (HEMT), indium phosphide double heterojunction bipolar transistor (HBT), and Si MEMS. In addition, this bonding method breaks the chip size reduction barrier for compound semiconductor technology.

Moore's law is approaching the physical limit [5]. Before the von Neumann architecture has not changed, the contradiction between the slowdown of chip performance improvement and the geometric progression of data demand became increasingly prominent. Under the circumstance that the chip volume cannot be further reduced effectively, multi-layer and three-dimensional heterogeneous integration (including new materials, thermal management, modeling, circuit/system design) technology is an important research topic that will promote the understanding and expectations of active array antennas.

1.3.3 Hybrid Integrated Circuit Technology

Hybrid Integrated Circuit (HIC) technology uses thick/thin-film technology, micro-assembly technology, and packaging technology to integrate semiconductor chips, passive components, etc., and is one of the main approaches to achieve miniaturization and light weight of electronic equipment. Figure 1.5 is a schematic diagram of the relationship between monolithic and hybrid integration. Monolithic integration is an eternal pursuit, and hybrid integration is a further and higher stage of monolithic integration. Hybrid integration technology involves a complex multi-level and multi-disciplinary technology system. It can be divided into a series of basic theories, manufacturing practices, and application technologies such as design technology, multi-layer interconnect substrate technology, micro interconnects technology, high air-tight packaging technology, and reliability evaluation and applications, involving electromagnetism, materials science, mechanics, physics, chemistry, and microelectronics, and many other disciplines.

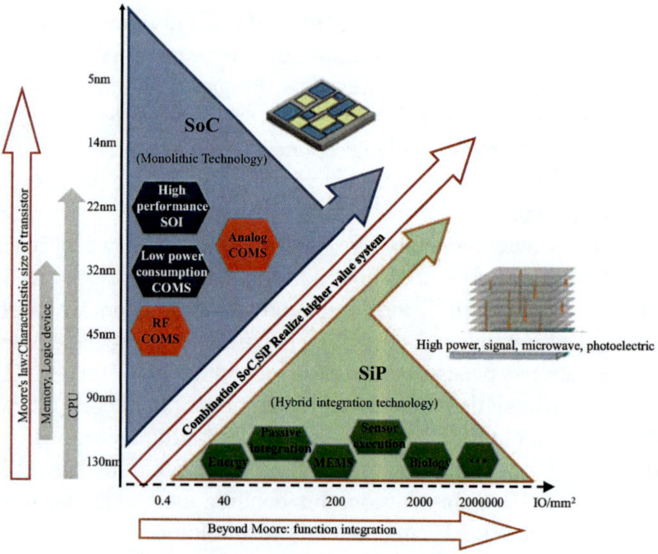

Fig. 1.5 Hybrid and monolithic integrations

The evolution of hybrid integration technology has gone through several significant changes. These changes include the transition from through-hole insertion technology to surface mount technology, from peripheral interconnection to area array interconnection, from single-chip to multi-chip, and from two-dimensional structure to three-dimensional structure. Each of these changes has facilitated the development of new hybrid integration technologies, improved assembly efficiency, and pushed HICs towards the direction of "four high, one small, and one light." The term "four high" refers to high assembly density, high frequency, high-power density, and high reliability, while "one small" refers to smaller size, and "one light" refers to reduced weight.

The improvement of the assembly density is reflected in the continuous reduction of the thickness of the package, the continuous reduction of the lead pitch, and the development of the lead arrangement from the two sides of the package to the surrounding and area arrays of the package. The three-dimensional heterogeneous integration technology can realize the integrated integration of new multi-functional devices (such as CMOS circuits, GaAs circuits, SiGe circuits or optoelectronic devices, MEMS devices, and various passive components), that is, the new connotation of "Hybrid Integration," which can increase the assembly density by more than 200%, has become an effective technical way to improve the function of electronic information systems. In terms of high frequency, high-density circuit components continue to expand to microwave and millimeter-wave applications. New satellite payloads, new seekers, mobile communications, and wireless local area networks will be the most active areas for high-density assembly applications in the future. The hybrid integrated power circuit is an indispensable key device in weapons and equipment. High thermal conductivity substrates are preferred in hybrid integrated power circuits, such as Direct Bond Copper (DBC), Aluminum Nitride (ALN), and Aluminum/Silicon Carbide (AL/SiC) substrates. Copper or aluminum tape bonding replaces the bonding process with the component welding process. It is also the key to ensuring high reliability and the long life of spacecrafts such as satellites. In terms of volume and weight, light weight packaging materials are used to reduce the weight of components, and integrated multi-layer substrate packaging technology is used to reduce the volume to overcome the common failure of glass insulator metal sealing to improve the reliability of components.

From the perspective of the development history of hybrid integrated circuits, the rise of hybrid integrated circuits first benefits from the urgent requirements of high performance, multi-function, miniaturization, and high reliability of military electronic equipment. It was subsequently developed by the demand for computers, communications, and automotive electronics. HICs are characterized by their structure and design flexibility, miniaturization, light weight, high reliability, shock and vibration resistance, and radiation resistance. It has been widely used in airborne communications, radar, fire control systems, missile guidance systems, communications, remote sensing, and telemetry systems for satellites and various space vehicles. With the continuous development of microelectronics technology, HICs have made significant progress in design technology, interconnection technology, packaging technology, substrate manufacturing technology, and materials, especially with the development of advanced HIC technology, such as Low Temperature Co-fired multi-layer Ceramic (LTCC) substrate, Aluminum Nitride (ALN) substrate, Multi-chip Module (MCM), three-dimensional MCM (3D-MCM), and System-in-package (SIP), HIC technology develops to a more advanced stage SIP technology. Its product assembly density is further improved with subsystem and even system-level functions, expanding hybrid integrated circuits' application space. Because the hybrid integration technology and products have the above-mentioned unique advantages, they are still irreplaceable by monolithic integrated circuits and other technologies and products, especially in applications with highly harsh environments or high-reliability requirements, such as space exploration, military applications, intelligent vehicle networking, and medical electronics.

1.4 Technology Development and Prospect of Active Array Antennas

In microwave imaging, high-resolution Earth observation tasks are mainly ground/sea stationary target reconnaissance, surveillance, and ground/air moving target detection. In addition, concealing hidden targets on the ground is one of the important tasks of reconnaissance and surveillance. Figure 1.6 shows the mission profile of a microwave imaging radar. The reconnaissance of stationary targets on the ground and sea and monitoring ground, sea, and low-altitude moving targets are the basic requirements for microwave imaging radars. For the core antenna system of a microwave imaging radar, the two characteristics of frequency and polarization should be taken as the starting point, and the related technologies such as multi-band, multi-polarization, and multimode of active array antennas should be studied around the actual requirements of the loading platform (Fig. 1.7).

As shown in Fig. 1.7, focusing on broadband, multi-band, and multi-polarization technologies, a microwave imaging radar "sees and distinguishes clearly" targets. The azimuth high resolution of a microwave imaging radar is achieved by relying on the synthetic aperture of a radar. Generally, the azimuth high resolution is half of the antenna aperture. The range resolution of a microwave imaging radar is achieved by bandwidth. Generally, the wider the frequency bandwidth, the higher the range resolution.

Different radar emission signal wavelengths correspond to different target echo information, and the ability to describe the scattering characteristics of the target is also different. A low-band microwave imaging radar can penetrate leaf clusters and stealth camouflage materials to detect hidden targets. Still, its ability of describing target contour and texture information is weak. A high-band microwave imaging radar corresponds to a shorter wavelength. It can clearly describe the contour and texture of a target, but its low penetration performance limits its ability to detect hidden targets. A multi-band SAR can obtain microwave images of different bands simultaneously, which can not only clearly describe the target's appearance but also see through to the hidden target and obtain richer and more reliable target information

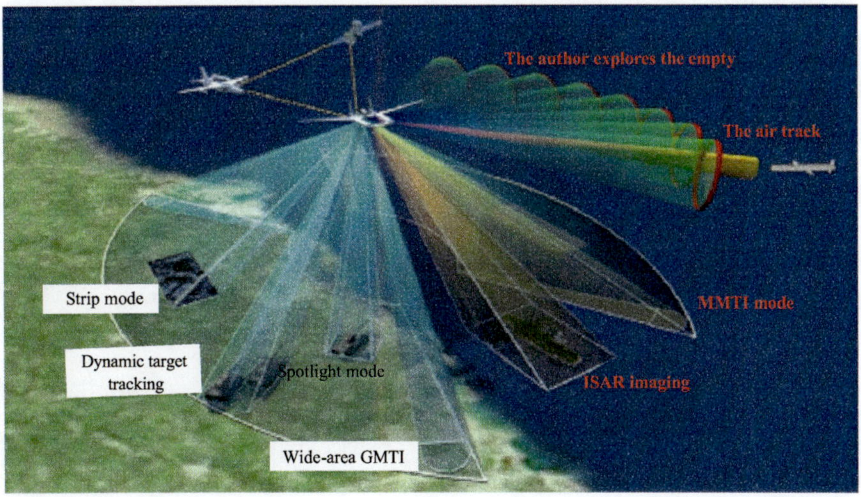

Fig. 1.6 Microwave imaging radar mission profile

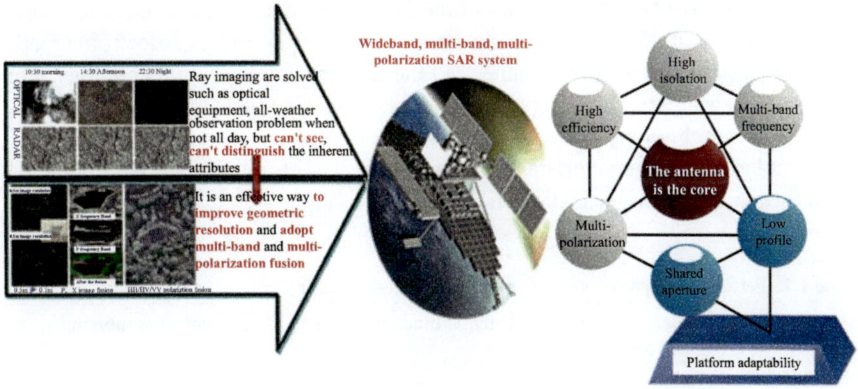

Fig. 1.7 Microwave imaging radar function and performance

than a single-band SAR system. Therefore, a multi-band microwave imaging radar is increasingly widely used in resource remote sensing, disaster assessment, and battlefield reconnaissance and surveillance.

Different electromagnetic wave polarizations of a microwave imaging radar for a specific target have different scattering characteristics. Multi-polarization fusion will enrich the scattering characteristics of a target but sometimes still cannot describe all the target details. This is because the conventional linear polarization cannot achieve the balanced excitation of all the target scattering points, and the target's partial scattering information cannot be obtained. In this case, fusion alone cannot increase the amount of effective information. On the other hand, a circularly polarized microwave imaging radar can obtain the complete details of the target as much as possible, especially the high-resolution circularly polarized microwave imaging radar has excellent imaging characteristics for aircraft, ships, and other targets. Compared with linear polarization, the structure of the target contour is complete and clear, and the layers are rich, which is an effective way to solve the problem of incomplete target contour information obtained from microwave imaging radar images and easy saturation of strong scattering point imaging.

1.4.1 Relationship Between an Imaging Radar and an Antenna

A high-resolution imaging radar can provide information on the structure and shape of ground stationary targets and ground, air, or space moving targets (vehicle, ship, aircraft, missile, satellite, etc.). Radar target imaging is equivalent to give the distribution of the scattering centers of the target. In theory, multimode radar imaging, such as multi-band, multi-polarization, variable viewing angle, and variable beam, is the distribution function of the spatial scattering center of the detection and perception target. Table 1.1 lists the descriptions of the performance parameters of a microwave imaging radar system, the interaction between the target and the electromagnetic wave, the basic theory of the antenna, and the antenna engineering technology. The performance parameters are related and restricted to each other, constituting an active array antenna technology system.

The polarization characteristic of a radar is the basis for applying radar polarization information. The polarization characteristics of radar systems are closely related

Table 1.1 Performance parameters of microwave imaging radar

System capability requirements	Objectives and electromagnetic wave actions	Antenna fundamental theory	Antenna engineering practice
High geometric resolution	Frequency	Broadband/ultra-wideband	Multi-band, shared-aperture
High radiation resolution	Polarization	Antenna configuration	Low profile, light weight
Dielectric penetration capability	Incidence angle	Radiation efficiency	Multi-polarization, high isolation
Spurious wave rejection capability	Resolution	Radiation characteristics	Dynamic radiation stability
Interference suppression capability	...	Coupling characteristics	Star, aircraft, and bomb platform adaptability

to the polarization characteristics of electromagnetic waves and antennas, the polarization characteristics of radar targets and active jammers, the optimal polarization theory of targets, and the theory of polarization target decomposition. It can be said that the research on the polarization characteristics of a target runs through all aspects of the radar system research. The basic idea of polarization target decomposition is to decompose the polarization scattering of a target into a combination of several basic scattering characteristics and then classify according to the similarity between the classification unit and the basic scattering mechanism or directly use the extracted new features [6, 7]. The advantage is that the classification result can better reveal the scattering mechanism of a target, which is helpful for people to understand the image and does not require training data during classification. Hence, it has a wide range of applications.

High-resolution and multi-polarization imaging characterize the scattering properties of targets from different aspects. High-resolution imaging technology improves the resolution of radar targets in the directions of distance, azimuth, and altitude, characterizes the subtle features of radar targets, and reduces the ambiguity of polarization description models. In addition, the polarization technology makes the structural information described by the high-resolution imaging technology more comprehensive because the polarization information of a target is intrinsically related to its shape. Through the extraction of polarization information, information that is difficult to characterize by other parameters such as surface roughness, symmetry, and orientation of a target can be obtained, which is indispensable to characterize the characteristics of the target completely. Therefore, a high-resolution imaging radar can analyze the target, which is of great significance for obtaining more target structure information and improving the intelligent processing ability of a radar detection system. The application efficiency of a high-resolution imaging radar mainly depends on the amount of information of its target echo. The richer the amount of echo information, the better the subsequent application performance. The amount of information in the echo depends on the radar's resolution, the echo's dynamic range, and the measurement degree of the polarization scattering performance of the observed target. Among them, the dynamic range of the echo mainly depends on the analog-to-digital sampling conversion module in the radar hardware, the storage space, and the size of the computing processing capacity of the system. Therefore, the spatial resolution and polarization measurement capability of an imaging radar system is the two most important indicators of a high-resolution imaging radar.

Resolution is one of the eternal themes pursued by imaging systems. There are usually two approaches of improving imaging resolution, e.g., one is to enhance and update hardware equipment to enable it to transmit broadband signals and synthesize large apertures while improving measurement accuracy; and the other one is to improve the imaging resolution of SAR/Inverse SAR (ISAR) by establishing physical and mathematical models and using new signal processing techniques. Still, the improvement and update of hardware have long cycles, high costs, and are limited by technological development. Therefore, it is urgent and important to use new signal processing technology to improve the resolution of radar imaging, which has become an important research direction of imaging radar processing.

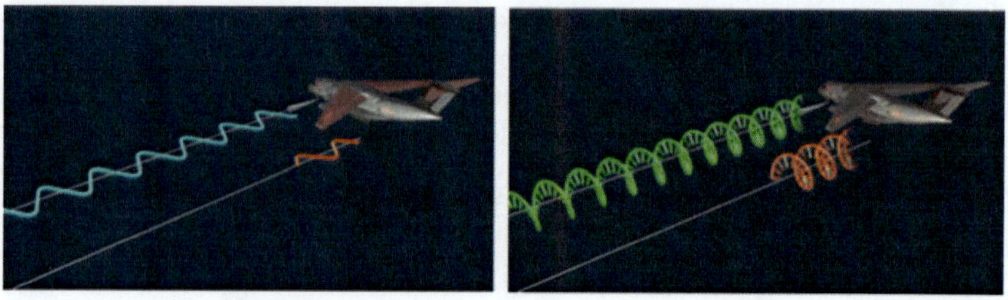

(a) (b)

Fig. 1.8 Electromagnetic wave scattering from the target, **a** linearly polarized, **b** circularly polarized

The polarization information of a radar target is closely related to its structure, material, shape, and attitude orientation. As shown in Fig. 1.8, through the extraction of polarization information, information that is difficult to characterize by other parameters such as surface roughness, symmetry, and orientation of the target can be obtained, which is an indispensable technical method to characterize the characteristics of the target completely. Polarimetric radar imaging organically integrates the high-resolution characteristics of a target and the full-polarization scattering characteristics. The high-resolution imaging technology dramatically reduces the ambiguity of the polarization description model. In contrast, the polarization technology makes the structural information described by the high-resolution imaging technology more comprehensive. The combination of the two can complement each other and combine polarization with high-resolution imaging. The combination of technologies is the most potential development direction of radar target recognition, which has received more and more attention.

In summary, the performance parameters of a high-resolution microwave imaging radar are closely related to the active array antenna's frequency, bandwidth, and polarization characteristics, as listed in Table 1.2. In order to improve the efficiency and reduce the size of a satellite antenna array, the multi-band and multi-polarization antenna shared-aperture technology is adopted to alleviate the contradiction between high resolution and wide swath and increase the observation bandwidth. Therefore, multi-channel technology is adopted. Although the application of shared-aperture and multi-channel technologies has dramatically improved the design difficulty of active array antennas, they are all important means to achieve broadband, multi-polarization, and multi-band, improve efficiency, reduce profile, and reduce weight. Therefore, it is necessary to compromise, analyze and optimize each relevant parameter from the level of an active array antenna system. Moreover, the following technologies must be studied from the theoretical and design methods.

(1) **Analysis and Optimization of Active Array Antenna Systems**

The first one is the sparse array optimization method. Based on the limited field of view requirements of microwave imaging and the slow optimization of large-scale phased array antennas, it is necessary to study the array factor spatial grating

Table 1.2 Main performance parameters of active array antenna

Performance parameter	Purpose and function
Broadband	Improve the imaging resolution of a radar
Multi-polarization	Improve target detection capability and information integrity
Multi-band	Improve target detection capability and information integrity
Multifunction	Improve the working ability of a radar in multiple modes
High isolation	Improve the signal quality for target imaging
High efficiency	Effective use of satellite platform power to reduce costs
Low profile	Improve the suitability of a radar to the satellite platform
Light weight	Reduce satellite platform launch costs

lobe optimization and sub-array arrangement of an active array antenna to solve the contradiction between the slow optimization speed, the reduction of the number of active channels, and the high grating lobe of the scanning beam. At the same time, engineering convenience should be considered.

The second one is the accurate modeling and computing technology. Accurate computing techniques of radiation characteristics can be developed based on antenna element and active channel measurements. The problem of mass testing and evaluation in the space/frequency domain of two-dimensional phased array antennas can be addressed to reduce the test wave number of large-scale active array antenna pattern from tens of millions to more than a thousand to facilitate engineering implementation.

The third one is the calibration method for a large aperture antenna. It is necessary to study the theory and method of online multi-channel amplitude and phase accuracy measurement of large-diameter active array antennas and to explore the amplitude and phase calibration technology and method of each channel of a high-efficiency antenna by using the optimized configuration of the antenna and the reciprocity of the antenna radiation field, such as the traveling wave calibration theory, based on the Fast Fourier Transform (FFT).

(2) **Active Array Antenna Theory and Design Method**

The first one is the antenna's high-efficiency technology. Referring to the idea of no cross-polarization and parasitic sidelobes, the new active array antenna structure and radiating element are studied; the mode control technology of an antenna element is studied, and the theory and method of the internal and external field matching of an antenna element in the active array environment are studied. Finally, ways of improving radiation efficiency can be explored.

The second one is the antenna polarization high-isolation technology. Based on the traditional dual-polarization symmetrical internal structure that causes asymmetric radiation of the antenna aperture, which deteriorates the basic principle of cross-polarization and isolation, new multi-polarization excitation methods and technologies and methods for reducing cross-polarization and improving isolation are studied, such as symmetrical dual-mode orthogonal conversion technology.

The third one is the antenna out-of-band spurious suppression technology. Based on the electromagnetic field entanglement and mutual interference effect caused by multi-band and multi-polarization shared-aperture nested superposition, the technology solves the in-band/out-of-band isolation reduction and cross-polarization deterioration caused by unwanted mode parasitic radiation is studied. The strong induction area control by shared-aperture nested overlay and differential cancellation technology can be studied, such as different frequency high-order mode suppression and symmetrical multimode quadrature technology.

The fourth one is circular polarization technology with a low axial ratio. It is necessary to further study and reveal the mechanism of circularly polarized scattering in microwave imaging and to study the theoretical model of circularly polarized scattering and penetration characteristics; for broadband, low-axis ratio circularly polarized antennas, to study the hybrid excitation of odd and even common mode/differential mode to achieve phase orthogonal extension. Theoretical methods include parity check, common mode/differential mode matching technology, etc.

The above discussion analyzes the characteristics of the active array antenna system required by a microwave imaging radar. At the same time, the parameters are closely related to the performance of a microwave imaging radar are:

(1) **Antenna aperture size**

As an example, a spaceborne microwave imaging radar shows that the antenna aperture's size is closely related to the ambiguity and the power aperture product, azimuth resolution, and observation bandwidth. From the perspective of ambiguity, the minimum unambiguous area of a SAR antenna is

$$A_{\min} = \frac{4v_s\lambda R}{c}\tan\theta \tag{1.5}$$

where c is the speed of light; λ is the working wavelength; v_s is the speed of the satellite (7540 m/s). If the altitude of the satellite is 632 km, the minimum antenna area of an X-band phased array antenna will increase with the increase of the antenna angle of view under the condition that the ambiguity design requirements are met, as shown in Fig. 1.9.

A large size can be chosen for a certain power aperture product to obtain the maximum antenna round-trip gain, reduce the transmit power, and save satellite energy resources. When a spaceborne microwave imaging radar system is analyzed, the power aperture product is usually calculated using the largest viewing angle (the farthest working distance).

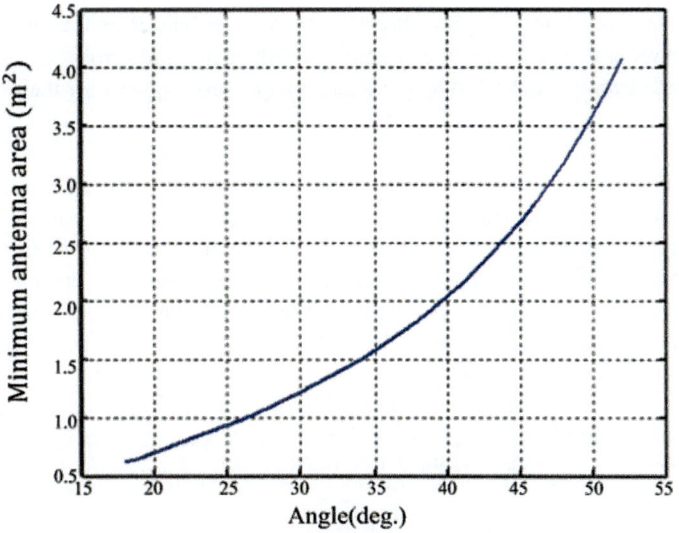

Fig. 1.9 The minimum unambiguous area of the antenna under different viewing angles

In the case of spaceborne, since the moving speed (ground speed) of the antenna beam in the irradiated area is slower than that of the platform, and a factor is usually multiplied when the azimuth resolution is calculated. Nevertheless, the azimuth resolution of a spaceborne SAR is still very close to half the azimuth aperture of an antenna. The antenna azimuth size should be less than 6 m to obtain the strip's 3 m azimuthal resolution. To achieve the azimuth 1 m resolution of the strip, the azimuth size of the antenna should be less than 2 m. However, in the actual analysis and design of the antenna aperture, the size of an antenna aperture should not be selected too small, and the effective aperture size after weighting the antenna aperture must be considered.

The size of the antenna's across aperture also needs to meet the requirements of the observation bandwidth, which depends on the antenna's across beamwidth, echo window width, number of range samples, and A/D sampling frequency. After the timing relationship of a radar system is determined, the width of the observation band, the incident angles of different viewing angles, near and far distances, the required range beam angle width, the required echo window time width, Pulse Repetition Frequency (PRF), etc. are analyzed and calculated according to the timing relationship diagram of radar signal transmission and reception, as well as the corresponding geometric relationship diagram of radio wave propagation, scattering, and echo.

(2) **Antenna beamforming**

A spaceborne SAR system is required to operate in various modes, including strip mode (Strip SAR), spotlight mode (Spotlight SAR), scanning mode (Scan SAR), ground moving target indication (GMTI), and others. To achieve the necessary azimuthal resolution and imaging observation zones under different viewing angles, the system requires the design of different azimuth beams and across beams for each mode. These beams are used to form the antenna beams required for the multimode operation of the spaceborne SAR system. Therefore, the design and optimization of these beams play a crucial role in the performance and efficiency of the system. Since the antenna front is a rectangular aperture and a rectangular grid, in addition to the beam width, beam shape, sidelobe level, and gain requirements, the range ambiguity requirements should also be considered in the design of the across directional beam. Azimuth beam design considers beam shape, width, sidelobe level, and gain in addition to the multiple modes of operation required to implement a system. Different working modes require antennas to have different beamwidths. For example, for a spaceborne SAR system must achieve strip SAR resolutions of 1 and 3 m, the azimuth beamwidth of the SAR antenna with a 1 m resolution is three times that of the SAR antenna with a 3 m resolution. The antenna is in different viewing angles. If the observation bandwidth is the same, the antenna across beam also needs to be widened.

(3) **Antenna beam scanning**

In a spaceborne SAR system, the stripe mode limits the length of the synthetic aperture due to the azimuth antenna beam width, and its azimuthal resolution will not be better than half of the antenna length. On the other hand, the spotlight working mode is a working mode suitable for small areas and high resolution. By controlling the azimuth direction of the satellite-borne SAR antenna beam, the same imaging area is continuously illuminated to increase the coherence time of the echo signal and the synthetic aperture's length. As a result, the antenna beam width no longer limits the azimuth resolution. A better azimuthal resolution than the stripe mode can thus be obtained. The key to the spotlight mode is to have an antenna with an azimuth beam that can be scanned, and by adjusting the azimuth direction of the beam, it can illuminate the area to be imaged for a long time.

The scanning SAR working mode is the wide-observation-band working mode of the spaceborne SAR, which is obtained at the expense of the azimuth resolution. In general, in the strip SAR, spotlight SAR, scanning SAR, and GMTI working modes commonly used in spaceborne SAR systems, it is desirable to scan the antenna beam in the range direction to achieve variable viewing angle imaging of the ground. Only the beamforming mode requires the beam to scan at a slight angle in the azimuth direction. However, in a high-resolution microwave imaging radar system, if the new imaging mode is adopted, the scanning angle of an active array antenna may be further increased.

(4) **Antenna instantaneous bandwidth**

The geometric resolution of SAR includes azimuth resolution and range resolution. The range resolution is mainly determined by the instantaneous signal bandwidth of a SAR system, the radar antenna viewing angle, and the ground processing weighting coefficient. The intrapulse phase error of the signal is one of the main factors affecting the range resolution. The slant range resolution of a spaceborne SAR is

$$\rho_{\text{gr}} = \frac{k_T \times k_I \times c}{2B \sin\theta} = \frac{\rho_r}{\sin\theta} \tag{1.6}$$

where B is the instantaneous signal bandwidth; k_T is the range-weighted broadening coefficient; k_I is the broadening coefficient caused by the system's amplitude and phase frequency characteristics; θ is the incident angle. When analyzing and designing, the values of the two coefficients k_T and k_I, are fundamental. Figure 1.10 shows the instantaneous signal bandwidth corresponding to different resolutions under different viewing angles. The images obtained by a spaceborne SAR system with the same signal bandwidth have different resolutions in the range direction. The smaller the viewing angle, the lower the range resolution; the larger the viewing angle, the higher the range resolution. The range resolution is poor at short distances, and the range resolution is good at long distances.

The instantaneous bandwidth of an antenna is different from the working bandwidth. For a spaceborne active phased array antenna, the maximum instantaneous bandwidth is limited by the transit time of an antenna aperture, and the instantaneous signal bandwidth should generally meet:

$$\Delta f \leq \frac{c}{4L_a \sin\theta} \tag{1.7}$$

where c is the speed of light; L_a is the length of an antenna aperture; θ is the incident angle.

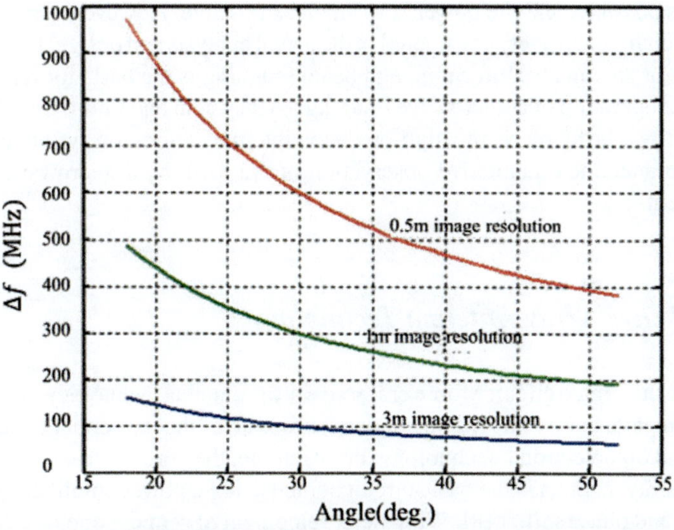

Fig. 1.10 The relationship between incident angle and instantaneous signal bandwidth

(5) **In-antenna calibration**

Due to the instability of many parameters of a microwave imaging SAR system, there will be errors in the measurement of the backscattering coefficient of a target. Therefore, in order to realize the quantitative measurement of ground targets by a microwave imaging SAR system, the system is usually calibrated internally and externally. The external calibration mainly completes the test of the transfer function of the SAR radar system and the radar antenna pattern and can also take into account the measurement of the SAR transmit power. The internal calibration mainly completes two functions: gain and power calibration, and monitoring the radar system's main working state. Internal calibration is the correction performed by the radar system to complete the quantitative measurement, mainly concerned with the absolute value of the channel change. It completes the calibration of the receiving channel gain of the SAR system, the gain of each T/R module, the transmit power of each T/R module, and the phase shifter of the T/R module. At the same time, the signal characteristics of the SAR system (such as the width of the range-compressed main lobe, the peak side lobe, and the integral side lobe) can be monitored. The internal calibration also has the function of antenna channel correction, which monitors the transmission characteristic changes of the active antenna system multi-channel and provides real-time compensation so that the amplitude and phase of each transmission channel maintain the required relationship. The amplitude and phase distribution of each channel of the feeder network can be obtained by using the system calibration data or special working states (such as the detection of components one by one) to realize the detection and compensation of the aging and failure of components.

A certain antenna power aperture product is the prerequisite for a microwave imaging SAR system to work normally. In principle, the antenna aperture size is as large as possible, and the power is as small as possible. However, various factors should be comprehensively considered in the actual analysis and design. For example, the ability of antenna beamforming and beam scanning is the basis for realizing the multimode operation of a microwave imaging SAR system and improving the observation ability. Furthermore, the high-precision internal calibration system can effectively guarantee the quantitative observation of the earth by a microwave imaging SAR system.

1.4.2 Active Array Antenna Technology

During the integrated circuit Moore era, active array antenna technology has emerged as a fundamental one by integrating modern phased array theory, semiconductor technology, and optoelectronic technology. For example, the core device T/R module of an active array antenna is the basic circuit composed of a power amplifier, low noise amplifier, and phase shifter [8]. With the development of semiconductor technology, Monolithic Microwave Integrated Circuit (MMIC) technology, Radio Frequency Microelectro Mechanical System (RF-MEMS) technology, and integrated packaging technology provide a technical approach to realize high performance, high reliability, miniaturization, and low cost of T/R modules. Integrated circuit technology is developing from narrow-band single-function to broadband multi-function, from Monolithic Integrated Circuit (MIC) to System on Chip (SOC), and from MCM to multi-function SIP [9]. In addition, the structural form of T/R modules has been developed from brick to tile, which has greatly promoted the development of active array antenna technology.

Traditional active array antennas typically consist of a "brick" structure, which is a combination of various functional modules and antennas. However, in response to the increasing demands for miniaturization, multi-functionality, high performance, high speed, low power consumption, and low cost in the new generation of high-resolution microwave imaging radar systems, new antenna technologies have emerged. These include Antenna on Chip (AOC), Antenna in Package (AIP), and System in Package (SIP) technologies, which have significantly advanced the architecture and integration methods of active antennas. These new antenna technologies have been made possible by the development and progress of semiconductor technology and advanced packaging techniques, and they are expected to pave the way for even more advanced and sophisticated radar systems in the future. AOC and AIP belong to the concepts of SOC and SIP, respectively. Based on these several antenna forms, the "Tiled Antenna" has emerged. The relationship between them is shown in Fig. 1.11.

AOC integrates an antenna and other circuits on the same chip through semiconductor materials and processes and is an on-chip antenna based on silicon-based technology [10]. AOC technology can improve system integration at a lower system cost. Still, due to the use of the same materials and processes, there is no way to optimize the performance of each type of circuit, resulting in reduced system performance and increased system power consumption. Due to the silicon substrate's low resistivity and high dielectric constant, a large part of the antenna's energy is concentrated in the silicon substrate when radiating, resulting in low radiation efficiency and gain of the antenna. The gain of conventional silicon-based technology on-chip antenna is generally less than −5 dBi, and the radiation efficiency is only 5% or even lower. If technologies such as proton injection, micromachining, artificial magnetic

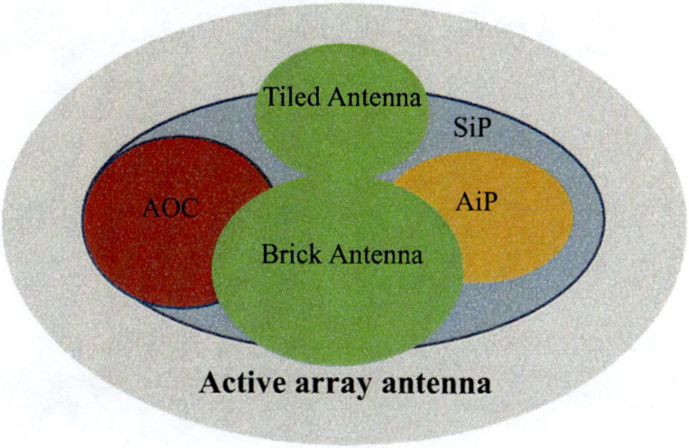

Fig. 1.11 Schematic diagram of the relationship between several active array antennas

conductors, and dielectric lenses are used, the gain or radiation efficiency of the antenna will be improved to a certain extent.

AIP integrates an antenna in the package that carries the chip through the packaging material. The packaged antenna technology inherits and develops the integrated concept of microstrip antennas, multi-chip circuit modules, and tiled phased array antenna configuration and extends the antenna to the fields of integrated circuits, packaging, materials, and processes [11]. Compared with AOC, AIP integrates a variety of devices and circuits in one package, implements complex functions and specific package-level systems that on-chip antennas cannot achieve, and effectively avoids the problem of gain loss caused by the low resistivity of semiconductor substrates, and the radiation efficiency generally reaches more than 80%. Figure 1.12 shows an example of the research results of a packaged antenna. The antenna array realized by the thick film technology and the radio frequency chip is packaged into a QFN package by gold wire bonding, and the packaged antenna with a center frequency of 122 GHz, a bandwidth of 12 GHz, and a maximum gain of 11.5 dBi is realized [12].

SIP adopts CMOS-SOI (Silicon-On-Insulator) process and QFN packaging technology, combines an on-chip antenna and a package antenna, and achieves a maximum antenna gain of 8 dBi in the frequency range of 54.5–63.4 GHz [13]. As shown in Fig. 1.13, flip-chip technology connects the LTCC antenna with the RF chip, and a phased array packaging system with 4 units is realized [14].

The structural feature of the tile array antenna is that the circuit board acts as the main body of the encapsulation shell simultaneously, and high-frequency and low-frequency connectors are often not used or rarely used, and the circuit board is parallel to the antenna array. As a result, using a tiled array can greatly reduce the antenna system's thickness and the number of connectors and cables used. In addition, commercial microwave packaging and manufacturing techniques are available for T/R modules, further reducing T/R module costs. The T/R modules of this architecture are packaged in an industry-standard Quad Flat No-lead (QFN) package, i.e., the QFN package is soldered directly to an inexpensive PCB, which is then soldered

Fig. 1.12 A packaged antenna based on QFN package design

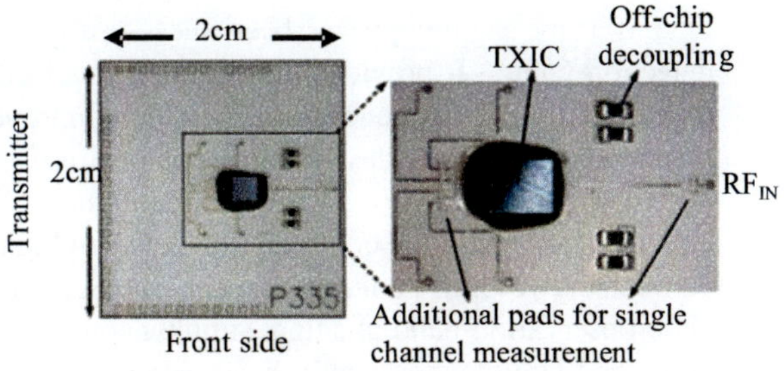

Fig. 1.13 LTCC antenna with TX/RX IC

directly to the back of the tile. The schematic diagram of the tiled T/R module array is shown in Fig. 1.14.

The above is analyzed and discussed from active array antenna integration and research. From the perspective of system and architecture technology, it can be divided into active phased array antennas, digital array antennas, and microwave photonic array antennas.

(1) Active phased array antennas are generally known as phased array antennas that are phase-controlled at radio frequency. With the successful application of active phased array antenna technology in microwave imaging radars, microwave imaging technology has developed rapidly, has developed from the original side-looking single beam strip mode (Strip SAR) to the scanning mode (ScanSAR), spotlight mode (Spotlight SAR), interferometric mode (INSAR), large viewing angle, multi-beam, ground moving target display (GMTI), and other working modes. At the same time, the requirements of low cost, high efficiency, and light weight for microwave imaging radars have promoted the development of active phased array antenna technology.

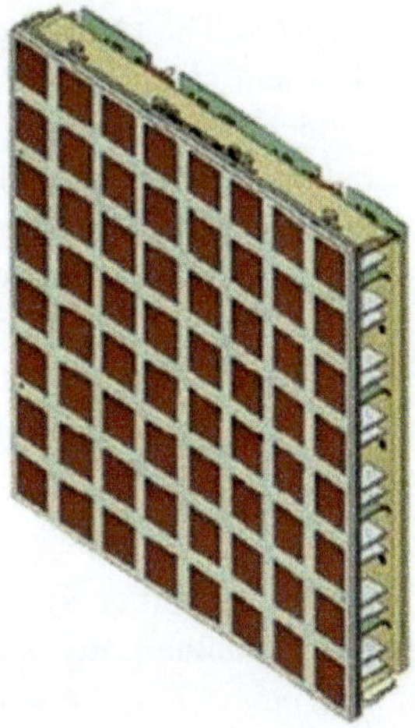

Fig. 1.14 Schematic diagram of tiled T/R module array

(2) Digital array technology has been successfully applied in early warning and detection radars. It provides multi-dimensional information such as airspace, time domain, and frequency domain, which improves the radar system's performance and simplifies the system's composition, and has become one of the development trends of radar. If it is combined with SAR technology, it can effectively alleviate the contradiction between high resolution and wide observation bandwidth in traditional spaceborne SAR. It can realize high-resolution wide-observation-band imaging, simultaneous multimode and multi-task operation, and simultaneous multi-beam adaptive forming. It can also change the transmit waveform during beam scanning, therefore it has the ability to have a low probability of interception.

The development of digital array SAR theory and technology has experienced the evolution process from receiving Digital Beam Forming SAR (DBF-SAR), the evolution process of receiving/transmitting DBF-SAR to DBF MIMO SAR. That is the gradual integration of digitization, software technology, transceiver distribution systems, and distributed system structures. A digital array SAR system has obvious performance and functional advantages compared to a conventional SAR system. However, if we want to make full use of these advantages, there are still some technical challenges to be solved, mainly including high-density integration of radar systems, broadband data acquisition, transmission, and processing technology. Nevertheless, with these problems to be solved, the improvement of the performance and function of the digital array SAR systems will be realized.

(3) The combination of microwave photonic technology and phased array technology produces optically controlled phased array technology, which has potential application prospects in generating, transmitting, and processing microwave signals. Modulating microwave signals onto optical carriers and using optical fibers for long-distance transmission of microwave signals have already been applied in communications, which has attracted the attention of the radar industry. The researchers have realized the use of optical fiber as the radar's data and signal transmission line in the radar system and the optical beamformer formed by the optical real delay line. In the optical phased array SAR, the radio frequency signal is modulated on the optical carrier and transmitted to the antenna element through different optical paths. Beamforming and control in the optical domain will have many desired effects. Therefore, applying microwave optoelectronic technology in SAR is one of the important research directions.

1.4.3 Antenna Array Microsystems

An antenna array microsystem is an advanced stage of active array antenna development. An antenna array microsystem integrates antenna array, active transceiver channel, power distribution/combination network, beam steering, power supply, and heat conduction structure in a narrow space by means of high-density three-dimensional package integration, thereby shortening the interconnection to result in smaller loss and better matching [15]. It can replace the traditional "brick" T/R module as the core of the antenna array. The basic technologies of antenna array microsystems include three-dimensional stacked packaging, chip-level heat dissipation microchannels, and Through Silicon Via (TSV)/Through Glass Via (TGV) vertical interconnection.

With the continuous development of Moore's Law, basic technical capabilities such as microelectronics, optoelectronics, and micromachines have developed rapidly, but the pace of further development to nano-scale integration is increasingly constrained by technology and cost. At the same time, with the rise of cross-border system architecture and software algorithms, cross-border integration to form new capabilities (beyond Moore) to meet potential needs has become an innovation hotspot. The development of the ultra-molecular era requires the close integration and fusion innovation of system technology and micro-nano electronic technology. Therefore, the antenna array microsystem technology is a multi-disciplinary frontier emerging technology, which is the product of the post-Moore's Law development.

3D integration technology will become mainstream in the post-Moore era. Three-dimensional integration technology can solve two significant core problems. One is to develop Moore's Law to double the density of transistors and improve chip performance. Only three-dimensional heterogeneous integrated circuits can surpass Moore's Law. The second one is to realize the multi-functional integration pursued by post-Moore's Law and integrate heterogeneous devices/modules.

In the future, the shape boundaries of active array antennas will be blurred, and more active and passive circuits will be integrated to develop in the direction of active array antenna microsystems. Still, the logical boundaries will become clearer and clearer. Therefore, achieving integration is an inevitable result. Furthermore, with the continuous development and improvement of civil and military network information infrastructure, active array antennas are bound to develop in the direction of integration, digitization, and multi-functional integration, which will profoundly affect everything from multi-platform high-resolution earth observation to cosmic exploration.

Antenna array microsystems will first integrate functional microsystems of modules and then integrate microsystems of different functional levels into larger system-level microsystems. There are two basic areas in the research of antenna array microsystems including implementation of microsystems on a single chip and integration of system-level microsystems through hybrid integration technology and three-dimensional (3D) heterogeneous technology. Hybrid integration technology is the basis for antenna array microsystems. The connection via cable connectors between the traditional array antenna and the RF channel and baseband is the main bottleneck restricting the miniaturization of the system, and the mismatch and loss caused by the connector and the transmission line will cause a great loss of performance. The scientific and technical issues involved in the antenna array microsystem include the following aspects.

(1) Architecture and topology technology under multi-physics constraints

The most significant differences between antenna array microsystems and other disciplines are architecture and topology technology under multi-physics constraints. The antenna array microsystem's architecture breaks through the microelectronics technology's scope. It cannot be divided into simple units regarding function and performance. Instead, it has layouts in disciplines such as force, light, material, electronics, and information and realizes the close relationship between light, machine, electricity, magnetism, sound, and other elements. The antenna array microsystem architecture has the characteristics of system-level architecture, performance, function, algorithm, etc., and component-level electrical, thermal, and material property. Under the constraints of multi-physics, antenna array microsystems' architecture is interdisciplinary, and the ambiguity and intersection of disciplines, functions, and performance interfaces [16] bring great difficulties to research antenna array microsystems.

(1) Multiphysics coupling mechanism. The large-scale antenna and the micro-scale chip are integrated into the same package. There are microscopic entanglement effects of electromagnetic fields radiated by large-scale antennas and different small-scale chips, crosstalk effects of RF signals and analog/digital signals in the package, and skin effects of RF signals at microscopic scales. Therefore, multi-physics coupling mechanisms need to be studied. With the entry point of multi-physics coupling, the time-domain and frequency coupling mechanisms of RF integration, high-density heterogeneity, high-precision transformation, and high-speed signal transmission interconnection at the micro-scale are analyzed. Furthermore, it guides the decomposition and assembly of system indicators and provides scientific guarantees for building a reasonable and effective antenna array microsystem architecture and topology.

    By extracting the optical, mechanical, electrical, magnetic, acoustic, and other multi-parameter characteristics of the antenna array microsystem architecture, combined with structure, fluid, mechanics, electromagnetism, etc., the coupling and mutual interference of multi-physical quantities at the micro-scale are carried out. The problems of complex phenomena caused by the interaction between multi-parameters can be solved and the coupling of multi-physics can be analyzed.

    Considering the transmission characteristics of complex signals such as radio frequency, analog, and digital analog in the three-dimensional micro-scale, from the reliability and Design for Manufacturing (DFM) analysis of the antenna array microsystem and the systematic research of continuous iterative improvement, solving the long-term stability and reliability of the antenna array microsystem can be focused, the standard model library can be established and improved, and the standard process of systematic electromechanical and thermal multi-physics simulation can be sorted out and established.

(2) Model of the electromagnetic properties of isomers. In the antenna array microsystem package, the three-dimensional micro-scale interconnection produces electromagnetic field discontinuities, resulting in electromagnetic field

mutual interference, harmful mode parasitic radiation, antenna polarization mismatch, and changes in the isolation characteristics within and outside the operating frequency band. Therefore, in order to obtain better working bandwidth, high efficiency, and low cross-polarization performance of the antenna, it is necessary to study the pattern matching technology of the antenna aperture field and the mutual coupling characteristics between the radiating elements of the array antenna under the specific boundary, especially under the condition of wide bandwidth scanning angle.

In the micro-scale environment, through the research and analysis of the electromagnetic characteristics of isomers, the electromagnetic distribution model of each functional unit in the antenna array microsystem is established, and the model is applied to the complex system design process to construct the multi-port characteristic model of the system.

By analyzing the conduction transformation, space radiation, impedance matching, and high-quality signal transmission of internal complex signals, the three-dimensional electromagnetic field enhancement modeling, and time-frequency domain algorithm analysis are carried out. The models of system-level functional units, interconnecting units, and encapsulation units are studied, and a three-dimensional multivariable function and a model library of electromagnetic characteristics of ultra-high-speed signals are obtained when signals are transmitted and transformed in a heterogeneous body. Build an equivalent model for simulation to solve key problems such as electromagnetic interference and crosstalk errors. On this basis, the distribution of heterogeneous materials and isomers in the antenna array microsystem is optimized, and the characteristic mode with the optimized electromagnetic performance is further obtained.

(3) Isomer Multidimensional Matching Tolerance Adaptation. In the case of large scale, the discontinuity of the microwave transmission line will generate high-order modes, which require a certain length of the transmission line to attenuate and eliminate. On the micro-scale, the interconnection between microwave devices and transmission lines and between transmission lines are both planar and three dimensional. The number of discontinuous points of the microwave transmission line is significantly increased, and the characteristics of the eigenmodes have changed. Therefore, studying the excitation mode matching theory under the strong constraints of boundary conditions and simulating and analyzing the parasitic effects of micro-scale interconnections is necessary.

Aiming at the problems of function and performance adaptability brought about by mechanical, electrical, thermal, and other matching in high-density packaging of microsystems, as well as the influence of process manufacturing precision on devices and systems, research on the multi-dimensional matching adaptability of heterogeneous isomers is carried out. On this basis, the tolerance evaluation of complex signal transmission and transformation in multiple dimensions (including mechanical, electrical, thermal, etc.) in heterogeneous isomers of antenna array microsystems is formed.

Based on the parameter deviation range and parasitic parameter variables in multiple dimensions of the microsystem, the mechanism, link, and effect of the functional circuit affected by it are further analyzed. It focuses on analyzing the increased power density due to the reduced thickness of the antenna array microsystem profile and the microscopic changes in dimensional accuracy or stress caused by heat, which affect performance and functionality. Multidimensional functions that affect the complex signal quality and multi-physics coupling are established, and system integration of active circuits/passive components/packages based on different functions under compatible process conditions is realized. This guides the robust design of 3D heteroisomers in microsystems.

(2) **Multiphysics matching hybrid integration technology**

HIC technology uses thick/thin-film, micro-assembly, and packaging technology to integrate semiconductor chips, passive components, etc., to achieve a given function. It is one of the main approaches to realize antenna array microsystems. Hybrid integration technology involves a complex multi-level and multi-disciplinary technology system. It can be divided into fundamental theories, manufacturing practices, and application technologies such as architecture design technology, multi-layer interconnect substrate technology, micro interconnects technology, high air-tight packaging technology, and reliability evaluation and applications.

The research on multi-physics matching hybrid integration technology is to meet the requirements of miniaturization, light weight, high density, and multi-function of antenna array microsystems. It is based on the research on hybrid integration frontier common technologies in electricity, light, magnetism, force, and other multi-physics fields. Complex signal transmission interconnect modeling breakthroughs, integrated passive components, three-dimensional heterogeneous micro-assembly, re-wafer, and other technologies are focused on. Technical challenges such as electromagnetic compatibility, high-speed signal transmission and crosstalk, thermal management, stress matching, and optoelectronic interference in antenna array microsystems are addressed.

Connectors connecting a traditional array antenna and RF channels, and the mismatch and insertion loss caused by the connectors will cause performance degradation, which is also the main bottleneck restricting the miniaturization of the system. Therefore, attention should be paid to the three-dimensional interconnection problem in researching antenna array microsystems.

(1) 2.5D/3D vertical interconnect technology. The 2.5D/3D vertical interconnection technology realizes the three-dimensional heterogeneous integration of different materials, different structures, different processes, different functional components, and aims to break through the limit of Moore's Law of graphic design [17]. Solving the problem of ultra-high-density interconnection under high-speed, high-frequency, and high-power transmission in the antenna array microsystem is focused on. 2.5D/3D interconnection through substrate via hole metallization vertical interconnection technology and bump technology are used to carry out electrical vertical interconnection. The difficulties such as crosstalk, delay, and

energy consumption that may occur in microsystems can be solved by studying the influence of various composite conductors and media on the adaptability and matching of complex signal transmission and shielding. At the same time, in the study of process, the parameter matching of thermodynamic and electrical properties is fully considered to avoid thermal mismatch and mechanical stress between different materials.

A typical stacked 3D package can be a stack of bare chips, and a stack of MCMs can even be a stack of wafers. Moreover, 3D-MCM can integrate chips of different process types (such as analog, digital, and radio frequency functional chips) in a single package structure to realize the integration of mixed signals. Under the circumstance that the mechanical integrity requirements of the antenna array microsystem module are met, and the module size, weight, and power consumption are extremely limited, the thickness of the antenna array microsystem package is reduced and the packaging density is maximized by optimizing the thickness of the multi-functional circuit board. Furthermore, the core of 3D integrated circuits and 3D silicon wafer integration is Through-silicon Via (TSV) technology, which is used to interconnect stacked chips, enhancing performance, shortening signal transmission time, and solving problems such as signal delay [18].

(2) 3D heterogeneous micro-assembly technology. Heterogeneous chip integrated fan-out technology is an advanced technology different from system-on-chip and wafer-level packaging. Solving the high-density interconnection between heterogeneous chips through the integration of heterogeneous chips through wafer reengineering and rewiring technology is the key to integrating antenna array microsystem functional unit modules.

Heterogeneous chip integration fan-out technology is used to integrate heterogeneous chips with different optical, electrical, magnetic, and other functions through advanced semiconductor technology to form a wafer [19], and high-density wiring is performed through thin films to form the multi-functional chips integrated technology. This technology can meet the requirements of reducing the thickness and volume of the antenna array microsystem functional unit module. Heterogeneous chip Fan-out Wafer-level Packaging (FOWLP) is thin and low cost, does not require a substrate, and does not require bumping on the wafer, flip-chip reflow, and flux cleaning. As a result, it can improve electrical and thermal performance and is easier for SIP, and 3D integrated circuit packaging.

Three-dimensional heterogeneous micro-assembly technology is a new type of micro-assembly technology. The integrated three-dimensional heterogeneous hybrid integration of different materials, structures, and functional components are realized based on multi-disciplinary system design and a micro-nano integrated manufacturing process. The electromechanical, thermal, mechanical, and other mismatches of heterogeneous materials are solved, and the system functions are improved simultaneously.

From the level of scientific research, it is necessary to study the limitations of semiconductor technology and the research direction of hybrid integration, such as which types of chips, structures, and materials can be hybrid integrated, and to extract universal laws and methods.

From the perspective of technical research, the three-dimensional heterogeneous micro-assembly technology is based on the system architecture; the highly integrated IC devices, microstructures, and other components are three-dimensionally integrated through the hybrid integration technology such as micro-soldering interconnection and micro-packaging. It is assembled into the package to form a high-density, high-reliability antenna array microsystem module, which bridges the chip function to the system function.

(3) High-density heterogeneous multi-layer substrate technology. The research of antenna array microsystems is usually based on three-dimensional heterogeneous hybrid integration technology, and typical technologies are MCM and SIP technology. Further, most of them are interconnect substrate technologies using Laminate (MCM-L), thin-film Deposition (MCM-D), and Co-fired ceramic (MCM-C) [20].

The research of high-density heterogeneous multi-layer substrate is to integrate substrate preparation technology and film integration technology through multi-layer substrate collaborative design and multi-physics coupling analysis. A high-density passive integrated heterogeneous multi-layer substrate with built-in resistance-capacitance and inductive elements is prepared by adopting a reasonable process method for matching and compatibility. Thick film passive component integrated substrate technology adopts advanced microelectronic technology and materials, built-in resistors, capacitors, inductors, and other components in the LTCC multi-layer substrate, which can shorten the interconnection length of discrete devices by more than 99%, reduce parasitic effects, and reduce interconnection. At the same time, system problems such as multipath weakening, spectrum congestion and noise interference are beneficial to be solved.

The hybrid multi-layer substrate integrates two or more substrates of different materials into a multi-layer substrate. Based on the physical parameters and characteristics of the different material substrates, the performance, wiring density, assembly efficiency, and cost of the multi-layer substrate are further improved. An example of this is the Multi-chip Module-Ceramic/Deposited Thin Film (MCM-C/D), where the use of thin-film multi-layer substrate enables the arrangement of high-speed signal lines, grounding lines, and welding areas with high wiring density and low signal transmission delay. On the other hand, power lines, ground lines, or low-speed signal lines are arranged on the co-fired ceramic substrate by making full use of its characteristics that it is easier to achieve more wiring layers and is suitable for large currents.

(3) **Packaging and thermal management technology**

Maximal functionalization, micro- and nano-scale, multi-scale structures, multi-functional materials, and active and passive embedded thick thin-film elements are important features of antenna array microsystems. With the development of antenna array microsystems toward miniaturization, high performance, and high-density integration, the power consumption of multi-functional devices (such as GaN and SoC chips) continues to increase. As a result, chip heat dissipation has grown from a few hundred milliwatts of small-scale integrated circuits to hundreds of watts. This will cause power chips and passive components to become non-uniformly distributed heat sources, increasing the heat flux density. Packaging and thermal management aims to dissipate heat through various methods to maintain the temperature inside the package within an allowable range, to avoid the internal temperature of the antenna array microsystem exceeding the limit value to result in creep of bonding materials, parasitic chemical reactions, diffusion of dopants, device stress rise, structural damage, and even melting, evaporation, and burning, leading the antenna array microsystem to stop working or lose its physical properties. Therefore, the package provides a heat dissipation channel for the antenna array microsystem, mechanical support, sealing protection, and internal and external signal interconnection for the internal chips, components, and substrates.

(1) Multi-intrinsic parameter adaptation material technology. Multi-intrinsic parameter adaptation material technology is interdisciplinary research covering structural design, material system, packaging technology, signal interconnection, environmental adaptability, reliability, and other fields. The study focuses on functional materials such as substrates, wirings, frames, interconnect conductors, inter-layer dielectrics, sealing materials, and encapsulation shells. The composite functional material is prepared for combining metals, ceramics, polymer matrix composite materials, metal matrix composite materials, ceramic matrix composite materials, and various reinforcements and material bodies. The requirements of antenna array microsystem package lightweight, miniaturization, low loss, and high thermal conductivity are realized.

    In response to the requirements of miniaturization and multi-functionalization of antenna array microsystem packaging, the combination and fusion technology of new substrate materials, conductive paste, substrate preparation technology, and film integration technology is the basis for realizing high-density heterogeneous multi-layer substrate technology. For example, the medium-temperature ceramic hole filling tungsten copper paste technology can realize high-speed DSP signal transmission. Single-chip fan-out technology can realize the interconnection of high-density micro-pitch chips and ceramic substrates. The vertical interconnection technology of aluminum nitride copper-filled pillars can realize high current transmission and meet the heat dissipation requirements of high-power devices. With the advent of large-scale commercialization of Wide Bandgap (WBG) semiconductor technology, the development of new packaging technologies is imminent.

(2) Embedded thermal management technology. Cooling devices based on micro-nano technology have played an increasingly important role in the thermal management of conventional microsystems. The cooling of electronic systems has developed from traditional natural convection, metal heat conduction, and forced air cooling to liquid cooling and heat pipe heat dissipation. The micro-channel cooling in the liquid cooling method is the most effective and convenient cooling method to meet the requirements of the antenna array microsystem.

The embedded microfluidic liquid-cooled substrate made by LTCC technology has the characteristics of small volume, large heat dissipation area, low-power consumption, and low cost of mass production. The runner cooler absorbs the heat on the chip and transfers the heat to the outside through the liquid circulation to achieve the purpose of cooling. The LTCC-embedded 3D micro-channel system is divided into multi-row straight groove type, meander type, and fractal flow channel. Taking advantage of the advantage that a single LTCC green ceramic sheet can be processed separately, a two-dimensional microfluidic channel is fabricated on a single LTCC green ceramic sheet by a punching process, and all the green ceramic sheets are laminated, hot-pressed, and sintered to form a complete 3D microfluidic channel.

(3) Ceramic-metal integrated packaging technology. Ceramic-metal Integral Substrate Package (ISP) technology is to use the multi-layer substrate as the carrier of the package, install the cavity wall of the package shell on the top surface of the substrate, wire the multi-layer substrate to constitute an integral part of the shell, and directly lead out the package on the substrate. The outer lead is a hermetic package and does not need to be packaged with an all-metal shell. While increasing packing density, reducing package thickness, and reducing weight, this technology also benefits microwave signaling and thermal management. In recent years, the emergence of three-dimensional heterogeneous hybrid integration technology is based on the LTCC process, which consists of two metal-ceramic modules connected through a metal transition board.

According to the boundary conditions such as environment, structure, and size the temperature field distribution and the influence of different conditions on the temperature field, thermal resistance and heat dissipation path, mechanical load and structural stress, electromagnetic field, and other microstructure multi-physics coupling analysis and design simulation optimization are carried out.

1.5 Chapter Summary

This book provides a comprehensive discussion and introduction to the analysis, optimization, and design of active array antennas because of the critical position of the active array antenna in the high-resolution microwave imaging radar and the eternal requirements for the low profile, high efficiency, and light weight of the active array antenna on its loading platform. This book takes the two essential factors "frequency and polarization" as the main line. Based on discussing the basic theory and design technology of active array antennas, the realization method of low profile, high efficiency, and light weight of active array antennas is studied. The architecture, analysis method, and engineering practice of active array antennas to realize broadband, multi-band, multi-polarization, and shared-aperture are systematically elaborated. Active packaged antennas are studied to further reduce the low profile of active array antennas, improve efficiency and achieve light weight. Hot research directions such as digital and microwave photonic antennas are studied. The main contents include:

(1) Introduction (this chapter). The characteristics of high-resolution microwave imaging radar and active array antenna, the development history of antennas, the development of semiconductor integrated circuit and hybrid integrated circuit technology and its application in active array antennas, and the technical development of active array antennas are introduced. A new concept of "antenna array microsystem" is proposed, and the new technology and development direction of active array antennas are discussed and prospected.
(2) Array antenna analysis and optimization (Chap. 2). Two kinds of characteristic parameters of antenna port and radiation are introduced, starting from the basic concept. Then, based on the specific application of microwave imaging, the design methods from linear arrays to planar arrays, sparse arrays, and their elements, as well as beamforming optimization techniques with only phase weighting and amplitude/phase weighting, are discussed and analyzed.
(3) Array antenna errors and compensation (Chap. 3). From the engineering design perspective, the error and compensation method of active array antennas, as well as the influence on the radiation characteristics of the antenna, are analyzed. In addition, measurement technology, rapid measurement, and accurate modeling technology of two-dimensional phased array antennas for microwave imaging radar are discussed.
(4) Broadband active array antennas (Chap. 4). The mechanism of limiting the instantaneous bandwidth of active array antennas is analyzed from the aspects of antenna beam pointing deviation, aperture transit time, and signal frequency modulation rate. In addition, the real-time delay line configuration method is analyzed and discussed from the perspective of one-dimensional and two-dimensional active array antenna systems. Finally, the basic principle, classification, performance parameters, and characteristics of microwave delay components are detailed. And standard microwave delay components' design methods and experimental results are given.
(5) Active array module integration (Chap. 5). Active array module integration is the basis for low profile, high efficiency, and light weight of active array antennas. Based on analyzing and discussing the integrated architecture of the active array, the composition, fundamental principle, design method, and integration technology of the "tile-type" active array module are elaborated in detail. In addition, challenging problems such as "tile-type" active array modules, miniaturization of transceiver components, and three-dimensional heterogeneous integration methods are addressed, and typical research results of active array modules are given.
(6) Shared-aperture array antennas (Chap. 6). The requirements and implementation methods of broadband multi-band and multi-polarization shared-aperture

antennas are introduced, and the design technologies of two types of double-line/circular-polarized shared-aperture antennas of microstrip patch and waveguide slot are introduced. On this basis, the latest research achievements of the three-band dual-polarized shared-aperture antenna are presented.

(7) Active packaged array antennas (Chap. 7). From the perspective of technological development, an active array packaged antenna is an antenna form between the active array antenna and the antenna array microsystem. Based on introducing the classification of packaged antennas and broadband packaged antenna elements, multiphysics' coupling and mutual interference mechanism on a micro-scale are analyzed, and the parasitic effects caused by the interaction of multi-parameters are studied. The multi-layer vertical interconnection technologies include inter-board hair button interconnection, inter-board BGA interconnection, inter-board LGA interconnection, intra-board inter-layer interconnection, and inter-chip TSV interconnection are mainly analyzed and studied. The technical methods of realizing the miniaturization, light weight, and high integration of microwave passive devices with embedded devices are explored. Three kinds of packaged antenna materials and technology, LTCC, HTCC, and organic matter, are introduced in detail. Finally, the research results of a millimeter-wave 64-element active array packaged antenna are given.

(8) Digital array antennas (Chap. 8). Based on the outstanding advantages of digital array microwave imaging radar system with high dynamics, low antenna sidelobes, and high beam scanning accuracy, the basic principles of digital array antennas including phase accumulator, phase/amplitude converter, and direct digital frequency synthesis (DDS) are introduced. At the same time, the digital signal generation technology of DDS spectrum characteristic analysis, DDS spurious suppression method, digital sampling, digital down-conversion, and other digital receiving technologies are introduced. Furthermore, the technical approach of reducing the noise figure of a digital array antenna system and improving the system's dynamic range are studied, and the design idea of the distributed frequency source that surpasses the traditional frequency synthesizer is proposed. At the same time, the research results of application examples are given.

(9) Microwave photonic array antennas (Chap. 9). Microwave photonic technology is an emerging technology that integrates microwave technology and photonic technology. According to the functional differences of microwave photonic technology in antenna arrays, microwave photonic digital array antennas, and optically controlled phased array antennas are studied and analyzed. The basic principles and implementation methods of solving the optical real-time delay, microwave signal modulation and demodulation, optical analog-to-digital conversion, and microwave photonic filtering of broadband active array antennas are elaborated. The basic characteristics of microwave photonic devices commonly used in microwave photonic links are introduced. The main performance parameters such as noise source, noise figure, dynamic ransge, isolation, and insertion loss in microwave photonic links are studied and

discussed. At the same time, an experimental system of broadband optically controlled phased array antenna is introduced.

References

1. Li D, Tong Q, Li R et al (2012) Some forefront scientific problems of high resolution earth observation. Sci China: Earth Sci 42(6):805–813
2. Lu J (2017) Design technology of synthetic aperture radar. National Defense Industry Press, Beijing
3. Rahmat Samii Y. How Should we excite non-engineers about our professions as antenna engineers and researchers? [EB/OL]. https://aps.ieee.org/images/pdfs/featureart050609. Accessed 16 May 2019
4. Lu J (2015) The technique challenges and realization of space-borne digital array SAR. In: IEEE 5th asia-pacific conference on synthetic aperture radar (APSAR). IEEE, Singapore, pp 1–5
5. Moore GE (1965) Cramming more components onto integrated circuits. Electronics 38(8):114–117
6. Wang Y, Lu J, Wu X (2006) Application of polarimetric target decomposition in target classification. J Anhui Univ Nat Sci Edn 30(5):33–36
7. Wang Y, Lu J, Zhang F (2007) Target classification algorithm based on polarization synthsis. Comput Sci Dev 17(3):30–32
8. Lu J (2015) Applications of microwave techniques in spaceborne SAR. In: Proceedings of national conference on microwave and millimeter wave 2015 (NCMMW2015). Chinese Institute of Electronics (CIE), Hefei, pp 10–15
9. Yangyuan W, Yongwen W (2012) The history and development law on micro-nano electronics. Scientia Sinica (Inf) 42(12):1485–1508
10. Fischer A, Tong Z, Hamidipour A et al (2014) 77 GHz multi-channel radar transceiver with antenna in package. IEEE Trans Antennas Propag 62(3):1386–1394
11. Zhang YP (2002) Integration of microstrip antenna on ceramic ball grid array package. Electron: Lett 38(1):14–16
12. Beer S, Gulan H, Rusch C, Zwick T (2012) Coplanar 122 GHz antenna array with air cavity reflector for integration in plastic packages. IEEE Antennas Wirel Propag: Lett 11:160–163
13. Luna JZ, Dussopt L, Siligaris A (2013) Hybrid on-chip/in-package integrated antennas for millimeter-wave short-range communications. IEEE Trans: Antennas Propag 61(11):5377–5384
14. Kuo JL, Lu YF, Huang TY et al (2012) 60 GHz four-element phased-array transmit/receive system-in-package using phase compensation techniques in 65 nm flip-chip CMOS process. IEEE Trans: Microw Theory Tech 60(3):743–756
15. Gupta KC, Hall PS (2000) Analysis and design of integrated circuit–antenna modules. Wiley, New York
16. Lu J, Wang W, Lu X, Zhang H et al (2020) Research on three matching problems in waveguide slot antenna. Radar Sci Technol 18(2):115–123
17. Santagata F, Sun FW (2018) System in package (SiP) technology: fundamental, design and applications. Microelectron Int 4:231–243
18. Su YF, Chiang KN (2017) Design and reliability assessment of novel 3D-IC packaging. J Mech 33:193–203
19. Lau JH, Li M, Li QQ et al (2018) Design, materials, process, fabrication, and reliability of fan-out wafer-level packaging. IEEE Trans Compon Packag Manuf Technol 8:991–1002
20. Merkle T, Götzen R, Choi J-Y et al (2015) Polymer multichip module process using 3-D printing technologies for D-band applications. IEEE Trans Microw Theory Tech 63:481

Chapter 2
Array Antenna Analysis and Optimization

Reflector antennas and array antennas are the two main types of antennas used commonly. A reflector antenna is composed of a feed structure and a reflector, which achieves a high antenna gain by generating an equal-phase radiation field on a reflector with a large area. On the other hand, an array antenna comprises a certain number of element antennas arranged in a certain configuration in space. This type of antenna achieves high-gain radiation in a specific direction in space through amplitude and phase weighting. Therefore, the analysis of the radiation characteristics of the reflector antenna mainly focuses on the antenna gain, sidelobe, polarization performance, and efficiency characteristics. As for the array antenna, due to the large freedom of each element's amplitude, phase, and array configuration, compared with the reflector antenna, the array antenna can achieve a higher gain and complex beam functions, such as electronically controlled scanning, directional null, and beamforming. Therefore, array antennas, especially phased array antennas, are increasingly widely used because array antennas are easy to realize patterns with special requirements.

The array antenna research is mainly divided into analysis and synthesis. Firstly, array antenna analysis is to analyze and predict performance parameters such as the pattern and impedance of the array antenna according to the known configuration, amplitude and phase excitation, and other conditions of the array antenna, and then obtain the parameters such as bandwidth, gain, polarization, and pattern of the antenna. The synthesis optimization of the array antenna is the process of determining the array configuration and the excitation of the amplitude and phase of the array elements according to the desired pattern characteristics. Then, the antenna performance parameters (such as gain, sidelobe level, and beam shape) are taken as the optimization goal, the array antenna configuration, excitation amplitude, and phase of the antenna element are comprehensively optimized. In this chapter, the basic theory and related contents of the array involved in the application of microwave imaging active array are discussed in detail.

J. Lu et al., *Active Array Antennas for High Resolution Microwave Imaging Radar*,
https://doi.org/10.1007/978-981-99-1475-3_2

2.1 Basic Parameters

As a device for transmitting or receiving electromagnetic waves in an electronic information system, the antenna's working characteristics are mainly characterized by two types of parameters. One is port parameters, mainly including the antenna's input impedance, and reflection coefficient, which characterize the conversion between the guided wave energy and the free-space energy in the system, and affect the antenna's efficiency. The other is radiation parameters, mainly including gain, sidelobe, and cross-polarization, which characterize the ability and characteristics of transmitting or receiving electromagnetic waves through the antenna, and are the goals of antenna design.

2.1.1 Port Parameters

(1) **Input impedance**

The input impedance of an antenna is a parameter reflecting the characteristics of the antenna circuit, defined as the impedance presented by the antenna at its input. The antenna impedance is equal to the ratio of its input terminal voltage U_{in} to the input terminal current I_{in}, that is

$$Z_{\text{in}} = \frac{U_{\text{in}}}{I_{\text{in}}} = R_{\text{in}} + jX_{\text{in}} \tag{2.1}$$

where R_{in} and X_{in} are the resistive and reactive components of the antenna impedance, respectively.

The antenna is generally directly connected to the feeder, and the input impedance determines the feeder's standing wave state. The relationship between the voltage reflection coefficient Γ of the feeder terminal and the antenna input impedance Z_{in} is

$$\Gamma = \frac{Z_{\text{in}} - Z_c}{Z_{\text{in}} + Z_c} \tag{2.2}$$

$$Z_{\text{in}} = Z_c \frac{1 + \Gamma}{1 - \Gamma} \tag{2.3}$$

where Z_c is the characteristic impedance of the transmission line of the feeder, Z_{in} is the input impedance of the antenna. The reflection coefficient $\Gamma = 0$ is the ideal working state of the antenna, which means that the antenna and the feeder are perfectly matched, and the signal has no energy loss during transmission. It shows that the input impedance Z_{in} of the antenna is equal to the characteristic impedance Z_c of the feeder.

In the field of antennas, another parameter that reflects the matching degree of the antenna is usually used, namely, the voltage standing wave ratio (VSWR), which is the ratio of the adjacent maximum voltage amplitude to the minimum voltage amplitude on the feeding transmission line, that is

$$\mathrm{VSWR} = \frac{|U_{\max}|}{|U_{\min}|} = \frac{1 + |\Gamma|}{1 - |\Gamma|} \tag{2.4}$$

where Γ is the reflection coefficient of the feeder. When the antenna is perfectly matched, $\Gamma = 0$ and VSWR = 1 and it means all the power transmitted on the feeder enters the antenna configuration to generate the maximum radiated power. At the same time, perfect matching indicates that no power is reflected in the feeder and enters the front-end microwave devices, such as amplifiers and oscillators, to avoid amplifiers and oscillators entering an unstable working state, causing the oscillation frequency and output power to change. Non-ideal matching will lead to standing waves on the feeder, correspondingly resulting in an impedance matching efficiency of less than 1, which is expressed as

$$\eta_z = 1 - |\Gamma|^2 \tag{2.5}$$

where Γ is the reflection coefficient of the feeder, the parameter η_z is usually expressed in dB, also known as standing wave loss. The relationship between typical standing wave ratio and reflection coefficient and impedance matching efficiency is shown in Table 2.1.

In phased array antennas, especially in the case of broadband and wide-angle scanning, it is more meaningful to investigate the active standing wave ratio of the antenna. The active standing wave ratio is defined as the ratio of the maximum voltage to the minimum voltage amplitude of the voltage wave transmitted to the input/output port of an antenna element when all elements in the array are excited. Different from the active standing wave ratio defined here, the VSWR is called the passive standing wave ratio. The passive standing wave ratio only considers the feeding of one antenna element under investigation, and the other elements are connected to the matched load, that is, there is no mutual coupling between the elements in the antenna array. The active standing wave ratio considers the situation in which all array elements are fed. In addition to considering the reflected signal caused by the impedance mismatch

Table 2.1 Typical values of VSWR and reflection coefficient

VSWR	Γ/dB	η_z/dB
1	$-\infty$	0
1.2	−20.8	0.04
1.5	−14.0	0.18
2	−9.5	0.51
3	−6.0	1.25
10	−3.5	4.81

Note $\Gamma(\mathrm{dB}) = 20\lg\Gamma$, $\eta_z(\mathrm{dB}) = 10\lg\left(1 - |\Gamma|^2\right)$

of the antenna element itself, the active standing wave ratio also considers the signal coupled to the antenna element by all elements in the array through spatial coupling. The active standing wave ratio is

$$\mathrm{VSWR}_a = \frac{|U_{\max,a}|}{|U_{\min,a}|} = \frac{1 + |S_a|}{1 - |S_a|} \tag{2.6}$$

where VSWR_a is the active standing wave ratio of the antenna element, $U_{\max,a}$ and $U_{\min,a}$ are the maximum voltage and minimum voltage of the total voltage wave transmitted on the antenna port, respectively, and S_a is the active reflection coefficient when all elements of the array are considered to be excited. The active reflection coefficient of the element is related to the amplitude and phase of the excitation of other elements, that is, under the premise of different weighting conditions and scanning states, the active reflection coefficient is different.

Because the active standing wave ratio is related to the amplitude and phase of the excitation of each element in the array, it brings difficulties to the port test. The active standing wave ratio can be directly tested using the power division network to excite the multi-antenna elements and the directional coupler to test the echo. For testing the performance of the weighted-shaped scanning state, additional phase shifters and attenuators are required in the power division network, which further increases the complexity of the test system. Therefore, in order to obtain the active standing waves of the antenna array under various excitation states, the passive standing waves of the antenna elements and the mutual coupling coefficients between the array elements are tested, and then the amplitude and phase weighting calculations are performed to obtain the active standing waves. The calculation formula is expressed as

$$S_{i,a} = \frac{\sum_{n=1}^{N} a_n e^{j\phi_n} S_{\mathrm{in}}}{a_i e^{j\phi_i}}, i = 1, 2, \ldots, N \tag{2.7}$$

where $S_{i,a}$ is the active reflection coefficient of the antenna element i, a_n and ϕ_n are the excitation amplitude and phase of the nth antenna element, S_{in} is the S-parameter between the antenna elements i and n, which is the reflection coefficient of the antenna element i when $i = n$. It can be known from Eq. (2.7) that after obtaining the reflection coefficient S_{ii} of the antenna element itself and the coupling S-parameter between the element and other elements in the array, by changing the excitation amplitude and phase of each element, the antenna element's active standing wave ratios under different excitation states in the array can be obtained. The biggest advantage of this method is that after one test, the active standing wave ratio can be obtained under any weighted shaping and scanning angle in the working frequency band. Figure 2.1 shows an example of a phased antenna element's active reflection loss test result. The coordinate y-axis represents the scanning angle, and the coordinate x-axis represents the operating frequency. According to the above method, it can be seen that high-resolution active return loss results in both frequency and scanning domains can be obtained if the sampling of the element S-parameter test frequency and calculation

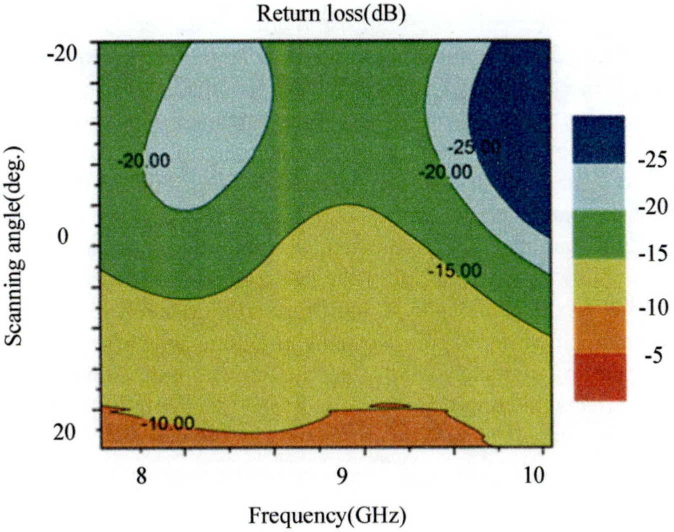

Fig. 2.1 Antenna scanning state active return loss

scan angle is sufficient. In the case of the two-dimensional scanning, it is necessary to set the determination value of the active standing wave exceeding the standard due to a large amount of data. During the calculation process, the frequency domain and scanning domain positions of the active standing wave ratio exceeding the index and sharply deteriorating blind spots are obtained. Obviously, this method is also applicable to the design optimization of phased array antenna elements.

(2) **Bandwidth**

The performance parameters of the antenna, such as input impedance, gain, and axial ratio, all change with the change of the operating frequency, and the working state of the electronic information system allow these performance parameters to change within a certain range. Therefore, when the performance parameters of the antenna vary within the allowable range, the frequency range between the corresponding upper and lower frequency limits is defined as the working frequency band, and the difference between the upper and lower frequency limits is called the bandwidth. Bandwidth is usually divided into absolute bandwidth and relative bandwidth. Absolute bandwidth is referred to the actual value of bandwidth measured in frequency elements, that is, the difference between the upper and lower frequency limits, while relative bandwidth refers to the ratio of the actual bandwidth value to the frequency band center frequency. Relative bandwidth is often used to evaluate the bandwidth performance of an antenna, and an antenna with a bandwidth of less than 10% is generally called a narrow-band antenna. Broadband antennas are generally referred to antennas whose frequency upper and lower limit ratios are greater than 2:1. When the ratio between the upper and lower bandwidth of the antenna is greater than 3:1, it is generally called an Ultrawide Band (UWB) antenna [1].

The input impedance, gain, axial ratio, and other characteristic parameters of the antenna vary within the allowable range, and the corresponding frequency ranges are not the same. Generally, the narrowest bandwidth value is regarded as the antenna bandwidth. Usually, the input impedance of the antenna is a parameter that is more sensitive to frequency changes, and the corresponding tolerance range is generally small. Therefore, in general, the antenna bandwidth is referred to the impedance bandwidth of the antenna. However, on some special occasions, the requirements for other antenna parameters are more stringent than the impedance. At this time, the antenna's bandwidth is referred to the corresponding parameter's bandwidth, usually indicated as the gain bandwidth or the axial ratio bandwidth. In addition, in high-resolution microwave imaging, the gain amplitude and phase characteristics in the antenna operating frequency band affect the pulse compression pattern sidelobe, which in turn affects the range imaging quality, so the amplitude/phase frequency-dependent performance is required.

2.1.2 *Radiation Parameters*

(1) **Directivity**

The directivity and gain of the antenna are the main performance parameters that characterize the radiation characteristics of the antenna and are used to quantitatively describe the distribution of the radiated power of the antenna along various directions in space. The directivity (D) of the antenna is defined as the ratio of the radiation intensity of the antenna along a specified direction to the average radiation intensity of the antenna in all directions. The average radiation intensity of the antenna in all directions is equal to the total power radiated by the antenna divided by the solid angle of the full space 4π. If no specific direction is specified, the directivity is referred to the directivity along the maximum radiation direction of the antenna. This definition can be expressed equivalently as the directivity of a non-isotropic radiating antenna which is the ratio of the radiated intensity of the antenna in a particular direction to the radiated intensity of an isotropic radiating antenna in that direction, that is

$$D = \frac{U}{U_0} = \frac{4\pi U}{P_{\text{rad}}} \tag{2.8}$$

where U is the radiation intensity (unit: watt/steradian), P_{rad} is the total power radiated by the antenna. If no specific direction is specified, the default is referred to the directivity coefficient along the maximum radiation direction, that is

$$D_{\max} = \frac{U_{\max}}{U_0} = \frac{4\pi U_{\max}}{P_{\text{rad}}} \tag{2.9}$$

where D is the directivity (dimensionless), $U_{\max}$ is the maximum radiation intensity, and U_0 is the radiation intensity of the isotropic antenna.

It can be seen from Eq. (2.9) for an isotropic antenna that its directivity is element 1 because U, $U_{\max}$, and U_0 are equal.

To sum up, the physical meaning of the directivity coefficient is that under the condition of the same radiation power, the radiation intensity of a directional antenna in the maximum radiation direction is D times that of an isotropic antenna. Therefore, $P_{\text{rad}} \cdot D$ is also called the antenna's Equivalent Isotropic Radiated Power (EIRP) in this direction. The antenna's function of amplifying the radiation intensity is that the antenna concentrates part of the power radiated by the isotropic antenna in other directions in a specific direction. The narrower the main lobe beam of the antenna, the more obvious the accumulative effect, and the higher the directivity coefficient of the antenna.

For the maximum radiation direction, the antenna is an amplifier of the radiated power, which is achieved by spatially distributing the radiated power. According to the antenna radiation intensity distribution in the whole space $U(\theta, \varphi)$, the directivity coefficient of the antenna can also be obtained. The total power radiated by the antenna is the integral of the radiation intensity of the antenna in the whole space, that is

$$P_{\text{rad}} = \oiint_{\Omega} U(\theta, \varphi) \sin\theta d\theta d\varphi \tag{2.10}$$

where U is the radiation intensity of the antenna, Ω is the solid angle range corresponding to the whole space, θ and φ are the coordinate variables of the spherical coordinate system. The antenna directivity coefficient is expressed as

$$D = \frac{4\pi U}{P_{\text{rad}}} = \frac{4\pi U_{\max}(\theta, \varphi)}{\oint_{\Omega} U(\theta, \varphi) \sin\theta d\theta d\varphi} \tag{2.11}$$

where $U_{\max}$ is the maximum radiation intensity.

In practical applications, the directivity coefficient of the antenna can be estimated under the condition of known beamwidth. The antenna directivity coefficient can be approximately expressed as the ratio of the space solid angle to the main beam solid angle, that is

$$D \approx \frac{4\pi}{\Omega_m} \tag{2.12}$$

where Ω_m is the solid angle of the main beam, which can be expressed as the product of the half-power beamwidth of the main beam along two mutually perpendicular main planes. When using the angle unit system, the directivity coefficient can be expressed as

$$D \approx \frac{4\pi}{\text{HPBW}_E \cdot \text{HPBW}_H} = \frac{4\pi(180/\pi)^2}{\text{HPBW}_E^\circ \cdot \text{HPBW}_H^\circ} = \frac{41253}{\text{HPBW}_E^\circ \cdot \text{HPBW}_H^\circ} \tag{2.13}$$

where HPBW_E° and HPBW_H° are the half-power beamwidths of the two main radiating planes in degrees.

For an antenna with orthogonally polarized components, the polarization directivity coefficient of polarization is the ratio of the antenna's radiated intensity component along a given polarization direction to the average radiant intensity in all directions. According to this definition, the directivity coefficient of the antenna in a specific direction is equal to the sum of the two orthogonal polarization directivity coefficients of the antenna in this direction, and the directivity coefficient of the antenna can be expressed as

$$D = D_\theta + D_\varphi \tag{2.14}$$

$$D_\theta = \frac{4\pi U_\theta}{(P_{\text{rad}})_\theta + (P_{\text{rad}})_\varphi} = \frac{4\pi U_\theta}{P_{\text{rad}}} \tag{2.15}$$

$$D_\varphi = \frac{4\pi U_\varphi}{(P_{\text{rad}})_\theta + (P_{\text{rad}})_\varphi} = \frac{4\pi U_\varphi}{P_{\text{rad}}} \tag{2.16}$$

where U_θ and U_φ are the radiation intensities in the θ and φ directions of the antenna along the specified direction, respectively, and $(P_{\text{rad}})_\theta$, $(P_{\text{rad}})_\varphi$ are the radiated powers corresponding to the θ component and the φ component in all directions of the antenna, respectively.

(2) **Polarization**

The antenna's polarization is referred to the polarization state of the electromagnetic wave radiated by the antenna in the specified direction. Generally, it is referred to the polarization state in the main lobe in the maximum radiation direction. Because the vibration of the electric field and the magnetic field vector in the electromagnetic wave may be located in any direction of the cross-section of the propagation direction, and there is a situation where the vibration changes with the propagation, that is, the motion trajectory of the field vector in the cross-section may be linear or elliptical. As a result, different polarization states of electromagnetic waves are formed.

The polarization state of the electromagnetic wave is distinguished by the trajectory formed by the end of the electric field vector of the electromagnetic wave. Generally, the vibration direction of the electric vector is used to describe the polarization state of the electromagnetic wave. It can be divided into three types: linear polarization, circular polarization, and elliptical polarization. If the electric field vector of the point under consideration is always on a straight line, the polarization state of the electric field at this point is linear polarization. More generally, the trajectory curve formed by the end of the electric field vector is an ellipse, and the polarization state of the field is elliptical polarization at this time. Finally, if the major and minor axes of the trajectory ellipse are equal, the polarization state of the field at this time is circular polarization.

For the circular polarization state of the electromagnetic wave, considering that the electric field vector rotates clockwise or counterclockwise, a circular trajectory

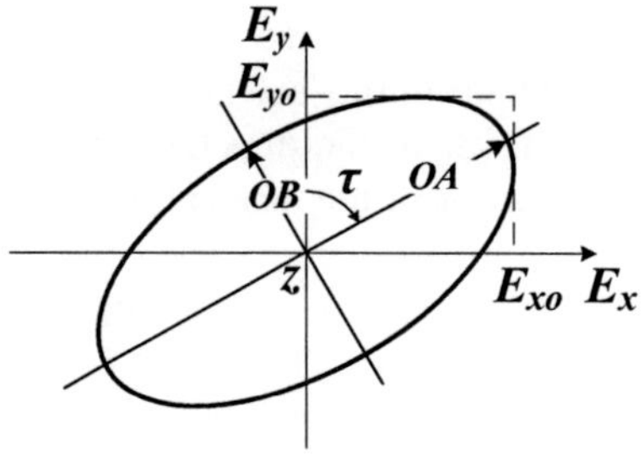

Fig. 2.2 Axial ratio definition diagram

can be formed, and the rotation direction of the electric field vector is observed along the direction of electromagnetic wave propagation. If it conforms to the right-hand rule, the polarization state is called right-handed circular polarization, otherwise, it is left-handed circular polarization. Note that the definition of left-handed and right-handed in the antenna domain is opposite to that in the optical domain [2].

For elliptical polarization, the trajectory formed by the end of the electric field vector is an inclined ellipse, as shown in Fig. 2.2. In order to describe the distribution of the electric field vector in all directions in the polarization plane, the parameter of the axial ratio is defined, which is defined as the ratio of the electric field intensities along the long and short axes of the ellipse, which can be expressed as

$$\mathrm{AR} = \frac{\mathrm{OA}}{\mathrm{OB}}, 1 \leq \mathrm{AR} \leq \infty \tag{2.17}$$

where OA is the length of the long axis of the electric field vector trajectory ellipse, OB is the length of the short axis of the trajectory ellipse.

Both linear and circular polarization can be regarded as special cases of elliptical polarization. According to the definition of the axial ratio, it can be calculated that the axial ratio of the linearly polarized wave is ∞, and the axial ratio of the circularly polarized wave is 1. In practical applications, the axis ratio expressed in dB is usually used, and the relationship between the dB value and the above definition is

$$\mathrm{AR(dB)} = 20\,\mathrm{lg}\,\mathrm{AR} \tag{2.18}$$

where AR is the polarization axis ratio.

Suppose the platform carrying the antenna is moving, including translation and rotation. In that case, the electromagnetic waves transmitted or received by the antenna often have a certain orientation deviation, which will cause the polarization state of the incident wave to be inconsistent with the polarization state of the antenna. And this will lead to a decrease in the efficiency of the antenna. The phenomenon that the antenna efficiency is reduced due to incomplete polarization matching is called the polarization mismatch phenomenon, and the corresponding reduction ratio of antenna efficiency is called the Polarization Loss Factor (PLF). The polarization mismatch factor can be obtained by multiplying the polarization vectors of the two antennas. Assume that the polarization vector of the radiated electromagnetic wave of the transmitting antenna is $\boldsymbol{\rho}_t$, and the antenna polarization vector of the receiving antenna along the direction of the incident wave is $\boldsymbol{\rho}_r$, that is, the electric field polarization of the incident wave and the receiving antenna can be expressed as

$$E_t = \rho_t \cdot E_t, E_r = \rho_r \cdot E_r \tag{2.19}$$

then the PLF can be expressed as

$$\text{PLF} = |\rho_t \cdot \rho_r|^2 = |\cos\psi_p|^2 \tag{2.20}$$

where $\boldsymbol{\rho}_t$ is the polarization vector of the transmitting antenna along the transmitting direction, $\boldsymbol{\rho}_r$ is the antenna polarization vector of the receiving antenna along the incident wave direction, E_t is the electric field vector of the transmitting antenna, and $\boldsymbol{E}_r$ is the electric field vector component of the receiving antenna. In the case of linear polarization, and ψ_p corresponds to the angle between the two polarization directions, while in the state of general circular polarization and elliptical polarization, it can be considered as the generalized angle between the two polarization states. Taking the transmission between a circularly polarized antenna and a linearly polarized antenna as an example, the polarization transmission efficiency between the two can be calculated as

$$\rho_t = \frac{1}{\sqrt{2}}x + \frac{1}{\sqrt{2}}y, \quad \rho_r = x \tag{2.21}$$

$$\text{PLF} = |\rho_t \cdot \rho_r|^2 = \frac{1}{2} = -3\ \text{dB} \tag{2.22}$$

Equation (2.22) shows that a 3 dB power loss will be caused by a polarization mismatch for transmission between a circularly polarized antenna and a linearly polarized one.

In the application of multi-polarization microwave imaging, polarization information is usually used. According to the international standard IEEE Std145-1993, the main polarization is the polarization component in the specified direction, and the cross-polarization is the component orthogonal to the main polarization. Circular polarization is only related to the beam propagation direction and is unique. However, there is no clear linear polarization definition, and the main polarization and cross-polarization proposed by A. C. Ludwig are usually adopted [3], as shown in Fig. 2.3. Among the three definitions, the first reference polarization is the plane wave polarization in the rectangular coordinate system, the second reference polarization is the polarization of the fundamental electric dipole, and the cross-polarization is the polarization of the magnetic dipole on the same axis, The three reference polarizations are the polarizations of the Huygens source, and the cross-polarization is the polarization of the Huygens source whose aperture field is rotated by 90°. The last

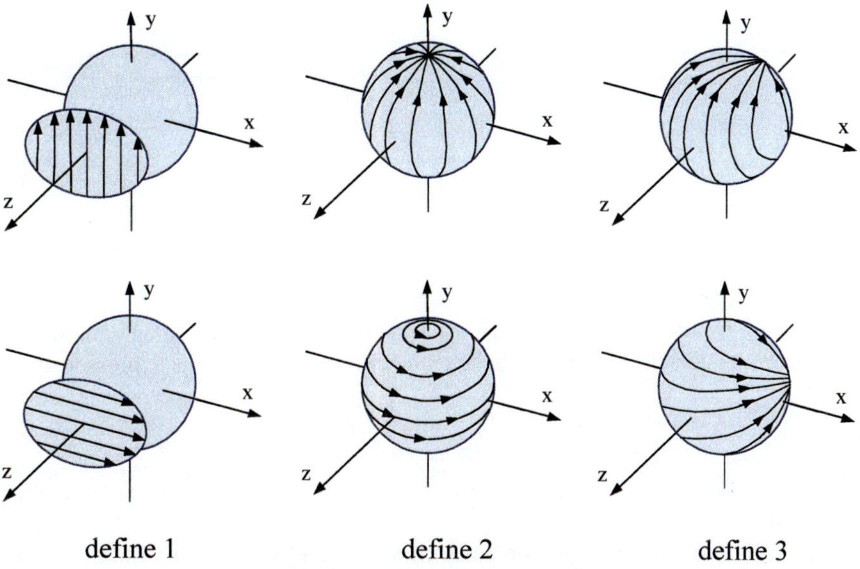

Fig. 2.3 Ludwig linear polarized antenna main/cross-polarization definition

one fits the application scenario and is easy to use. Therefore, the third definition is usually applied in the field of radar.

(3) **Radiation efficiency**

The radiation efficiency of an antenna is used to characterize the ratio of the energy actually fed into the antenna to be radiated out, which is defined as the ratio of the radiated power of the antenna to the actual power fed into the antenna. The difference between the power fed into the antenna and the power radiated by the antenna is the loss caused by the energy radiation process of the antenna, including high-order mode loss, medium and metal loss, etc. When the electromagnetic wave propagates in the antenna configuration, the dielectric and metal materials will generate heat loss. As shown in Fig. 2.4, the radiation efficiency of the antenna is

$$\eta_{\text{rad}} = \frac{R_r}{R_r + R_L} \tag{2.23}$$

where R_r is the antenna's radiation resistance, and R_L is the antenna's resistance.

According to the formation process of the antenna's radiation efficiency, improving the antenna's radiation efficiency can be achieved by reducing the loss tangent of the dielectric material and increasing the electrical conductivity of the metal material.

In practical applications, the parameter "Antenna Efficiency" is sometimes used, which is defined similarly to the radiation efficiency of an antenna. The difference between the two is that the antenna efficiency is based on the total power fed into

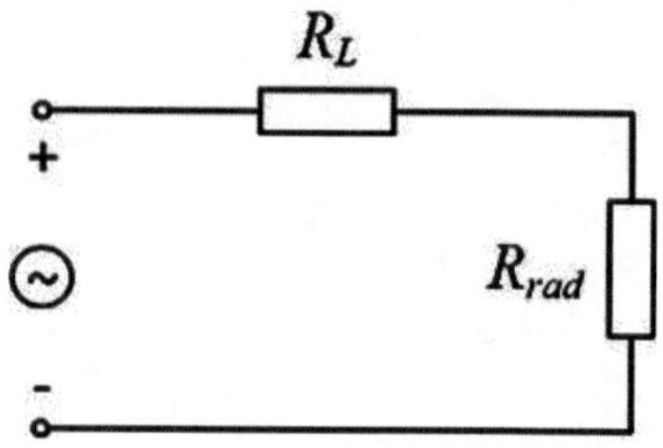

Fig. 2.4 Antenna radiation equivalent circuit diagram

the antenna by the feeder, including the power reflected from the antenna due to impedance mismatch. Therefore, the antenna efficiency η_{ant} can be expressed as

$$\eta_{\text{ant}} = \eta_Z \eta_{\text{rad}} = \left(1 - |\Gamma|^2\right)\eta_{\text{rad}} \tag{2.24}$$

where Γ is the reflection coefficient of the antenna, η_{rad} is the antenna's radiation efficiency.

The difference between the antenna efficiency and the antenna radiation efficiency is the standing wave loss, that is, the loss caused by the standing wave due to impedance mismatch. Usually, the efficiency of the antenna is required to be maximized. Therefore, the impedance matching efficiency and antenna radiation efficiency of the antenna should be improved as much as possible, and the standing wave loss and dielectric and metal loss of the antenna should be reduced.

2.2 Linear Arrays

Limited by the aperture of a single antenna element, the gain of the antenna element is generally small. However, in many applications, the antenna has high directivity or gain to meet the ground's long-range, high-resolution microwave imaging needs. In order to improve the gain of the antenna, it is necessary to arrange the antenna elements in an array, and the antenna composed of multiple antenna elements is called an array antenna.

The total field radiated by the array antenna is the vector superposition of the fields radiated by the individual elements in the array. Usually, the array antenna is composed of antenna elements with the same configuration. Therefore, the factors affecting the radiation pattern of the array antenna are mainly the distance between the array elements and the geometric layout of the array (such as linear array, circular array, rectangular array, and spherical array), the excitation amplitude, and phase of the antenna element, the radiation pattern of the antenna element itself, etc.

2.2.1 Linear Arrays

The linear array is the simplest and most commonly used array form among the array antennas. The parameters affecting the array configuration in the linear array are mainly the element pattern, the number of array elements, and the array elements' spacing. The more commonly used linear array form is the equidistant linear array. After the antenna elements are formed into an array, the radiation characteristics of the array are determined by the element pattern, the array configuration, and the excitation amplitude of the antenna elements. Figure 2.5 shows a two-element array. Now we analyze and discuss the relationship between the radiation characteristics of the array and the antenna array.

It is assumed that in the two-element array, the antenna elements are infinitely small dipoles, and the two dipoles are arranged in a direction perpendicular to the direction of the dipoles. The radiation field of the two-element array is the superposition of the respective radiation fields of the two array elements. Assuming that the excitation amplitudes of the two array elements are equal, then have

$$E_t = E_1 + E_2 = \alpha_\theta j\eta \frac{kI_0 l}{4\pi}\left\{\frac{e^{-j[kr_1-(\beta/2)]}}{r_1}\cos\theta_1 + \frac{e^{-j[kr_2+(\beta/2)]}}{r_2}\cos\theta_2\right\} \quad (2.25)$$

where β is the excitation phase difference between the two elements. Under the condition that the observation point is located in the far field, the phase in Eq. (2.25) can be approximated, that is

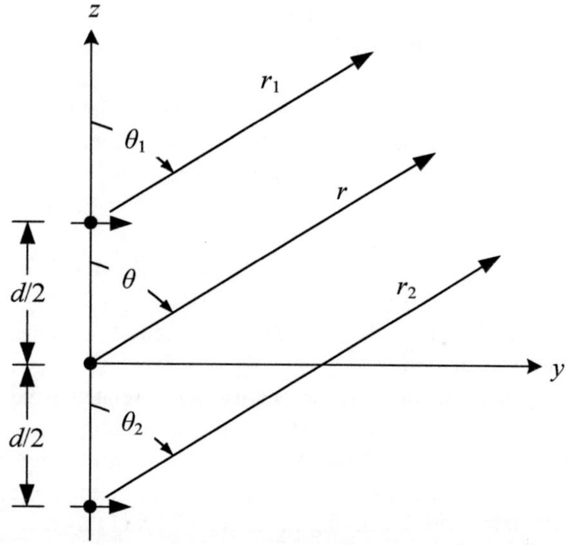

Fig. 2.5 Schematic diagram of two-element array

$$\theta_1 \simeq \theta_2 \simeq \theta$$

$$\text{Phase term}\begin{cases} r_1 \simeq r - \frac{d}{2}\cos\theta \\ r_2 \simeq r + \frac{d}{2}\cos\theta \end{cases}$$

$$r_1 // r_2 // r$$

The total radiated field of the array can be simplified as

$$\begin{aligned} E_t &= \alpha_\theta j\eta \frac{kI_0 l}{4\pi}\cos\theta\left[e^{j(\mathrm{kd}\cos\theta+\beta)/2} + e^{-j(\mathrm{kd}\cos\theta+\beta)/2}\right] \\ &= \alpha_\theta j\eta \frac{kI_0 l}{4\pi}\cos\theta\left\{2\cos\left[\frac{1}{2}(\mathrm{kd}\cos\theta + \beta)\right]\right\} \end{aligned} \tag{2.26}$$

where η is the free-space wave impedance, I_0 is the antenna excitation current.

It can be seen from Eq. (2.26) that the total radiation field of a two-element array can be expressed as the product of the field of a radiating element located at the center of the array and a factor independent of the element pattern. This factor, independent of the element pattern, is called the array factor. For a binary array fed with equal amplitude, the array factor is

$$\mathrm{AF} = 2\cos\left[\frac{1}{2}(\mathrm{kd}\cos\theta + \beta)\right] \tag{2.27}$$

where d is the distance between the antenna elements, k is the free-space wave number, β is the excitation phase difference between the two array elements.

The array factor is a function related to the array geometry and the magnitude and phase of the array element excitation. Therefore, the array factor and antenna pattern can be changed by adjusting the array element spacing, excitation amplitude, and phase.

The above analysis points out that the far-field electric field of an equal-amplitude two-element array is equal to the product of the radiation field of the antenna element located at the center of the array and the array factor of the array. Extending the above expression to the general array pattern, it can be expressed as

$$E_t = E_i \cdot \mathrm{AF} \tag{2.28}$$

where E_i is the radiation field of the element antenna, and AF is the array factor of the array antenna.

In general, the array factor is a function related to the number of array elements, the distance, and geometry between the array elements, and the relative magnitude and phase of the feeds. Since the array factor has nothing to do with the array element's pattern, it is usually assumed that the array element is an isotropic point source when analyzing and discussing the array factor. In this case, the antenna array's radiation field is the antenna array's array factor. According to the multiplication principle of the pattern, the radiation pattern of a real array antenna can be obtained by multiplying the radiation direction of the antenna element and the antenna array factor.

2.2.2 *Equal-Amplitude Linear Arrays*

The equal-amplitude linear array is referred to a linear array with the same excitation amplitude of each antenna element in the array, which is relatively simple and commonly used in linear arrays, as shown in Fig. 2.6.

Each array element is replaced by an isotropic point source, and the excitation factor of the i-th element is set as

$$M_i = I_i e^{j(i-1)\beta} \tag{2.29}$$

where β is the phase difference between adjacent elements. Since the amplitude of each array element is the same, then $I_i = 1$, and its array factor is

$$\begin{aligned} \mathrm{AF} &= 1 + e^{j(\mathrm{kd}\cos\theta+\beta)} + e^{j2(\mathrm{kd}\cos\theta+\beta)} + \cdots + e^{j(N-1)(\mathrm{kd}\cos\theta+\beta)} \\ &= \sum_{n-1}^{N} e^{j(n-1)(\mathrm{kd}\cos\theta+\beta)} \end{aligned} \tag{2.30}$$

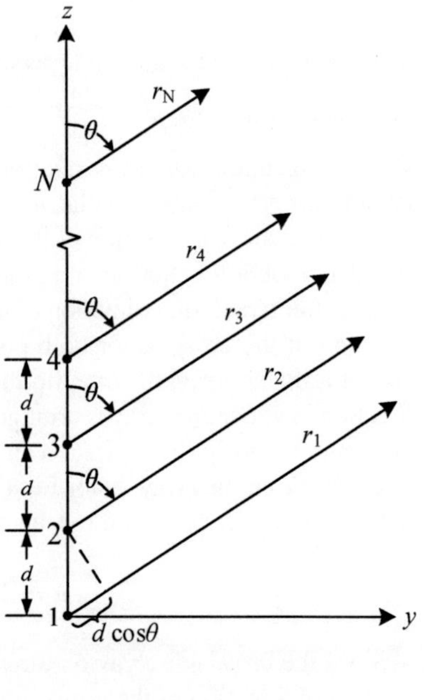

Fig. 2.6 Schematic diagram of uniform linear array

where d is the distance between antenna elements, k is the free-space wave number, β is the phase difference between adjacent elements, N is the number of antenna elements.

In this case, the array factor is usually abbreviated as

$$\mathrm{AF} = \sum_{n-1}^{N} e^{j(n-1)\psi}, \ \psi = \mathrm{kd}\ \cos\theta + \beta \tag{2.31}$$

According to the summation rule of geometric series, Eq. (2.31) can be further simplified as

$$\mathrm{AF} = \frac{e^{jN\psi} - 1}{e^{j\psi} - 1} = e^{j[(N-1)/2]\psi} \frac{\sin \frac{N}{2}\psi}{\sin \frac{1}{2}\psi} \tag{2.32}$$

If the phase reference point is taken as the center point of the array, the array factor can be simplified to

$$\mathrm{AF} = \frac{\sin \frac{N}{2}\psi}{\sin \frac{1}{2}\psi} \tag{2.33}$$

In summary, the array factor relates to the array element spacing d and the progressive feeding phase difference β between the array elements. When the array element spacing and feeding phase difference β are different, the radiation conditions of the array are also different. The more typical linear arrays include broadside arrays, end-fire arrays, and scanning arrays.

(1) **Broadside arrays**

Many application scenarios require that the maximum radiation direction of the antenna array be perpendicular to the arrangement plane of the array, that is, the maximum radiation direction of the array points to the normal direction of the array. In order to obtain better antenna radiation performance, it is usually required that the maximum radiation direction of the antenna element and the maximum radiation direction of the array factor both point to the normal direction of the array. In the case of a linear array, the direction is referred to the direction perpendicular to the direction in which the array is arranged. According to the analysis of the array factor, the requirements of the excitation amplitude and phase of the broadside array can be obtained. From the array factor Eq. (2.31) of the equal-amplitude linear array, it can be known that the condition for the array factor to obtain the maximum value is

$$\psi = \mathrm{kd}\cos\theta + \beta = 0 \tag{2.34}$$

Since the broadside array requires the maximum radiation direction to appear in the normal direction of the array arrangement direction, that is, $\theta = 90°$, there is

$$\psi = \text{kd}\cos\theta + \beta|_{\theta=90^\circ} = \beta = 0 \tag{2.35}$$

Therefore, to make the linear array's maximum radiation direction face the broadside direction, the excitation phases of all array elements are required to be the same. In the above analysis process, the array element spacing d can take any value, but in order to ensure that the array factor does not appear grating lobes in other directions, the array element spacing is required to be less than one wavelength ($d < \lambda$).

For an antenna array with many elements, the antenna array's pattern shape mainly depends on its array factor's pattern. The half-power beamwidth of the array factor can be obtained according to the half-power point angle $\theta_{1/2}$ of the main lobe of the array factor pattern, so that

$$\text{AF}_n = \frac{\sin\frac{N}{2}\psi}{N\sin\frac{1}{2}\psi} = \frac{\sin\left(\frac{N}{2}\text{kd}\cos\theta\right)}{N\sin\left(\frac{1}{2}\text{kd}\cos\theta\right)} = \frac{\sqrt{2}}{2} \tag{2.36}$$

can obtain

$$\frac{N}{2}\text{kd}\cos\theta = 1.391 \tag{2.37}$$

Since the maximum direction of the array factor appears in the direction of $\theta = \pi/2$, the half-power beamwidth of the array factor is

$$\text{HPBW} = 2\arcsin\frac{2.782}{\text{Nkd}} = 2\arcsin 0.443\frac{\lambda}{\text{Nd}} \approx 0.886\frac{\lambda}{\text{Nd}} \tag{2.38}$$

where N is the number of antenna elements, d is the distance between antenna elements, and λ is the working wavelength of the antenna.

It can be seen from Eq. (2.38) that the half-power beamwidth of a broadside array is inversely proportional to the number of array elements N. The more the number of array elements, the narrower the lobe.

When the array factor is set to 0, the null direction of the array factor can be obtained. At this time, have

$$\frac{N}{2}\text{kd}\cos\theta = \pm n\pi, n = 1, 2, 3, \ldots \tag{2.39}$$

$$\theta = \arccos \pm \frac{n\lambda}{\text{Nd}} \tag{2.40}$$

When the antenna array is used for reception, no interference signal will be received in these null directions, so in the antenna design, the null direction is often aligned with the interference direction. From the null distribution of the antenna array factor, it can be known that the First-Null Beamwidth (FNBW) of the broadside array is

$$\mathrm{FNBW} = 2\arcsin\frac{\lambda}{\mathrm{Nd}} \approx \frac{2\lambda}{\mathrm{Nd}} \tag{2.41}$$

The ratio of the maximum value of the sidelobe of the antenna array to the maximum value of the main lobe is defined as the sidelobe level (SLL), which is usually expressed in decibels. From Eq. (2.36) of the array factor, it can be known that the first sidelobe of the broadside array is approximately

$$\mathrm{AF}_n = \left.\frac{\sin\frac{N}{2}\psi}{N\sin\frac{1}{2}\psi}\right|_{\frac{N}{2}\psi\approx\frac{3\pi}{2}} = 0.212 = -13.46\,\mathrm{dB} \tag{2.42}$$

That is, the first-SLL of the broadside array is about −13.46 dB.

(2) **End-fire arrays**

In addition to the broadside array, another typical radiation direction of the antenna array is the end-fire. That is, the array's maximum radiation direction is the linear array's arrangement direction. Substitute $\theta = 0°$ in the end-fire direction into Eq. (2.31), and according to the maximum radiation direction condition $\psi = 0$ of the equal-amplitude linear array, we can get

$$\psi = \mathrm{kd}\cos\theta + \beta|_{\theta=0°} = \mathrm{kd} + \beta = 0 \tag{2.43}$$

Therefore, the radiation direction of the antenna array points to the end-fire direction, and the requirement for the feed phase is $\beta = -\mathrm{kd}$. In the end-fire direction, since the radiation fields of each element are superimposed in-phase in this direction, the resulting maximum is synthesized.

Beam width and null orientation. Under the condition of $\beta = -\mathrm{kd}$, the array factor of the end-fire array can be expressed as

$$\mathrm{AF}_n = \frac{\sin\frac{N}{2}\psi}{N\sin\frac{1}{2}\psi} = \frac{\sin\left[\frac{N}{2}\mathrm{kd}(\cos\theta - 1)\right]}{N\sin\left[\frac{1}{2}\mathrm{kd}(\cos\theta - 1)\right]} \tag{2.44}$$

where N is the number of array elements, d is the distance between array elements, k is the free-space wave number.

Let the array factor be $\sqrt{2}/2$ half-power state, the half-power point angle can be obtained, namely,

$$\mathrm{AF}_n = \frac{\sin\frac{N}{2}\psi}{N\sin\frac{1}{2}\psi} = \frac{\sin\left[\frac{N}{2}\mathrm{kd}(\cos\theta - 1)\right]}{N\sin\left[\frac{1}{2}\mathrm{kd}(\cos\theta - 1)\right]} = \frac{\sqrt{2}}{2} \tag{2.45}$$

can obtain

$$\cos\theta = 1 - \frac{1.394}{\mathrm{Nkd}} = 1 - \frac{0.444\lambda}{\mathrm{Nd}} = 1 - 2\sin^2\frac{\theta}{2} \tag{2.46}$$

So the HPBW is

$$\mathrm{HPBW} = 2\theta = 4\arcsin\sqrt{\frac{0.222\lambda}{\mathrm{Nd}}} \approx 1.88\sqrt{\frac{\lambda}{\mathrm{Nd}}} \tag{2.47}$$

Equation (2.47) shows that, unlike the case of the broadside array, the half-power beamwidth of the end-fire array is inversely proportional to the square root of the number N of array elements. Under the same conditions, the beamwidth of the end-fire array is significantly wider than that of the broadside array, and its gain is correspondingly lower than that of the broadside array.

The null direction of the end-fire array can also be obtained by the expression of the array factor, that is

$$\cos\theta - 1 = -\frac{n\lambda}{\mathrm{Nd}}, n = 1, 2, 3, \ldots \tag{2.48}$$

$$\theta = \arccos\left(1 - \frac{n\lambda}{\mathrm{Nd}}\right) = 2\arcsin\sqrt{\frac{n\lambda}{2\mathrm{Nd}}} \tag{2.49}$$

Its nth zero beamwidth is

$$\mathrm{FNBW} = 2\theta = 4\arcsin\sqrt{\frac{n\lambda}{2\mathrm{Nd}}} \tag{2.50}$$

where N is the number of antenna elements, λ is the working wavelength, d is the distance between antenna elements.

From the array factor Eqs. (2.36) and (2.44) of the end-fire array and the broadside array, it can be known that the sidelobe distribution position of the end-fire array is different from that of the broadside array, but its SLL value is the same with that of the broadside array, and the first sidelobe is also −13.46 dB, that is

$$\mathrm{AF}_n = \left.\frac{\sin\frac{N}{2}\psi}{N\sin\frac{1}{2}\psi}\right|_{\frac{N}{2}\psi\approx\frac{3\pi}{2}} = 0.212 = -13.46\,\mathrm{dB} \tag{2.51}$$

(3) **Scanning arrays**

It can be seen from the previous analysis that when the phase difference between the adjacent elements of the N-element equal-amplitude linear array $\beta = 0$, the radiation direction of the array is the broadside direction, when the phase difference between adjacent elements $\beta = -\mathrm{kd}$, the radiation direction is the end-fire direction. It can be seen that the maximum radiation direction can be changed by controlling the phase difference β of adjacent elements. Assuming that the maximum radiation direction of the expected array occurs at $\theta = \theta_0$, then

$$\begin{aligned} \psi = \mathrm{kd}\cos\theta + \beta|_{\theta=\theta_0} = \mathrm{kd}\cos\theta_0 + \beta = 0 \\ \beta = -\mathrm{kd}\cos\theta_0 \end{aligned} \tag{2.52}$$

Therefore, by adjusting the array element's progressive phase difference, the antenna array's maximum radiation direction can be controlled to point to any specified direction, which is the basic principle of the phased array antenna.

According to the periodicity of the array factor function $\frac{\sin\frac{N}{2}\psi}{N\sin\frac{1}{2}\psi}$, when $\psi = \pm 2n\pi$, the array factor takes the maximum value. Therefore, grating lobes will be generated when the array factor takes multiple maximum values. Therefore, to suppress the generation of grating lobes, the spacing of the array elements needs to meet the following conditions:

$$|u|_{\max} < 2\pi \tag{2.53}$$

$$\frac{2\pi d}{\lambda}|\cos\theta - \cos\theta_0|_{\max} < 2\pi \tag{2.54}$$

Therefore there is

$$d < \frac{\lambda}{1 + |\cos\theta_0|} \tag{2.55}$$

θ in Eq. (2.55) is calculated from the axis direction of the array. If θ is calculated from the normal direction of the array, $\cos\theta_0$ should be changed to $\sin\theta_0$, that is

$$d < \frac{\lambda}{1 + \sin\theta_0} \tag{2.56}$$

The maximum direction of the grating lobe corresponding to Eq. (2.55) does not appear in the visible area, but the part of the lobe from the maximum direction of the grating lobe to its nearest null can still fall within the visible area. The strict requirement is that this part of the grating lobe does not appear in the visible area, then there is

$$\frac{2\pi d}{\lambda}|\cos\theta - \cos\theta_0|_{\max} < 2\pi - \frac{\pi}{N} \tag{2.57}$$

So

$$d < \frac{\lambda}{1 + |\cos\theta_0|}\left(1 - \frac{1}{2N}\right) \tag{2.58}$$

If θ is calculated from the normal direction of the array, then the requirement is

$$d < \frac{\lambda}{1 + |\sin\theta_0|}\left(1 - \frac{1}{2N}\right) \tag{2.59}$$

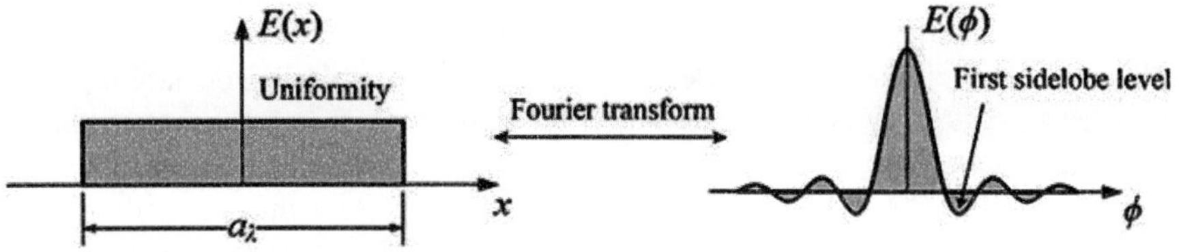

Fig. 2.7 Relationship between linear array aperture distribution and radiation pattern

2.2.3 Unequal-Amplitude Linear Arrays

The SLL of the aforementioned equal-amplitude linear array is all about −13.46 dB. In some applications, the SLL is too high, which cannot meet the application requirements. For example, in radar applications, the SLL is often required to be lower than −30 dB to reduce signal interference from other directions, and there are similar requirements in high-resolution microwave imaging. For the low SLL requirement proposed in this type of application, the equal-amplitude distribution cannot make it, so it is necessary to make corresponding improvements to the equal-amplitude linear array to reduce the SLL. The analysis shows that the SLL of the antenna array is mainly related to the spatial distribution of the current amplitude of the antenna array. When the array current changes with the spatial position of the array element, an unequal-amplitude linear array is produced. In order to control the SLL of the array to reach a specific value, it is necessary to do a detailed study on the relationship between the amplitude distribution of each array element in the unequal-amplitude linear array and the SLL of the array.

Due to the Fourier transform relationship between the radiation pattern of the antenna array and its aperture field distribution, the pattern of the constant-amplitude linear array has a shape similar to the sinc function. Therefore, the SLL of the equal-amplitude linear array is the first-SLL of the sinc function, as shown in Fig. 2.7. By adjusting the amplitude distribution of the array elements in the antenna array, the SLL of the antenna array can be controlled.

2.2.4 Unequally Spaced Linear Arrays

An unequally spaced antenna array is referred to an antenna array in which the spacing between adjacent elements in the array is not a fixed value. There are two main types of common unequally spaced antenna arrays. One is an array in which the spacing of the array elements increases symmetrically from the array's center to both sides with certain rule, the other is an equally spaced linear array based on a certain rule. An array is formed by extracting a certain percentage of the array elements. Of course, there is also a completely irregular random element spacing array form, which is rarely used because it is difficult to achieve large-scale implementation in engineering. Unequally spaced arrays can be called sparse arrays and can also be called density-weighted arrays. The first type of array is also called density-tapering array because its array density distribution decreases from the center to both sides.

The advantage of the unequally spaced linear array is that the number of cells is reduced. In order to control the parameters, such as the SLL of the unequally spaced linear array, it is necessary to analyze the parameters, such as the array factor. Unlike the equally spaced linear arrays, the pattern characteristics and sidelobe characteristics of unequally spaced linear arrays generally cannot be obtained analytically and can only be solved by numerical methods. Therefore, the analysis and synthesis process of the unequally spaced linear array is realized by analyzing and optimizing the related numerical problems.

For the first type of unequally spaced array, the array element spacing is not limited to an integer multiple of a specific value and can take any value. Common analysis methods include matrix, perturbation, equivalent current, and statistical methods [4]. The matrix method is a method of approximating the radiation lobe of the antenna with a finite-term Fourier series. The inverse matrix is required at this time, called the matrix method. The perturbation method obtains a density-weighted array by modifying the arrangement of the antenna elements of the equally spaced linear array, thereby deriving the array factor of the antenna array. The equivalent current method is to equivalently express the spacing occupied by the antenna elements as the aperture current distribution and then derive the antenna array factor. Finally, the statistical method uses random numbers that satisfy a certain probability density function to determine the position of the antenna element. In this case, the statistical values of antenna gain, beamwidth, and SLL can be obtained through relevant statistical theory.

For the second type of unequally spaced array, the array element spacing is an integer multiple of a primary spacing. It is usually a uniform equidistant array after removing some lattice points. Common analysis methods include enumeration, optimization, and statistical methods. The enumeration method calculates the corresponding pattern for all possible arrangements of the missing element array and selects the arrangement with the lowest first-SLL. The result obtained by this method is the global optimal solution, but when the number of cells is large, the amount of calculation will be huge, and the time and computational cost will be high, making it difficult to achieve, so it is not suitable for the design of large-scale arrays. The optimization method is also called the dynamic programming method. This method starts from a certain array layout, forms a new array layout by adjusting the positions of individual cells, selects the layout with better performance by comparing the sidelobe performance of the new layout and the old layout, and iterates the above process until an array layout that meets the requirements of sidelobe performance is obtained, or the sidelobe performance cannot be further improved. The statistical method obtains an aperture current distribution that can produce a lower SLL through theoretical analysis, uses this distribution as a probability density function, and then is referred to the function statistics to determine whether to place or not place cells at each position. This results in a density-weighted matrix where the distribution density of the antenna elements is the same as the expected current distribution. For each possible layout optimized by this process, the first-SLL is higher or lower than the ideal current distribution, but the statistical average of the various arrangements should be consistent with the expected SLL of the aperture current distribution. When the reference aperture current distribution is appropriately selected, an antenna array with a lower SLL can generally be obtained.

For large linear arrays with random element spacing, the beamwidth of the random array converges to that of the average pattern. Therefore, this performance parameter is less affected by the number of array element reduction, so it is more advantageous to use a sparse array with random spacing to obtain narrow beams [5]. However, its directivity coefficient D is still proportional to the total number of array elements N. At the same time, a certain element density is required to obtain a given low SLL, that is, there is a minimum limit on the number of array elements N.

2.2.5 Effect of Element Pattern on Array Pattern

For the actual antenna, its pattern is anisotropic, and the actual radiation characteristic of the antenna array is affected by the pattern of the antenna element. In order to examine the effect between the antenna element pattern and the array factor, the principle of pattern multiplication is described here. The radiation pattern of an antenna array is the product of the radiation pattern of the antenna element and the radiation pattern of an isotropic point source array excited at the same position, the same relative amplitude, and phase as the array, namely,

$$E(\theta, \varphi) = f(\theta, \varphi) g_a(\theta, \varphi) \tag{2.60}$$

where $f(\theta, \varphi)$ is the array factor, $g_a(\theta, \varphi)$ is the pattern of the antenna element, θ and φ are the coordinate variables of the spherical coordinate system.

In general, forming an array of antenna elements aims to improve the antenna gain and achieve higher directivity. Therefore, the antenna element's maximum radiation direction and the array factor's maximum radiation direction are usually designed to be in the same direction to obtain the highest possible gain.

In order to investigate the effect between the element antenna pattern and the array factor, we first discuss a common coaxial two-element dipole array. Since the pattern of the dipole itself is a Fig. 2.8, the array factor of the two-element array (equal amplitude and phase) is also a Fig. 2.8, and the product of the two will form a Fig. 2.8 pattern with a narrower beam, as shown in Fig. 2.8a.

If the dipoles in the array are rotated 90°, they point perpendicular to the direction of the array layout. As a result, the element pattern maximum of the dipole will be aligned with the null of the array factor, and the null of the element pattern will be aligned with the maximum of the array factor. At this time, the pattern will generate 4 nulls, and the maximum radiation direction of the antenna will point to an oblique direction, as shown in Fig. 2.8b. The electric field strength in this direction is also lower than the maximum value of the pattern shown in Fig. 2.8a. By comparing

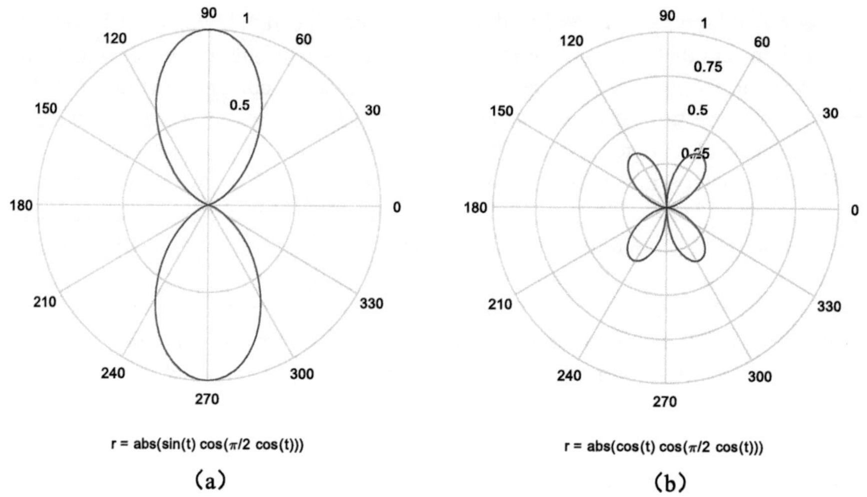

Fig. 2.8 Influence of dipole spatial orientation on array pattern. **a** The dipole orientation is consistent with the array direction, **b** the dipole orientation is perpendicular to the array direction

Fig. 2.8a, b, it can be found that the element pattern greatly influences the antenna array's overall radiation pattern.

In practical application, the antenna element's maximum radiation direction and the array factor's maximum radiation direction are usually aligned to obtain the highest gain possible. At this time, the shape of the element pattern will affect the sidelobes of the antenna array. Figure 2.9 shows the change of the normalized direction coefficient of the antenna array with the beamwidth when the width of the antenna element pattern is different. Among them, Fig. 2.9a is the array pattern with isotropic radiation source as the array element, and Fig. 2.9b is the array pattern when the element pattern is $\cos\theta$, $\cos^2\theta$, $\cos^3\theta$, $\cos^4\theta$, respectively. It can be seen from Fig. 2.9b that when the gain of the element antenna is higher and the beam is narrower, the far-field sidelobe of the array is lower. Therefore, in the scanning array, not only the alignment of the maximum radiation direction of the element pattern with the direction of the array factor but also the coverage of the element pattern during phased scanning of the array factor should be considered to avoid a sharp drop in the antenna scanning gain caused by the element pattern during scanning. Therefore, in some wide-angle scanning phased arrays, measures need to be taken to expand the beamwidth of the element pattern to obtain a relatively flat gain characteristic within the scanning range.

To sum up, the pattern of the antenna element has a significant influence on the overall pattern of the array. Therefore, when designing the antenna array, the antenna element and the antenna array layout should be considered as a whole, and the appropriate antenna element should be selected according to the actual engineering needs.

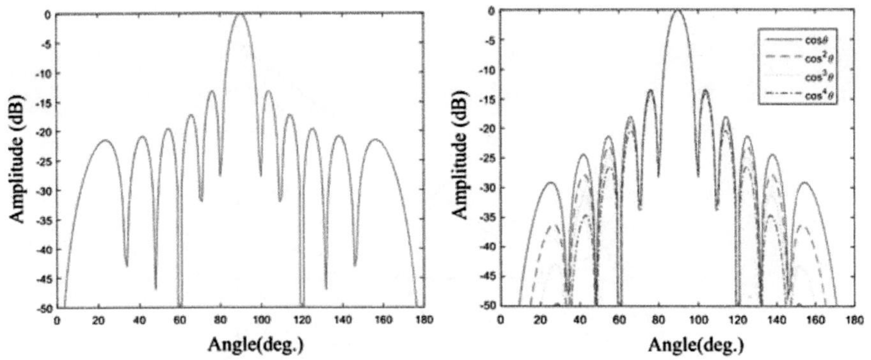

(a) (b)

Fig. 2.9 Influence of the pattern of the antenna element on the side lobes in the far area of the array. **a** Pattern of the array factor, **b** pattern of the antenna array in the case of element modulation

2.3 Planar Array Antennas

When an antenna array is formed, in addition to arranging the antenna elements along a one-dimensional linear direction to form a linear array, a rectangular array, also known as a planar array, can also be formed along a two-dimensional rectangular grid. Planar arrays have one more dimension than linear arrays and one more dimension in controlling beam shape and pointing adjustment than linear arrays. Using a planar phased array can make the antenna gain higher and realize two-dimensional flexible scanning, which makes the planar phased array antenna widely used in radar, communication and electronic warfare, and other fields.

2.3.1 Array Layout

Since the planar array generally contains many array elements, the layout of the planar array generally adopts a relatively regular configuration to obtain better engineering practicability. There are two commonly used antenna element layouts: rectangular grids and triangular grids.

(1) **Rectangular grid layout**

All array elements are arranged on a rectangular grid point, and their configuration are shown in Fig. 2.10.

Assuming that the number of array elements along the x-direction is M, the spacing is d_x, the number of array elements in the y-direction is N, and the spacing is d_y, then the array factor of the linear array along the x-direction is

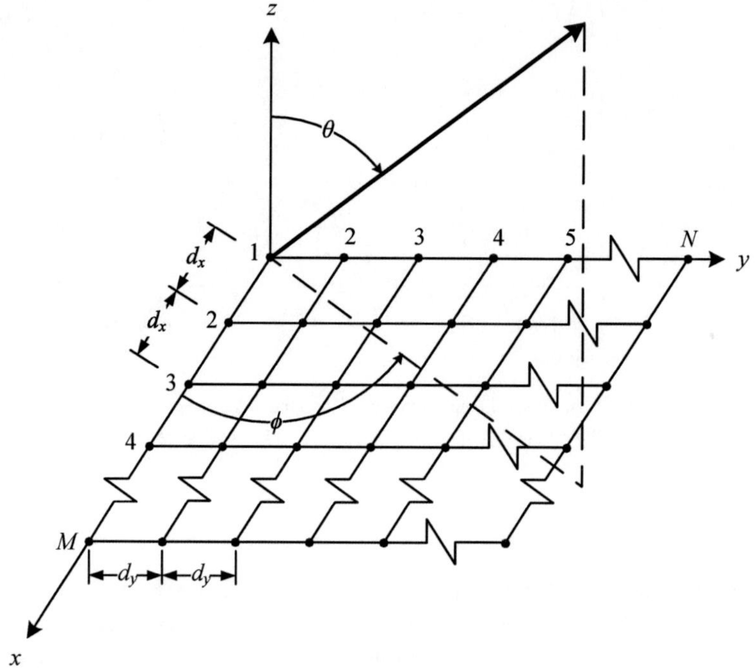

Fig. 2.10 Schematic diagram of rectangular array configuration

$$\mathrm{AF}_x = \sum_{m=1}^{M} I_{m1} e^{j(m-1)(\mathrm{kd}_x \sin\theta\cos\varphi + \beta_x)} \tag{2.61}$$

where I_{m1} is the excitation amplitude of the array element, d_x and β_x are the distance and excitation phase difference between adjacent elements along the x-direction, m represents the element number in the x-direction.

If the linear sub-array is replicated N times along the y-direction, a rectangular array is formed, and the array factor of the planar array can be expressed as

$$\mathrm{AF} = \sum_{n=1}^{N} I_{1n}\left[\sum_{m=1}^{M} I_{m1} e^{j(m-1)(\mathrm{kd}_x \sin\theta\cos\varphi + \beta_x)}\right] e^{j(n-1)\left(\mathrm{kd}_y \sin\theta\sin\varphi + \beta_y\right)} \tag{2.62}$$

where d_x and d_y are the distances between adjacent elements along the x and y directions, β_x and β_y are the excitation phase differences between adjacent elements along the x and y directions, m and n are the element numbers in the x and y directions. Equation (2.62) can be simplified as the product of array factors in two directions, namely,

$$\mathrm{AF} = \mathrm{AF}_x \mathrm{AF}_y = \sum_{m=1}^{M} I_{m1} e^{j(m-1)(\mathrm{kd}_x \sin\theta \cos\varphi + \beta_x)} \cdot \sum_{n=1}^{N} I_{1n} e^{j(n-1)(\mathrm{kd}_y \sin\theta \sin\varphi + \beta_y)} \tag{2.63}$$

If the excitation amplitudes of all elements in the array are equal, the array factor can be expressed as

$$\mathrm{AF} = I_0 \sum_{m=1}^{M} e^{j(m-1)(\mathrm{kd}_x \sin\theta \cos\varphi + \beta_x)} \cdot \sum_{n=1}^{N} e^{j(n-1)(\mathrm{kd}_y \sin\theta \sin\varphi + \beta_y)} \tag{2.64}$$

Then, the normalized array factor can be obtained, namely,

$$\mathrm{AF}_n(\theta, \varphi) = \frac{\sin \frac{M}{2}\psi_x}{M \sin \frac{\psi_x}{2}} \cdot \frac{\sin \frac{N}{2}\psi_y}{N \sin \frac{\psi_y}{2}}$$

$$\psi_x = \mathrm{kd}_x \sin\theta \cos\varphi + \beta_x,\ \psi_y = \mathrm{kd}_y \sin\theta \sin\varphi + \beta_y \tag{2.65}$$

When the spacing between the array elements is greater than λ, the array will generate grating lobes. In order to avoid the grating lobes, the spacing between the array elements is required to satisfy $d_x < \lambda, d_y < \lambda$. Similar to a linear array, the maximum radiation direction of the array can be scanned to a specific direction when the elements in the array are excited with progressive phases. If the array is expected to scan in the direction of $\theta = \theta_0, \varphi = \varphi_0$, the progressive phase difference between the array elements is

$$\beta_x = -\mathrm{kd}_x \sin\theta_0 \cos\varphi_0,\quad \beta_y = -\mathrm{kd}_y \sin\theta_0 \sin\varphi_0 \tag{2.66}$$

Conversely, if the phase difference between the x and y directions is known, the beam scan angle of the array is

$$\tan\varphi_0 = \frac{\beta_y d_x}{\beta_x d_y},\quad \sin^2\theta_0 = \left(\frac{\beta_x}{\mathrm{kd}_x}\right)^2 + \left(\frac{\beta_y}{\mathrm{kd}_y}\right)^2 \tag{2.67}$$

The beamwidth of the rectangular array is

$$\begin{cases} \theta_h = \sqrt{\dfrac{1}{\cos^2\theta_0\left(\frac{\cos^2\varphi_0}{\theta_{x0}^2} + \frac{\sin^2\varphi_0}{\theta_{y0}^2}\right)}} \\ \varphi_h = \sqrt{\dfrac{1}{\frac{\sin^2\varphi_0}{\theta_{x0}^2} + \frac{\cos^2\varphi_0}{\theta_{y0}^2}}} \end{cases} \tag{2.68}$$

where θ_h is the half-power beamwidth of the array along the elevation plane, φ_h is the half-power beamwidth of the array along the azimuth plane, θ_{x0} is the half-power beamwidth of the broadside array with equal spacing and equal element number with the x-direction sub-arrays of the array. The definition of θ_{y0} is similar to that of θ_{x0}, and the two are expressed as

$$\begin{cases} \theta_{x0} = \arccos\left(-\frac{2.782}{\text{Mkd}_x}\right) - \arccos\left(\frac{2.782}{\text{Mkd}_x}\right) \\ \theta_{y0} = \arccos\left(-\frac{2.782}{\text{Mkd}_y}\right) - \arccos\left(\frac{2.782}{\text{Mkd}_y}\right) \end{cases} \tag{2.69}$$

For the analysis of a planar array, its beam solid angle Ω_A is a commonly used physical quantity, which is equal to the product of the half-power beamwidths of the array on the two vertical planes in the maximum direction, namely,

$$\Omega_A = \theta_h \varphi_h = \frac{\theta_{x0}\theta_{y0}\sec\theta_0}{\left(\sin^2\varphi_0 + \frac{\theta_{y0}^2}{\theta_{x0}^2}\cos^2\varphi_0\right)^{1/2}\left(\sin^2\varphi_0 + \frac{\theta_{x0}^2}{\theta_{y0}^2}\cos^2\varphi_0\right)^{1/2}} \tag{2.70}$$

The directivity of the rectangular array is

$$D_0 = \pi\cos\theta_0 D_x D_y \tag{2.71}$$

where D_x is the directivity coefficient of the broadside array with equal spacing and the same number of elements in the x-direction of the array, D_y is the directivity coefficient of the broadside array with equal spacing and the same number of elements in the y-direction of the array.

$$D_x = \frac{\text{Mkd}_x}{\pi},\ D_y = \frac{\text{Nkd}_y}{\pi} \tag{2.72}$$

Given the known half-power beamwidth, the directivity coefficient of the planar array is estimated as

$$D_0 \approx \frac{\pi^2}{\Omega_A(\text{rad})} = \frac{32400}{\Omega_A(\text{deg})} = \frac{32400}{\theta_h\varphi_h(\text{deg})} \tag{2.73}$$

(2) **Triangular grid layout**

In addition to the rectangular grid arrangement, the common element arrangement in the array antenna also has a triangular grid arrangement. In a large-scale two-dimensional active phased array, triangular grids can reduce the number of antenna elements and active channels, effectively reducing costs. For example, for a conventional two-dimensional phased array antenna, as shown in Fig. 2.11, when an equilateral triangular grid is used, the number of radiating elements can be reduced by up to 13% compared to a rectangular grid. When the absolute number of elements is

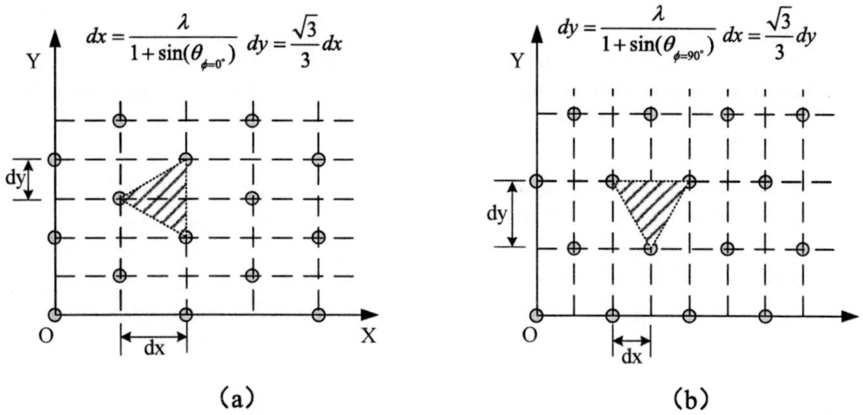

Fig. 2.11 Schematic diagram of the triangular grid configuration. **a** First determine the side triangle grid of d_x, **b** first determine the regular triangle grid of d_y

large, reducing the number of radiating elements is quite beneficial for system cost reduction.

When the triangular grid is used to arrange the array, the grid spacing must also be limited to prevent grating lobes during array scanning. When the maximum scanning angle of the array is θ_s, the array element spacing of the triangular grid array should meet the conditions

$$\begin{cases} d_x \le \frac{1}{\sin\alpha}\frac{\lambda}{1+|\sin\theta_s|} \\ d_y \le \frac{1}{\cos\alpha}\frac{\lambda}{1+|\sin\theta_s|} \end{cases} \frac{\pi}{6} \le \alpha \le \frac{\pi}{3} \tag{2.74}$$

where α is the angle between the hypotenuse of the triangular mesh and the x-axis.

Whether a rectangular grid or a triangular grid array, the condition of no grating lobes can be analyzed intuitively, that is, on any straight line of the antenna array, the element projection spacing satisfies the maximum scanning state, and no grating lobes appear.

2.3.2 *Planar Array Synthesis*

The synthesis and optimization of an array antenna involve a variety of parameters, mainly the amplitude and phase of the antenna elements, as well as the array shape limited by the structural envelope, the grid layout of the elements, and the element spacing when grating lobes are tolerated. The goal is to obtain the required radiation pattern, including parameters such as the main lobe, side lobe, and beamwidth. The main lobe parameters can be divided into pencil beam, difference beam, and shaped beam requirements, while the side lobe parameters can be divided into low side lobes, extremely low side lobes, low side lobes in local areas, and pattern null requirements. The synthesis of the antenna array generates the optimal excitation coefficients of the expected antenna pattern shape. In the specific synthesis, the amplitude and phase of the antenna elements are controlled to approximate these optimal excitation coefficients [6–8]. In practical engineering, the antenna array shape, element grid layout, and element spacing are usually designed first according to the limiting boundary conditions, and then the target pattern is synthesized and optimized. The previous step is empirical and essential in engineering, especially in large phased array antennas, the choice of element spacing directly affects the number of antenna active channels and the cost of the phased array. This design process is mainly determined by the scanning range and tolerance to grating lobes. In some applications, the maximum scanning angle can be reduced by selecting an appropriate installation angle of the antenna array [9]. A well-developed design usually requires multiple iterations between these two steps.

For simple two-dimensional separable variable arrays, the synthesis process of planar arrays can be simplified into two simpler one-dimensional linear array synthesis problems. Common linear array synthesis methods include the binomial method, Chebyshev array method, Fourier transform method, Shekunov polynomial method, and Taylor distribution method.

(1) Binomial method

The binomial array cleverly uses the binomial expansion's properties to synthesize the antenna array's array factor pattern into a shape without sidelobes [10]. Therefore, to analyze the radiation characteristics of the binomial array, it is necessary to study the array factor of the binomial array. The basic configuration of the binomial array is shown in Fig. 2.12.

Assuming that the element spacing of the array is d, when observed along the θ direction, the phase of each array element is 0, kdcosθ, 2kdcosθ, ..., $(N-1)$kdcosθ, and the amplitude of each array element in the binomial array is C_{N-1}^{0}, C_{N-1}^{1}, C_{N-1}^{2}, ..., C_{N-1}^{N-1}, the pattern function of the composite field can be obtained as

$$\begin{aligned}\mathrm{AF}(\theta) =& C_{N-1}^{0}\cdot 1 + C_{N-1}^{1}\cdot e^{j\mathrm{kd}\cos\theta}\\ &+ C_{N-1}^{2}\cdot e^{j2\mathrm{kd}\cos\theta} + \cdots + C_{N-1}^{N-1}\cdot e^{j(N-1)\mathrm{kd}\cos\theta} = \left(1+e^{j\mathrm{kd}\cos\theta}\right)^{N-1}\end{aligned} \tag{2.75}$$

Considering the magnitude characteristic of the array pattern, its expression is

$$|\mathrm{AF}(\theta)| = \left|1+e^{j\mathrm{kd}\cos\theta}\right|^{N-1} = 2^{N-1}\cos^{N-1}\left(\frac{\mathrm{kd}}{2}\cos\theta\right) \tag{2.76}$$

It can be known from Eq. (2.76) that when kd $< \pi$, the distance between the array elements is less than half wavelength, the array factor will have no side lobes. As the number of array elements increases, the main lobe gradually narrows, but the array factor has no side lobes.

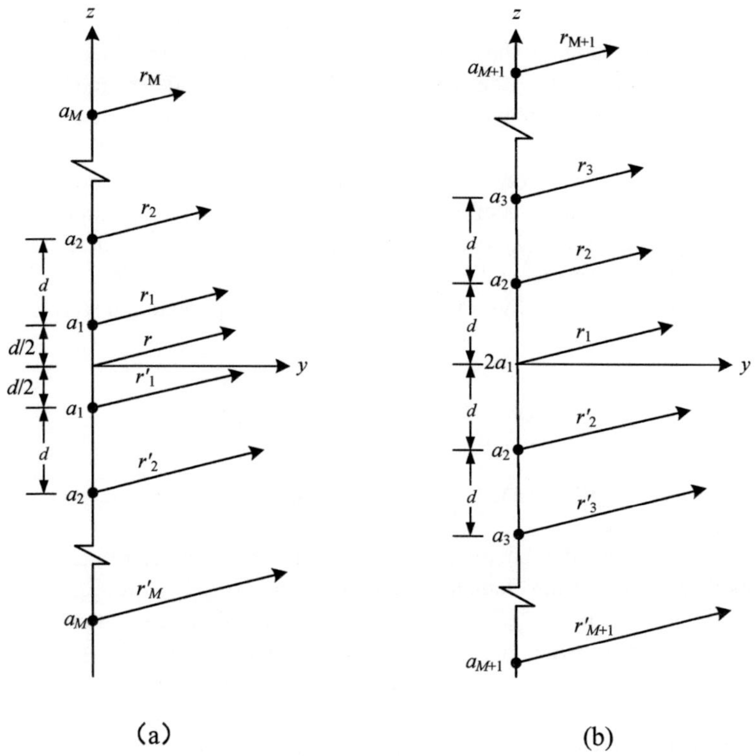

Fig. 2.12 Basic configuration of binomial array, **a** when the number of array elements is even, **b** when the number of array elements is odd

The advantage of the binomial array is that the side lobes of the array factor are completely eliminated, but it has obvious disadvantages: the beam of the antenna array is wider, the gain is low, and the utilization efficiency of the antenna aperture is low. In practical applications, it is usually not necessary to completely eliminate the side lobes of the antenna array but only to control the SLL within a specified range.

(2) **Dolph-Chebyshev method**

Compared with the simple constant-amplitude linear array, the array element excitation amplitude of the unequal-amplitude linear array has a higher freedom in design so that the aperture field of the antenna array can show a gradually decreasing trend, resulting in the effect of reducing the antenna side lobes [11]. In most cases, controlling the SLL to a certain level can improve the gain of the antenna array and obtain high aperture utilization. In addition, the engineering feasibility of amplitude and phase control must be considered.

The Dolph-Chebyshev array is one of the methods to control the sidelobe level of the antenna array at a specified level. This method uses the equal-amplitude fluctuation characteristics of the Chebyshev polynomial to generate an array factor of equal-ripple sidelobes. The difference between the Chebyshev array and the binomial array lies in generating the excitation amplitude of each array element. Among them, the design idea of the Chebyshev array is to use the undetermined coefficient method to fit a Chebyshev polynomial whose order is related to the number of array elements and then use the undetermined coefficient method to obtain the coefficients as the amplitude of each array element. Then the excitation amplitude needs to be solved by the undetermined coefficient method [8]. The excitation amplitude of the binomial array is determined. After the number of array elements is determined, the amplitude of each array element can be directly given.

The array factor of a Chebyshev array is a Chebyshev polynomial with respect to θ, that is

$$\mathrm{AF} = T_{N-1}\left[z_0\cos\left(\frac{\mathrm{kd}}{2}\cos\theta\right)\right] \tag{2.77}$$

where $N-1$ is the order of the Chebyshev polynomial, z_0 is a constant determined by the expected SLL.

When the number of elements of the Chebyshev array is determined, the array factor is determined. If the element spacing d of the array is adjusted, the SLL of the array factor will not be affected, but the antenna gain and the number of side lobes will be affected, when the element spacing is too large, grating lobes may be generated. Therefore, in most applications, under the condition that the SLL of the array meets the requirements, the element spacing of the antenna array should be as large as possible to obtain a higher gain. According to the properties of Chebyshev polynomials and the correspondence between array factors and polynomials, when the array element spacing d increases, the grating lobes of the array will appear in the end-fire direction. Therefore, the condition that the antenna array does not generate grating lobes with SLL better than the specified value is the independent variable of the Chebyshev polynomial should not exceed -1 for the θ value which is corresponding to the end-fire direction, that is

$$z_0\cos\left(\frac{\mathrm{kd}}{2}\cos\theta\right)\bigg|_{\theta=0,\pi} \le -1 \tag{2.78}$$

where z_0 is the independent variable shrinkage factor determined by the expected SLL. From Eq. (2.78), the maximum allowable spacing under this condition is

$$d_{\max} = \frac{2}{k}\arccos\left(-\frac{1}{z_0}\right) = \frac{\lambda}{\pi}\arccos\left(-\frac{1}{z_0}\right) \tag{2.79}$$

The synthesis process of the Chebyshev array is to determine the number of array elements according to the expected antenna array gain and select different array factors as the basis according to the number of array elements being odd or even. Using the properties of Chebyshev polynomials, the array factor's higher order term $\cos(m\psi)$ is expanded. According to the expected SLL of R_0, the independent variable shrinkage factor z_0 of the Chebyshev polynomial is calculated, and the calculation condition is based on $T_{N-1}(z_0) = R_0$. Taking the Chebyshev polynomials of various orders as the basis, the array factor $AF = T_{N-1}\left(z_0\cos\frac{\psi}{2}\right)$ is expanded to this group of basis, and the coefficient of each basis function is obtained by the undetermined coefficient method.

Each coefficient obtained through the above process is the excitation amplitude of each array element to be solved. A careful examination of the solution process shows that the synthesis process is a process of fitting a higher-order polynomial with a lower-order polynomial, in which the lower-order polynomial refers to the contribution of each array element to the array factor. The closer the array element is to the center of the array, the faster the pattern changes with the θ angle and the higher the corresponding polynomial order. The higher-order polynomial as the fitting target refers to the expected realization of the array factor pattern. The polynomial is a Chebyshev polynomial with a shrinkage ratio z_0. The higher the shrinkage ratio z_0, the smaller the relative fluctuation of the ripple in the polynomial, and the lower the SLL on the pattern.

After the above synthesis process, the excitation amplitude of each array element can be obtained. The analytical expression of the amplitude of each array element of the Chebyshev array is as follows:

$$a_n = \begin{cases} \sum\limits_{q=n}^{M} (-1)^{M-q}(z_0)^{2q-1}\frac{(q+M-2)!(2M-1)}{(q-n)!(q+n-1)!(M-q)!}, & M \text{ is even} \\ \sum\limits_{q=n}^{M} (-1)^{M-q+1}(z_0)^{2(q-1)}\frac{(q+M-2)!(2M)}{\epsilon_n(q-n)!(q+n-2)!(M-q+1)!}, & M \text{ is odd} \end{cases}$$

$$\epsilon_n = \begin{cases} 2, n = 1 \\ 1, n \neq 1 \end{cases} \tag{2.80}$$

where M is the number of array elements, a_n is the excitation amplitude of the nth array element, z_0 is the independent variable shrinkage factor determined by the SLL.

For Chebyshev arrays with SLLs between −20 and −60 dB, the approximate directivity coefficient is

$$D_0 = \frac{2R_0^2}{1 + \left(R_0^2 - 1\right)f\frac{\lambda}{L+d}} \tag{2.81}$$

$$f = 1 + 0.636\left\{\frac{2}{R_0^2}\cosh\sqrt{\left(\cosh^{-1}R_0\right)^2 - \pi^2}\right\}^2 \tag{2.82}$$

where f is the beam broadening factor, which can be expressed as the ratio of the beamwidth of the Chebyshev array to the beamwidth of the same number of linear arrays with the identical spacing and equal amplitude, R_0 is the voltage ratio of the main lobe to the side lobe calculated from the SLL. Correspondingly, the half-power beamwidth of the Chebyshev array is

$$\mathrm{HPBW} = f\Theta_h = f\left[\arccos\left(\cos\theta_0 - \frac{2.782}{\mathrm{Nkd}}\right) - \arccos\left(\cos\theta_0 + \frac{2.782}{\mathrm{Nkd}}\right)\right] \quad (2.83)$$

where Θ_h is the beamwidth of the equal-amplitude linear array with the same element spacing as the Chebyshev array.

(3) **Fourier transform method**

The principle of the Fourier transform method is to use the Fourier transform relationship between the current distribution of the linear array and the corresponding radiation pattern and obtain the desired current distribution by performing inverse Fourier transformation on the expected radiation pattern.

For a line source whose current distribution is $I(z)$, its radiation pattern can be expressed as

$$f(\theta) = \frac{1}{\lambda}\int_{-L/2}^{L/2} I(z)e^{j\beta z\cos\theta}dz \quad (2.84)$$

where $I(z)$ is the current amplitude and phase at each position in the line source, Eq. (2.84) is not a standard Fourier transform relationship. To convert it into a Fourier transform relationship, the variables need to be transformed as follows:

$$w = \cos\theta,\ s = \frac{z}{\lambda} \quad (2.85)$$

After the variable transformation, the relationship between the radiation pattern and the current distribution is obtained as

$$f(w) = \int_{-L/2\lambda}^{L/2\lambda} I(s)e^{j2\pi ws}ds \quad (2.86)$$

Equation (2.86) is the standard Fourier transform from $I(s)$ to $f(w)$, which can be obtained by inverse transform

$$I(s) = \int_{-\infty}^{+\infty} f(w)e^{-j2\pi ws}dw \quad (2.87)$$

Therefore, through the expected pattern $f(w) = f(\cos\theta)$, the current distribution $I(s)$ that realizes the pattern can be obtained, but the $I(s)$ obtained by this transformation is usually not the current distribution of finite length. Its distribution range is infinite, and it is not the desired distribution in the range of $|s| \le L/2\lambda$. It is impossible to realize an infinite-length linear array. The usual way to obtain a finite-length linear array is to truncate the above current distribution, that is

$$I(s) = \begin{cases} I(s), |s| \leq L/2\lambda \\ 0, |s| > L/2\lambda \end{cases} \tag{2.88}$$

This truncation will cause a certain deviation between the pattern produced by the linear array and the expected pattern, which can be reduced by expanding the truncation range.

(4) **Sekunov polynomial method**

The principle of the Shekunov polynomial method is to control the position of the pattern's null for array synthesis. After determining the pattern's null by a heuristic method, the excitation coefficient of the array element can be obtained [12]. The Shekunov polynomial refers to the polynomial that can be obtained after the array factor of the antenna array is transformed to a certain extent, and then the shape of the pattern corresponding to the polynomial can be changed by changing the null distribution of the polynomial.

The array factor of the N-element equidistant linear array is

$$\mathrm{AF} = \sum_{i=0}^{n-1} I_i e^{jiu}, u = \mathrm{kd}\cos\theta + \psi \tag{2.89}$$

where ψ is the phase difference between adjacent elements, I_i is the excitation amplitude of the ith array element. To make $z = e^{ju} = e^{j(\mathrm{kdcos}\theta + \psi)}$, then the array factor can be rewritten as a polynomial about z, that is

$$\mathrm{AF} = \sum_{i=0}^{n-1} I_i z^i = I_0 + I_1 z + I_2 z^2 + \cdots + I_{N-1} z^{N-1} \tag{2.90}$$

Equation (2.90) is a polynomial of degree $N - 1$ with respect to z. According to algebraic theory, a polynomial of degree $N - 1$ has $N - 1$ roots and can be expressed as the product of $N - 1$ factors, namely,

$$\mathrm{AF} = I_{N-1}(z - z_1)(z - z_2)(z - z_3) \cdots (z - z_{N-1}) \tag{2.91}$$

These $N - 1$ roots are the nulls of the corresponding antenna array pattern. By adjusting the position of these $N - 1$ roots relative to the maximum radiation direction of the antenna corresponding to $z_0 = e^{j(\mathrm{kdcos}\theta_0 + \psi)}$, the distance between each null point and maximum radiation direction can be adjusted. The closer the nulls are to the maximum radiation direction, the lower the SLL of the antenna. Therefore, the position of the null of the Shekunoff polynomial is adjusted by a heuristic method to reduce the antenna SLL. After the position of each null is determined by the heuristic method, these nulls are substituted into the polynomial, and each factor is expanded to obtain the corresponding coefficients of each power z^i of z in the polynomial, and then the excitation amplitude and phase of each source can be obtained.

Through the above analysis, this method needs to go through many trial processes and cannot directly obtain comprehensive results like the Fourier transform method, but this method reveals the influence of the null position of the directional function of the antenna array on its pattern. To suppress the SLL of the antenna array, the roots of the polynomial need to be close to each other on the unit circle, at the cost of widening the main lobe. At the same time, when the distribution of the polynomial roots is adjusted to be close to each other, the coefficients of the polynomial will change, which also shows that adjusting the excitation amplitude and phase of the array elements can change the SLL of the antenna array.

(5) **Taylor distribution method**

The Taylor distribution in the antenna array synthesis is a modification of the Chebyshev linear array[13], which can be directly used for the synthesis of continuous line sources, called Taylor linear sources, or it can be discretized for linear arrays [14]. Both Taylor distribution and Chebyshev distribution commonly use narrow beam low sidelobe synthesis methods. However, the difference between the Taylor distribution and the Chebyshev linear array is that the pattern obtained after synthesis only maintains several near-field sidelobes with the same level value, while the other sidelobes decrease in turn. In addition, the width of its main lobe is slightly larger than that of the Chebyshev array, similar to the quasi-optimal pattern of the Chebyshev array.

Compared with the Chebyshev array, the advantage of the Taylor line source is that it avoids the sharp rise of the current distribution of the Chebyshev array. When the target SLL is low, the current distribution synthesized by the Chebyshev array will rise sharply at both ends of the array, and the current at both ends significantly influences the sidelobe, which increases the difficulty of engineering realization to a certain extent. On the contrary, the Taylor line source avoids the phenomenon of the current steep rise, and thus the realization difficulty of the Taylor distribution array is lower than that of the Chebyshev array.

2.4 Sparse Arrays

The antenna arrays discussed above are all uniform arrays, i.e., the array elements are periodically arranged with fixed spacing on a regular grid point. The grating lobes usually need to be avoided in radar and communications. Therefore, the characteristics of grating lobes limit the maximum element spacing of a uniform array, i.e., the distribution density of elements in the array cannot be lower than a lower limit. However, in practical applications, there are often some requirements to reduce the distribution density of array elements as much as possible. At this time, it is necessary to improve the configuration of the uniform array so that it can still avoid the occurrence of grating lobes under the condition of low distribution density of array elements, or the grating lobe is suppressed to the extent allowed and a certain side lobe performance is maintained.

Usually, the purpose of reducing the number of array elements and the distribution density is to reduce the cost and the beamwidth. Cost reduction is an essential consideration for larger-scale antenna arrays. In some application fields, the narrower beam of the antenna can improve the angular resolution. Therefore, it is of great significance to increase the array element spacing and reduce the distribution density of the array elements. After the array element spacing is increased, the antenna array is called a sparse array, and the design of the sparse array also requires that no grating lobes appear in the visible space. To sum up, related array sparse technologies mainly focus on eliminating or suppressing array grating lobes, reducing array side lobes, and improving array gain.

The first problem to be considered in designing a sparse array is the grating lobe problem caused by increased array element spacing. The array produces grating lobes because of the increase in the spacing of the array elements and the periodic arrangement between the array elements. The grating lobes can be suppressed to a certain extent by performing an aperiodic arrangement at the unit or sub-array levels. Through the aperiodic arrangement at the unit level, the non-periodization of all the array elements can be realized, which is easy to be realized when the array scale is small. Through the non-periodic arrangement of sub-array levels, the grating lobes of each sub-array appear in different directions in space and thus cannot be superimposed to form high grating lobes.

2.4.1 Random Sparse Array Layout

The random sparse array usually refers to the unit-level array sparse technology. The basic idea is to use the array elements' distribution density to simulate the full array's amplitude weighting to achieve a lower SLL with a smaller number of array elements. Whether the array elements are restricted to appearing on the periodic grid points or not during the design process, the random array can be divided into two forms: the density sampling with equal spacing, and the spacing of the array elements can be randomly arranged in a continuously variable manner.

(1) **Equally spaced density sampling**

The basic idea is based on a full equally spaced array. The elements in the array are extracted according to the probability determined by the expected amplitude distribution, and the remaining elements form a density sampling array. The probability distribution density of the array elements in the antenna array becomes the expected amplitude distribution through the density sampling process. This spatial density weighting is equivalent to amplitude weighting and can also achieve the effect of reducing the SLL of the antenna array.

The specific synthesis of the density sampling array is divided into two parts: determining the configuration and amplitude distribution of the full array and determining the position of the active array element. The configuration and amplitude distribution of the full array are based on the requirements of the antenna, such as scanning coverage, array inclination, and structural parameters, to determine the grid configuration and array element spacing of the full array. And further through the analysis of the directivity coefficient, beamwidth, SLL, and other parameters, the antenna aperture, the number of array elements, and the amplitude weighted distribution function are determined. After determining the configuration and amplitude distribution of the full array, it is necessary to determine the position of each active array element. Then, according to the spatial distribution density function of the array element, a random 0/1 lattice is generated, and the lattice element is placed in the position where the lattice is 1, and the position where the lattice is 0 is set as a vacancy element.

A more direct method of determining the spatial distribution density of array elements is to make the spatial distribution density directly equivalent to the amplitude weighted distribution function. At this time, the array sampled according to the distribution density function can achieve the expected low side lobe effect under the condition of equal-amplitude feeding. An improvement of this method is to perform unequal-amplitude feeding to the array after sampling so that the product of the array density distribution and the feeding amplitude distribution is equal to the expected amplitude weighted distribution. After this improvement, the SLL can be further reduced with the same number of array elements, or the number of array elements can be reduced without raising the SLL.

After these two steps, an equally spaced density sampling array with a specific amplitude can be obtained. The radiation performance of the randomly distributed array generated by one generation process may be different from the expected value. By repeating the above generation process and selecting the result with the best performance from the multiple generation results, the radiation performance of the array can be optimized to meet the expected random array requirements.

(2) **The spacing of the array elements can be randomly arranged in a continuously variable manner**

The equal-spaced density sampling array is based on the equal-spaced array and produces a specific weighting effect by extracting a certain ratio of cells. At this time, the spacing between the array elements is an integer multiple of the array element spacing of the original equal-spaced array, which is a discrete distribution. The random array with continuously variable array element spacing removes this limitation, makes the value of the array element spacing continuous, increases the randomness of the array element distribution, and further improves the radiation performance of the antenna array. However, compared with the equal-spaced density sampling array, the array with continuous distribution is more difficult to realize in engineering, which puts forward higher requirements for designing the antenna feeding network.

The synthesis process of the continuously variable random arrangement of the array elements is similar to that of the density sampling array, and the configuration and amplitude distribution of the reference full array must be determined. The difference between the two is that after the amplitude distribution is determined, the continuous-spaced random array determines the position of the array elements by generating random numbers on the entire array aperture.

2.4.2 Sub-array Layout

The phased array antenna based on sub-array technology refers to the phased array in one or two dimensions. Multiple antenna element groups or sub-arrays exceed the conventional antenna element size and reach the size of several wavelengths. The advantage of this phased array antenna configuration is that the number of active channels of the antenna is greatly reduced, and it can still meet the requirements of the scanning angle of the microwave imaging radar in various working modes. But its disadvantage is obvious. A large number of periodic grating lobes appear in the scanning. In SAR, this shortcoming will affect microwave imaging radar's imaging quality and limit the observation bandwidth. Therefore, according to the considerations of volume, weight, and cost, it is necessary to focus the research on the following topics for the application of microwave imaging radar, including (1) the optimal compromise between the sub-array size and the scanning capability, (2) to suppress the grating lobe level caused by beam scanning.

In practical applications, the aperture of the antenna array is often large, and the number of elements may reach tens of thousands. Although the unit-level aperiodic array can obtain relatively good radiation characteristics, it will greatly increase the difficulty of processing and manufacturing the antenna array. And it makes other parts of the phased array antenna system (such as the realization of feeder and beam control) extremely complicated, so the unit-level aperiodic array is difficult to apply to the design of large-scale sparse arrays. On the other hand, the sub-array-level random array retains the periodicity of the local sub-arrays of the array to a large extent under the condition of low periodicity of the overall configuration of the array, thereby reducing the difficulty of processing and manufacturing and having high engineering application value.

Common sub-array-level aperiodic array configurations include ring grid array, irregular sub-array grid array, sub-array center position aperiodic array, sub-array dislocation aperiodic array, and aperiodic array formed by rotating sub-arrays (shown in Fig. 2.13).

The basic configuration of the ring grid array is shown in Fig. 2.13a. It changes the spacing between each array element based on the circular array so that the spacing between adjacent array elements is equal, which is not the case under normal circumstances, where the central angles are equal. From the radial direction, each antenna element is no longer in the same direction, from the circumferential direction, the number of array elements on each circumference is no longer equal. This change in

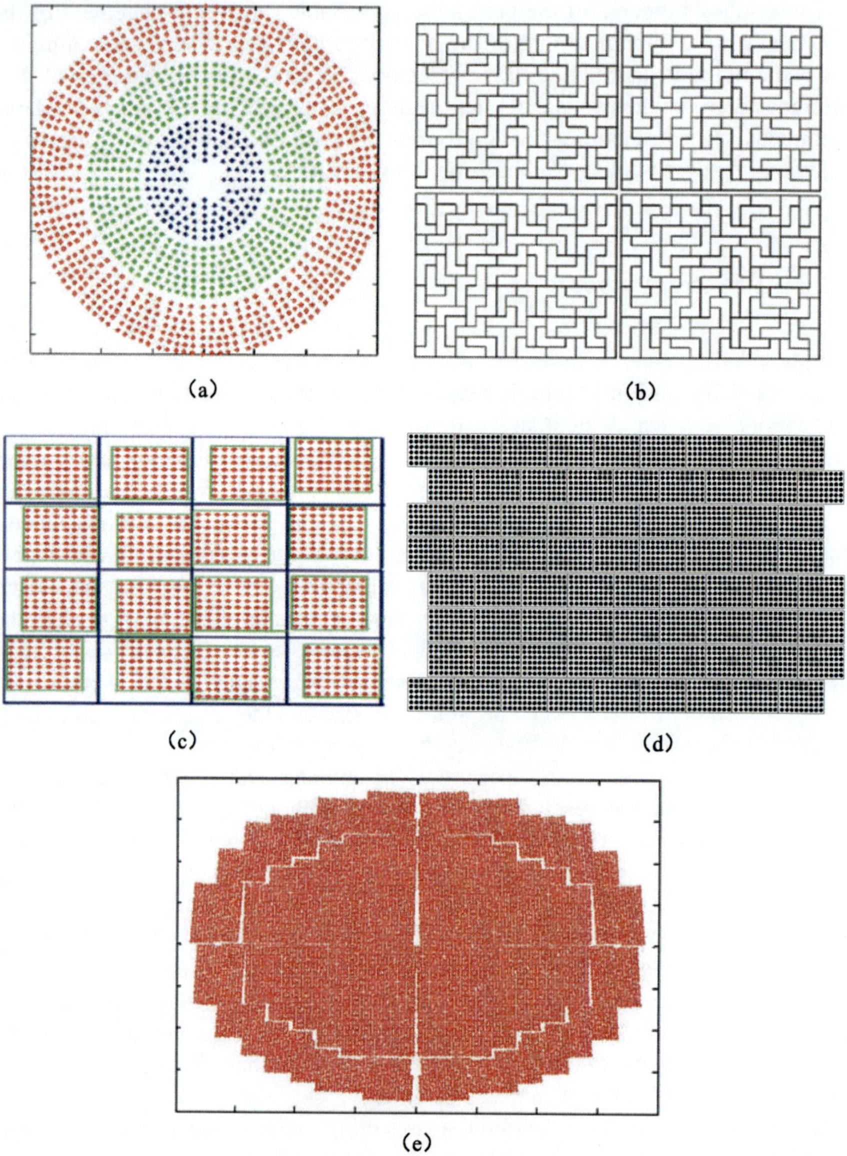

Fig. 2.13 Sub-array-level aperiodic array arrangement, **a** ring grid array, **b** irregular sub-array random array, **c** sub-array center position aperiodic array, **d** sub-array dislocation aperiodic array, **e** rotates the sub-array non-periodically

the array configuration makes the configuration no longer periodic, thereby reducing the grating lobes generated by the configuration's periodicity. As a result, the grating lobe suppression performance of the ring grid array is better, but the variety of sub-arrays is more difficult to process and assemble, and the outer contour size of the circular array is generally larger than that of the rectangular array with the same number of array elements.

The configuration of the random array of irregular sub-arrays is shown in Fig. 2.13b. By using irregular sub-arrays or irregular deformed sub-arrays to form the array, the phase centers of the sub-arrays in the array are randomly distributed, which can avoid superposition of quantized lobes and weaken the beam scanning grating lobes of the large-spacing array. Compared with the rectangular conventional array, the grating lobe suppression is better than 14 dB in the large-angle scanning range [15]. When forming an array, all the antenna elements of the sub-array share a T/R channel, and the array is filled with the array radiation aperture by changing the layout. This arrangement method not only randomizes the phase center spacing of each sub-array to reduce the grating lobe level but also avoids the reduction of aperture efficiency caused by increased sub-array spacing. As a result, the grating lobe suppression and gain performance are good, but the layout and assembly of the back-end power divider and transceiver components are difficult due to strong randomness.

The configuration of the aperiodic array at the center of the sub-array is shown in Fig. 2.13c. The array arrangement is to randomize the spacing of the sub-arrays in the two directions of the row and column of the array, which destroys the in-phase superposition condition of the array grating lobes and suppresses the level of the grating lobes. This arrangement method's structure and assembly relationship are relatively simple, and it is easy to realize engineering. However, to obtain good grating lobe suppression, it is necessary to use two-dimensional random distribution with large spacing, resulting in a large number of quantized lobes in the principal section of the pattern, low aperture efficiency, and large scan gain loss.

Figure 2.13d shows the configuration of the aperiodic array with sub-array dislocation. The array arrangement is a one-dimensional simplified form of the aperiodic array at the sub-array center. The basic operating principle is the same as that of the former. The full array is used in one dimension, and sub-array dislocation is used in another dimension to achieve grating lobe suppression. The aperture efficiency of this array is higher than that of sub-array center aperiodic array, and it also has the advantages of simple assembly relationship and easy implementation. However, because it is distributed in a full array in one dimension, more transceiver components need to be used in the back end, so the system complexity and cost cannot be substantially reduced.

The non-periodic array configuration formed by the rotating sub-array is shown in Fig. 2.13e. It divides the entire array into a certain number of sub-arrays, and rotates each sub-array to different degrees, so that the grating lobes of each large sub-array rotate in different directions in space, leading to the grating lobes without overlapping each other. Therefore, the effect of suppressing grating lobes is produced. Since each sub-array is still periodic, it is less difficult to implement in engineering, and the aperture efficiency and gain loss caused by the dislocation of sub-arrays are also less heavier. The disadvantage of this array arrangement is that each sub-array has a small angle of rotation, which increases the difficulty of designing the feed network.

Among the above-mentioned sub-array-level arrangements, the non-periodic array technology based on sub-array dislocation and rotation [16–18] is less challenging to implement in engineering, so it has higher application prospects.

In the sub-array phased array antenna used in microwave imaging radar, a linear array with equal amplitude and a uniform distribution is usually used, and the radiation pattern is fixed. The antenna array realizes array factor scanning by controlling the phase of each sub-array in the array, thereby realizing antenna beam scanning. Since the sub-array spacing is far more than one free-space wavelength, the array factor appears to be more periodic grating lobes in the visible space.

According to the principle of pattern product, the main lobe of the array factor of the normal direction coincides with the maximum point of the pattern of the sub-array to form the main lobe of the antenna array, while the grating lobes of other array factors coincide with the null of the pattern of the sub-array, as shown in Fig. 2.14. During the scanning process, the main lobe of the array factor deviates from the maximum point of the sub-array pattern, while the grating lobe of the array factor deviates from the null of the sub-array pattern and enters the grating lobe, and the grating lobe appears when the antenna array pattern is synthesized. With the increase of the scanning angle, the main lobe, after combining the array factor main lobe and the sub-array pattern, gradually decreases, while the synthesized grating lobe gradually increases. When the array factor's scanning angle is half the grating lobe spacing, the two peaks of the array factor appear symmetrically in the main lobe of the sub-array pattern, and the maximum grating lobe has the same level as the main lobe.

Sub-array-level antenna array pattern optimization, especially grating lobe suppression, can be carried out by sub-array separation and recombination [17],

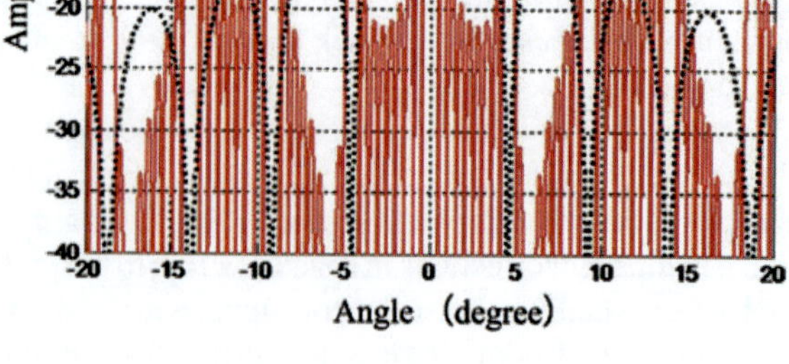

Fig. 2.14 Schematic diagram of formation and suppression of grating lobes of sub-array phased array

which is fast and requires little computation. For the sub-array phased array antenna, due to the modulation effect of the sub-array pattern, the large-angle grating lobes of the array factor far from the main lobe region of the sub-array can be well suppressed. Therefore, the essence of grating lobe suppression is to suppress the first grating lobe level in the array factor. Figure 2.14 intuitively illustrates the principle of grating lobe suppression when a large antenna array is composed of sub-arrays, in which the dotted line is the pattern of uniformly distributed sub-arrays, and the solid line is the array factor that suppresses the level of the grating lobe adjacent to the main lobe. When the beam is scanned, the pattern of the sub-array remains unchanged, the main lobe of the array factor deviates from the maximum value of the main lobe of the sub-array, and the composite value decreases while the first grating lobe enters the main lobe, and its level value gradually increases with the increase of the scanning angle. Since the level value of the first grating lobe is suppressed by a certain order, the level of the first grating lobe can be controlled compared with the conventional sub-array level antenna array scanning. In contrast, the sub-array side lobes suppress the other grating lobes.

Assuming that the multi-element uniform sub-arrays are arranged along the y-axis direction, the number of wavelengths of the sub-array length is $L_\lambda(\gg 1)$, and the number of wavelengths of the sub-array spacing in the x-axis is $dx_\lambda(<1)$. After scanning the two-dimensional planar array expanded and arranged by this sub-array, the beam direction is (θ_B, φ_B). According to the $dx_\lambda(<1)$ array factor pattern, the grating lobe closest to the main lobe in the scanning state is directed (θ_g, φ_g) as

$$\theta_g = \arcsin\left(\frac{\sin\theta_B \cos\varphi_B}{\cos\varphi_g}\right) \tag{2.92}$$

$$\varphi_g = \arctan\left(\tan\varphi_B \pm \frac{1}{L_\lambda \sin\theta_B \cos\varphi_B}\right) \tag{2.93}$$

There are many ways of choosing sub-array dislocation. According to the scale, it is divided into single sub-array or multi-sub-array parallel arrangements of sub-blocks. The dislocation size can be divided into the same and different dislocation sizes, as shown in Fig. 2.15.

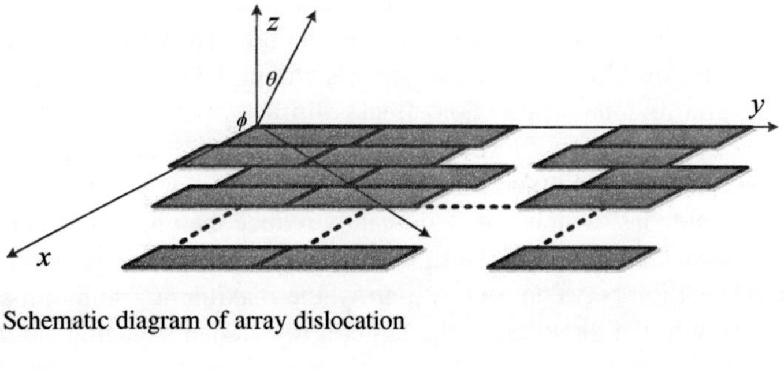

Fig. 2.15 Schematic diagram of array dislocation

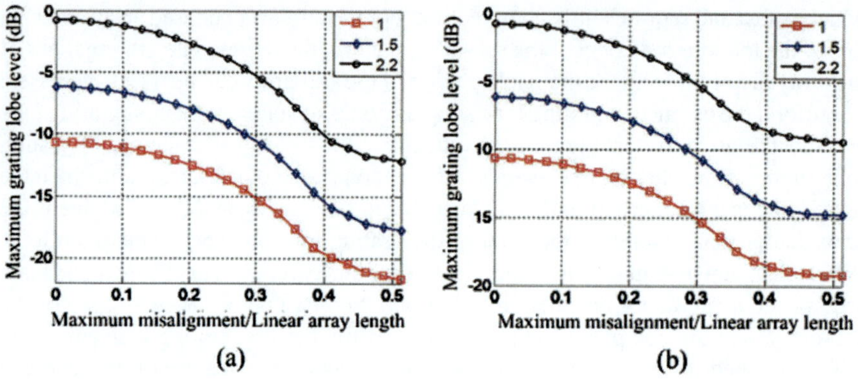

Fig. 2.16 The relationship between the maximum grating lobe level and the maximum amount of misalignment, **a** 8-row scale sub-array, **b** 16-row scale sub-array

Taking a sub-array phased array antenna with 128 rows and 9 columns as an example, the size of the grid array is $89.6\lambda \times 112.3\lambda$, the scale along the x-direction is 128 rows, and the maximum scanning angle is 20°. The length of the direction element is $L_\lambda = 2.5$, which is the direction of the larger size of the sub-array, and there are 9 sub-arrays along the y-direction.

The optimization of sub-array dislocation can be realized by Particle Swarm Optimization (PSO). Figure 2.16 shows the relationship between the maximum grating lobe level of the antenna array composed of sub-blocks and the maximum amount of misalignment between sub-blocks in the case of 8 and 16-row sub-arrays, where the amount of misalignment is different for each block. It can be seen from the calculation results that when the misalignment is 0, that is, when there is no misalignment, the maximum grating lobe levels when the antenna array is scanned in the azimuth direction of 1°, 1.5°, and 2.2° are about −10.5, −6 and −1.5 dB.

As shown in Fig. 2.16a, it is an 8-row scale sub-array. With the increase of the maximum dislocation amount between sub-blocks, the maximum grating lobe level of the same scanning angle gradually decreases. When staggered by half the length of the sub-array, the grating lobe level is suppressed to the maximum, and when the scanning reaches 2.2°, the grating lobe level is about 12 dB. As shown in Fig. 2.16b, it is a 16-line scale sub-block. When scanning to 2.2°, the maximum suppression of the grating lobe is close to 8 dB. It can be seen from Fig. 2.16 that when the maximum dislocation of the sub-blocks of the two sizes is within 0.35 times the length of the sub-array, the grating lobe suppression effect is similar.

In summary, the sub-array-level phased array grating lobe suppression technology is a very effective method to broaden the scanning capability of microwave imaging radar antennas, which can significantly reduce the number of active channels. Furthermore, after adopting the sub-array-level phased array technology, optimizing the dislocation between each sub-array, the maximum grating lobe level of the antenna array in the direction of the larger array element spacing can be effectively suppressed. Therefore, the method can be applied to microwave imaging radar antenna systems and other antenna systems with small scanning ranges and no low sidelobe requirements.

2.4.3 Sparse Array Elements

Compared with a conventional array, the element density of a sparse array is reduced, so the spacing between elements is larger, and the effective area of the low-gain antenna element is smaller, resulting in lower efficiency of aperture utilization in the antenna array. In order to increase the gain of the antenna array and fully fill the large-spacing area, a high-gain antenna is usually used as the array element. Commonly seen high-gain antenna elements are horn antennas, helical antennas, Yagi antennas, and sub-array antennas.

(1) **Horn Antennas**

The horn antenna has a simple configuration and feeding advantages, high controllability of beamwidth and gain, good frequency characteristics, and small loss. As shown in Fig. 2.17, horn antennas can be divided into rectangular and circular horn antennas according to their basic shapes. Among them, rectangular horn antennas are further divided into H-plane fan-shaped horns that expand along the broad side of the waveguide, E-plane fan-shaped horns that expand along the narrow side of the waveguide, and cone horns that expand along both the broad and narrow sides of the waveguide.

The radiation gain of the pyramid horn antenna is lower than the antenna gain corresponding to its physical area. The reason is that the aperture field distribution of the horn antenna is not an ideal equal-amplitude and in-phase distribution, and its aperture field has different amplitudes and unequal phase problems everywhere due to the field distribution. Its radiation gain is

$$G = \frac{4\pi \mathrm{ab}}{\lambda^2}\eta_a \tag{2.94}$$

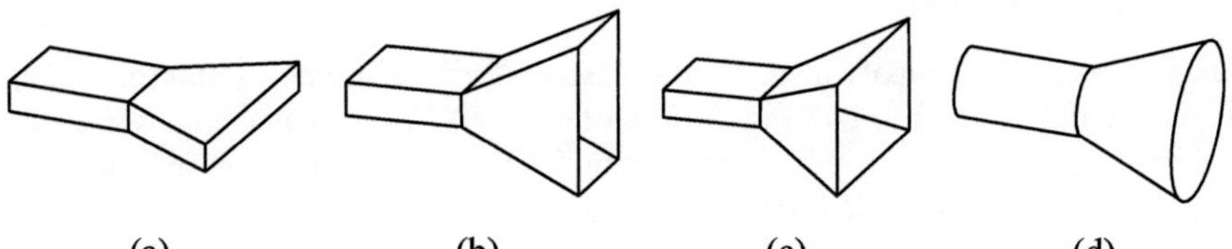

(a) (b) (c) (d)

Fig. 2.17 The basic form of horn antenna. **a** H-plane sector horn, **b** H-plane sector horn, **c** pyramid horn, **d** conical horn

$$\eta_a = \frac{\left|\int_{S_0} E_a dS\right|}{A_0 \int_{S_0} |E_a|^2 dS} = \eta_t \eta_{\text{ph}}^H \eta_{\text{ph}}^E \tag{2.95}$$

$$\begin{cases} \eta_t = \dfrac{\left|\int_{-a/2}^{a/2} \cos\frac{\pi x}{a} dx\right|^2}{a\int_{-a/2}^{a/2} \left|\cos\frac{\pi x}{a}\right|^2 dx} \\ \eta_{ph}^H = \dfrac{\left|\int_{-a/2}^{a/2} \cos\frac{\pi x}{a} e^{-j\frac{\pi x^2}{\lambda R_H}} dx\right|^2}{\left|\int_{-a/2}^{a/2} \cos\frac{\pi x}{a} dx\right|^2} \\ \eta_{ph}^E = \dfrac{\left|\int_{-b/2}^{b/2} e^{-j\frac{\pi x^2}{\lambda R_E}} dy\right|^2}{b\int_{-b/2}^{b/2} dy} \end{cases} \tag{2.96}$$

where a and b are the aperture sizes of the pyramid horn along the broad and narrow sides of the waveguide, respectively, η_a is the aperture efficiency of the horn, η_t is the aperture efficiency caused by the aperture field distribution, η_{ph}^H is the aperture efficiency caused by the square-law phase distribution of the H-plane, η_{ph}^E is the aperture efficiency caused by the square-law distribution of the E-plane phase.

There is an optimal value for the opening angle of the pyramid horn. The gain corresponding to the horn aperture's physical area increases with the opening angle. As the opening angle increases, the phase efficiency of the aperture caused by the phase distribution of the horn decreases. Therefore, when the opening angle is at an intermediate value, the two product is the largest, and the gain of the corresponding pyramid horn is the largest. When both the E and H surfaces of the pyramid horn have the optimal opening angles, the aperture efficiency is about

$$\eta_t = 0.81, \eta_{\text{ph}}^H = 0.79, \eta_{\text{ph}}^E = 0.80, \eta_a = \eta_t \eta_{\text{ph}}^H \eta_{\text{ph}}^E = 0.51 \tag{2.97}$$

Therefore, for a high-gain pyramid horn, its aperture efficiency is low. Therefore, in the design and analysis process, the influence of aperture efficiency on the performance of the antenna system needs to be considered. In order to improve the electrical performance of horn antennas, researchers have developed various forms of special horn antennas, including multimode horns, corrugated horns, ridged horns, dielectric-loaded horns, and grid-loaded horns.

(2) **Helical antennas**

The basic configuration of a helical antenna is a uniform helix with constant elevation angle and radius, as shown in Fig. 2.18. Different sizes of helical antennas work in different modes. Common radiation modes include normal, axial, beam splitting, etc. The specific working mode is determined by its electrical size.

The helical antenna is used as a high-gain antenna and generally works in the axial mode state. In the working state of axial mode, the main feature of the helical antenna is that the maximum radiation is along the axis direction, the radiation field is a circularly polarized wave, and the traveling wave is approximately transmitted along the helix.

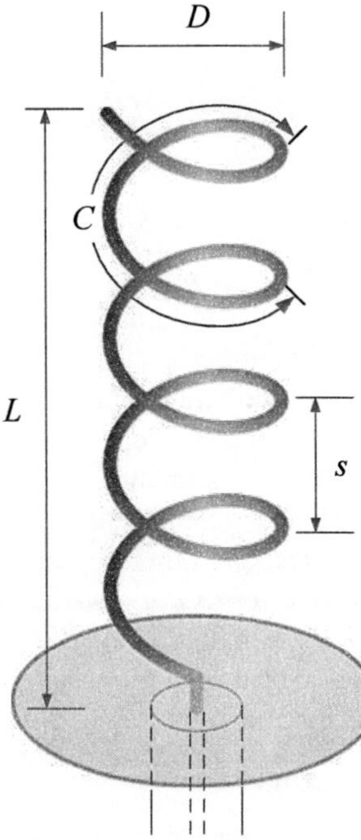

Fig. 2.18 Schematic diagram of helical antenna configuration

The pattern of an axial mode helical antenna can be approximated as the product of the pattern of a single-turn helix and the array factor of a Hansen-Wood end-fire array. The pattern of a single-turn spiral can be approximated by the current element in its axis-containing plane pattern cosθ. After calculating the array factor in combination with the relevant theory, the antenna pattern can be obtained as

$$F(\theta) = \cos\theta \frac{\sin\frac{\pi}{2N}\cos\left[\frac{N\pi S}{\lambda}(1-\cos\theta)\right]}{\sin\left[\frac{\pi S}{\lambda}(1-\cos\theta)+\frac{\pi}{2N}\right]} \tag{2.98}$$

where N is the number of turns of the helix, S is the pitch, λ is the free-space wavelength.

The half-power beamwidth of an axial mode helical antenna is

$$\text{HPBW} = \frac{52^\circ}{(C/\lambda)\sqrt{\text{NS}/\lambda}} \tag{2.99}$$

The directivity coefficient is

$$D = 12(C/\lambda)^2\text{NS}/\lambda \tag{2.100}$$

The axial ratio of the maximum radiation direction is

$$|r_A| = \frac{2N+1}{2N} \tag{2.101}$$

The input impedance is almost purely resistive and can be approximated as

$$R_{\text{in}} = 140\frac{C}{\lambda}\Omega \tag{2.102}$$

where N is the number of turns of the helix, S is the pitch, λ is the free-space wavelength.

For the helical antenna, the optimum helix angle with an excellent directional pattern, impedance, and axial ratio characteristics is 12°–14°, which can be used as an essential reference for the design of helical antennas.

Helical antennas have been widely used in satellite communications, aerospace vehicles, and other occasions requiring circularly polarized antennas and are also suitable for use in low-frequency microwave imaging radar systems. Its circularly polarized radiation characteristics can avoid the problem of polarization mismatch caused by the Faraday rotation effect of electromagnetic waves when passing through the ionosphere.

(3) **Yagi antennas**

A Yagi antenna consists of a driven element and several parallel parasitic elements. In the passive elements, there is a parasitic element that acts as a reflector and a plurality of parasitic elements that act as directors, as shown in Fig. 2.19. By adjusting the length and spacing of each parasitic element, the amplitude and phase of the induced current on the parasitic element can be changed, and a good end-fire pattern and higher gain can be obtained. Yagi antennas are widely used in the metric wave and decimeter wave bands in communication and radar systems. Its advantages are simple configuration, convenient feeding, and high gain, but the antenna profile is relatively high when forming an array.

The gain and half-power beamwidth of the Yagi antenna can be estimated as

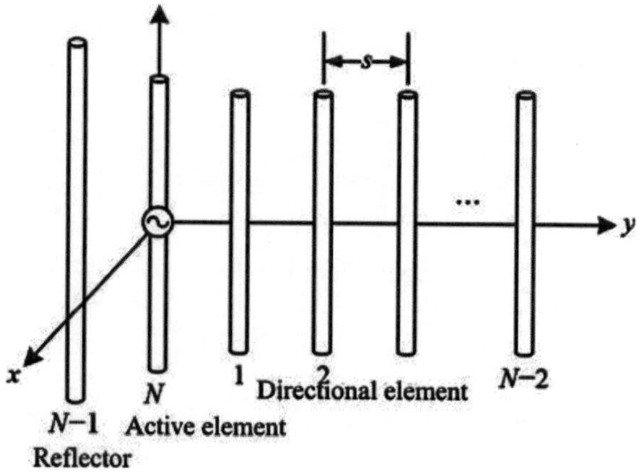

Fig. 2.19 Schematic diagram of Yagi antenna configuration

$$G \approx 10^{(N-1)s}/\lambda \tag{2.103}$$

$$\text{HPBW} = 55^{\circ}\sqrt{\lambda/L} \tag{2.104}$$

where s is the distance between adjacent elements, λ is the working wavelength, N is the number of elements of the antenna.

The more the number of elements of the Yagi antenna, the longer the array length and the higher the array gain. However, when there are too many directors, the gain of the Yagi antenna cannot be increased infinitely. Generally speaking, the reasonable length limit of the Yagi antenna is $L = (3 \sim 3.5\lambda)$. Under the condition of not exceeding this length, the number of directors of the Yagi antenna is generally 6–14, and the antenna gain is generally 13–18 dBi. To improve the antenna's gain further, you can use multiple Yagi antenna arrays such as 1×2, 2×2, and 2×4 to obtain higher gain.

(4) **Sub-array antenna**

The sub-array antenna is a plurality of antenna elements synthesized into a small antenna array with centralized feeding through a passive network. The antenna elements can be microstrip patches, slots, etc. Because of its scalability and uniformity of amplitude and phase, the aperture efficiency is higher than that of the above-mentioned high-gain element. Therefore, in the case of a small scanning angle and high element gain in the phased array antenna, the sub-array antenna is selected as the finite-scan phased array. According to the characteristics of the scanning range of the phased array, the sub-arrays are divided into linear arrays, rectangular arrays, and hexagonal arrays.

2.5 Array-Shaped-Beam Synthesis

The array mentioned above antenna pattern is generally a pencil beam directed at a single point, and the beam shape and coverage area are relatively regular. However, many applications place more complex requirements on the beam shape of the array. For example, some high-resolution imaging radar antennas form a specified shaped beam with the expected coverage on the ground, and some air-to-air radar antennas need to form a cosecant square beam in the elevation plane. Fixed beamforming can be realized through network amplitude and phase control [19] and real-time controllable beamforming can also be achieved by using a phased array system. Based on function requirements, most modern radar systems choose a phased array system to achieve rapid switching of antenna patterns in different beam shapes. In general, beamforming produces a specific irregular pattern to achieve coverage of a specific area. Usually, in engineering applications, the one-way pattern is optimized for beamforming. The transmitting pattern and the receiving pattern are independently formed according to the required coverage, but sometimes the two-way pattern is also optimized according to the power coverage.

The array-shaped-beam synthesis problem of the array antenna is usually transformed into the optimization problem of the pattern, and the classical optimization algorithms such as Chebyshev, Taylor, and Wardwood are aimed at a particular type of characteristic problem. On the other hand, more general optimization algorithms are usually used in research for synthesis with specific constraints, such as genetic algorithm, simulated annealing method, PSO, and steepest descent method [20–24]. For example, the genetic algorithm encodes the parameters to be optimized to generate the initial population of chromosomes, performs genetic breeding operations to generate the next generation population, designs the fitness function, selects the next generation population according to the fitness function, sets the control parameters, and forms an iterative search process of breeding plus detection. On the other hand, PSO is a global optimization algorithm that initializes a group of random particles in the global space and then finds the optimal solution through continuous iteration. In practical applications of the above algorithms, it is usually necessary to make certain improvements to the parameters and details of the algorithm according to different objectives to improve the execution efficiency and optimization results.

2.5.1 Phase Weighting

There are many different methods of beamforming for antenna arrays. For phased array antennas, only changing the phase distribution of the array elements, that is, only phase weighting can achieve beamforming without losing power. This method is usually applied in the transmission state of active array antennas.

The phase-weighted beamforming problem is treated as an optimization problem in the study, and there are usually two choices for the optimization goal: one is the antenna array pattern that is expected to be realized, the other is the specific parameters about the pattern, such as SLL and beamwidth. Among them, the direct optimization of the pattern is more common, and the corresponding mathematical description is as follows

$$\min \Delta f = \| f(\theta, \beta_1, \beta_2, \ldots, \beta_N) - f_0(\theta) \|, \beta_1, \beta_2, \ldots, \beta_N \in [0, 2\pi] \quad (2.105)$$

where Δf is the difference between the actual pattern in optimization and the expected pattern, $f(\theta, \beta_1, \beta_2, \ldots, \beta_N)$ is the pattern when the phases of each element of the antenna array are $\beta_1, \beta_2, \ldots, \beta_N$, $f_0(\theta)$ is the pattern of the expected realization, $\|\cdot\|$ is the second-order norm of the function, which is specifically expressed as the root mean square value of the function in the definition domain.

After the optimization objective is selected, the phase of each element of the antenna is used as the optimization variable, and the appropriate optimization algorithm is selected according to the nature of the problem itself. Finally, the optimal or reasonable solution to the problem is obtained by adjusting the parameters in the optimization process.

In phase-weighted beamforming research, commonly used comprehensive methods include interpolation, genetic algorithm, steepest descent, minimum value, quasi-Newton method, adaptive algorithm, and PSO algorithm.

It can be seen from the above analysis that the general solution of only phase-weighted beamforming is to treat the beamforming problem as an extreme value optimization problem and solve the corresponding optimization problem through various algorithms. In the optimization, according to the specific characteristics of the problem, the parameters of the algorithm are adjusted, or the algorithm is improved to some extent, and then the purpose of beamforming the antenna array is achieved.

2.5.2 Amplitude and Phase Weighting

The weighting methods in which amplitude and phase are variables simultaneously are primarily used in receiving arrays. In the application process, the received signal of each element is usually amplified first, and then the amplitude and phase are controlled by the phase shifter and attenuator in the T/R module. Due to the introduction of degrees of freedom in the amplitude distribution simultaneously, the amplitude-phase-weighted beamforming will obtain superior pattern characteristics.

The amplitude and phase-weighted beamforming problem can be described as determining the excitation amplitude and phase of the antenna array elements so that the radiation pattern can be as close to the expected pattern as possible. According to different applications, this expected pattern may vary significantly. For example, in synthetic aperture radar, the range scanning angle is different, that is, the range beam coverage position is different, and the position and width of the ambiguity area are different. Therefore, the optimization objectives include not only gain, beamwidth, and flatness, as well as side lobes and blur area suppression, etc. [21], as shown in Fig. 2.20a, in high-resolution imaging, there are also special requirements for broadband and other beamwidths, as shown in Fig. 2.20b.

With the rapid improvement of computer power, this type of beamforming is usually obtained by an optimization algorithm. The corresponding mathematical description is similar to Eq. (2.105), except that the magnitude term is added to the variable.

2.5.3 Applications

For array antennas, it is a complex nonlinear optimization problem to achieve the desired beam shape by synthesizing the amplitude and phase of the radiating elements of the antenna array. Usually, the number of antenna array elements is significant, and many parameters need to be adjusted. The genetic algorithm and PSO algorithm are

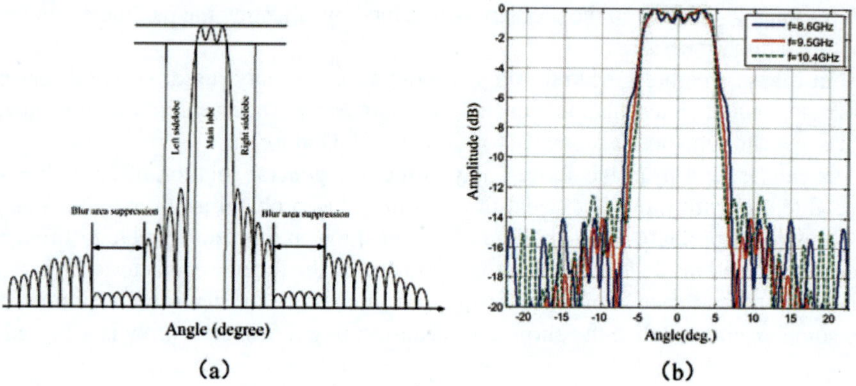

Fig. 2.20 Beamforming, **a** local area suppression of side lobes, **b** equal beamwidth

efficient algorithms to solve such problems. In engineering, the amplitude and phase weight distribution of the antenna array is obtained through optimization, and the desired pattern is achieved through the control of the network in the passive centrally fed antenna array or the phase shifter and attenuator control in the phased array.

Take the PSO algorithm as an example, it is a global optimization algorithm with few adjustable parameters, simple, fast running, easy to implement, and powerful. The algorithm's principle is to initialize a group of random particles and then find the optimal solution through iteration. In each iteration, the particle updates itself by tracking two "extremes." One is the optimal solution the particle finds, that is, the individual extremum. The other is the optimal solution found by the entire population, the global extremum.

After the particle finds the above two extreme values, it updates its speed and position, that is

$$\begin{cases} V = w \cdot V + c_1 \cdot \text{rand} \cdot (p\text{best} - \text{present}) + c_2 \cdot \text{rand} \cdot (g\text{best} - \text{present}) \\ \text{present} = \text{present} + V \end{cases} \tag{2.106}$$

where V is the speed of the particle, present is the current position of the particle, rand is a random number between 0 and 1, c_1 and c_2 are the learning factors, usually $c_1 = c_2 = 2$, w is the weighting coefficient, the value is in between 0.1 and 0.9. If w decreases linearly with the algorithm iteration progress, the algorithm's convergence performance will be significantly improved. Let $w_{\max}$ be the maximum weighting coefficient, $w_{\min}$ be the minimum weighting coefficient, iter is the current number of iterations, and iter_m is the total number of iterations of the algorithm, then we have

$$w = w_{\max} - \text{iter} \cdot (w_{\max} - w_{\min})/\text{iter}_m \tag{2.107}$$

Through continuous learning and updating, the particle finally solves the position of the optimal solution in the space, and the search process ends. The final output *g*best is the optimal global solution. During the update process, the maximum velocity of each dimension of the particle is limited to V_{max}, and the coordinates of each dimension of the particle are also limited within the allowable range. Figure 2.21 shows the basic flow of the PSO algorithm [24].

To check whether the algorithm converges, it is usually necessary to select a fitness function. The fitness function is different, indicating that the optimization problem

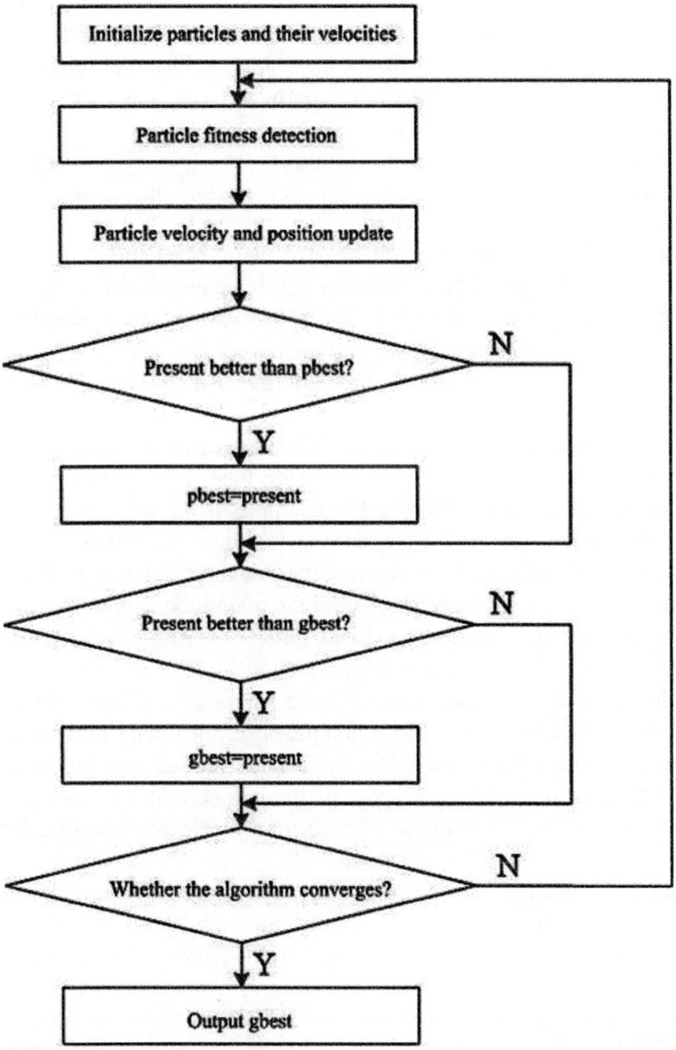

Fig. 2.21 PSO algorithm workflow

is different. In the integrated beamforming of the antenna array, the target pattern is often used as the optimization target for the main lobe. For beam broadening, the fitness function of the PSO algorithm can be set as

$$F = A\left|\frac{\mathrm{SLL} - \mathrm{SLL}_0}{\mathrm{SLL}_0}\right| + B\left|\frac{\theta_{3\,\mathrm{dB}} - \theta_{3\,\mathrm{dB0}}}{\theta_{3\,\mathrm{dB0}}}\right| + C\left|\frac{\delta}{\delta_0}\right| \tag{2.108}$$

where SLL is the current pattern side lobe level, SLL_0 is the maximum allowable side lobe level, $\theta_{3\,\mathrm{dB}}$ is the −3 dB width of the main lobe, $\theta_{3\,\mathrm{dB0}}$ is the target main lobe width, δ is the maximum ripple in the main lobe, δ_0 is the maximum ripple allowed in the main lobe, A, B, and C are the weighting coefficients. Different weighting coefficients will result in different SLLs, main lobe widths, and main lobe ripples. Properly increasing a specific weighting coefficient can make the corresponding parameters closer to the target value.

When only the phase-weighted beamforming is used, the element amplitude distribution is uniform, and the element phase is controlled by the phase shifter, thereby realizing the antenna array beamforming.

The optimal design object is a linear array with 24 elements and an element spacing of 0.5 wavelengths. The amplitude of each element is the same, and through phase weighting, the shape of the pattern is 0–30° cosecant square shaping, the level of all side lobes is lower than −10 dB, and the coverage is above 30°.

When only the phase is weighted, the main lobe usually has a larger fluctuation relative to the cosecant squared target, especially when the side lobe is required to be lower, and the fluctuation is larger. Therefore, in the optimization process, the weighting coefficients of the main and side lobe areas are different, and the optimization results are also different. For example, if the weighting coefficient of the main lobe area is increased, the cosecant square shaping is better, if the weighting coefficient of the side lobe area is increased, the side lobe will become lower, and the fluctuation of the cosecant square shaping of the main lobe area will become larger. Therefore, the weighting coefficients should be set according to the actual needs to achieve the parameters that meet the requirements.

Through the simulation optimization of the example, it is found that when the side lobe level is −8 dB, the main lobe and the cosecant square forming goal curve are in good agreement, as shown in Fig. 2.22a, and the SLL is required to reach −8 dB. At 12 dB, the 30° coverage airspace pattern has certain fluctuations. However, it is easier to achieve SLL lower than −15 dB for high elevation angles, as shown in Fig. 2.22b.

Cosecant beamforming is realized under the condition of amplitude and phase weighting. The optimal design object is a linear array, the number of elements is 16, and the element spacing is $0.64\lambda_0$. The shape of the pattern is a cosecant square of 0–30°, all sidelobes are below −35 dB, and the coverage is above 30°.

The optimized target pattern is shown in Eq. (2.107), that is, the dotted line part shown in Fig. 2.23, that is

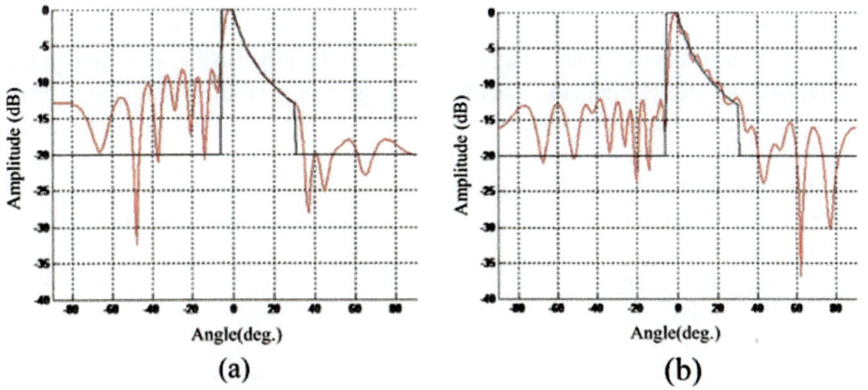

Fig. 2.22 Realization of antenna phase-weighted cosecant square shape based on PSO algorithm, **a** when the main lobe is smooth, **b** when the side lobe is low

$$f_0(\theta) = \begin{cases} 20\lg[\csc(\theta + 8°)/\csc 8°], & 0 \leq \theta \leq 30° \\ -37\,\text{dB}, & \text{Others} \end{cases} \quad (2.109)$$

The PSO fitness function is set to

$$F = A\left|\frac{\text{SLL} + 37}{37}\right| + B \cdot \max_{\theta \in [0, 30°]} |f(\theta) - f_0(\theta)| \quad (2.110)$$

The simultaneous weighting of amplitude and phase is adopted, the number of iterations is set to 500, the particle swarm size is set to 40, the range of amplitude value is [0, 1], and the value range of phase is also [0, 1] after normalizing with 360°. The weighting coefficients A and B of the degree function are finally optimized to

Fig. 2.23 Realization of antenna amplitude-phase-weighted cosecant square shape based on PSO algorithm

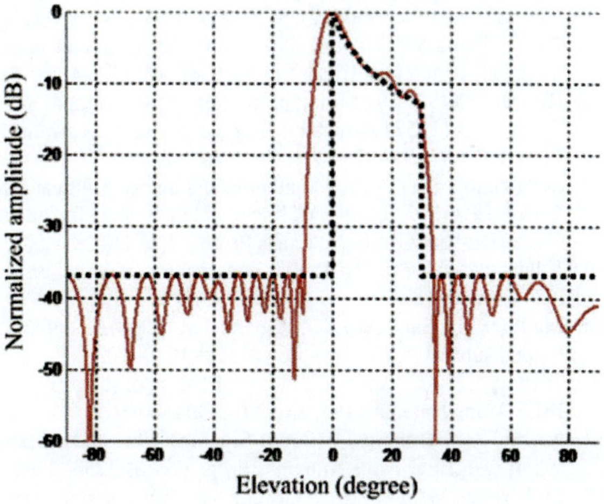

obtain each antenna array element's amplitude and phase coefficients. As shown in Fig. 2.23, the antenna's main lobe coincides with the target pattern. Although the number of elements is small, the side lobe level also reaches −36.6 dB. It can be seen that the particle swarm optimization algorithm can be well used for the optimization of radiation pattern beamforming.

The above two examples illustrate that by optimizing the feeding amplitude and phase of each element in the array antenna by the PSO algorithm, the beamforming of the main lobe and the suppression of the SLL can be realized at the same time, or the beamforming of the main lobe can be realized by only phase weighting, which can be obtained good cosecant square shaping effect. Therefore, the PSO algorithm has the advantages of simple theory, few parameters, and easy implementation and is more suitable for solving the beamforming problem of phased array antennas.

References

1. Zhong S (2011) Antenna theory and techniques. Publishing House of Electronics Industry, Beijing
2. Balanis CA (2005) Antenna theory-analysis and design, 3rd edn. Wiley, New Jersey
3. Ludwig AC (1973) The definition of cross polarization. IEEE Trans Antennas Propag 21(1):16–19
4. Nie Z (2014) Handbook of antenna engineering. UEST (University of Electronic Science and Technology) Press, Chengdu
5. Lo YT, Lee SW (1982) Antenna handbook: theory, applications, and design. Van Nostrand Reinhold, New York
6. Zhang H, Wang W, Jin M, Zou Y (2017) The design of broadband low side-lobe flat waveguide slots antenna. In: Proceedings of National conference microwave and millimeter wave 2017. Chinese Institute of Eleconics (CIE), Hangzhou, pp 850–852
7. Wang W, Fu D, Chen S (2001) Conductance computation of waveguide traveling-wave array. J Xidian Univ 28(2):234–237
8. Fang Z, Jin M (2012) The design of an L-band double-ridged waveguide slot array antenna with low sidelobe. Radar ECM 32(4):43–46
9. Zou Y, Wang W, Wu R (2008) Geometric parameter optimization for tilted-back planar 2D digital array antenna. Radar Sci Technol 6(6):481–485
10. Ricardi LJ (1972) Radiation properties of the binomial array. Microw J 15(12):20–21
11. Barbiere D (1952) A method for calculating the current distribution of tschebysheff arrays. Proc IRE 40(1):78–82
12. Schelkunoff SA (1943) A mathematical theory of linear arrays. Bell Syst Tech J 22:80–87
13. Taylor TT (1955) Design of line-source antennas for narrow beamwidth and low side lobes. Trans IRE Prof Group Antennas Propag 3(1):16–28
14. Villeneuve AT (1984) Taylor patterns for discrete arrays. IEEE Trans Antennas Propag 32(10):1089–1093
15. Mailloux RJ, Santarelli SG, Roberts TM (2006) Wideband arrays using irregular(polyomino) shaped subarrays. Electron Lett 42(18):1019–1020
16. Toyama N (2005) Aperiodic array consisting of subarrays for use in small mobile earth stations. IEEE Trans Antennas Propag 53(6):2004–2010
17. Kerby KC, Bernhard JT (2006) Sidelobe level and wideband behavior of arrays of random subarrays. IEEE Trans Antennas Propag 54(8):2253–2262
18. Lu J, Wang W, Qi M (2013) Grating lobes suppression in phased array antenna for space-borne SAR applications. J Microw 29(5/6):135–138

19. Zhang HT, Wang W, Zheng Z et al (2018) Design of a beamforming circular-polarization waveguide antenna array. In: Proceedings of the 2018 IEEE 7th asia-pacific conference on antennas and propagation (APCAP), New Zealand, pp 66–67
20. Hao L, Ming Z, Desen F, Jiaguo L (2002) Synthesis of antenna arrays shaped-beam using genetic algorithm. Chin J Radio Sci 17(5):539–548
21. Sun H (2012) Study on SAR antenna wideband and optimizing technologies. Chinese Academy of Sciences
22. Xu H, Li J, Hu M (2007) Research on beam widening of spaceborne SAR. J Electron Inf Technol 29(3):540–543
23. Jiao Y, Yang K, Chen S, Zhang F (2006) Application of particle swarm optimization in antenna array pattern synthesis. Chin J Radio Sci 21(1):16–20
24. Qi M, Wang W, Jin M (2008) Synthesis of antenna arrays using particle swarm optimization. Radar Sci Technol 6(3):231–234

Chapter 3
Array Antenna Error and Compensation

3.1 Introduction

As we all know, an antenna is a device that converts the guided electromagnetic fields into space radiation electromagnetic waves. For microwave imaging radar antennas, the radiation performance, e.g., gain, beamwidth, SLL, and beam pointing, are directly related to the imaging performance. In the antenna design, these radiation characteristics are usually analyzed and optimized under idealized conditions. As a result, it is difficult to consider the effects with some of the practical engineering parameters, such as the mechanical structure force, the thermal response, the channel consistency, etc. There is always some certain deviation between the analysis and the experimental test. Therefore, the radiation characteristics of the antenna need to be calibrated, corrected, and measured through experiments.

The spatial radiation distribution of the array antenna is mainly determined by four parameters, i.e., the location of the radiating element in the array antenna, the element pattern, and the exciting amplitude and phase of the element. Based on the uniqueness principle [1], the radiation pattern of the antenna array is determined when the four parameters are determined.

The research of the active array antenna is based on the optimization, design, engineering implementation, and control of the above four parameters. It also includes structure, thermal control, power supply, beam steering, and other factors such as environmental suitability and reliability that serve the four parameters. Considering a variety of influencing factors, the radiation performance is ultimately attributed to two parameters, amplitude and phase distribution, as listed in Table 3.1. Generally, when the analysis and design of an active array antenna are completed, whether it is a planar array or a conformal one, as shown in Fig. 3.1, the geometrical position of the element and its radiation pattern has been determined. The radiation pattern of the array is only determined by the amplitude and phase of the element, which is the basic operation principle of a phased array antenna. Of course, with the development of the high-integration and multi-function of the active array antenna, the position of the antenna element can be applied to improve the performance of the active

J. Lu et al., *Active Array Antennas for High Resolution Microwave Imaging Radar*,
https://doi.org/10.1007/978-981-99-1475-3_3

Table 3.1 Parameters influencing factors

	Element location	Element pattern	Element excitation
Parameter	Phase	Amplitude/phase	Amplitude/phase
Object	3D space radiation field	3D space radiation field	Port matching and 3D Space radiation field

(a) (b)

Fig. 3.1 Typical active array antennas, **a** planar array, **b** conformal array

array antenna. For example, a three-dimensional (3D) antenna array with mobile and reconfigurable spatial positions of antenna elements can be achieved based on a swarm of unmanned aerial vehicles.

Compared with the conformal array, the design and implementation of planar arrays are easier, because of the lack of one-dimensional spatial position variables. Modular design, processing, assembly, and expansion of the active array antenna can greatly simplify its analysis and design, such as pattern calculation and optimization, beam steering, pattern measurement, etc. Meanwhile, due to the uniformity of the radiation pattern of the antenna elements in the planar array, it is easy to achieve the ideal radiation characteristics of the antenna array. Of course, its disadvantages are also obvious. The gain of the antenna is related directly to its radiation area. In the case of large-angle scanning, the decrease of the antenna gain is inevitable. Therefore, for a planar array antenna, its coverage space is limited, and it is necessary to combine multiple planar arrays or mechanical scanning to solve this problem.

In the applications of microwave imaging, the beam scanning range of the antenna is usually relatively narrow. For example, in spaceborne imaging radar, the beam scanning range is generally about $\pm 20^{\circ}$, and the azimuth beam scanning is only a few degrees. In the case of high resolution and wide observation area, the azimuth coverage will extend to about $\pm 45^{\circ}$. On the airborne platform, due to the low flight height of the aircraft, the beam scanning range of the antenna needs to be increased. However, there is no application scenario with the requirement of 360° azimuth coverage in the early warning radar or communication system, and therefore the planar array antenna with limited scanning range is usually preferred. Figure 3.2 shows the typical structure of the planar phased array antenna [2] applied to microwave imaging radar. First, multiple antenna modules form a planar array

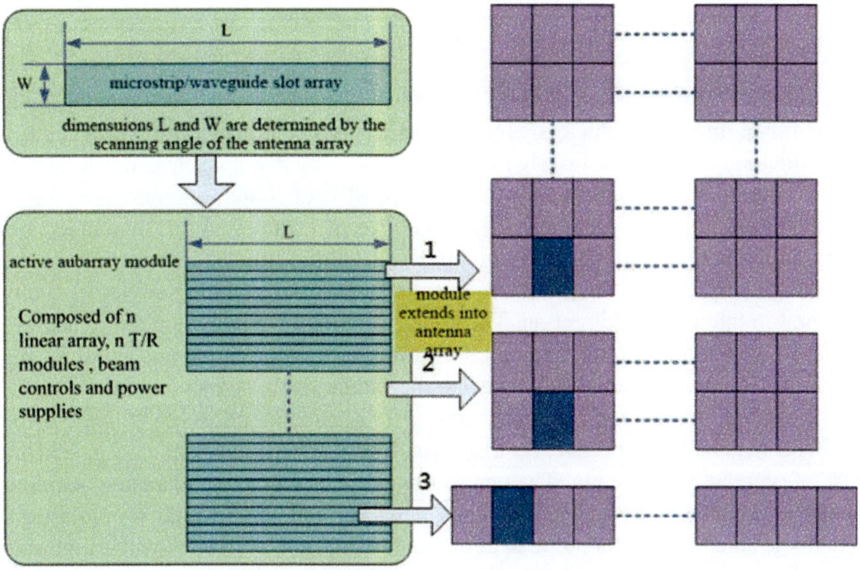

Fig. 3.2 Modular configuration of microwave imaging antenna

sub-array, and then multiple sub-arrays form a planar array antenna. This planar array antenna configuration is very suitable for satellite platforms. When the satellite is in the launch state, the antenna is in a folded state, and after the satellite enters orbit, the unfolding mechanism unfolds the planar array antenna.

3.2 Radiation Characteristic Parameters

The far field pattern of the antenna array is formed by the coherent spatial superposition of the electromagnetic fields radiated by each antenna element in the array. For a conventional planar array antenna, the distribution of the antenna elements is fixed. Therefore, the only factors that affect the radiation pattern of the antenna array are the element's pattern and the amplitude and phase of the excitation element. In essence, which is essentially the modulation of the antenna array factor superposition element pattern. Therefore, when analyzing, optimizing, designing, and evaluating the performance of the antenna array, the antenna array can be separated and reorganized to the sub-array. The modulation of the antenna element pattern directly affects the scanning range and the beam pointing and also has the characteristics of reducing the sidelobe level of the antenna array. The antenna array factor pattern is the main factor affecting the beam shape and beam scanning of the array antenna.

3.2.1 Sidelobe Level

The microwave imaging radar phased array antenna usually has a large scanning angle in the range direction, which is generally achieved by element-level phase control. The Sidelobe-Level (SLL) design of the antenna usually considers the influence of the ambiguous area and tries to reduce the SLL of the ambiguous area. There are many optimization methods for this kind of SLL suppression. A simpler and more direct method is to perform amplitude weighting through an attenuator in the receiving state to obtain the required low SLL. A small scanning angle is usually required in the azimuth direction. Thus, the sub-array is usually applied to achieve the beam scanning, i.e., the excitation in terms of the amplitude and phase of the antenna elements inside the sub-array are the same.

In wide-angle scanning antenna arrays, it is usually necessary to select the elements with wide-beam coverage to alleviate the deterioration of antenna gain in the case of wide-angle scanning. The conventional method is to add parasitic coupling configurations between adjacent antenna radiating elements, which is similar to a monopole antenna. The radiation pattern is in the shape of an "apple" with a high side beam direction and a recessed zenith direction. It complements the "steamed bread"-shaped radiation distribution of the main antenna to form a wide-beam coverage, as shown in Fig. 3.3. However, the radiation level of the wide-beam antenna element in the large-angle direction, which is far from the broadside direction, is still high. Therefore, the suppression effect on the effect of the sidelobe in the far region of the antenna array is weak. Thus, the ideal wide-angle scanning array element should construct its beam shape according to the beam scanning range so that it covers the scanning range of the antenna array. Therefore, compared with the conventional wide beamwidth element, the ideal antenna element can better suppress the far-region sidelobes of the array antenna and, simultaneously, reduce the gain drop of the scanning state.

In the sub-array phased array antenna, the azimuth far-region sidelobe is greatly affected by the sub-array modulation. For the sub-array antenna, the element spacing in the array directly affects the far-region sidelobe. Under the premise of realization, the small-spacing array can suppress the far-region sidelobes better than the large

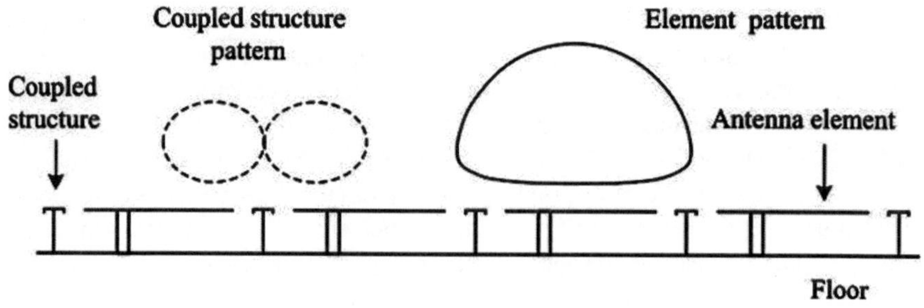

Fig. 3.3 Wide-angle scanning antenna array with parasitic configuration added

one. Figure 3.4 shows the calculation example of the influence of the element spacing in the sub-array on the far-region SLL. In the calculation, the number of sub-arrays in the antenna array is the same, and the length of the sub-array is the same. As shown in Fig. 3.4a, the element spacing inside the sub-array is 0.5λ, and when scanning 1.5°, the peak value of the SLL in the farthest area of the sub-array is −24 dB. The peak value of the grating lobe is −27.5 dB, as shown in Fig. 3.4b, the element spacing in the sub-array is 0.8λ, and the calculated corresponding values are −17 dB and −19.6 dB, respectively. The results show that the sub-array antenna formed with the 0.5λ element spacing improves the level of the grating lobe in the far region of the antenna array by 8 dB compared with the 0.8λ spacing. As a result, the peak level of the first grating lobe in both antenna distributions is the same at −11.1 dB. In the sub-array phased array design, on the premise of satisfying the requirement of Equivalent Isotropically Radiated Power (EIRP), the sub-array scale can be selected to be larger to further reduce the number of active channels. However, in this case, it will result in a higher grating lobe level in the scanning state. The grating lobe suppression technology is adopted to destroy the periodic structure through modular dislocation, rotation, and irregular block, and then it can suppress the level of the grating lobe on the outer of the main beam.

In addition, the antenna sub-array pattern reconfigurable technology can solve the problem of high grating lobes. The basic principle is that the maximum beam pointing of the sub-array pattern can be switched through the switch circuit. When the beam scanning angle increases, causing the grating lobe to increase continuously, the sub-array pattern is switched in time so that the null direction of the sub-array pattern is as close as possible to the grating lobe direction to reduce the grating lobe. This technology can better solve the problem of high grating lobes, but the loss caused by switching devices in reconfigurable circuits cannot be ignored. Therefore, this grating lobe level suppression method needs to solve the high loss problem of reconfigurable circuits.

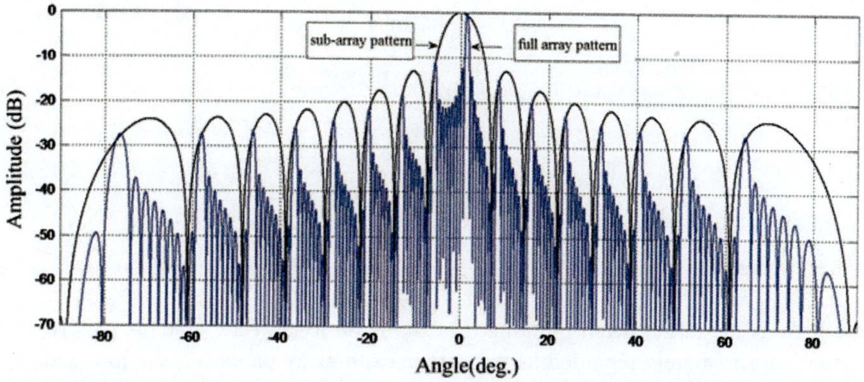

Fig. 3.4 The influence of different element spacings in the sub-array on the sidelobe in the far region of the antenna array (The element spacing is 0.5λ)

3.2.2 Beam Pointing

For microwave imaging radar, the antenna beam pointing accuracy directly impacts the swath width, image resolution, ambiguity, and image positioning accuracy. Beam pointing errors are divided into fixed and random errors. The former can carry out systematic compensation, while the latter is difficult to predict, and the error varies randomly and is unstable over time. Such errors need to be controlled in analysis, design, and engineering. The modulation of the antenna element pattern in the antenna sub-array phased array is also an essential factor affecting the beam pointing [3].

Due to the modulation of the antenna element pattern, the beam pointing deviation is inevitably caused when the antenna array is scanned. The larger the beam scanning angle, the greater the beam pointing deviation, and this phenomenon is particularly evident in sub-array phased array antennas. This is because the beam width of the sub-array pattern is narrow, and the slope at the −3 dB drop point of the beam is large. Therefore, when the antenna array factor and the sub-array pattern are combined in space, the maximum point of the scanning angle is not on the maximum value of the array factor pattern.

The relationship between the sub-array scale of the phased array antenna and the beam pointing error is analyzed and discussed below. Suppose in a sub-array phased array antenna, the number of radiating elements in the azimuth sub-array is N_1, the number of sub-arrays is N_2, the element spacing is d, and the sub-array spacing is $N_1 \times d$. The antenna elements in the sub-array are evenly distributed, the excitation amplitude and phase of each sub-array are same, and the antenna array factor controls the beam scanning through the excitation phase of each sub-array.

The antenna sub-array pattern is expressed as

$$f_1(x) = 10\lg\left(\left|\frac{\sin(N_1 x)}{N_1 \sin x}\right|\right) \tag{3.1}$$

$$x = \frac{1}{2}\text{kd}\sin\theta$$

The array factor pattern is expressed as

$$f_2(x) = 10\lg\left(\left|\frac{\sin(N_1 N_2(x - x_b))}{N_2 \sin N_1(x - x_b)}\right|\right) \tag{3.2}$$

$$x_b = \frac{1}{2}\text{kd}\sin\theta_b$$

where θ_b is the target beam pointing. For the main lobe of the antenna pattern, f_2 can be approximately regarded as the antenna sub-array pattern f_1 compressed N_2 times along the coordinate x-axis and then translated by x_b. So, there is the following relationship:

$$f_2(x) = f_1(N_2(x - x_b)) \tag{3.3}$$

$$f_2'(x + x_b) = N_2 f_1'(N_2 x) \tag{3.4}$$

The slope of the tangent line where the antenna array factor deviates from its maximum direction x is equal to N_2 times the slope of the tangent line where the sub-array deviates from the normal direction $N_2 x$. According to the principle of pattern multiplication, the pattern of the final antenna array is obtained by $f_1 \cdot f_2$. The antenna beam pointing is related to the change in both slopes. For the convenience of calculation, the pattern is simplified as

$$f(x) = \ln\left(\left|\frac{\sin Nx}{\sin x}\right|\right) \tag{3.5}$$

The pattern derivative is

$$f'(x) = \frac{N}{\tan(Nx)} - \frac{1}{\tan(x)} \tag{3.6}$$

In the vicinity of the maximum radiation direction $x = 0$, $x \ll 1$, Eq. (3.6) can be approximated as

$$f'(x) = \frac{\left(1 - N^2\right)x}{3} \tag{3.7}$$

The physical meaning of Eq. (3.7) is that in the vicinity of the maximum radiation direction of the uniformly distributed antenna array pattern, the variation of the pattern derivative with x is linear. Therefore, when the main lobe of the antenna array factor pattern is close to x_b/N_2^2 on the maximum side of the sub-array pattern, the tangent slopes of the two patterns are approximately opposite numbers to each other, which is the beam pointing of the pattern. Considering $\sin x \approx x$, the beam direction of the composite pattern obtained by scanning θ_b of the antenna array factor is $\left(1 - 1/N_2^2\right)\theta_b$, that is, the deviation of the beam direction relative to the direction of the array factor is about $-\theta_b/N_2^2$.

Accordingly, when the antenna beam scanning phase is controlled, the preset scanning angle needs to be increased by $\left(N_2^2 - 1\right)/N_2^2$ times, that is, the preset scanning angle is

$$\theta_b' = \frac{\left(N_2^2 - 1\right)\theta_b}{N_2^2} \tag{3.8}$$

Taking the one-dimensional linear array antenna with the number of sub-array elements $N_1 = 16$ and the number of sub-arrays $N_2 = 9$ as an example, the element spacing $d = 0.78\lambda$, then the sub-array spacing is 12.48λ. The sub-array beamwidth

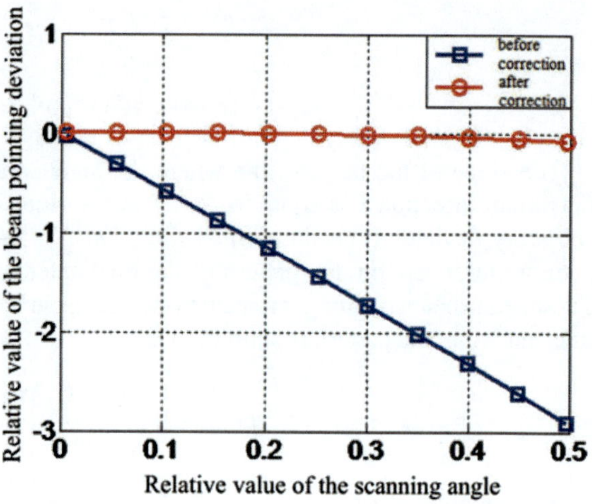

Fig. 3.5 Scan angle and beam pointing deviation in a small angle range

is 4°, and the antenna array beamwidth is 0.44°. Figure 3.5 shows the calculated comparison curve. The x-axis is the ratio of the scanning angle to the beam half power width of the sub-array, and the y-axis is the percentage of the beam pointing deviation relative to the beam width of the antenna array. For example, when the antenna array is scanned to a quarter of the 3 dB beam width of the sub-array, the uncorrected beam pointing deviation is −2.9% relative to the beam width of the antenna array. According to the above calibration method, when calculating the beam scanning compensation phase, the antenna beam pointing premultiplied $\left(N_2^2 - 1\right)/N_2^2 = 81/80$ times, then the final composite beam pointing deviation is reduced to −0.083%.

In Eq. (3.8), correlation approximation is carried out when calculating small-angle scans, and simplified results are obtained. However, when the antenna array factor is scanned at a large angle, such as scanning to the beam width of 3 dB of the sub-array antenna, the beam pointing deviation obtained by Eq. (3.8) is relatively large. In order to reduce the beam pointing deviation, it is necessary to compensate for the excitation phase of the sub-array and let the slope of the tangent at the sub-array pattern θ_b be

$$K_b = f_1'(x_b) = \frac{N_1}{\tan(N_1 x_b)} - \frac{1}{\tan(x_b)} \tag{3.9}$$

When the number of sub-arrays is $N_2 \gg 1$, the slight difference between the derivative of the sub-array pattern at x_b and $\Delta x + x_b$ can be ignored, and we get

$$f_2'(\Delta x + x_b) = N_2 f_1'(N_2 \Delta x) = -K_b \tag{3.10}$$

$$f_1'(N_2 \Delta x) = -\frac{K_b}{N_2} \tag{3.11}$$

$$c\tan x = \frac{1}{x} - \frac{1}{3}x - \frac{1}{45}x^3 - \cdots \tag{3.12}$$

$$\begin{aligned} f_1'(N_2\Delta x) &= \frac{N_1}{\tan(N_1 N_2 \Delta x)} - \frac{1}{\tan(N_2\Delta x)} \\ &\approx \frac{N_2\left(1 - N_1^2\right)}{3}\Delta x + \frac{N_2^3\left(1 - N_1^4\right)}{45}(\Delta x)^3 \end{aligned} \tag{3.13}$$

Combining Eqs. (3.10) and (3.11), the unary cubic equation is obtained as

$$(\Delta x)^3 + p\Delta x + q = 0 \tag{3.14}$$

$$p = \frac{15}{N_2^2\left(1 + N_1^2\right)}, q = \frac{45K_b}{N_2^4\left(1 - N_1^4\right)}$$

In general, Eq. (3.14) has one real root and two imaginary roots, and the real root is taken as the solution to the equation

$$\Delta x = \sqrt[3]{-\frac{q}{2} + \sqrt{\left(\frac{q}{2}\right)^2 + \left(\frac{p}{3}\right)^3}} + \sqrt[3]{-\frac{q}{2} - \sqrt{\left(\frac{q}{2}\right)^2 + \left(\frac{p}{3}\right)^3}} \tag{3.15}$$

Then the beam pointing deviation is obtained as

$$\Delta\theta = \arcsin\left(\frac{2x_b}{\text{kd}}\right) - \arcsin\left[\frac{2(x_b + \Delta x)}{\text{kd}}\right] \tag{3.16}$$

Therefore, when the beam of the antenna array is scanning, to obtain the composite beam pointing θ_b, the beam scanning angle should be preset as

$$\theta_b' = \frac{\theta_b^2}{\theta_b + \Delta\theta} \tag{3.17}$$

For the above example, when the beam scans to the 3 dB width of the sub-array, the beam pointing deviation before calibration is −6.5% relative to the beam width of the entire antenna array, and the relative value of the beam pointing deviation after calibration is 0.04%, as shown in Fig. 3.6.

For the element-level phased array antenna, because the element beam width is wide, the beam drop slope is gentle at the 3 dB beam width, and the pointing deviation caused by the element pattern is small, so it can usually be ignored.

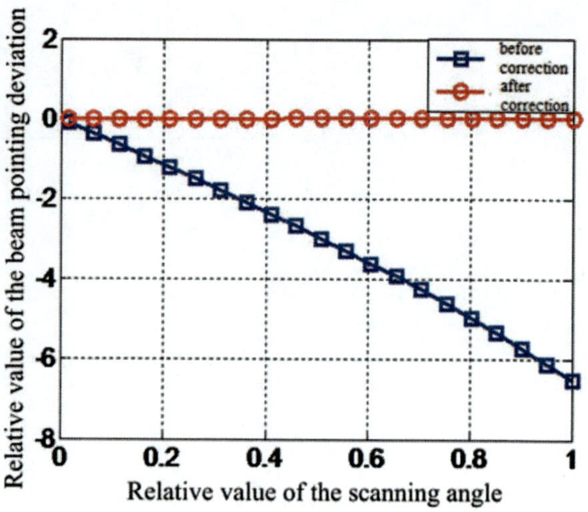

Fig. 3.6 Beam pointing deviation scanned within 3 dB beam width of sub-array

3.2.3 Antenna Gain

In microwave imaging radar, the quality of the image is closely related to the radiated power and gain of the antenna and is also closely related to the radar scattering cross section of the target. Usually, the noise Equivalent Sigmazero (NEσ_0) is applied to represent the sensitivity of the system, an important parameter that reflects image quality. According to the microwave imaging radar equation [4], that is

$$\mathrm{NE}\sigma_0 = \frac{2(4\pi)^3 R^3 kT F_n v_{\mathrm{st}} L_s}{P_{\mathrm{av}} G^2 \lambda^3 \rho_{\mathrm{rg}}} \tag{3.18}$$

where P_{av} is the average transmit power of the radar, G is the one-way gain of the antenna, ρ_{rg} is the range resolution of the radar, R is the distance between the radar and the target, L_s is the system loss, λ is the working wavelength, v_{st} is the velocity of platform, T is the noise temperature, F_n is the noise figure, and k is the Boltzmann constant, $k = 1.38054 \times 10^{-23}$ J/K.

Two major factors, namely the directivity and efficiency of the antenna, affect the gain of antenna. The former represents the energy radiated by the electromagnetic energy in space through the antenna, and the latter represents the efficiency of the feed-guided wave converted into the radiated wave.

The directivity and efficiency of the antenna have been described in Chap. 2. For an active array antenna, the transmitting state is usually selected for saturation amplification, the antenna aperture field is generally uniformly distributed, and the directivity coefficient can be simplified according to the area, which is

$$D = \frac{4\pi A}{\lambda^2} \tag{3.19}$$

where A is the effective area of the antenna array and λ is the working wavelength. When a non-uniform weighting is adopted, the effective aperture of the antenna is relatively reduced, the beam broadens, the SLL decreases, and therefore the directivity coefficient decreases. For example, when a Taylor weighting distribution is adopted on the antenna aperture, the decrease iin directivity coefficient can be evaluated by the beam broadening coefficient [5].

The relationship between the antenna gain and the directivity coefficient is

$$G = \eta_a D \tag{3.20}$$

where η_a is the antenna efficiency, the factors affecting the antenna efficiency include the ohmic loss of the antenna element, the dielectric loss, and the reflection loss.

In the antenna design, the main way to reduce the ohmic loss is to select appropriate materials and reduce the length of the feeder. The main way to reduce dielectric loss is to choose a substrate with a low-loss tangent, control the feed length, and reduce the discontinuous parasitic radiation and surface wave radiation of the feedline. The reflection loss mainly affects the active standing wave in the antenna design. In order to ensure the scanning performance in all angles, the active standing wave must be controlled within a certain range.

Printed dipole antennas have the advantage of wideband, and the bandwidth can reach about 20% [6], but the profile of the antenna element is relatively high, and the thickness of the antenna will be further increased when the sub-array design is combined with the power division network. In addition, due to the inherent defects of the transmission line form, the antenna efficiency is low, the structural characteristics of the printed dipole antenna make it impossible to realize the integrated realization of the planar array, and it needs to be processed and assembled line by line, due to the high profile, this kind of antenna has poor mechanical properties and large load space. Therefore, printed dipole antennas are usually not used as spaceborne SAR antennas.

Microstrip antennas have the advantages of low profile, small size, low mass, and easy integration with active devices. At present, microstrip antennas have solved the bandwidth problem through multi-layer [7] and cavity-backed structures [8, 9]. The cavity-backed single-layer microstrip patch element can achieve a bandwidth of 20%. However, like printed antennas, the loss of feeding network becomes the limiting factor for their application in the sub-array phased array antennas. In general, the efficiency of the X-band 16-element sub-array is only about 50%, and the efficiency of the 8-element antenna sub-array is 60–70%.

The waveguide slot array antenna is disadvantageous in size, mass, and bandwidth, but it's very low feed loss makes it have obvious advantages in higher frequency bands, especially in the X-band and higher frequency bands. As a result, the bandwidth of the waveguide slot array antenna has been greatly expanded, and the relative bandwidth exceeds 13% [10, 11], which meets the requirements of high-resolution microwave imaging, and its efficiency is usually higher than 80%. In addition, the appropriate surface oxidation treatment of the aluminum alloy waveguide slot array antenna can control the antenna surface's solar absorption rate and emissivity, which is beneficial to the implementation of thermal control of the antenna array.

The open-end waveguide antenna can obtain 18% bandwidth through matching [12], and 40% bandwidth can be obtained through stepped ridge matching. At the same time, due to the use of metal materials, the antenna efficiency in the broadband range can reach about 90%. However, the antenna profile is high due to its structural characteristics, and the additional network further increases the profile height. Therefore, this antenna is more suitable for element-level phased array antennas but not for sub-array antennas.

The comparison of the above antennas, electrical and thermal performance is listed in Table 3.2, in which the efficiency of the antenna array includes the loss caused by the amplitude/phase error in the sub-array, and the scope of application is used to indicate whether it is suitable for spaceborne microwave imaging active array antenna.

As a phased array antenna system, especially the antenna applied to the spaceborne platform, the requirements for efficiency are not only the efficiency of the passive radiating surface but also strict requirements on the active module. Such as T/R modules, delay amplifier modules, and secondary power supplies.

In the spaceborne active array antenna, the high power, high efficiency, light, and miniaturization of T/R modules are one of the important research points. Among them, the high efficiency of the T/R module is more important, and it is the crucial technical parameter of the spaceborne T/R module. The efficiency level directly affects the energy distribution, consumption, and thermal control of the spaceborne payload. As far as the efficiency of modules is concerned, it is important to improve the chip efficiency. Obviously, the advancement of Monolithic Microwave Integrated

Table 3.2 Antenna element/array comparison

	Printed dipole antenna	Microstrip patch element	Open-end waveguide antenna	Microstrip antenna array	Waveguide slot array antenna
Bandwidth (%)	20	20	40	13	13
Efficiency (X-band)	Same as microstrip antenna	85%	85%	16 elements: ~50% 8 elements: ~65%	16 elements: ~80% 8 elements: ~85%
Mechanics	Rely on structural parts	Rely on structural parts	High self-strength	Rely on structural parts	High self-strength
Thermal control difficulty	Difficult	Difficult	Mature technology	Difficult	Mature technology
Applicable scope	Not recommended	Large swath width	X-band and above, large swath width	Below C-band	C-band and above

Circuits (MMIC) technology is difficult to enhance chip efficiency significantly. Therefore, the scope of this book will not be repeated here. In addition, it is also crucial to reduce the circulator, isolator, and feeder insertion loss. In general, the actual composition of T/R modules varies slightly with different radar systems, and the complexity of specific circuits is also different, but their basic principles are almost the same. It mainly comprises a transmitting channel, receiving channel, shared channel, power modulation circuit, and drive control circuit. The efficiency of the module is

$$\eta_{T/R} = \frac{P \times D}{V_{D1} \times I_1 + V_{D2} \times I_2 + V_{D3} \times I_3} \tag{3.21}$$

where P is the output pulse power of the module, D is the pulse duty ratio of the module, generally taking 10–30%, V_{D1} is the working voltage of the transmitting power amplifier, including the driving amplifier and the final amplifier, generally about 8.5 V, I_1 is the working current of the transmitting power amplifier, generally in the order of amperes, V_{D2} is the voltage of the receiving low-noise amplifier and the drive control circuit, generally 5 V, I_2 is the current of the receiving low-noise amplifier and the drive control circuit, generally in the order of 100 mA, and V_{D3} is the voltage of the drive control circuit, generally −5 V, I_3 is the current of the drive control circuit, generally in the order of tens of mA.

It can be seen from Eq. (3.21) that the main factors affecting the module efficiency are the output power and the magnitude of each current value. Therefore, reducing the loss between the output of the final stage power amplifier and the output port of the T/R module and increasing the output power P of the component are the keys to improving the whole efficiency. Among them, improving the efficiency of the final-stage power amplifier and the power supply modulation circuit and reducing the top drop of the RF output pulse power are the auxiliary means to improve the efficiency of the modules. Therefore, selecting high-efficiency final-stage power amplifiers is the primary task of designing high-efficiency T/R modules, and high-efficiency final-stage power amplifiers can reduce the working current I_1 value of the transmit power amplifier, which has a significant effect on improving the efficiency of modules. Reasonable selection of low-power, low-noise amplifiers and reducing the switching loss of receiving power modulation can effectively reduce the I_2 value, and also have a certain effect on improving the efficiency of modules. The drive control circuit current I_3 is smaller than I_1 and I_2, and its proportion in the module efficiency can be ignored. Placing the power modulation circuit inside the module can supply power to the power chip nearby, reduce the current consumption I_2, and improve the conversion efficiency of the secondary power supply. The power supply modulation of the transmitting channel drain is usually realized by a high-power P-channel MOS FET tube, which reduces the on-resistance of the *P*-channel MOS FET tube, improves the power efficiency of the pulse modulation circuit, and thus achieves the purpose of improving the efficiency of the modules.

Typical efficiency parameters that can be achieved by T/R modules of conventional devices in different bands are listed in Table 3.3.

Table 3.3 Comparison of typical indicators of T/R modules in different frequency bands

	L-band	C-band	X-band	Ku-band
Typical efficiency (%)	45	28	25	20
Test output power (W)	200	80	20	12
Test bandwidth	200 MHz	700 MHz	1 GHz	2 GHz
Device mode	MMIC + discrete device (final power amplifier)	MMIC	MMIC	MMIC

Note Output power refers to the use of monolithic microwave integrated circuits without power combine

It is also crucial to reduce the insertion loss of the connection between the T/R module and the passive radiating antenna. Excessive insertion loss will have two effects on the array antenna as a whole: First, the excessive loss will reduce the overall efficiency of the array antenna, making the total radiated power of antenna decreases, secondly, insertion loss will cause heating of the feeder, and connection problems will generate electromagnetic leakage. Both of these effects will have adverse effects on the operation of the system. The heat will accelerate the aging speed of the devices at the connection and reduce the reliability of the connection. The electromagnetic leakage will cause interference between the channels and reduce the signal quality of each channel. Therefore, when analyzing and designing the feeding configuration between the T/R module and the passive antenna, it is generally necessary to reduce the insertion loss of the feeding configuration as much as possible while ensuring a good mechanical connection to obtain better overall electrical performance.

3.3 Error Analysis

The far-field radiation pattern of the antenna array depends on the pattern of the antenna element, the amplitude, and phase of the excitation of the antenna element, the location of the element, etc. In essence, the geometric position is ultimately reflected in the phase. Therefore, the array antenna error has only two factors, i.e., amplitude and phase. Based on the principle of "uniqueness", the source and the field are in one-to-one correspondence under certain boundary conditions. Meanwhile, the determined source can calculate the unique radiated electromagnetic field. Therefore, in the antenna design, the source distribution of the antenna array is usually optimally designed according to the expected radiation field distribution, and the required pattern can be obtained by realizing the target source distribution. Error analysis, calibration, and control is the process of making the actual antenna array radiation source amplitude and phase distribution infinitely close to the target distribution value.

3.3.1 Array Antenna Error Sources

The errors that cause the difference between the theoretical design value and the actual value of the phased array antenna mainly come from the phased array antenna error, the test system, and the interaction between the two during the test process.

Phased array antenna errors are usually divided into three categories: systematic errors, gradual errors, and random errors. The system error mainly comes from the asymmetry of the antenna configuration, the misalignment of the array mechanical and electrical axis of the array, and the stability of the platform. The gradual error usually comes from the thermal deformation and the deformation under the action of external force. Random error sources include manufacturing tolerances, assembly and installation tolerances, antenna element failures, and amplitude/phase errors caused by the nonuniform temperature distribution of the antenna aperture. Such errors are random and difficult to analyze and calculate strictly in advance. The amplitude and phase distribution of the electric field at the aperture of the array antenna changes randomly, which can be analyzed by statistical theory.

Test system errors include the stability of the transmitting signal source and receiver, calibration errors, mismatch effects, test accessories errors, and test environment errors. During the antenna measurement, scattered scattering on the ground and surrounding objects make the measured results and the actual situation inaccurate. The error caused by the site has a significant influence on measurement accuracy. In order to improve the measurement accuracy, microwave-absorbing materials can be laid in the area where the scattering effect is more significant.

The errors associated with the interaction of the antenna with the test system are mainly calibration errors. Whether it is a traditional analog phased array antenna or a digital array antenna, due to various reasons such as device inconsistency, manufacturing tolerances, and assembly errors, there must be amplitude and phase errors in each channel of the phased array antenna. Therefore, before testing the performance of the phased array antenna, it is necessary to perform the initialization calibration of the phased array antenna, that is, to accurately measure the initial amplitude and phase characteristics of each channel, and obtain the difference from the ideal amplitude and phase distribution. Then compensate by adjusting the amplitude and phase of each channel so that the entire antenna can reach the best state required by design.

The precondition for single-channel calibration of the antenna is to assume that the pattern of each element is the same as that of the element in the array, and the radiation characteristics of the antenna element will change slightly with its position in the array, e.g., the pattern of the antenna element near the edge of the array and the array assembly slot will be affected, and its pattern will deviate from the pattern of the antenna element in the array. At the same time, it is also necessary to assume that the position of the source antenna is accurately known during the calibration and there is an error in the actual erection. In addition, it is also assumed that the operational characteristics of each active channel are the same as the actual situation. During the actual calibration, each channel has instability, and when the channels work one by one, the characteristics of the full antenna array will be different. Aperture field inversion requires the antenna to perform full array calibration according to Nyquist sampling constraints, so the calibration time is long. Especially in the case of large antenna arrays, the stability of the test system, nonuniform temperature distribution caused by the heating of the active channel of the antenna array, and changes in ambient temperature will all introduce errors.

3.3.2 Error Source Analysis

Due to manufacturing and installation, the unevenness of the Z-axis will affect the phase error of the antenna element. As shown in Fig. 3.7, the difference between the elements in the Z-direction is Δz, that is, the normal spatial phase difference is $2\pi \Delta z/\lambda$. Without considering other factors, it is assumed that the antenna array completes the initial state calibration in the near field of the plane and obtains equal phase distribution. When the antenna scans to the θ direction, the spatial phase difference caused by Δz becomes the ΔL phase difference, where $\Delta L = \Delta z/\cos\theta$, which is an additional increase of $2\pi (L - \Delta z)/\lambda$ compared with the equal phase plane after the array compensation phase difference. This error is directly related to the Z-direction position error and the scanning angle. Therefore, in the design process of the antenna array, it is necessary to control the flatness of the antenna aperture.

Figure 3.8 shows the schematic diagram of the lateral coordinate position error for the antenna element. The phase error caused by the lateral position error of the antenna element position mainly comes from the antenna beam scanning phase calculation. It is difficult to obtain the exact spacing between each antenna element in antenna engineering, especially in large-phased array antennas. Therefore, the design element spacing is usually used as a reference when calculating the scanning phase. Taking a 1D phased array antenna as an example, it is assumed that the design element spacing is dx, while the actual element spacing is $dx + \Delta x$, and the beam deviates from the normal direction to scan to θ. Obviously, after analysis, the difference between the calculated compensation phase and the actual space phase is $2\pi \Delta x \sin\theta/\lambda$.

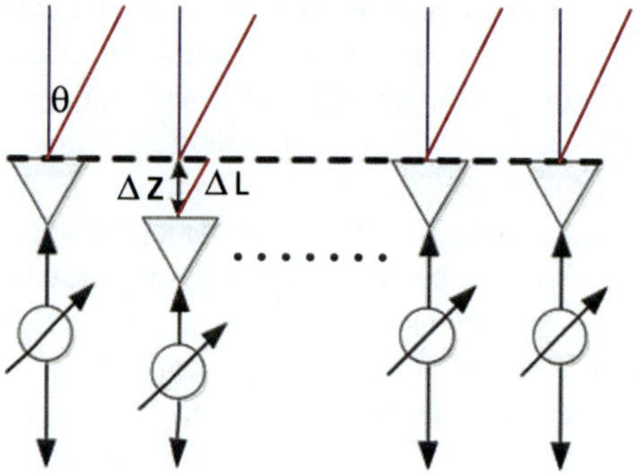

Fig. 3.7 Antenna array flatness error

Fig. 3.8 Antenna array element lateral position error

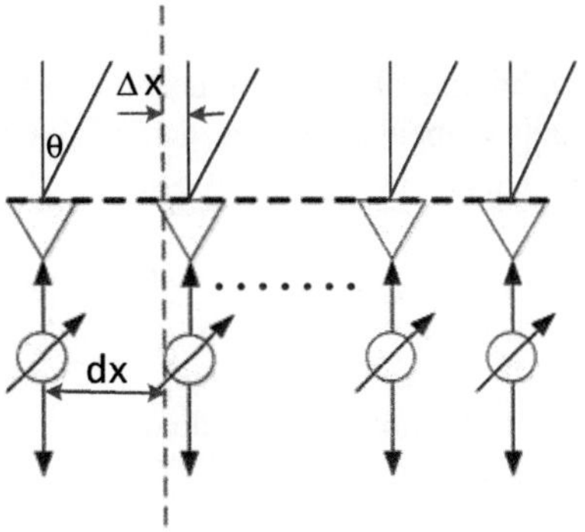

Based on the position error factors of the elements, when designing an antenna array, it is necessary to analyze and allocate the antenna element position error, constrain and guide the antenna array design and manufacture according to the specific requirements of the array antenna, such as the antenna beam scanning range, SLL, and beam pointing, etc.

For the amplitude and phase errors of the passive network of the array antenna, in the transmitting state of the phased array antenna, the discreteness of the network insertion loss of each port is controlled within a certain range so that the excitation signal can pass through the output port of each network to the T/R module and still ensure its final stage. The power amplifier works in the saturation amplification region, so the amplitude error between each channel of the passive network has little effect on the amplitude of the output signal of each channel. At this time, the amplitude error of each channel is determined by the inconsistency of the component transmit power of each channel. The phase shifter compensates for the phase error of the passive network and the active channel, and the final phase error depends on the control error of the phase shifter of the T/R module and its in-band signal amplitude flatness, in the receiving state, the amplitude and phase are using the attenuator and phase shifter of the T/R assembly to compensate.

The antenna element excitation amplitude and phase errors caused by the non-uniformity of the array surface temperature can theoretically be compensated. That is, by testing the curve of the operational characteristics of each channel changing with temperature, a temperature sensor is installed in the antenna array, and the phase shifter is controlled to compensate for the phase difference caused by the difference between the working and the calibration temperatures. However, this method greatly increases the system complexity in engineering, so the antenna array temperature control system is usually used to control the temperature difference of the entire array surface within a certain tolerance range when the antenna is working.

There are many types of amplitude and phase errors in T/R modules and delay amplifier modules, such as the initial difference of amplitude and phase of each module, control errors, in-band flatness, parasitic amplitude modulation error, parasitic phase modulation error, etc. However, based on the design idea of simplifying the phased array antenna calibration system, most of these errors are strictly controlled in the manufacturing process so that the operational characteristics are within the tolerance range, and the introduction of more complex error measurement and compensation measures is avoided as much as possible.

Due to the stress deformation of the antenna aperture, such as gravity, heat, wind, and other factors, the gradual error is usually unavoidable, so it needs to be controlled from the aspects of the material, structural rigidity, and temperature control design.

The influence of systematic error on the antenna performance is weakened by the quality of design, fabrication, integration, and control of the test links, which can sometimes be compensated by on-orbit calibration. During the testing of the array antenna, in addition to its error, the antenna under test will also introduce errors due to the calibration state of the test system, as shown in Fig. 3.9. Due to the limitation of the calibration accuracy of the antenna under test, there is a certain angle between the parallelism of the plane array radiation surface and the probe sampling surface, as shown in Fig. 3.9a–c will cause the electrical and mechanical axes of antenna array to misalign and affect the test accuracy of the beam pointing. Furthermore, when there is a horizontal error in the antenna array, the antenna array rotates on the Z-axis, as shown in Fig. 3.9b, which leads to the increase of the cross-polarization component and affects the measurement accuracy of the antenna cross-polarization. At the same time, the rotation of the antenna causes the relative relationship between the probe and each element to change, which affects the calibration accuracy of the base state of the antenna element. The pointing error can be eliminated when the microwave imaging radar performs on-orbit pointing calibration.

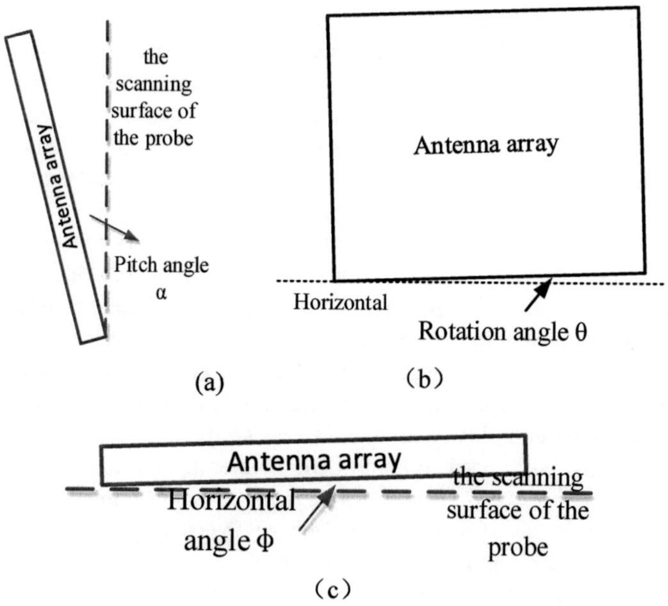

Fig. 3.9 Schematic diagram of array antenna test setup error

Based on the above analysis, it can be found that there are many error sources in the phased array antenna design. The system error can be corrected and compensated by various methods, and the measurement system can calibrate each channel's amplitude and phase errors inside the antenna array. Therefore, the final error term only includes the control error of the T/R module, the parasitic amplitude/phase modulation error, in-band fluctuations, measurement errors, channel amplitude/phase errors caused by temperature differences, and phase errors caused by mechanical deformations. These errors are unpredictable and difficult to control.

3.3.3 Error Acquisition

The characteristics of each channel of the active array antenna include two categories: the inherent amplitude and phase of the passive network and the amplitude and phase of the active channel. In high-resolution microwave imaging radar systems, the microwave network is usually designed with a parallel feed isoelectric length to obtain superior wide instantaneous bandwidth performance. Therefore, when the temperature difference of the antenna aperture is controlled within a small range, it can be considered that the amplitude and phase between the channels in the passive network are fixed. Therefore, the deviation between them can be compensated by channel calibration during the test in the anechoic chamber, and the compensation value is fixed. The difference between the active channels is that the T/R modules and the delay amplifying modules work in different states. While the difference is large, the active channel is more sensitive to the temperature change characteristics than the passive network. Therefore, the temperature distribution on the antenna aperture is a factor affecting the amplitude and phase differences between the channels.

The channel parameters of the antenna can be obtained at multiple stages in the development process, mainly divided into the stand-alone, module, and full array. Under modern information management and automated testing conditions, when every single unit (such as power divider/combiner, cable, T/R modules, delay amplifier modules, etc.) is accepted, all kinds of port characteristics have been tested and stored. Most of the phased array antennas for microwave imaging radars use modular assembly. Therefore, it can go through state inspection and performance testing after each module is integrated, and the performance parameters of each channel of the module can be obtained at this stage. After the antenna array is assembled and integrated, the antenna channel calibration is first performed, and on this basis, the tests of various states of the array antenna are completed.

Active array antenna full-array channel calibration and measurement are usually performed synchronously with antenna measurement. Therefore, the antenna array measurement method selection is directly related to the test content.

3.3.4 Error Analysis

The statistical method is generally used to calculate the random error response of the antenna SLL, the beam pointing, and the gain of antenna to analyze the influence of the active array antenna error on the antenna radiation performance. Based on the above analysis of the error source, the random error source is identified, the basic error parameters are obtained, and the antenna radiation performance is calculated according to the statistical calculation method.

The statistical relationship between the magnitude of the amplitude and phase error and the probability of a certain SLL SL_P parameter can be achieved as follows:

$$P(\mathrm{SL} < \mathrm{SL}_P) = \int_0^{\mathrm{SL}_P} \frac{\mathrm{SL}}{\sigma_R^2} \exp\left(-\frac{\mathrm{SL}^2 + \mathrm{SL}_T^2}{2\sigma_R^2}\right) \cdot I_0\left(\frac{\mathrm{SL} \cdot \mathrm{SL}_T}{\sigma_R^2}\right) d\mathrm{SL} \tag{3.22}$$

where SL_T is the theoretically designed SLL parameter, σ_R is the standard deviation of the SLL, and $I_0\left(\frac{\mathrm{SL} \cdot \mathrm{SL}_T}{\sigma_R^2}\right)$ is the first-class deformed zero-order Bessel function.

The relationship between the probability that the antenna can reach a certain SLL (SL_P) and the standard deviation of the SLL (σ_R) can be obtained by Eq. (3.22) under the condition of a given theoretical design SLL. As shown in Fig. 3.10, under the uniform distribution, the theoretical SLL of the array antenna beam is $\mathrm{SL}_T = -13.26$ dB. According to the theoretical and the required SLLs, it can be known by calculation that $\sigma_R < 0.024$ must be obtained to obtain the actual antenna SLL of −11 dB under a probability greater than 99.5%.

The antenna elements are assumed to be non-directional, and the amplitude and phase error variance is independent of the beam scan angle. According to the statistical distribution characteristics of the main lobe and sidelobe of the antenna pattern,

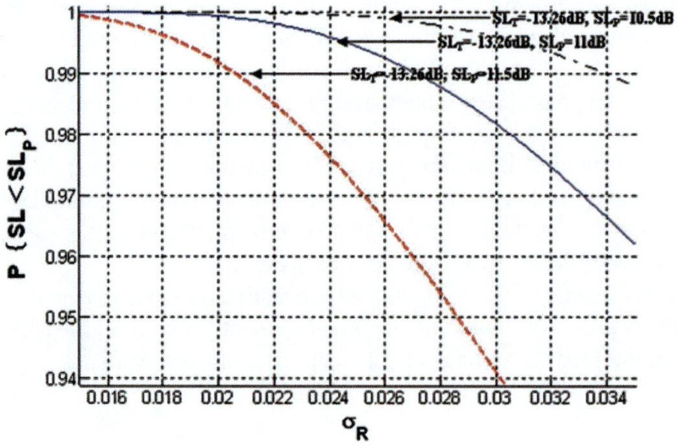

Fig. 3.10 Uniformly distributed sidelobe error curve

it can be known that the standard deviation σ_R of antenna SLL is related to the number of effective elements pN of the array, the element amplitude, and phase error ε^2 and the antenna efficiency η, that is

$$\sigma_R^2 = \frac{\varepsilon^2}{2\eta pN} \tag{3.23}$$

$$\varepsilon^2 = (1 - p) + \sigma_A^2 + p\sigma_P^2 \tag{3.24}$$

where η is the antenna efficiency, N is the number of antenna array elements, p is the proportion of the antenna elements that can work normally, ε^2 is the error term that characterizes the element amplitude and phase error, σ_A is the root mean square value of the amplitude error, and σ_P is the root mean square value of the phase error. Therefore,

$$(1 - p) + \sigma_A^2 + p\sigma_P^2 = 2 \cdot \eta pN\sigma_R^2 \tag{3.25}$$

In the case of obtaining σ_R, various random errors can be allocated according to Eq. (3.25), and conversely, the antenna SLL range can be obtained from the known errors.

Figure 3.10 shows the relationship between the probability of achieving the standard for the sidelobes of the three antenna arrays and the standard deviation of the SLLs. The meaning of the curve is the probability $P\,(\mathrm{SL} < \mathrm{SL}_P)$ that the SLL value SL of the antenna array is lower than the expected SLL value SL_P under the condition that the standard deviation of the SLL is equal to σ_R. The difference between the three curves lies in their respective expected SL_P. It can be known from the changing trend of the curve that the larger the standard deviation of the SLL of the antenna array, the lower the probability that the array sidelobe can reach the expected SL_P. Taking the same sidelobe compliance probability $P\,(\mathrm{SL} < \mathrm{SL}_P)$ as the standard, the lower the expected SL_P is, the smaller the SLL standard deviation σ_R is required, that is, the stricter the requirement for the SLL standard deviation.

The degradation of antenna gain performance due to random errors can be expressed as

$$\Delta G = 10\ \lg\left(\frac{1}{1 + \varepsilon^2/p}\right) \tag{3.26}$$

where ε is a parameter that integrates the magnitude of various amplitude and phase random errors and p is the proportion of the normal working antenna element.

The antenna beam pointing error can be expressed as

$$\sigma_{\Delta\theta} = \frac{\sqrt{12}\sigma_c}{\frac{2\pi}{2}d \cdot \mathrm{MN}} \tag{3.27}$$

$$\sigma_c^2 = \left(1 + \sigma_A^2\right)\sigma_P^2$$

where $\sigma_{\Delta\theta}$ is the beam pointing error, and M and N are the number of row and column elements of the antenna, respectively. It can be seen from Eq. (3.27) that the pure amplitude error has no effect on the beam pointing, but the effect of the phase error on the antenna pointing can be enhanced.

When calculating the antenna performance error, it is also possible to use the Monte Carlo method [13] to perform multiple calculations on the pattern to obtain the error envelope and determine the error effect based on the setting of the error range. Otherwise, the allowable range of error can be deduced by calculation result of the radiation performance error envelope. Figure 3.11 shows an example of the error response envelope of a uniformly distributed antenna array under different beam conditions. Among them: $\sigma_A \leq 1.0\,\text{dB}$, $\sigma_P \leq 10°$, and 800 active channels, after 100 calculations, the maximum error is -11 dB.

It can be seen from Fig. 3.11 that the Monte Carlo method can complete the error analysis of the antenna array sidelobe, gain, beam pointing, and other characteristics. Furthermore, according to this result, random error distribution can also be completed, that is, the allowable range of random errors such as feed amplitude error, phase error, and the element failure rate can be determined.

To sum up, the error analysis of the radiation performance of the phased array antenna can be realized by analytical calculation and Monte Carlo simulation. The advantage of analytical calculation is that the theoretical relationship between various factors in the error analysis can be obtained, and the computational complexity has nothing to do with the antenna array size. However, its disadvantage is that when the granularity of the error analysis problem is further refined, if it is necessary to consider that the phase error distribution is different due to the temperature difference at different positions of the array, the theoretical analysis will be difficult. The Monte Carlo method has made a more detailed simulation of the actual formation process of the overall pattern of the array. Therefore, additional parameters can be introduced to characterize the influencing factors with finer granularity. The simulation process can also give information about the direction of the antenna array. However, the disadvantage is that the amount of calculation increases with the size of the array. When the antenna array size is large and the problem granularity is finer, the simulation process is time-consuming.

3.4 Antenna Measurement

The measurement of antenna radiation characteristics is an important part of antenna research. After the active array antenna is integrated and assembled, it is first necessary to measure and calibrate each channel to achieve the initial error matrix of the channel and compensate it [14–17]. Then, a number of radiation performance parameters of the antenna are tested to evaluate the actual performance of the antenna.

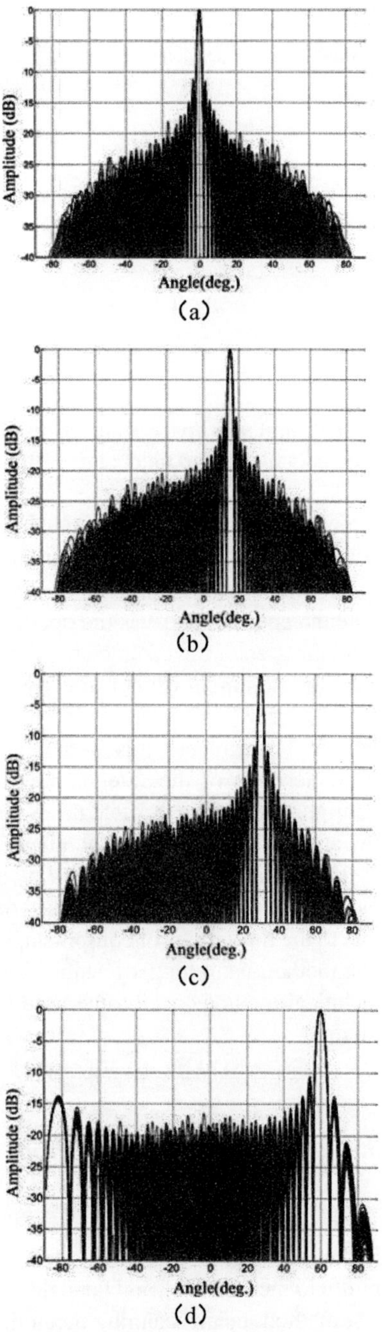

Fig. 3.11 Simulation results of the azimuth pattern with errors in the transmitting state, **a** normal direction, **b** scanning at 15°, **c** scanning at 30°, **d** scanning at 60°

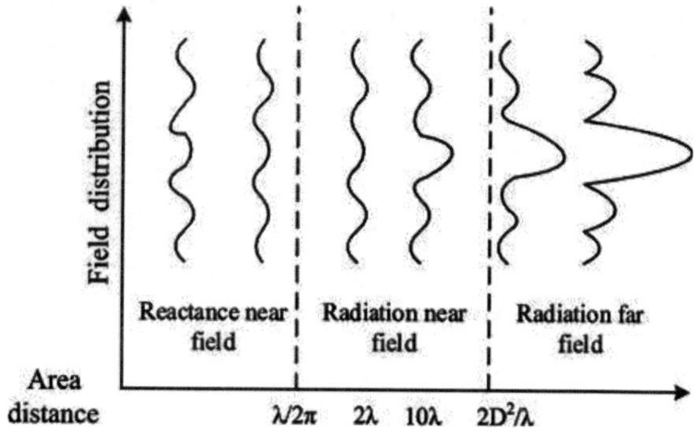

Fig. 3.12 Antenna field division

The antenna space field is divided into three regions, as shown in Fig. 3.12, the reactance near field region, the radiation near field region, and the radiation far-field region. Different measurement methods are used to measure different radiation regions of the antenna. Obviously, it is more intuitive to measure the radiation pattern in the far-field region, but the observation point is far away from the antenna under test. Generally, the test distance is greater than $2D^2/\lambda$. Here, D represents the antenna aperture and λ is the operating wavelength. The larger the antenna aperture size, the greater the distance required. However, the data obtained through the near-field measurement cannot directly reflect the characteristics of the antenna pattern and can only be obtained through the near-field and far-field transformation calculations. Nevertheless, this method has great advantages. Since the far-field radiation characteristics of the antenna are obtained by calculation, it is easy to obtain the configuration of the far-field three-dimensional pattern of the antenna under test, and the amount of information obtained is greater.

In general, for small active array antennas, the plane near-field method is usually used, or the far-field method is used for testing using a three-axis positioner. However, due to the limitations of equipment, site, time, and other factors, the antenna far-field test method is no longer feasible for large-phased array antennas, and some unique technical methods need to be used for far-field testing, such as the celestial source method.

3.4.1 Antenna Test Method

There are many antenna measurement methods. For example, from the sampling point in the antenna radiation field, there are the aperture field method, near-field method, midfield method [18], and far-field method [19]. In addition, there are focusing beam test method, beam scanning method, and so on.

(1) **Aperture field method**

The aperture field method uses a probe with known characteristics, places it in the reactance area of each radiating element in front of the antenna array, and moves the test probe successively according to the grid position of the element. In this way, the amplitude and phase distribution of each element channel in the antenna array can be directly measured, and then the far-field pattern of the antenna can be calculated from the amplitude and phase distribution.

(2) **Planar near-field method**

At a distance of 3–10 wavelengths from the antenna, the near-field amplitude and phase distribution of the antenna are measured, and the far-field radiation field is obtained by applying the more stringent mode expansion theory. Since the test distance is very close to the antenna array in the near-field measurement, it does not require a huge test space, and the antenna test can be completed in a limited space, which has the advantage of being unaffected by the weather environment. This method has been widely used, and the technology is mature.

(3) **Midfield test method**

Using the auxiliary coupling antenna that is several times the diameter of the antenna, the transfer function of the active array antenna is tested at several specific positions, the amplitude and phase characteristics of each antenna element are obtained through data correlation processing, and then the far-field pattern is calculated to obtain far-field pattern characteristics. The basic principle is shown in Fig. 3.13a–c are the three sampling points of the auxiliary antenna. The sampling point position relative to the antenna array does not meet the far-field conditions, but relative to the antenna element is the far-field region. In addition, the angle between the sampling point selection and each element in the array changes little to ensure the consistency of the amplitude/phase of each element. The disadvantage of this method is that the antenna pattern cannot be directly obtained, complex calculations are required, and it is difficult to exact measure the gain value of the antenna.

(4) **Focus test method**

The geometric position relationship between the test antenna and the array to be tested is obtained by the method of antenna aperture calibration, and then the wave path difference is calculated by the coordinates of each antenna element in the array and the coordinates of the auxiliary antenna, and converted into the corresponding phase, as shown in Fig. 3.14. During the test, the path difference phase calibration corresponding to each antenna channel of the array and the channel calibration phase is preset together in the beamforming system so that the antenna lobe is focused on the auxiliary antenna so that the auxiliary antenna point is equivalent to the infinite distance of the antenna under test, so as to achieve the far-field equivalent test purpose. Then use the beam scanning method to complete the test of the antenna pattern, that

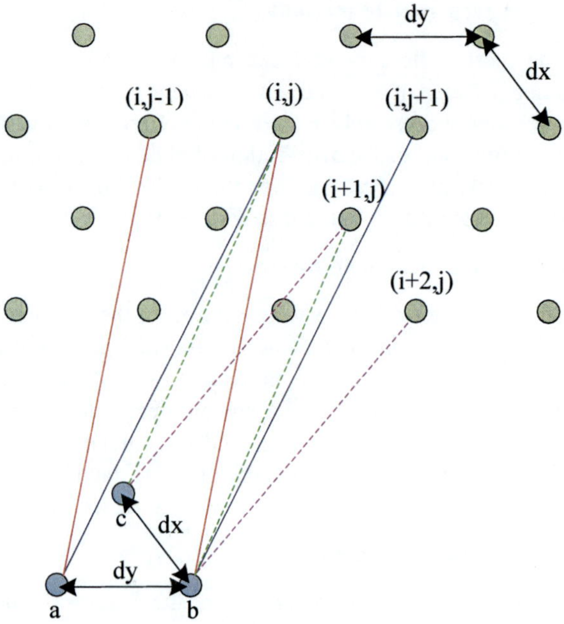

Fig. 3.13 Schematic diagram of the midfield test method

is, control the antenna beam in the azimuth direction within the range of $-\theta \sim +\theta$, scan at a certain angle interval, and record a signal level L_i for each angle scanned so that you can obtain the far-field pattern with the antenna pointing angles (α_0, β_0).

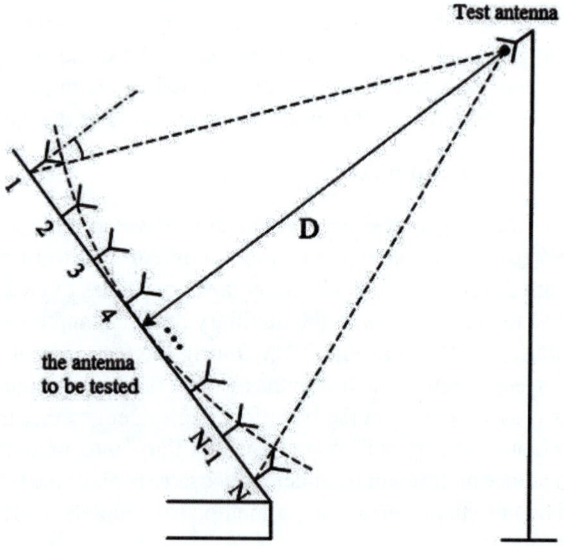

Fig. 3.14 Schematic diagram of the focus test method

(5) **Far-field method**

This method is relatively simple. The antenna to be tested is rotated horizontally or vertically through a one-dimensional or two-dimensional turntable. The amplitude and phase of the antenna to be tested on the rotating section are tested by the auxiliary antenna to obtain the radiation pattern of the section antenna. This method has drawbacks. It is also obvious that each time the pattern of a slice is measured, the rotation section of the antenna needs to be adjusted, and the test workload is large and time-consuming.

(6) **Beam scanning method**

The auxiliary antenna is fixed in the far-field region with the azimuth and elevation in front of the antenna array (α_0, β_0), and the beamforming system is used to align the maximum value of the main lobe of the antenna pattern to the antenna and control the antenna beam in the azimuth direction. In the range of $-\theta \sim +\theta$, scan at a certain angular interval, and record a signal level L_i for each scan of θ_i so that the far-field pattern with the antenna pointing angle (α_0, β_0) can be obtained. The principle is shown in Fig. 3.15. If you want to test the elevation plane, control the antenna to scan in the elevation plane. The theoretical basis of this method is based on the essence of array antenna beam scanning, the translation of the antenna pattern in the visible space. But this translation is not constant gain but has a modulation of cosθ, considering that the pattern of active elements in the array will also change under different scanning states. Therefore, the pattern tested by this method is only close to the actual pattern in the area near the main lobe. The advantage of the beam scanning method is that the transceiver antenna position is relatively fixed, the test process is simple, and the test result of the antenna pattern is relatively direct. The disadvantage is that the test error of the sidelobe in the far field region is relatively large, and the position of the auxiliary antenna must be adjusted every time a beam direction is tested.

In addition, far-field test methods such as the celestial source method, calibration star reflection method, and flying platform far-field measurement method are usually applied to the test of large antenna arrays.

Most of the microwave imaging array antennas are concentrated in the L ~ X-band, the antenna array aperture is moderate, and the antenna array measurement is mostly done in the indoor plane near field.

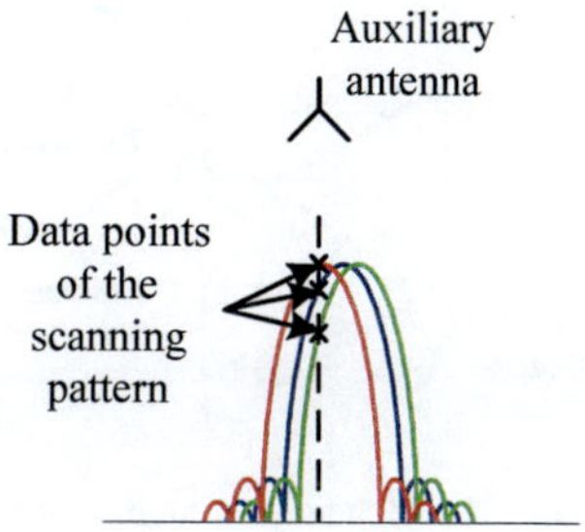

Fig. 3.15 Schematic diagram of the beam scanning test method

3.4.2 Near-Field Measurement

Antènna near-field measurement is to use a probe with known characteristics to scan in the near-field region of the antenna to be measured, measure the amplitude and phase distribution of the electromagnetic field on the scanning surface, and obtain the far-field characteristics of the antenna to be measured through near-field/far-field transformation. Near-field measurement can be divided into plane near-field, cylindrical near-field, and spherical near-field according to the characteristics of the sampling scanning surface. The plane near-field test method is shown in Fig. 3.16. The Antenna Under Test (AUT) is set up in the near-field distance area of the probe in the near-field measurement system, and the antenna surface is parallel to the probe scanning surface. According to the beam coverage of the array antenna and the level of the sidelobe in the far area of the beam, determine the sampling truncation angle of the probe, and calculate the size of the scanning surface. Then, the probe moves row-by-row or column-by-column under the condition that the Nyquist sampling limit is satisfied in the scanning plane, measures the S21 parameter between the antenna and the probe, and obtains the amplitude and phase values of each sampling point position.

The amplitude and phase of the probe output and the coordinate information of its corresponding position can be obtained by measuring, and the plane wave

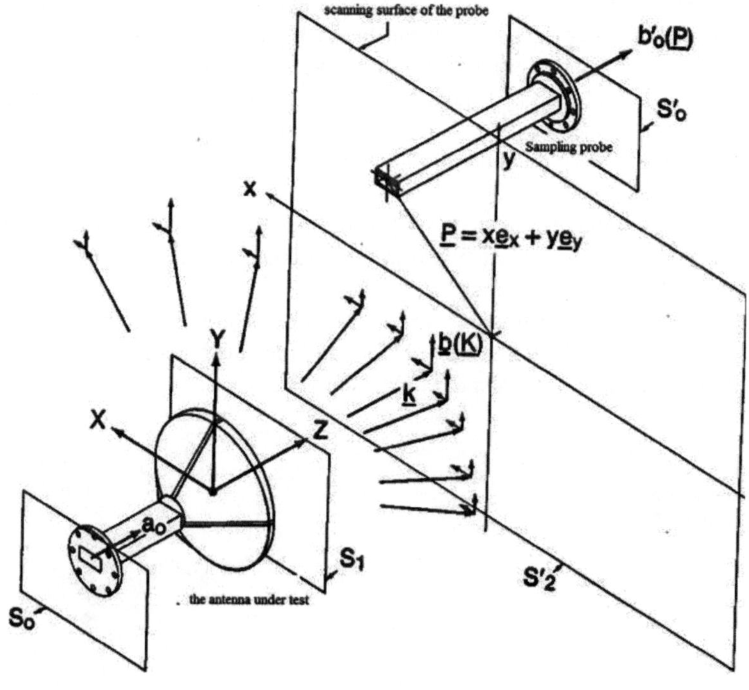

Fig. 3.16 Schematic diagram of the plane near-field measurement

spectrum of the antenna to be tested can be obtained through Fourier transform. The relationship between the probe output amplitude and phase and the antenna under test can be expressed as

$$B_0(x, y, z = d_0) = \int_{-\infty}^{+\infty}\int T(K_x, K_y) \cdot S(K_x, K_y) e^{iK_z d_1} e^{i(K_x x + K_y y)}\, \mathrm{d}K_x \mathrm{d}K_y \tag{3.28}$$

where B_0 is the amplitude and phase value of the probe measurement output, T is the plane spectrum of the antenna to be measured, that is, the value to be evaluated, S is the plane spectrum of the probe antenna, and the latter two are the phase factors of the probe on the sampling plane, d_0 is the distance between the probe and the antenna to be measured, K_x, K_y, and K_z are the propagation constants in three directions, and $K_z = \sqrt{K_0^2 - (K_x^2 + K_y^2)}$, $K_y = K \sin\theta \sin\phi$, $K_x = K \sin\theta \cos\phi$.

The plane spectrum of the antenna under test can be expressed as

$$T(K_x, K_y) = E(K_x, K_y) \sum_{m=1}^{M}\sum_{n=1}^{N} A(x_m, y_n) \cdot e^{-i(K_x x_m + K_y y_n)} \Delta x \Delta y \tag{3.29}$$

$$\begin{aligned} A(x_m, y_n) &= E(K_x, K_y) \sum_{-k_1}^{k_1}\sum_{-k_2}^{k_2} \frac{e^{-ik_z d_1} D(K_x, K_y)}{4\pi^2 E \cdot S} e^{i(K_x x_m + K_y y_n)} \Delta K_x \Delta K_y \\ &= \sum\sum \left\{ \frac{e^{-ik_z d_1}}{4\pi^2 E \cdot S} \sum\sum \begin{bmatrix} B_0(x, y) e^{-i(K_x x + K_y y)} \Delta x \Delta y \\ \cdot e^{i(K_x x_m + K_y y_n)} \Delta K_x \Delta K_v \end{bmatrix} \right\} \end{aligned} \tag{3.30}$$

The planar near-field test system contains nearly 20 error terms [20], mainly in the aspects of the probe and its position, receiver and transmitter, cables, and microwave anechoic chamber environment. For a qualified plane near-field measurement system, all kinds of errors have been guaranteed to be within a certain allowable range in construction and later maintenance.

3.4.3 Aperture Field Inversion Calibration

Aperture field inversion technology is one of the antenna array measurement and calibration methods. This method uses the near/far field patterns obtained by simulation or testing to obtain each element's excitation amplitude and phase through Fourier transform [21, 22]. The near-field inversion method has strong operability and is significant to the element fault diagnosis and amplitude and phase calibration. The principle and effect are shown in Fig. 3.17.

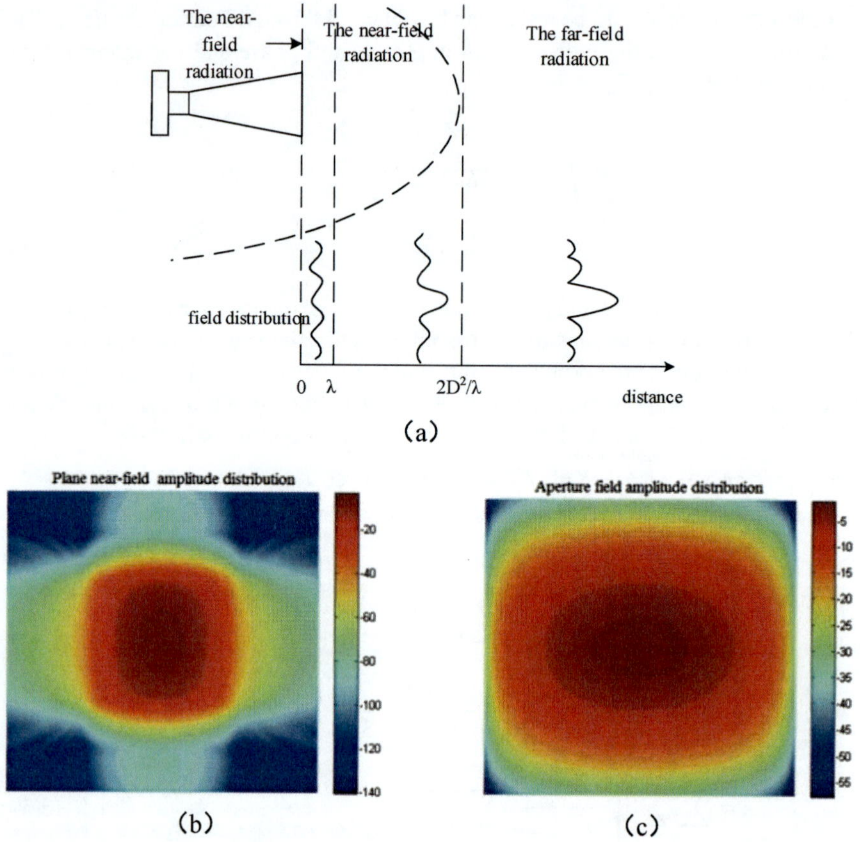

Fig. 3.17 Schematic diagram of aperture field inversion. **a** A schematic diagram of the radiation near-field formed by the aperture field, **b** the amplitude distribution of the plane near-field test, **c** the inversion result of the aperture field of the antenna array

Basic principle: The theoretical basis of the aperture field calibration algorithm is the plane wave expansion theory. Assuming that the propagation medium of the antenna is a linear, uniform, and isotropic passive and lossless medium, in the antenna propagation coordinate system shown in Fig. 3.18, in the space where $z \geq 0$, the electromagnetic wave propagation satisfies Helmholtz equation, that is

$$\nabla^2 \begin{bmatrix} E \\ H \end{bmatrix} + k^2 \begin{bmatrix} E \\ H \end{bmatrix} = 0 \tag{3.31}$$

$$E(r) = A(k)e^{-jkr}, \; H(r) = \frac{1}{\omega\mu} k \times A(k)e^{-jkr} \tag{3.32}$$

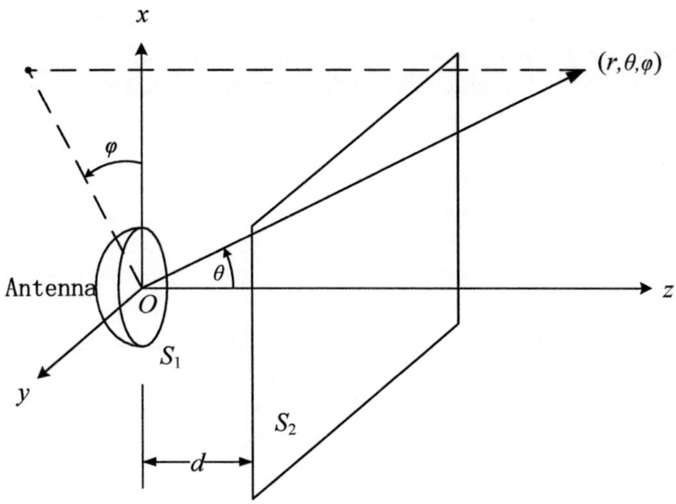

Fig. 3.18 Antenna propagation coordinate system corresponding to aperture field inversion theory

where E and H are the electric and magnetic field components of the electromagnetic wave, respectively, r is the spatial position vector, A is the complex amplitude vector of the electromagnetic wave, k is the wave vector of the electromagnetic wave, ω is the angular frequency of the electromagnetic wave, and μ is the permeability of the medium. Equation (3.32) is the plane wave solution of the Helmholtz equation and the plane wave spectrum $A(k)$ of the electromagnetic wave can be obtained through Eq. (3.32).

In the space where $z \geq 0$, the electromagnetic field distribution at any point can be regarded as a spatial combiner of electromagnetic waves in all directions, that is

$$E(r) = \iint_{-\infty}^{\infty} A(k)e^{-jkr}\,\mathrm{d}k_x\mathrm{d}k_y$$

$$H(r) = \frac{1}{\omega\mu}\iint_{-\infty}^{\infty} k \times A(k)e^{-jkr}\,\mathrm{d}k_x\mathrm{d}k_y \tag{3.33}$$

where x and y are the wave number in the x-direction and the wave number in the y-direction of the electromagnetic wave, respectively.

After obtaining the electromagnetic wave complex amplitude vectors in all directions, the electromagnetic field distribution at any point in the space where $z \geq 0$ can be obtained according to Eq. (3.33). The electromagnetic field distribution at infinity is the far-field antenna pattern, and the electromagnetic field distribution in the $z = 0$ plane is the antenna aperture field distribution. Usually, when $k_x^2 + k_y^2 > k^2$, the propagation of electromagnetic waves will decay rapidly, and this part of the component is called evanescent waves in the plane wave spectrum. In the near-field measurement, the distance between the sampling surface and the antenna aperture surface is in the order of several wavelengths so that the evanescent wave component can be ignored, that is, the upper and lower integral limits of Eq. (3.33) can be changed from $\pm\infty$ to $\pm k$.

Implementation method: In theory, the near-field diagnosis only needs to obtain the complex amplitude vector $A(k)$ of electromagnetic wave, but the near-field data obtained in the actual near-field test is the amplitude and phase information received by the probe. The real plane spectrum of the antenna array to be calibrated also needs to eliminate the influence of the probe.

In the near-field test, the applied probe generally has a known pattern, or the far-field pattern can be obtained by calculation, which provides convenience for probe compensation. The influence of the probe can be eliminated by considering the near-field coupling Eq. (3.34) of the probe, that is

$$\frac{b_0(r_0)}{a_0 F} = \iint\limits_{-\infty}^{\infty} \frac{K_z}{K} A(k) \cdot S(k) e^{-jkr_0}\, \mathrm{d}k_x \mathrm{d}k_y \tag{3.34}$$

where $b_0(r_0)$ is the signal received by the probe at position r_0, $A(k)$ and $S(k)$ are the plane wave spectrum of the antenna to be measured and the plane wave spectrum of the probe, respectively, a_0 and F are the mismatch factors between the input signal of the antenna under test and the probe and the near-field test system, respectively. By ignoring the constant factor and performing Fourier transform on the near-field data, the coupled $A(k) \cdot S(k)$ distribution can be obtained, that is

$$A(k) \cdot S(k) e^{-jk_z d} = \iint\limits_{-\infty}^{\infty} b_0(x, y, d) e^{j(k_x x + k_y y)} \mathrm{d}x \mathrm{d}y \tag{3.35}$$

where (x, y, d) represents the Cartesian coordinate component of the electromagnetic field position vector $\boldsymbol{r}$. In the actual near-field test, a linearly polarized probe is usually used, and then in the near-field range, two orthogonal polarization components are measured, and then the two sets of data are calculated. The two orthogonal polarization components are expressed as

$$\begin{cases} I_1(k) = \iint\limits_{S_2} B_{01}(x, y, d) e^{j(k_x x + k_y y)} \mathrm{d}x \mathrm{d}y \\ I_2(k) = \iint\limits_{S_2} B_{02}(x, y, d) e^{j(k_x x + k_y y)} \mathrm{d}x \mathrm{d}y \end{cases} \tag{3.36}$$

where $I_1(k)$ and $I_2(k)$ are the $A(k) \cdot S(k) e^{-jk_z d}$ distribution after considering the influence of probe cross-polarization coupling, $B_{01}(x, y, d)$ and $B_{02}(x, y, d)$ represent the probe received signals of the two polarization components at the (x, y, d) position, respectively. Therefore, it is possible to

$$\begin{bmatrix} f_E \cos\phi & -f_H \sin\phi \\ -f_E \sin\phi & -f_H \cos\phi \end{bmatrix} \begin{bmatrix} A_\theta \\ A_\phi \end{bmatrix} = e^{j2\pi \cos\theta \frac{d}{\lambda}} \begin{bmatrix} I_1 \\ I_2 \end{bmatrix} \tag{3.37}$$

where f_E and f_H are the probe pattern of the E and H planes, respectively. According to Eq. (3.35), the plane wave spectral components A_x and A_y in the rectangular coordinate system can be obtained, respectively, that is

$$\begin{cases} A_x = \left[I_1 \left(\frac{\cos\theta \cos\varphi^2}{f_E} + \frac{\sin\varphi^2}{f_H} \right) - I_2 \sin\varphi \cos\varphi \left(\frac{\cos\theta}{f_E} - \frac{1}{f_H} \right) \right] e^{j2\pi \cos\theta \frac{d}{\lambda}} \\ A_y = \left[I_1 \sin\varphi \cos\varphi \left(\frac{\cos\theta}{f_E} - \frac{1}{f_H} \right) - I_2 \left(\frac{\cos\theta \sin\varphi^2}{f_E} + \frac{\cos\varphi^2}{f_H} \right) \right] e^{j2\pi \cos\theta \frac{d}{\lambda}} \end{cases} \tag{3.38}$$

According to the plane wave spectrum components A_x and A_y, the electromagnetic field distribution of the aperture surface in the x-and y-directions are obtained, that is

$$E_{x,y}(x, y, 0) = \iint_{-k}^{k} A_{x,y}(k) e^{-j(k_x x + k_y y)} \mathrm{d}k_x \, \mathrm{d}k_y \tag{3.39}$$

where $E_x(x, y, 0)$ and $E_y(x, y, 0)$ are the electromagnetic field distribution on the aperture surface, $-k$ and k are the upper and lower limits of the sampling range.

After the above inversion process, the one-to-one amplitude and phase values corresponding to the antenna apertures can be obtained. By repeatedly controlling the amplitude and phase of the T/R modules in the active array antenna, the aperture distribution can be infinitely close to the required target value, and the compensation of the antenna array can be completed. In the transmitting state, the T/R modules of the antenna array usually work in a saturated amplification state, and the compensation can be accomplished only by controlling the phase shifter in the T/R modules. In the receiving state, the compensation of amplitude and phase can be completed by controlling the phase shifter and attenuator of the T/R module.

After the compensation and calibration of the normal beam of the antenna are completed, according to the scanning and weighting requirements of the antenna array, the beam control code of the active array antenna is calculated, and the scanning state and shaped beam measurement are completed in the near field of the plane. If the deviation from the target value is large, the beam calibration in this state can be completed through aperture field inversion, amplitude, and phase control iterative approximation again.

3.4.4 One-by-One Calibration

The single-channel one-by-one calibration technology utilizes the independent switching function of the T/R module design in the active array antenna. When calibrating each channel of the antenna array, through computer control, only one channel in the antenna array is working, and the other channels are not working. Then, the probe samples the channel to obtain the transmission coefficient parameter S21. The full array of the antenna is switched on and off one by one, and the probe is moved accordingly and sampled synchronously to obtain each channel amplitude and phase value. Compared with the ideal objective function, the attenuator and phase shifter in the T/R module are controlled to make the amplitude and phase values approach the target value, and the reference state calibration is completed, thereby obtaining the antenna array base state error matrix. Figure 3.19 shows the schematic diagram of the one-by-one calibration test method. By controlling the computer and the grid of the antenna array to be calibrated, the radiation fields of the antenna array elements are sampled one by one correspondingly, and the calibration amplitude and phase data of each element of the antenna array are obtained.

This technology is very suitable for sub-array-level active array antennas, and a sub-array antenna composed of multiple radiating elements corresponds to one transceiver channel. Only one point or several points need to be sampled on the sub-array antenna, and the probe is controlled so that the sampling position of each sub-array antenna is the same. As a result, there is no need to meet the Nyqnist sampling constraints, so the calibration speed is much faster than the aperture field inversion method.

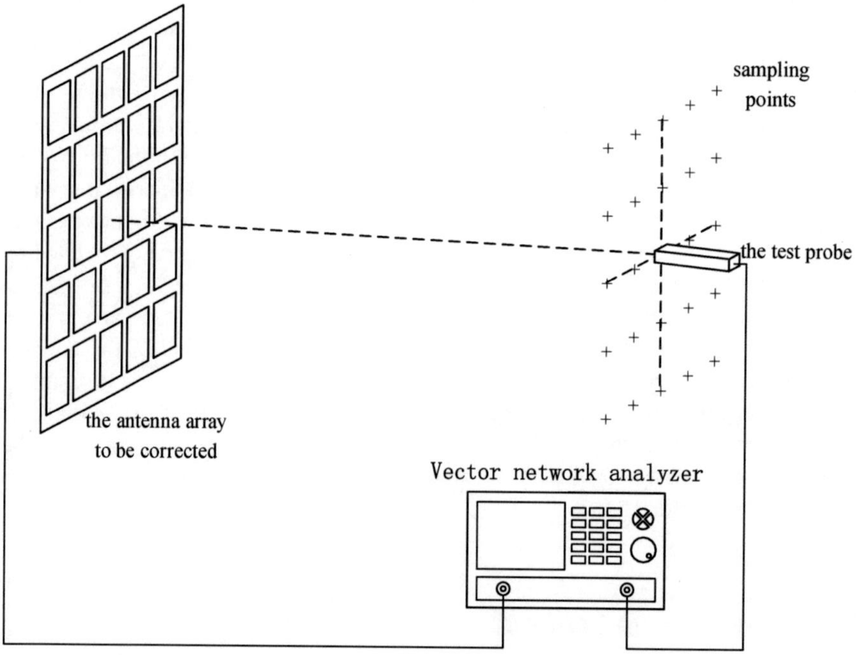

Fig. 3.19 Schematic diagram of the one-by-one calibration test system

3.5 Accurate Calculation of Radiation Performance

3.5.1 Principle

Microwave imaging radar sometimes cares about its radiation resolution. Therefore, the system must accurately obtain the gain and pattern characteristics of the phased array antenna. The multimode microwave imaging radar phased array antenna has a huge number of 2D scanning beam positions, which is difficult to measure completely. For high-resolution broadband antennas, it is necessary to accurately obtain in-band amplitude and phase parameters, which further increases the antenna measurement workload. Due to the limitations of the near-field measurement system probe scanning speed, instrument data acquisition, control exchange, frequency switching, etc., broadband 2D scanning phased array antenna pattern measurement has become a task with a huge workload. Therefore, it is necessary to seek fast measurement technology to solve the problem of massive pattern measurement of microwave imaging 2D phased arrays. Although using multi-probe sampling technology can improve the speed of antenna measurement. However, considering the factors such as switch switching, mechanical scanning speed, and the influence of probe array error, it is still impossible to guarantee the precise measurement of the radiation characteristics of microwave imaging radar 2D phased array under different scanning angles of broadband.

The plane near-field is based on the near-field electromagnetic measurement data of the antenna, and the far-field radiation characteristics are obtained through the near/far-field transformation calculation. Similarly, based on the antenna element or sub-array pattern and its excitation amplitude and phase data, the radiation performance of high-precision active array antennas can also be obtained through accurate modeling. Aiming at the accurate calculation of massive patterns of 2D active array antennas, it is verified by plane near-field measurement to solve the accurate calibration of the radiation performance of high-resolution microwave imaging radar broadband active array antennas.

3.5.2 Accurate Modeling

Based on the multiplication principle of the antenna array pattern, the far-field pattern of the antenna array can be constructed by the 3D pattern of each element and the array factor pattern. By obtaining the actual spatial distribution of the electromagnetic field of each antenna element in the active array antenna, combined with the channel amplitude and phase values obtained by accurate measurement, the far-field pattern of the antenna array can be accurately calculated. This method is suitable for quantitative calibration of the microwave imaging sub-array-level active array antenna pattern. The basic method is as follows:

(1) Measure and obtain the 3D pattern of each typical position element or sub-array in the antenna array;
(2) Accurately obtain the amplitude and phase values of each channel and each working state in the antenna array;
(3) Using the plane near-field measurement system to complete the base state calibration of the active array antenna, compensating the amplitude and phase values of each channel of the normal equal-amplitude and co-phase uniform distribution beams;
(4) Taking the amplitude/phase distribution in the case of normal beams as the benchmark, the theoretical values of the amplitude and phase of beam scanning and beamforming are objective functions. Call the amplitude and phase data of each working state obtained by accurate measurement of each channel, select the working state of each channel, and make the amplitude and phase value of the antenna array approach the objective function;
(5) After the working state of each active channel is selected, combined with the element pattern, the far-field pattern of the antenna array is obtained by calculation.

According to the working state of the selected active channel, determine the phase control code of the beam and use the plane near-field measurement system to measure and verify the modeling and calculation pattern. The far-field pattern of the antenna is

$$\mathrm{AF}(\theta,\varphi)=\sum_{m=0}^{M-1}\sum_{n=0}^{N-1}\Big(f_{\mathrm{sa,mn}}(\theta,\varphi)\cdot I_{\mathrm{mn}}\cdot E_{\mathrm{sa,mn}}\cdot e^{jk\cos\theta\sin\varphi\left(\frac{M-1}{2}+m\right)\Delta x}\,e^{\mathrm{jk}\cos\theta\sin\varphi\left(\frac{N-1}{2}+n\right)\Delta y}\Big) \tag{3.40}$$

where θ and φ are the space angles in spherical coordinates, M and N are the numbers of columns and rows of the antenna array sub-array, Δx and Δy are the antenna spacing of the sub-array, respectively, $k=\frac{2\pi}{\lambda}$ is the center frequency wave number, $E_{\mathrm{sa,mn}}$ is the base state error, which is obtained by zeroing the normal beam of the antenna array, and I_{mn} is the excitation factor of each sub-array, that is, the amplitude and phase values of the T/R modules and the corresponding delay amplification modules. It is obtained by testing on the T/R modules, and a database is established to correspond one to one with the assembly position of the antenna, $f_{\mathrm{sa,mn}}$ is the pattern of the element in the array, and it is the pattern of the sub-array in the sub-array-level phased array, which is obtained by testing.

When the number of active channels of the antenna array is small, due to the large proportion of edge sub-arrays, the sub-array antenna pattern needs to be divided into the edge, corner, and array middle conditions for separate testing and collection. For high-resolution microwave imaging radar applications, collecting sub-array antenna patterns in the broadband range is also necessary. The typical pattern results need to establish a database and correspond to the positions in the antenna array for accurate pattern calculation.

3.5.3 Calculation Example

According to the above method, an example illustrates the accurate modeling and calculation of a certain sub-array-level active array antenna pattern. In order to obtain an accurate calculated value of the antenna pattern, the pattern characteristics of the sub-arrays at each typical location are considered in the antenna array, such as horizontal and vertical plane edges, four corners, and sub-arrays in the environment of the array. The channel data is obtained through accurate measurement at the unit and module levels, respectively. The typical sub-array and channel data are numbered and correspond to one-to-one positions in the antenna array. The calculated radiation pattern of the 2D active phased array antenna for microwave imaging is realized through the compiled software of objective function optimization, channel data look-up table, sub-array data, position information embedding, pattern calculation, and beam control code automatic generation.

Figure 3.20 shows the comparison between accurate modeling and near-field measurement. It can be seen that both of them are in good agreement. On this basis, the antenna gain is calculated. In most cases, the difference between the two is within ±0.05 dB, as shown in Fig. 3.20a–c. In some special beam broadening cases, the difference in the pattern is also controlled within ±0.1 dB, as shown in Fig. 3.20d.

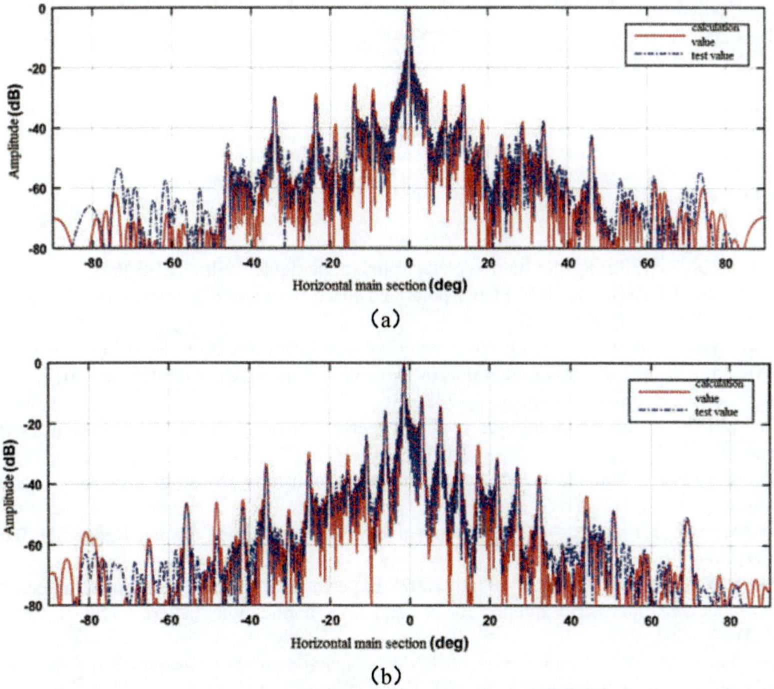

Fig. 3.20 Comparison of accurate modeling and testing patterns, **a** normal pattern in the horizontal plane, **b** scanning pattern in the horizontal plane, **c** pattern in the evenly distributed vertical plane, **d** beam broadening pattern in the vertical plane

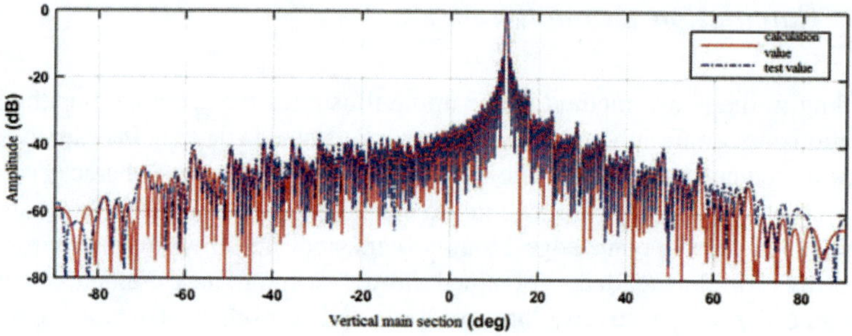

(c)

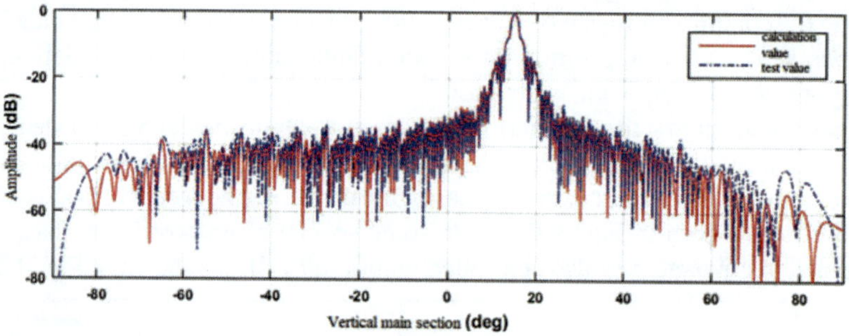

(d)

Fig. 3.20 (continued)

References

1. Harrington RF (2001) Time-hamonic electromagnetic fields. Wiley-IEEE Press
2. Wei W, Lei LI, Zhihui Z (2007) Dual-polarized space-borne SAR antenna array. Remote Sens Technol Appl 22(2):166–172
3. Ji M, Wei W, Jin M (2011) Discussion on antenna beam direction deviation of spaceborne SAR. In: Proceedings of National microwave and millimeter wave conference 2011. Chinese Institute of Electronics, Qingdao, pp 733–736
4. Lu J (2017) Design technology of synthetic aperture radar. National Defense Industry Press, Beijing
5. Zhong S (2011) Antenna theory and techniques. Publishing House of Electronics Industry, Beijing
6. Xiaopeng L, Yan L, Li A (2006) A novel BALUN for printed dipole. Radar Sci Technol 4(4):236–239
7. Wang W, Liang X, Zhang Y et al (2007) Experimental characterization of a broadband dual-polarized microstrip antenna for X-band SAR applications. Microw Opt Technol Lett 49(3):649–652
8. Hongtao Z, Wei W, Zhihui Z et al (2011) A broadband single layer edge-fed cavity-backed microstrip antenna. Microw Opt Technol Lett 53(2):2831–2834
9. Xiao-peng LU, Wei WA, Chu GA (2012) Microstrip line fed wideband single-layer patch antenna. Radar Sci Technol 10(1):112–116

10. Wang W, Zhang H, Zhang Z et al (2014) Broadband antenna array for SAR applications. In: IEEE antennas and propagation society international symposium (APSURSI). IEEE, Memphis, TN, pp 138–139
11. Wang W, Zhong S, Zhang Y et al (2006) A broad bandwidth slotted ridged-waveguide antenna array. IEEE Trans Antennas Propag 54(8):2416–2420
12. Fang Z, Lu J (2008) A wide-band, wide-angle compact rectangular waveguide antenna. Radar & ECM (2):16–17
13. Lee J, Lee Y, Kim H (2005) Decision of error tolerance in array element by the Monte Carlo method. IEEE Trans Antennas Propag 53(4):1325–1331
14. Ausmann HM, Fenn AJ, Willwerth FG (1989) Phased array antenna calibration and pattern prediction using mutual coupling measurements. IEEE Trans Antennas Propag 37(7):844–850
15. Zou Y, Zhang Y, Cao J (2003) Calibration and compensation for active phased antenna using series travelling-wave matrix-switch built-in test system. Radar Sci Technol 1(1):60–64
16. He C, Zhang Y, Chen S (2005) Analysis of one by one calibration and diagnosis method for receiver channels of active phased array antennas. Radar Sci Technol 3(3):185–188
17. Li L, Wang W, Feng W (2009) The application of unitary matrix in monitoring phased array antennas. Radar ECM 4:38–42
18. Zheng X, Gao T (2005) Study on mid-field calibration technique of phased array antenna. J Microw 21(5):22–25
19. Mao N, Xinde J (1987) Antenna measurement handbook. National Defense Industry Press, Beijing
20. Zhang F (1999) Planar near-field measurement errors and compensation techniques of ultra-low sidelobe antenna. Xidian University, Xian
21. Jin J, Wan X, Wang W, Jin M (2007) Design of planar waveguide slotted antenna arrays. Radar Sci Technol 5(3):232–235
22. Liu H, Yang T, Chen X (2015) Aperture inversion of array antenna based on pattern multiplication. Radar Sci Technol 13(3):305–309

Chapter 4
Broadband Active Array Antennas

The fine resolution of the observation target is a prominent feature of broadband radar, and its fine resolution capability is proportional to the signal bandwidth of the radar. When the range resolution of the radar is smaller than the linear size of the target of interest, the radar echo of the target can form a one-dimensional (1D) range image. When this broadband radar is combined with synthetic aperture technology or inverse synthetic aperture technology, two-dimensional (2D) imaging of the target can be achieved. If the radar is combined with coherent measurement technology, three-dimensional (3D) imaging of the target can also be performed.

Most of the traditional electronic information systems (such as radar, communication, etc.) are narrowband, i.e., a narrowband signal is applied to modulate a sinusoidal carrier frequency signal. However, with the increase in signal bandwidth, the traditional radar theory and technology based on narrowband will encounter new challenges. For example, the antenna radiation characteristics, channel transmission characteristics, target reflection characteristics, signal processing, and detection methods are all quite different from the narrowband case.

Broadband radars have higher range resolution than narrowband radars. In general, synthetic aperture radar and inverse synthetic aperture radar are broadband signal radars with high-resolution target imaging, high-precision target recognition, non-cooperative target classification and recognition, and strong anti-jamming ability. Moreover, modern active array antenna systems have as large an instantaneous signal bandwidth as possible. Therefore, broadband active array antenna is an essential technical system of the radar system, especially microwave imaging radar system.

4.1 Instantaneous Bandwidth Limitation

The bandwidth of an active array antenna usually has two meanings: one refers to its operating bandwidth, and the other refers to its instantaneous bandwidth or instantaneous signal bandwidth. The so-called operating bandwidth refers to the

J. Lu et al., *Active Array Antennas for High Resolution Microwave Imaging Radar*,
https://doi.org/10.1007/978-981-99-1475-3_4

operating frequency range in which some performance indicators of the active array antenna system meet the specified requirements. Different indicators have different requirements for bandwidth. Usually, the indicators include input impedance or port reflection coefficient, gain or directivity, cross-polarization level, beamwidth, and SLL. Instantaneous bandwidth refers to the signal bandwidth that a radar antenna system can transmit and receive. In general, the operating bandwidth must be greater than the instantaneous bandwidth to realize the broadband function of the array antenna.

The bandwidth of the active array antenna is not only limited by the bandwidth of the antenna element but also restricted by factors such as the distance between the array elements, the amplitude, and phase of the excitation signal, the feeding network, and the active module.

The bandwidth of the antenna element is related to the element type, configuration, and feeding method. For example, the bandwidth of the monopole antenna, dipole antenna, slot antenna, and microstrip antenna element is usually related to its impedance characteristics, the bandwidth of the horn antenna, dielectric rod antenna, and surface wave antenna is determined by the primary and higher propagation modes, the edge diffraction effects limit the bandwidth of electrically small aperture antennas, the primary feed essentially limits the bandwidth of single or double reflector antennas, the bandwidth of any antenna with a fixed aperture will be limited by the beamwidth requirements that vary with frequency.

Suppose the element spacing of the array antenna is too small. In that case, the mutual coupling effect between the array elements will be enhanced, and the mismatch between the antenna and the feed network will be more serious. It is tough to design broadband active standing waves. If the element spacing is too large, undesired grating lobes will appear in the visible area when the beam is scanning. As we all know, antenna arrays mainly have two feeding configurations, i.e., serial and parallel feeding. The serially fed array will cause the beam pointing dispersion due to the dispersion effect of the transmission line. At the same time, the beamwidth of the parallel-fed array not only has the frequency response characteristics but also widens with the increase of the scanning angle. The SLL also varies with frequency and increases as the scan angle increases. The RF excitation signal will generate amplitude and phase quantization errors and periodic errors for array elements and sub-array antennas. In addition, the antenna will generate quantization side lobes in the frequency band, thereby deteriorating the antenna gain and beam pointing.

Modern radar systems have higher and higher requirements for target recognition accuracy, imaging resolution, detection distance, and anti-jamming capabilities, and the functions and working modes of radar are becoming more complex and diverse. Correspondingly, radar systems have higher and higher performance requirements than antenna systems. As a result, active array antennas with a large aperture, sizeable instantaneous bandwidth, and scanning angle are increasingly adopted by radar engineers.

For large phased array antennas with wide instantaneous signal bandwidth and wide-angle scanning, the instantaneous signal bandwidth is limited by spatial dispersion effects. Especially for array antennas with large instantaneous signal bandwidth, due to the use of non-dispersive phase shifters, the phase shift provided by the phase shifters cannot effectively compensate for the spatial phase difference with linear frequency response characteristics in a wide frequency band. As a result, the beam pointing shifts and beam broadening will occur, and the characteristics of the broadband signal spectrum will be correspondingly degraded.

In the case of wide-angle scanning, the instantaneous bandwidth of the broadband active array antenna will be affected by the antenna spatial dispersion effect [1], which is mainly manifested in beam pointing deviation, aperture fill time, chirp signal frequency modulation rate change, and grating lobe, etc. [2, 3]. On the other hand, in the scanning state of a broadband phased array antenna without real-time delay line compensation, it will affect the image quality of synthetic aperture radar [4].

4.1.1 Beam Pointing Deviation Limit

Beam pointing deviation refers to the dispersion phenomenon of an antenna beam pointing with frequency. In general, the beam direction of the array antenna is consistent with the normal direction of the array phase wavefront. If a phase shifter provides the phase of each element of the array antenna, the phase shifter is driven by the system beam control circuit. Since the phase of a phase shifter is usually independent of frequency, if the active array antenna controls the excitation phase of each antenna element separately so that the antenna array obtains a linear phase taper, the phase wavefront of the antenna array will deviate. In the normal direction of the array, the beam is pointed from the array normal to the direction perpendicular to the phase wavefront, so the scanning of the antenna beam occurs. When the beam scanning of the phased array antenna deviates from the normal direction, since the phase value provided by the digital phase shifter of the antenna array does not change with the frequency, the spatial phase difference generated by the different elements of the antenna array transmitted or received within a certain signal bandwidth cannot be generated by the digital signal. Instead, the phase shifters are compensated so that the beam pointing of the antenna array is shifted at those frequencies off-center. This phenomenon of the antenna beam pointing shifting as the signal frequency changes are called the "aperture effect" or "spatial dispersion effect" of the phased array antenna.

A one-dimensional linear array antenna composed of N antenna elements is shown in Fig. 4.1.

Assuming that the central operating frequency of the linear array antenna is f_0, the length of the linear array aperture is L. When the maximum direction of the array beam is θ_S, the phase difference required by the phase shifter provided by the antenna elements at both ends of the antenna array is

$$\Psi = \frac{2\pi L}{\lambda_0}\sin\theta_S = \frac{2\pi f_0 L}{C}\sin\theta_S = 2\pi f_0 T_C \tag{4.1}$$

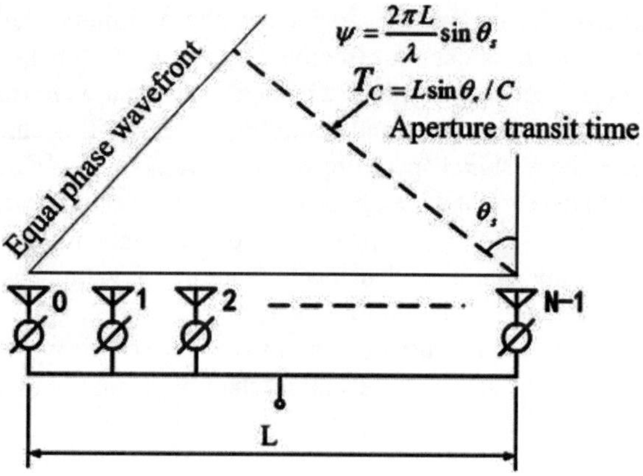

Fig. 4.1 Relationship between linear array configuration and beam pointing

$$T_C = \frac{L \sin \theta_S}{c} \tag{4.2}$$

where $\lambda_0 = \frac{C}{f_0}$ is the center frequency wavelength, C is the speed of light, and T_C is the aperture fill time of the array antenna.

The aperture fill time is determined by the size of the antenna array aperture and the beam scan angle. When the phase shifter is used to control the beam scanning, if the signal frequency is changed from f_0 to $f_0 + \Delta f$, and the same array antenna element phase is still given, the beam pointing will be shifted, and the beam direction will change from the θ_S direction to the $\theta_S + \Delta\theta$ direction, so there is

$$\Psi = \frac{2\pi f_0 L}{c} \sin \theta_S = \frac{2\pi (f_0 + \Delta f) L}{c} \sin(\theta_S + \Delta\theta) \tag{4.3}$$

A small frequency change Δf, the corresponding beam pointing change $\Delta\theta$ is also small, and there is

$$\sin(\theta_S + \Delta\theta) \approx \sin \theta_S + \Delta\theta \cos \theta_S \tag{4.4}$$

From Eqs. (4.3) and (4.4), we can get

$$\Delta\theta = \frac{\Delta f}{f_0} \tan\theta_S \tag{4.5}$$

where Δf is the frequency offset relative to the center frequency, $\Delta\theta$ is the beam pointing offset. Equation (4.5) shows that when the signal frequency changes from f_0 to $f_0 + \Delta f$, the antenna beam pointing will be shifted by $\Delta\theta$.

If the sine space coordinate system is used, the expression for the change of the beam pointing to the antenna array with the frequency is

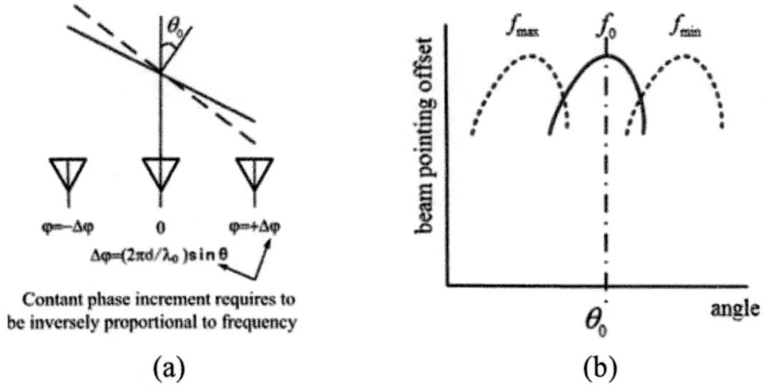

Fig. 4.2 Beam pointing diagram. **a** Relationship between Beam direction and phase, **b** relationship between beam direction and frequency

$$u = \frac{u_0 f_0}{f} \tag{4.6}$$

$$u = \sin\theta, u_0 = \sin\theta_0$$

where θ_0 is the beam direction corresponding to the center frequency, and the relationship between the beam pointing and the frequency expressed by Eq. (4.6) is shown in Fig. 4.2.

It can be seen from Fig. 4.2 that when the signal frequency deviates from the center frequency, the antenna beam pointing is skewed. The antenna beam pointing angle decreases when the signal frequency is higher than the center frequency and increases when the signal frequency is lower than the center frequency.

For an active array antenna that provides a wideband instantaneous signal with beam scanning phase adjustment by a phase shifter, each antenna element numerically controls the phase shifter to set the beam control phase code according to the center frequency to provide the same phase weighting for the wideband array antenna element. The excitation signal phase of each frequency point within the signal bandwidth is the same as the center frequency point. If the bandwidth is defined by the frequency limit where the gain is reduced by half, then for an antenna array beamwidth of θ_{3dB}, the fractional bandwidth is

$$\frac{\Delta f}{f} = \frac{\Delta u}{u_0} = \frac{\theta_{3dB}}{\sin\theta_0} = 0.886 B_b \frac{\lambda}{L\sin\theta_0} \tag{4.7}$$

where B_b is the beam broadening coefficient of the array, corresponding to different beam widening coefficients, the relationship between the length of the antenna array and the fractional bandwidth is shown in Fig. 4.3.

The relationship between the array length in meters (m) and the array bandwidth in megahertz (MHz) is shown in Fig. 4.4.

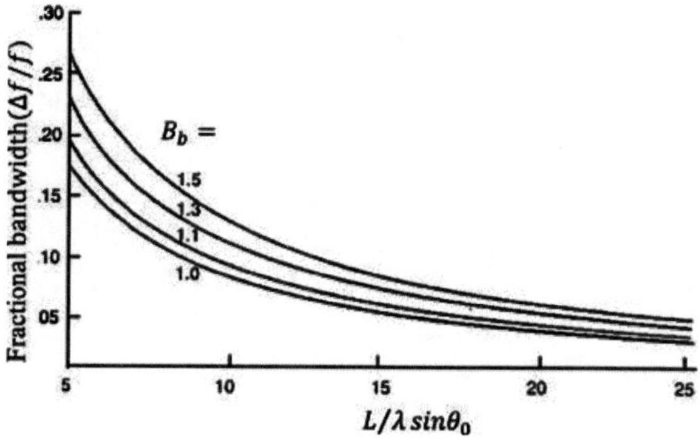

Fig. 4.3 Relationship between 3 dB bandwidth and array aperture $(L/\lambda)\sin\theta_0$

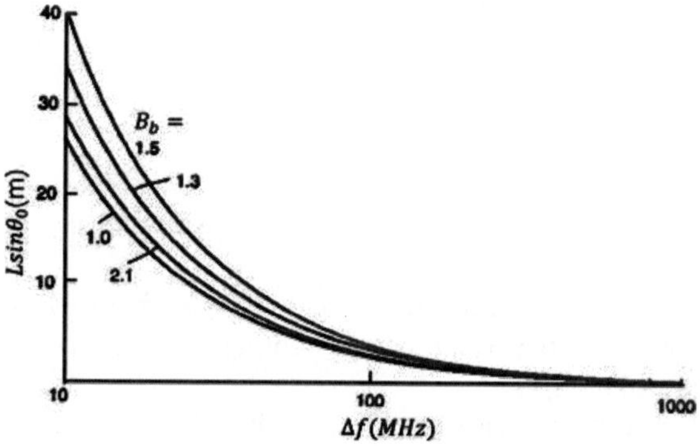

Fig. 4.4 Relationship between array aperture and bandwidth

The large beam pointing deviation of the array antenna leads to the rapid deterioration of the antenna beam characteristics in the broadband range, which will seriously affect the imaging quality of inverse synthetic aperture radar and synthetic aperture radar. The beam pointing deviation of the array antenna increases with the beam scanning angle and the signal bandwidth. The condition of the signal bandwidth limitation is formed by the constraint of the beam pointing deviation, also called the array antenna bandwidth criterion.

In general, a reasonable criterion for limiting the bandwidth is that the beam pointing with frequency deviation does not exceed $\pm 1/4$ of the beam width at this time [5], that is

$$\left|\frac{\Delta\theta_0}{\theta_B(\text{scanned})}\right| \leq \frac{1}{4} \tag{4.8}$$

From Eq. (4.5), we can get

$$\frac{\Delta f_{\max}}{f_0} \leq \frac{\theta_{3\text{dB}}}{2\sin\theta_S} = 0.443 B_b \frac{\lambda_0}{L\sin\theta_S} \tag{4.9}$$

Then the maximum bilateral bandwidth of the instantaneous signal is

$$\Delta f_{\max} \leq \frac{f_0 \theta_{3\text{dB}}}{2 \sin \theta_S} = 0.443 B_b \frac{C}{L \sin \theta_S} \tag{4.10}$$

where $\Delta f_{\max}$ is the bilateral maximum bandwidth with f_0 as the intermediate frequency, and $\Delta f_{\max} = 2\Delta f$. Under the conditions of beam pointing offset, maximum scanning angle, and weighting coefficient, the maximum instantaneous bandwidth of the array antenna is uniquely determined by the aperture size of the antenna.

For example, for a wide-angle scan ±60°, using the bandwidth criterion above, with 1/4 beamwidth as the band edge limit, which corresponds to a loss of about 3/4 dB, the array percent bandwidth limit is

$$\text{Bandwidth}\,(\%) = \text{Beamwidth}\,(^\circ)\ (\text{CW}) \tag{4.11}$$

For the ±60° wide-angle scanning array antenna operating in pulses, the 3 dB beamwidth is taken as the limiting condition of the band edge, which corresponds to the energy loss of about 0.8 dB when the beam impinges on the target. At this point, the array percentage bandwidth limit is

$$\text{Bandwidth}\,(\%) = 2 \text{ times beam width}\,(^\circ)\ (\text{Pulse Wave}) \tag{4.12}$$

For example, if an array antenna scans ±60° with a 1° beamwidth, the instantaneous signal bandwidth is about 1% for a continuous wave phased array radar. On the other hand, the instantaneous signal bandwidth is about 2% for the pulse wave phased array radar.

4.1.2 Aperture Fill Time Limit

For an active array antenna operating in a pulsed system, when the beam is scanned to the maximum scanning angle, if the spatial delay difference of the antenna element signals at both ends of the array exceeds a pulse width, it will appear that the transmitting beam cannot be synthesized or the receiving signals cannot be added and synthesized at the same time.

From the perspective of the aperture fill time of an active array antenna, a pulsed plane wave is incident on an antenna array from a direction at an angle θ from the normal direction. As a result, a very short pulse (large signal bandwidth) arrives at the antenna elements at different times at both ends of the array. If the array elements do not have corresponding delay compensation, it is impossible to add and synthesize the signals of each element and benefit from the antenna array gain. A schematic diagram of the aperture fill time of the narrow pulse incident antenna array is shown in Fig. 4.5.

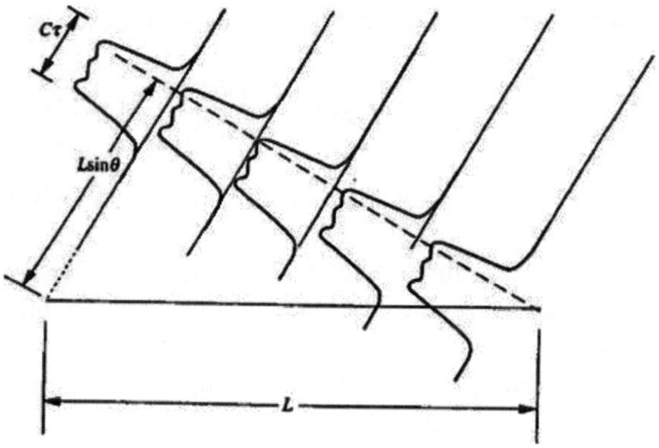

Fig. 4.5 Schematic diagram of aperture fill time of a narrow pulse incident antenna array

From the angle of the FM signal envelope, for a linear array with an aperture fill time of T_C, as shown in Fig. 4.1, if the target is in the direction of θ_S, the radiation signal of the $N-1$th element is ahead of the signal of the 0th element. T_C time to reach the target. Therefore, for a linear FM pulse compressed signal with a pulse width of T and a bandwidth of $\Delta f_{\max}$, the signal envelope synthesized by the signals radiated by each antenna element at the target position is no longer a rectangle, but a trapezoid, as shown in Fig. 4.6. The time for each antenna element signal to reach the target simultaneously for synthesis is equal to $(T - T_C)$.

After the target reflects the signal, it is received and synthesized by each antenna element of the array, and the signal sent to the pulse compression receiver is further broadened. The width of the entire received signal envelope is $T + 2T_C$, the time for the signals received by all antenna elements to be added and synthesized at the same time is $T - 2T_C$, and the time of the front and rear edges of the signal waveform increases by $2T_C$. This means that the main lobe of the output signal of the pulse compression receiver will be broadened, and the broadening factor is about $1 + 2T_C/T$, and the SLL of the output signal waveform will also increase. After the receiver performs pulse compression, the signal width is compressed by T to τ, $\tau = 1/\Delta f_{\max}$. The antenna aperture fill time T_C should be less than the signal pulse width $\tau = 1/\Delta f_{\max}$. Otherwise, the signals received by the antenna elements at both ends of the array will be completely separated in time after pulse pressure and cannot

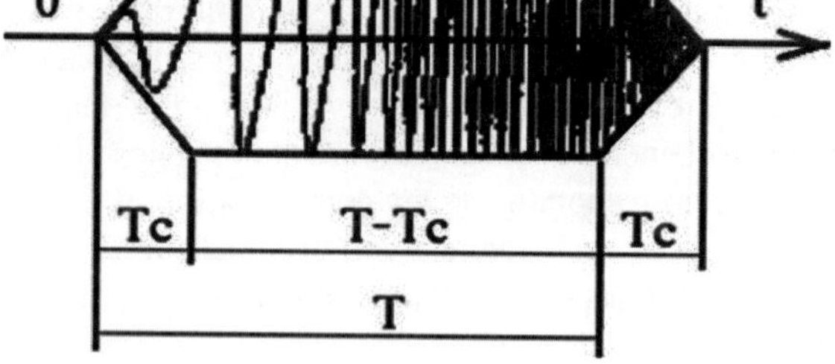

Fig. 4.6 Influence of aperture fill time on FM signal envelope

be synthesized. Therefore, under the condition of a particular array aperture fill time, the instantaneous bandwidth of the signal will be limited by the aperture fill time. The limitation of the propagation time of the array aperture on the instantaneous bandwidth of the signal is

$$\Delta f_{\max} \leq \frac{1}{\mathrm{kT}_C} \tag{4.13}$$

where k is a constant, usually between 2 and 10. From Eq. (4.13), it can be obtained that when no time delay line is added, the maximum signal bandwidth allowed by the array is

$$\Delta f_{\max} \leq \frac{1}{\mathrm{kT}_C} = \frac{1}{k}\frac{c}{L \sin\theta_s} \tag{4.14}$$

For an array antenna with a center frequency of 1300 MHz and an aperture size of 12 m, if the beam scanning of ±60° is realized, its instantaneous signal bandwidth is only about 2%.

4.1.3 Signal Frequency Modulation Rate Limit

When the array antenna scans a wide-angle beam, the beam is significantly pointed away from the normal direction of the antenna array. For a radar system using a broadband linear frequency modulation (LFM) signal, the signal transmitted or received by its antenna element introduces a "square phase error" due to pulse compression. At the same time, the signal transmitted by the array element reaches the target, deviating from the normal direction, and there will be a square phase error between the signals of each antenna array element. The squared phase error introduced by the pulse pressure signal between the ith element and the 0th reference element is

$$\Delta\varphi_i = \pi\frac{\Delta f_{\max}}{T}(i\Delta\tau)^2 = \pi k_\nu (i\Delta\tau)^2 \tag{4.15}$$

where $\Delta\tau$ is the time delay of the adjacent element signal of the array, k_ν is the frequency modulation rate of the LFM signal, T is the signal pulse width. It can be seen from Eq. (4.15) that the square phase error of the pulse pressure signal is proportional to the FM rate of the chirp signal and is related to the beam scanning angle and the size of the antenna aperture. Therefore, adverse effects such as antenna beam pointing offset, beam deformation, and SLL elevation will occur when the antenna array has a significant square phase error.

The limitation of the array antenna on the square phase error of the element signal has the same principle as the limitation of the antenna far-field test. For the square phase error introduced by the chirp signal, the limiting condition is that the two-way phase error of the antenna elements at both ends of the antenna aperture should be less than π/16. From this, the limit requirement for the chirp signal frequency modulation rate can be obtained, that is

$$k_v = \frac{\Delta f_{\max}}{T} \le \frac{C^2}{16(N-1)^2(d\ \sin\theta_s)^2} \tag{4.16}$$

where N is the number of elements, C is the speed of light, and θ_s is the beam scanning angle. Equation (4.16) shows that when the number of array elements, element spacing, maximum beam scanning angle, and signal pulse width are determined, the maximum instantaneous signal bandwidth allowed by the array antenna is also determined.

4.2 Delay Compensation Methods

It is straightforward to see from the above analysis that it is a challenge to achieve a large instantaneous signal bandwidth for a large wide-angle scanning array antenna driven by a conventional non-dispersive phase shifter. To obtain a large instantaneous bandwidth, the active array antenna must use a delay line with dispersion characteristics to compensate for the path difference of the spatial signal.

True time delay line (TTDL) technology is essential to improve the beam pointing accuracy of broadband array antennas and enhance the performance of broadband phased array radars. For the active array antenna of a high-resolution imaging radar system, to realize the wide-angle scanning of the beam of the large aperture large-signal bandwidth antenna, the time delay line is used for delay compensation, which is an effective method to eliminate the aperture effect of the wideband and wide-angle scanning active array antenna.

There are usually two delay line configuration methods for wideband and wide-angle scanning active array antennas: array element-level configuration and sub-array-level configuration.

4.2.1 Element-Level Delay Line Configuration

Taking a linear array antenna as an example, as shown in Fig. 4.7, in addition to a phase shifter, each element of the antenna array also inserts a delay line as the time delay compensation of each antenna element.

If the delay difference between adjacent elements is τ, and the total delay of the antenna array is $\tau_c = (N-1)\tau$, the antenna aperture fill time will be reduced to

$$\Delta T = T_c - \tau_c = T_c - (N-1)\tau \tag{4.17}$$

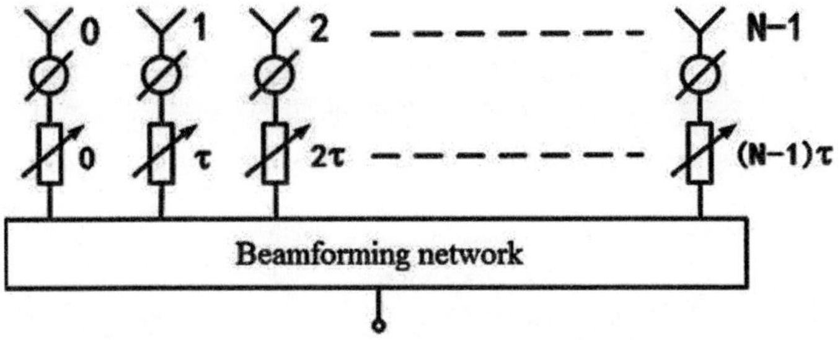

Fig. 4.7 Line array element-level delay line configuration

The relationship between the beam pointing offset and the frequency offset (relative bandwidth) after inserting the delay line can be obtained from Eq. (4.5), namely

$$\Delta\theta' = -\frac{\Delta f}{f_0}\left(1 - \frac{\tau_c}{T_c}\right)\tan\theta_s \tag{4.18}$$

It can be seen from Eq. (4.18) that after the time delay line is inserted at the element level, the beam pointing deviation of the array can be reduced accordingly. In particular, when the antenna elements at both ends of the array are inserted into the delay line, the total delay amount is equal to the aperture fill time. That is, $\tau_c = T_c$, the beam pointing deviation is 0, the array will have no aperture effect, and the antenna beam pointing will not be affected by the instantaneous bandwidth of the signal.

After applying element-level real-time delay line compensation, the limit on the instantaneous signal bandwidth of the antenna becomes

$$\Delta f'_{\max} \leq \frac{1}{k(T_C - \tau_C)} \tag{4.19}$$

The active array antenna can achieve an ideal wideband instantaneous signal bandwidth with an element-level delay line configuration. However, since the delay line components are relatively expensive, bulky, and heavy, it is also necessary to provide power supply and control management circuits, which is not conducive to reduction of system cost, volume, weight, and power consumption.

4.2.2 Sub-array-Level Delay Line Configuration

The element-level delay line configuration method is often unaffordable for large active array antennas. It is necessary to comprehensively consider factors such as cost, equipment quantity, system power consumption and reliability, antenna beam pointing dispersion, gain loss, SLL, grating lobe suppression, and other factors. An

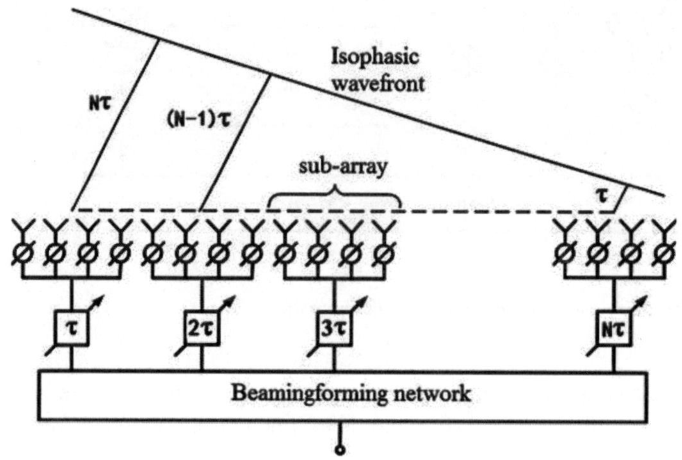

Fig. 4.8 Sub-array-level delay line configuration

acceptable compromise is configuring the delay line at the sub-array level of the array antenna [6–8], which is a practical method for most large active array antenna delay line configurations. Still taking the linear array as an example, the array of *N* elements is divided into *M* sub-arrays, and the delay components are set in the sub-arrays, as shown in Fig. 4.8.

In Fig. 4.8, the antenna can be regarded as an array composed of sub-arrays, and the sub-array pattern forms the element factor. The phase shifter still drives each element in the sub-array for beam scanning and its beam direction changes with the frequency. In contrast, the delay element drives the scanning of the array factor (sub-array level), and its scanning beam direction is independent of frequency.

After the delay line is configured at the sub-array level, the aperture effect in the sub-array still exists, the beam direction of the sub-array is still shifted, and the signals of each antenna element in the sub-array still have superimposed transient effects. That is, the signals of each antenna element cannot be added simultaneously, and the signal waveform will be broadened.

If the division of the sub-array is uniform, the aperture fill time in the sub-array will be reduced to 1/*M* of the aperture transit time of the large array, so the restriction on the instantaneous bandwidth of the signal can be relaxed by *M* times, that is

$$\Delta f_{\max} \leq \frac{M}{\mathrm{kT}_C} = \frac{M}{k}\frac{c}{L\sin\theta_S} \tag{4.20}$$

where *M* is the number of sub-arrays, *L* is the length of the antenna aperture, θ_s is the beam scanning angle, *c* is the speed of light. Under the condition that the aperture of the antenna array and the number of elements are determined, if the sub-array-level delay line configuration is adopted, in order to achieve the required instantaneous signal bandwidth, the number of sub-arrays of the array obtained from Eq. (4.20) should meet

$$M \geq k\Delta f_{\max} T_C = \Delta f_{\max} \frac{\text{kL} \sin \theta_s}{c} \tag{4.21}$$

Using the sub-array-level delay line configuration method can significantly reduce the number of delay line components while achieving the sizeable instantaneous signal bandwidth of the array. Still, at the same time, it will also bring adverse effects of reduced array sidelobe performance and gain degradation. For arrays configured with sub-array delay lines, the overall pattern of the array is the multiplication of the array factor and the element factor. A change in frequency will cause the appearance of grating lobes rather than a shift in the pointing of the main beam. As shown in Fig. 4.9, the sub-array pattern (the element factor of the array) at the center frequency f_0 has a null at the grating lobe (array factor) position, that is, the null of the sub-array pattern coincides with the position of the maximum value of the array factor. According to the principle of pattern multiplication, the overall pattern of the array has no grating lobes, when the frequency changes δ_f, the sub-array pattern (dashed line in Fig. 4.9) will scan an angle, and the null of the sub-array pattern and the array factor is the largest at this time. As a result, the value positions are no longer coincident, and grating lobes will appear in the overall pattern of the synthesized array.

Since the sub-array antenna pattern is scanned with frequency, the gain loss and grating lobe size of the array is functions of the sub-array bandwidth factor K (the beam width in the normal direction of the sub-array), which can be expressed by the bandwidth factor K. Figure 4.10 shows the relationship between the gain loss and the relative magnitude of the grating lobes and the sub-array bandwidth factor K for a sub-array-level delay array scanned with a ±60° beam.

For a sub-array-level delay line configuration antenna array, the beamwidth in Fig. 4.10 is the sub-array normal beamwidth. Therefore, if the aperture is divided into

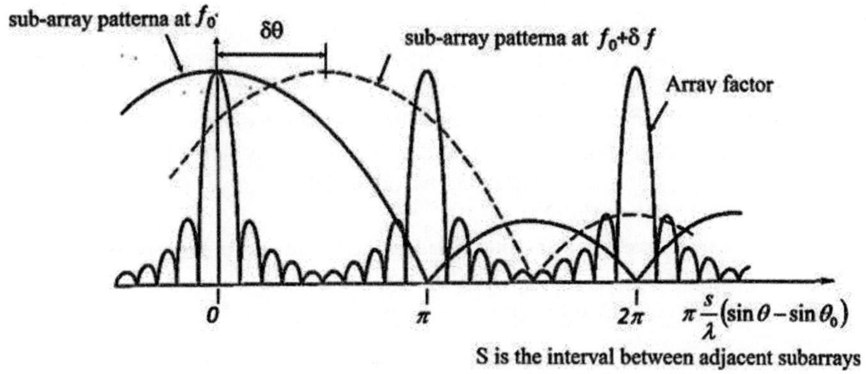

Fig. 4.9 Grating lobes due to frequency change

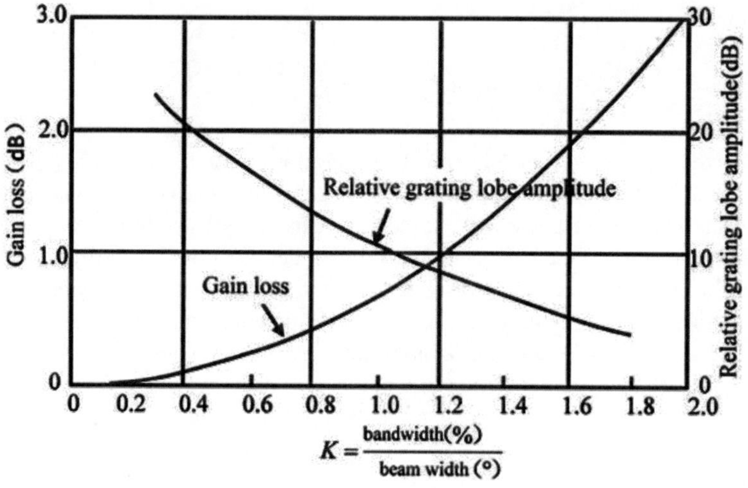

Fig. 4.10 Functional relationship between gain loss and grating lobe amplitude with bandwidth

M sub-arrays in a plane, and each sub-array is configured with a delay component, its bandwidth will increase *M* times.

For large-scale active array antennas, it has a large number of elements. If large-angle beam scanning and wide instantaneous signal bandwidth are achieved, the total time delay that the array needs to compensate for is very large, sometimes as high as several hundred wavelengths. Therefore, if a single-stage sub-array delay line configuration scheme is adopted, the loss of delay line components will be huge due to the many delay lines configured. To compensate for this link loss, a high-gain compensation circuit needs to be designed, which leads to a corresponding increase in power consumption, size, and weight of delay line components, as well as increased component complexity and reduced reliability. Therefore, this single-stage delay line configuration method is unacceptable for system design.

A multi-level TTDL configuration method is generally used for large-scale array antennas with large delay compensation. Figure 4.11 shows a three-stage delay line configuration antenna array schematic diagram.

In Fig. 4.11, the antenna array is configured with large-step, large-delay line components in the first-stage sub-array and a medium-delay delay line component in the second-stage sub-array. That is, a low-order delay line with a slight delay is configured in each T/R module.

For large-scale active array antennas, it is necessary to comprehensively evaluate the system performance, development difficulty, and cost of the delay line and select a feasible multi-level delay line configuration method.

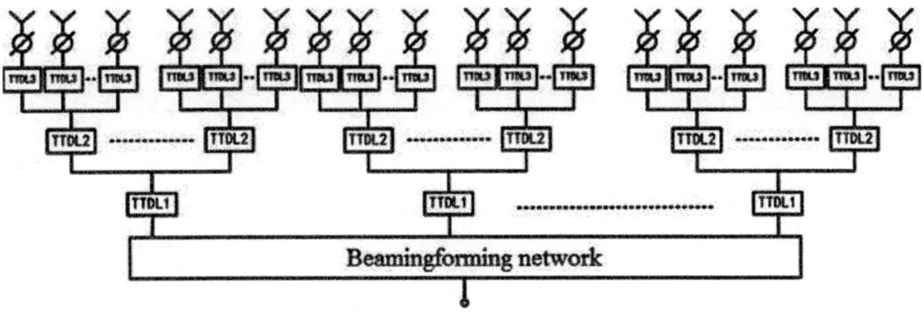

Fig. 4.11 Schematic diagram of a three-stage delay line configuration antenna array

4.2.3 Array Antenna Coordinate System

The configuration design of the active array antenna delay line first needs to solve the problem of calculating the maximum delay of the array. As discussed earlier, the maximum delay (aperture transit time) of an array antenna is related to the array aperture size and maximum beam scan angle. For the 2D active array antenna, the maximum time delay is also related to the maximum beam scanning angle of the two main planes of the array. However, the beam scanning angles of the two main sections in the azimuth and range directions of the 2D planar array antenna have different meanings due to different reference coordinate systems. There are usually three reference coordinate systems for a 2D planar array antenna: the $\theta - \varphi$ coordinate system (spherical coordinate system), the AOE coordinate system, and the EOA coordinate system. The three coordinate systems and their corresponding beam pointing angles are shown in Fig. 4.12.

In Fig. 4.12, the antenna array is located in the *X*–*Y* plane, and the coordinate origin is located in the array center. The *X* axis is parallel to the azimuth direction of the front, the *Y* axis is parallel to the range direction (elevation direction), the *Z* axis is parallel to the normal direction of the front, and *R* is the beam pointing vector. In Fig. 4.12a, (θ, φ) are the beam scanning angles in the elevation and azimuth directions of the antenna, respectively. In Fig. 4.12b, c, (α, β) is the beam scanning angles in

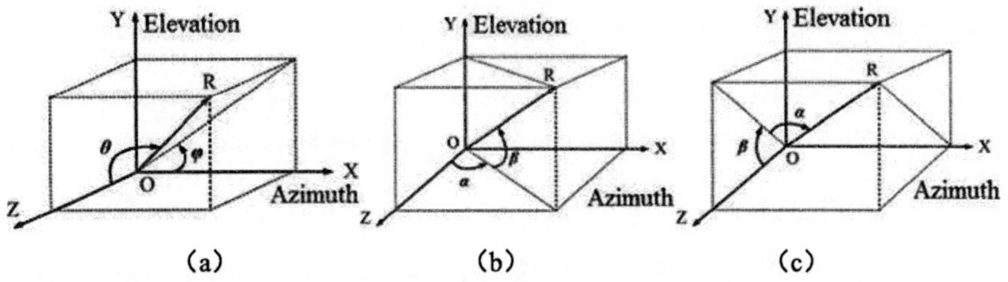

Fig. 4.12 Array antenna reference coordinate system and its corresponding beam pointing angle, **a** $\theta - \varphi$ coordinate system, **b** AOE coordinate system, **c** EOA coordinate system

the azimuth direction and the range direction (elevation direction) of the antenna, respectively.

Figure 4.12a is the $\theta - \varphi$ coordinate system (spherical coordinate system) of the antenna, θ is the angle between the Z-axis and the beam pointing, and φ is the angle between the projection of the beam pointing vector on the X–Y plane and the X axis. Figure 4.12b is the AOE coordinate system of the antenna. The azimuth angle α of the beam scanning is the angle between the beam pointing vector on the Z–X plane and the Z axis, and the elevation angle β is the angle between the beam pointing vector on the Z–X plane projection line and the beam pointing vector. Figure 4.12c is the EOA coordinate system of the antenna, the azimuth angle α of the beam scanning is the angle between the beam pointing vector on the Y–Z plane projection line and the beam pointing vector, and the elevation angle β is the angle between the beam pointing vector on the Y–Z plane projection line and the Z axis.

The $(\theta,\ \varphi)$ coordinate system of the antenna is suitable for the operating mode of beam conical scanning. In contrast, the AOE coordinate system and the EOA coordinate system are more appropriate for the operational mode of progressive beam scanning in azimuth or elevation (range direction). Therefore, the three azimuth and range beam scanning angles defined by the three coordinate systems of the antenna system are all used in the radar system. Obviously, for different coordinate systems, the meanings of antenna beam azimuth and range scan angle are entirely different. Therefore, when designing the antenna array, the antenna designer must specify the coordinate system corresponding to the scanning angle of the antenna beam, otherwise, it is easy to produce wrong results.

4.2.4 Delay Line Configuration Design

A large-aperture antenna is usually selected for a spaceborne microwave imaging radar system with a decimeter-level resolution to ensure sufficient power-aperture product. Sometimes the antenna beamwidth is about 0.1°, and the beam pointing dispersion is less than 5% of the beamwidth. However, the aperture size reaches hundreds of wavelengths, and the instantaneous signal bandwidth exceeds 1.5 GHz. In order to ensure the sizeable instantaneous signal bandwidth of the active array antenna, real-time delay line compensation methods must be adopted to achieve the performance of tiny beam pointing dispersion. In practical applications, a multi-level sub-array delay line configuration is usually applied for large-scale array antennas. For the array antenna whose parameters such as aperture size, operating frequency, maximum beam scanning angle, and instantaneous bandwidth have been determined, the typical method of delay line configuration design is as follows:

(1) According to the specified operating frequency and instantaneous signal bandwidth, calculate the maximum beam pointing deviation according to Eq. (4.5), or calculate the maximum instantaneous bandwidth that the array can achieve according to the specified beam pointing deviation according to Eq. (4.14). This

evaluates the beam pointing deviation or signal bandwidth performance that the array antenna can achieve without a delay line and confirms whether the array antenna needs delay compensation.

(2) If a delay line needs to be applied, calculate the maximum delay length of the array antenna.

(3) According to the composition and scale of the antenna array, formulate a preliminary delay line configuration strategy, and at the same time, consider the entire array delay block division.

(4) For the initial delay line configuration strategy, through the simulation and iterative optimization of the beam performance, an acceptable optimal delay line configuration method is formed.

(5) For the final optimized array delay line configuration, carry out a detailed simulation and evaluation of the antenna parameters, especially the beam pointing deviation, and the dispersion characteristics of the antenna SLL.

It can be known from the array antenna theory that if the pattern of each antenna element is the same, the pattern of the array is mainly determined by the array factor. For a 2D planar array antenna with a rectangular grid, its array factor can be expressed as

$$F(\theta, \phi) = \sum_{m,n} \exp\{\mathrm{jk_0}[\mathrm{md}_x(u - u_0) + \mathrm{nd}_y(v - v_0)]\} \tag{4.22}$$

where d_x and d_y are the element spacings in the azimuth and range directions of the array, respectively, and the (u, v) variables corresponding to the three antenna coordinate systems are

$$(\theta, \varphi)\text{ coordinate system: } \begin{cases} u = \sin\theta\cos\varphi \\ v = \sin\theta\sin\varphi \end{cases}$$

$$\text{AOE coordinate system: } \begin{cases} u = \sin\alpha\cos\beta \\ v = \sin\beta \end{cases}$$

$$\text{EOA coordinate system: } \begin{cases} u = \sin\alpha \\ v = \cos\alpha\sin\beta \end{cases}$$

where (α, β) is the beam scanning angles in the azimuth and range (elevation) directions in the AOE and EOA coordinate systems. In the three coordinate systems, the mutual conversion relationship between the azimuth and range beam scanning angles is listed in Table 4.1.

Table 4.1 Azimuth and range beam scan angle conversion relationship

Coordinate system	u	v	w
(θ, φ)	$\sin\theta\cos\varphi$	$\sin\theta\sin\varphi$	$\cos\theta$
AOE	$\sin\alpha\cos\beta$	$\sin\beta$	$\cos\alpha\sin\beta$
EOA	$\sin\alpha$	$\cos\alpha\sin\beta$	$\cos\alpha\cos\beta$

According to the array aperture size and the maximum beam scanning angle, the maximum extension degree $\mathrm{TTDL_{max}}$ of the array antenna can be derived as

$$\mathrm{TTDL_{max}} = uL_x + vL_y \tag{4.23}$$

where L_x and L_y are the aperture sizes in the azimuth and range directions of the array antenna, respectively, for the three coordinate systems, the corresponding formulas for calculating the maximum extension of the antenna array are as follows

$$(\theta, \varphi)\text{ coordinate system: }\mathrm{TTDL_{max}} = L_x \sin\theta\, \cos\varphi + L_y \sin\theta\, \sin\varphi \tag{4.24}$$

$$\text{AOE coordinate system: }\mathrm{TTDL_{max}} = L_x \sin\alpha\, \cos\beta + L_y \sin\beta \tag{4.25}$$

$$\text{EOA coordinate system: }\mathrm{TTDL_{max}} = L_x \sin\alpha + L_y \cos\alpha\, \sin\beta \tag{4.26}$$

where (θ, φ) and (α, β) are the maximum beam scanning angles of the array antenna. Usually, the time extension is normalized by the working wavelength of the antenna. That is, the number of wavelengths expresses the array time delay to facilitate the design and control management of delay line components.

4.2.5 1D Sub-array Delay Line Configuration Example

(1) **Basic array configuration**

Figure 4.13 shows an array antenna system configured with a 1D sub-array-level delay line. The antenna is a 2D planar active array, which adopts a rectangular grid array, works in the X-band and is composed of 96 column line sources. Each antenna element is followed by a T/R module, which is used for the amplitude and phase of the antenna element. The aperture size of the array is 1500 mm in azimuth and 240 mm in range. The maximum beam scanning angle is ±20° in azimuth and ±15° in range (AOE coordinate system). The maximum instantaneous signal bandwidth of the array antenna is 1.4 GHz.

(2) **Beam pointing deviation and signal bandwidth without delay compensation**

If the array does not add a delay line, according to the above parameters, the beam pointing angle deviation of the antenna when scanning the maximum beam in the azimuth and range directions can be obtained, as listed in Table 4.2.

If the required beam pointing deviation within the signal bandwidth is less than 1/4 of the beam width during the maximum scanning of the antenna beam, then without adding a delay line, it can be known from Eqs. (4.10) and (4.14) that when $B_b = 1, k = 2, \alpha = 20°, \beta = 15°$, the maximum instantaneous signal bandwidth that the array antenna can achieve is about 259 MHz and 292.3 MHz, respectively.

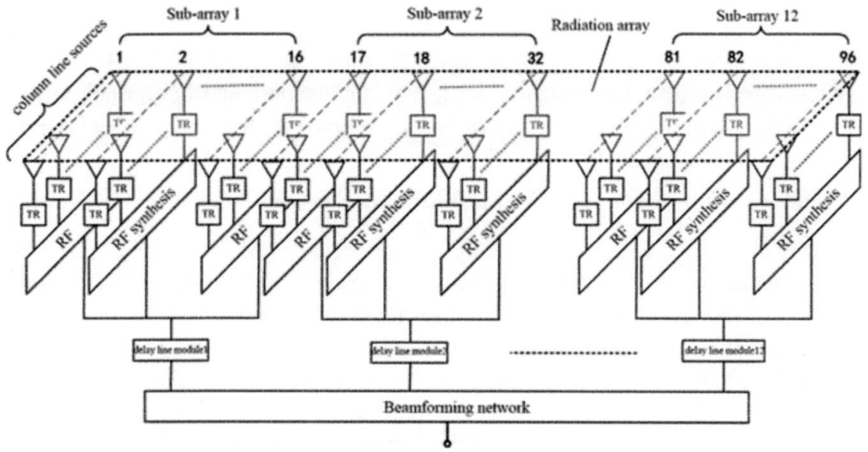

Fig. 4.13 Array antenna configuration diagram

Table 4.2 Beam pointing angle deviation for maximum beam scanning in azimuth and range of the 1D sub-array delay line configuration

Maximum scanning angle	Normal beamwidth/(°)	Maximum scanning beamwidth/(°)	In-band maximum beam pointing deviation/(°)	Beam pointing deviation/beamwidth/(%)
Azimuth (±20°)	1.06	1.13	1.54	134
Pitch (±15°)	6.72	6.96	1.13	16.2

If the delay line is not configured, the effective instantaneous signal bandwidth and beam pointing deviation performance cannot meet the requirements of the high-resolution synthetic aperture radar system.

(3) Calculation of the maximum delay of the array and the number of sub-arrays

According to the calculation Eq. (4.25) of the maximum delay of the array antenna in the AOE coordinate system, the maximum delay of the array antenna can be obtained, that is

$$\mathrm{TDL}_{\max} = L_x \sin\alpha \cos\beta + L_x \sin\beta \approx 557.7\,\mathrm{mm} \approx 17.7\lambda_0$$

That is, the total delay of the antenna array is about 18 wavelengths.

The array antenna is a medium-scale active phased array antenna. Due to a large number of elements, it is unacceptable to use element-level delay compensation from the perspective of development cost and system equipment. Instead, the use of sub-array-level delay compensation is a feasible option. The number of delay line sub-arrays in the two directions of the array is calculated by Eq. (4.21). When $k = 4$, the number of azimuth sub-arrays $M \geq 9.6$, and the number of range sub-arrays $N \geq 1.1$. Therefore, the array needs to be equipped with at least 10 delay line sub-arrays in the azimuth direction.

(4) **Delay compensation method and simulation results**

From the above calculation results of the number of delay line sub-arrays, it can be known that more than one delay line sub-array needs to be configured in the range direction of the array. In order to simplify the system complexity, the number of delay line sub-arrays is reduced as much as possible without increasing the amount of system equipment. The array antenna adds a 2-bit small delay line at the element level, which does not require additional equipment, and can solve the problem of increasing the delay line sub-array in the range direction. This simplifies the 2D delay sub-array configuration of a medium-sized 2D array antenna to a 1D delay line sub-array configuration.

In the azimuth direction, 12 delay line sub-arrays are designed, considering factors such as the specific equipment composition, operation mode, and sub-aperture configuration of the antenna system.

After comprehensive optimization, the array antenna adopts a two-stage delay compensation method: a hybrid delay compensation method of sub-array-level delay compensation and element-level low-order delay compensation. As a result, the maximum equivalent of the delay line of the antenna is $18\lambda_0$, and the specific delay line configuration is as follows.

Element-level (integrated into T/R modules): 2-bit delay line.

Sub-array-level: 12 sub-arrays, each sub-array adds a 4-bit delay line, $[1, 2, 4, 8]\lambda_0$, the maximum delay is $15\lambda_0$.

If a 2-stage delay line configuration design is adopted within the 1.4 GHz signal bandwidth and desired beam scanning range, the simulation results of the beam pointing deviation and SLL of the array antenna as a function of the scanning angle are shown in Figs. 4.14 and 4.15.

After adding delay line compensation, within the 1.4 GHz signal bandwidth and desired beam scanning range, the maximum offset of the active array antenna azimuth beam pointing is less than $\pm 0.02°$, and the maximum offset of the ranging beam pointing is less than $\pm 0.05°$, about 1.8% and 0.72% of the beamwidth, respectively.

Another advantage of this hybrid delay compensation method is that it can suppress the periodic grating lobes generated by the sub-array-level delay compensation. The azimuth scanning patterns of the antenna beams at the three frequency points before and after the optimal configuration of the delay line are shown in Fig. 4.16.

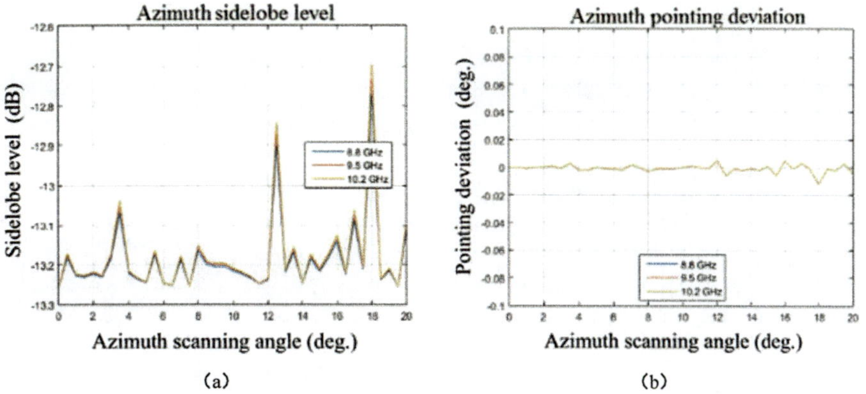

Fig. 4.14 Azimuth beam sidelobe and beam pointing deviation versus scanning angle. **a** Azimuth SLL, **b** azimuth pointing error

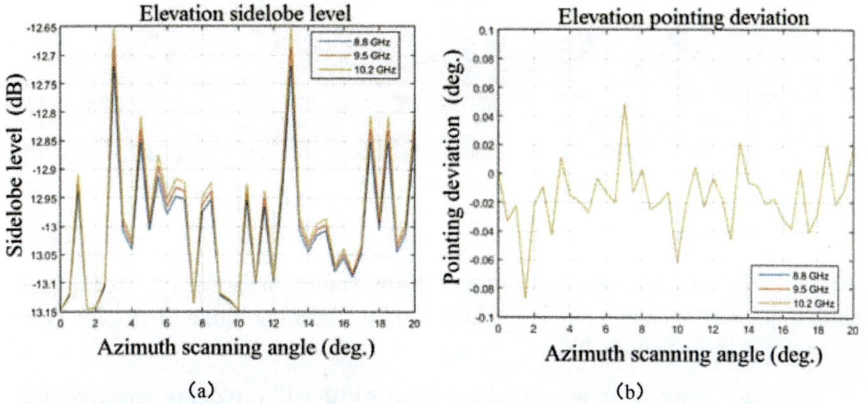

Fig. 4.15 Range-direction beam sidelobes and beam pointing deviation versus scan angle, **a** range-direction SLL, **b** range-direction pointing error

4.2.6 2D Sub-array Delay Line Configuration Example

(1) **Basic array architecture**

The delay line configuration of large 2D active array antennas mainly adopts a 2D multi-level sub-array configuration strategy, dividing multiple sub-arrays in the azimuth and rang directions, respectively, and forming a multi-level sub-array architecture. For example, a large 2D active array antenna array is shown in Fig. 4.17.

The antenna is a 2D planar active array using a rectangular grid arrangement, working in the X-band, the azimuth direction of the array is 15600 mm, the range direction is 2420 mm, and the total number of elements is about 120,000. The maximum beam scanning angle of the antenna is: azimuth direction ±42°, range

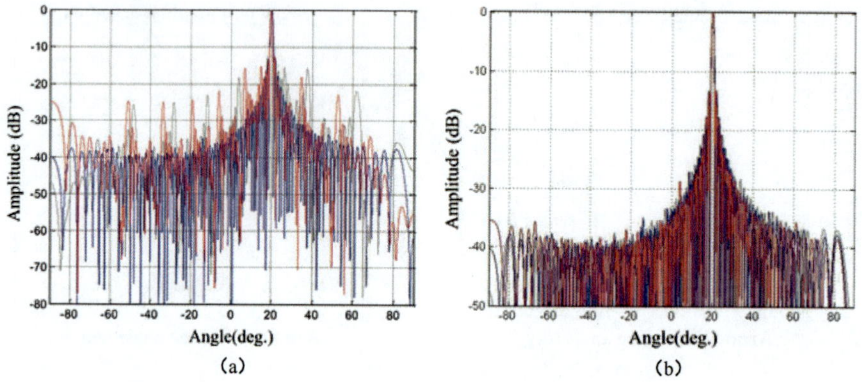

Fig. 4.16 Antenna scanning pattern before and after delay line configuration optimization, **a** without delay component, **b** with delay component

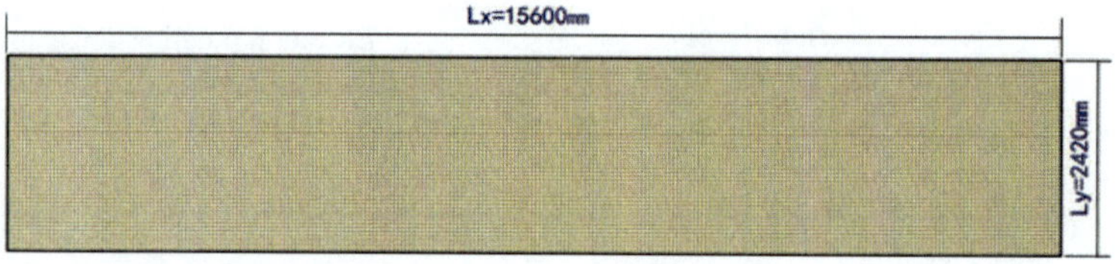

Fig. 4.17 Schematic diagram of the large 2D active array antenna

direction ±18° (AOE coordinate system), the maximum instantaneous signal bandwidth required by the antenna is 3.0 GHz, and the beam pointing dispersion is not greater than ±3% beamwidth.

(2) **Beam pointing deviation and signal bandwidth without delay compensation**

Suppose the antenna array does not add a delay line according to the antenna parameters mentioned above. The antenna beam pointing angle deviation when scanning the maximum beam in the azimuth and range directions can be obtained, as listed in Table 4.3.

Table 4.3 Beam pointing angle deviation for maximum beam scanning in azimuth and range of the 2D sub-array delay line configuration

Maximum scanning angle	Normal beamwidth/(°)	Maximum scanning beamwidth/(°)	In-band maximum beam pointing deviation/(°)	Beam pointing deviation/beamwidth/(%)
Azimuth (±42°)	0.10	0.135	0.14	104
Elevation (±18°)	0.66	0.694	0.051	7.3

If the required beam pointing deviation within the signal bandwidth is less than 1/4 of the beamwidth during the maximum scanning of the antenna beam, then without adding a delay line, it can be known from Eqs. (4.10) and (4.14) that when $B_b = 1$, $k = 2$, $\alpha = 42°$, $\beta = 18°$, the maximum effective instantaneous signal bandwidth that the array antenna can achieve is 12.7 MHz and 14.4 MHz, respectively.

It can be seen that if the delay line is not applied, the instantaneous signal bandwidth and beam pointing deviation that the antenna can achieve is far from the needs of high-resolution microwave imaging radar systems.

(3) **Calculation of the maximum delay of the array and the number of sub-arrays**

According to the calculation Eq. (4.25) of the maximum delay of the array antenna in the AOE coordinate system, the maximum delay of the array antenna can be obtained as

$$\text{TDL}_{\max} \approx 10675.4\,\text{mm} \approx 341.6\lambda_0$$

That is, the total delay line delay that needs to be configured for this antenna is about 342 wavelengths.

Since the aperture area of the active array antenna is nearly 40 m^2, the number of elements is more than 120,000, the array scale is huge, and the amount of equipment is large. Suppose the element-level delay line configuration scheme is adopted. In that case, at least 10 delay line components need to be configured $[1, 2, 4, 8, 16, 32, 64, 128, 128]\lambda_0$, as far as the technical level of delay line components is concerned, the development cost of such large delay components is high, and the volume is large and heavy. Considering the performance parameters of the active array antenna, the complexity of the system, the development cost, and the feasibility of the project, the active array antenna adopts a 2D multi-level delay line sub-array configuration scheme.

According to Eq. (4.21), when $k = 4$, the number of delay line sub-arrays in the two directions of the antenna array is calculated respectively, and the number of azimuth sub-arrays $M \geqslant 417.5$ and the number of range sub-arrays $M \geqslant 29.9$. Therefore, the array antenna needs to be equipped with at least 418 delay line sub-arrays in the azimuth direction and at least 30 delay line sub-arrays in the rang direction.

(4) **Delay compensation scheme and simulation results**

It can be known from the above calculation result of the number of delay line sub-arrays that the array antenna needs to configure delay line sub-arrays in the two dimensions of the azimuth and range directions. Therefore, to solve the problem of difficulty and cost in developing large-delay delay lines, it is a reasonable choice to use a 2D multi-level delay line sub-array configuration strategy.

According to the preliminary delay line configuration strategy, considering the operating mode and integrated architecture of the antenna system, through the simulation and iterative optimization of the beam performance, the active array antenna is divided into a total of 16,384 delay line sub-arrays of which 512 delay lines are

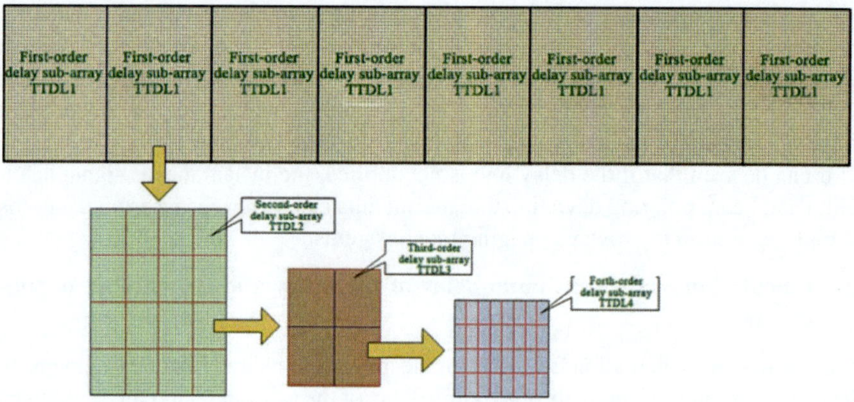

Fig. 4.18 Configuration of 2D multi-level delay line sub-array for large array antenna

divided in the azimuth direction. In addition, the line sub-array is divided into 32 delay line sub-arrays in the range direction. As a result, the antenna configuration of 2D multi-level delay line sub-array is shown in Fig. 4.18.

The specific configuration of the 2D multi-level delay line sub-array is listed in Table 4.4.

In order to achieve high-performance beamforming, broaden beam, and multi-mode beam management, in the multi-level sub-array delay line configuration architecture, the total delay amount of the delay line of the antenna system is larger than the theoretical calculating value.

According to the above-mentioned four-level configuration strategy of the antenna array delay line, within the 3.0 GHz signal bandwidth and desired beam scanning range, the optimization simulation results of the antenna beam pointing dispersion are shown in Fig. 4.19.

It can be seen that after the optimal configuration of the multi-level delay line, the directional dispersion of the antenna azimuth beam is less than $\pm 1.5\%$ of the beamwidth, and better than $\pm 1\%$ of the beamwidth in range deviation.

Table 4.4 2D multi-level sub-array delay line configuration

Sub-array level	Delay line	Number	Delay bits
First-level sub-array	TTDL1	8	[32, 64, 128, 128]λ_0
Two-level sub-array	TTDL2	128	[8, 16, 32, 32]λ_0
Three-level sub-array	TTDL3	512	[2, 4, 8, 108]λ_0
Four-level sub-array	TTDL4	16,384	[1, 2, 4]λ_0

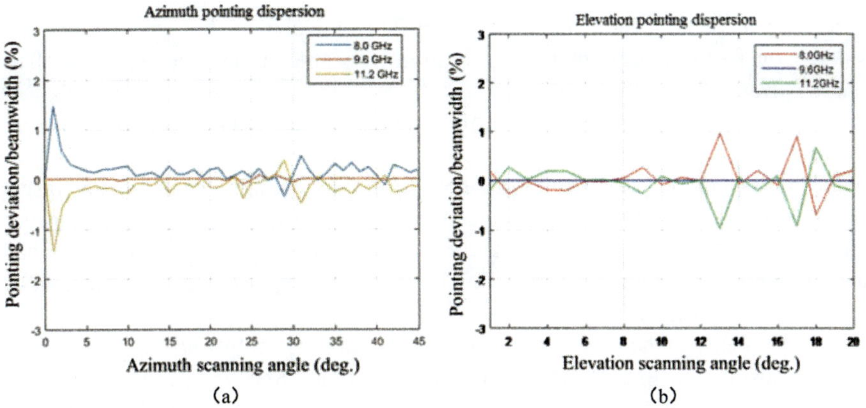

Fig. 4.19 Antenna beam pointing dispersion characteristics after delay line compensation, **a** azimuth beam pointing deviation, **b** range beam pointing deviation

4.3 RF Time Delay Components

4.3.1 Introduction

The RF delay line is a microwave component whose primary function is to realize the real-time delay of the RF signal. When a signal passes through the two-port time delay line device, the output signal group delay will generate a certain time delay relative to the input signal.

Although both the phase shifter and the time delay device can be used as the phase control device of the array antenna element, the beam scanning of the array can be realized by changing the phase/time delay of the excitation signal of each antenna element. However, due to the frequency response characteristics of the phase shifter and the time delay line being different, the scanning characteristics of the antenna array beam will be different. Most of the commonly used digital phase shifters are non-dispersive, and their phase shift value does not change with frequency, and the maximum phase shift amount is 2π. On the other hand, the digital time delay device is a dispersion type, and its phase shift value changes with the frequency in the operating frequency band.

Time delay lines have various structural forms, which can be roughly divided into three types according to different signal transmission and modulation methods: optical delay [9, 10], digital delay [11–13], and RF delay.

The optical delay is to change the transmission time of the optical signal in the medium by controlling the length of the optical fiber or the transmission characteristics (dispersion characteristics and frequency selectivity) of the optical fiber device to realize the time delay of the signal. Its advantages are slight fiber loss, large signal bandwidth, large delay compensation, and fine delay control. The disadvantage is that a photoelectric conversion device is required, which is large in size, low in conversion efficiency, heavy in weight, and high in cost. The optical delay has great application value in the fields of large delay compensation array antenna and optically controlled phased array.

The digital delay is the time delay of the signal in the digital domain by adjusting the initial frequency/phase of the signal and moving the sampling point. The digital delay has the characteristics of large delay compensation, high precision, low insertion loss, good phase nonlinearity, high amplitude consistency, fast switching speed, and good reliability. The disadvantage is that the cost is high. The digital delay has a good application in the high-precision large delay compensation array antenna of the digital array radar.

The RF delay is the time delay of the microwave signal by changing the length of the transmission line. Digitally controlled RF delay lines by means of switches mainly include switch lines and reflection types. The digitally controlled RF delay line mainly adopts the transmission line configuration, such as a microstrip line, strip line, or coaxial line. Its main features are simple structure, large bandwidth, low cost, and good reliability. The disadvantage is that the accuracy of a slight delay is poor, and the insertion loss of a large delay is large. Nevertheless, digitally controlled RF delay lines are widely used in active array antennas due to their simplicity, low cost, and practicality.

Other types of delay lines include acoustic wave delay lines, magnetostatic wave delay lines, ferrite delay lines, and high-temperature superconducting delay lines. This type of delay line mainly changes the electromagnetic characteristics of the carrier and the transmission mode of electromagnetic waves through additional conversion devices, e.g., electro-acoustic, and modulation of the transmission carrier. However, the essence is still to change the signal wavelength or propagation speed in the transmission carrier to realize the time delay of the signal.

The digital RF delay line adopts the digital quantization method and realizes various quantization delay states through the step of the minimum quantization bit. The digital delay line is also called a numerically controlled delay line because of its digital quantization, which is convenient for managing the array delay by computer control. The digital delay is conceptually completely different from the above-mentioned digital delay based on the digital domain delay. Usually, the time delay of the delay line is expressed by the time extension of the number of equivalent wavelengths. For example, a 5-bit delay line has 5 ground states $[1, 2, 4, 8, 16]\lambda$, the minimum quantization bit is 1λ, and the maximum extension is 31λ.

With the development of Monolithic Microwave Integrated Circuit (MMIC) technology, digitally controlled delay line chips based on monolithic microwave integrated circuits have achieved a delay of fewer than 5 bits. Furthermore, the planar assembly can be integrated with T/R modules or high-density array modules, so it has a good application prospect in high-integration active array antennas.

In order to compensate for the gain requirement of the transceiver link, the delay line is often integrated with the amplifier circuit of the transceiver signal to form a delay amplifier component. At the same time, to obtain good performance and high stability and reliability, other functional circuits will be added to the delay-amplifying component, such as amplitude and phase equalizers to improve in-band amplitude and phase performance, an ESC attenuation to improve amplitude and phase consistency, RF filters to improve anti-interference performance, as well as monitoring and protection circuits, temperature compensation circuits, power supply, and control circuits, etc.

4.3.2 Real-Time Latency Fundamentals and Classification

After a signal of wavelength λ (angular frequency ω) passes through a carrier of length l at speed v, the output signal time delay is

$$\Delta\tau = \frac{l}{v} = l\sqrt{\varepsilon\mu} \tag{4.27}$$

$$v = 1/\sqrt{\varepsilon\mu}$$

where ε is the permittivity, μ is the permeability. It can be found from Eq. (4.27) that the essence of the RF delay line to realize the signal time delay is to change the length of its signal transmission path or the material properties such as the magnetic permeability or permittivity of the transmission carrier. The latter is equivalent to changing the wavelength or propagation speed of the signal in the carrier. The digitally controlled RF delay line realizes the signal time delay by changing the path length of the transmission line.

Although both digitally controlled phase shifters and delay lines can be used in array antennas to achieve beam scanning of the antenna, there are specific differences in electrical characteristics between the two.

A true time delay line is a control device with a flat group delay frequency response within a particular frequency band, and the group delay value does not change with the change of the inserted phase. It has two characteristics [14]: (1) For the linear frequency response of the relative phase shift value, the slope changes with the change of the phase shift value, (2) For the flat group delay frequency response, the evolution of the delay value will cause the time characteristics of the input RF signal pulse envelope to change.

The relative group delay amount $\Delta\tau$ of the real-time delay line is defined as

$$\Delta\tau = \frac{\Delta\angle S_{21}(\omega)}{\omega} \tag{4.28}$$

where $\Delta\angle S_{21}(\omega)$ is the relative phase shift at a given signal angular frequency ω.

The frequency characteristics of a real delay line are shown in Fig. 4.20.

Since the delay line has a flat delay frequency response in different delay states, its relative phase shift has a linear frequency response and the phase shift gradient changes with the change of the phase shift value. As a result, the wavefront plane of

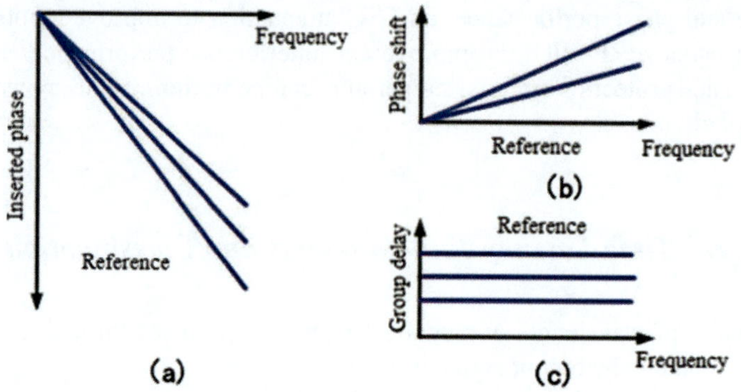

Fig. 4.20 Real delay line frequency characteristics, **a** insertion phase, **b** relative phase shift, **c** group delay

the array signal has different group times. The delay frequency response characteristic leads to changes in the timing of the pulse envelope of the input frequency signal.

Correspondingly, a digitally controlled phase shifter is defined as a control device with a group delay frequency response that does not change with the change of the inserted phase within a certain frequency band. Digitally controlled phase shifters have two characteristics: (1) The relative phase shift of each state has a flat phase shift frequency response and the phase shift value of each state does not change with frequency, (2) The time delay of each state is the same and does not change with frequency and the constant group delay causes the input RF pulse envelope to not change with time.

The frequency response characteristics of the phase shifter are shown in Fig. 4.21.

Since the phase shifter has a constant time delay in different phase-shifted states, the wavefront plane of the array antenna does not change with the relative phase change. Therefore, the phase shifter is unsuitable for the beam scanning of a

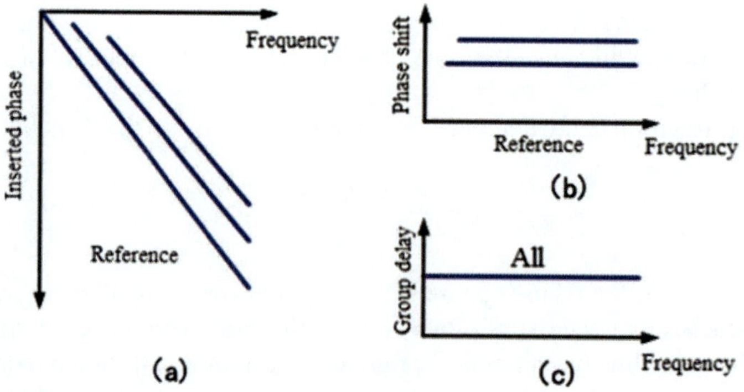

Fig. 4.21 Phase shifter frequency characteristics, **a** insertion phase, **b** relative phase shift, **c** group delay

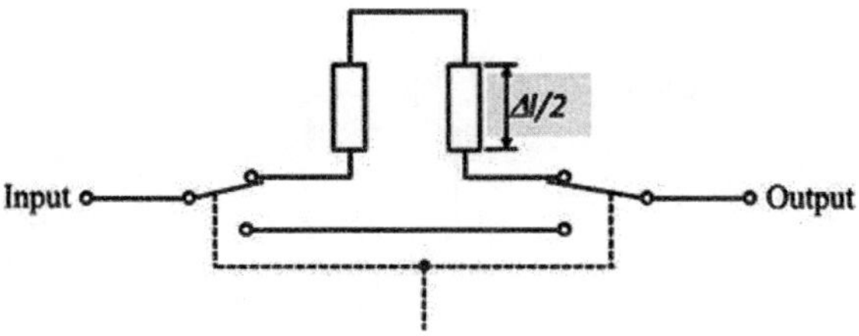

Fig. 4.22 Schematic diagram of the switch linear delay line

large aperture broadband array antenna to avoid beam pointing deviation and pulse broadening.

RF delay lines that use switches for state control include switch lines and reflection types mainly.

(1) **Switching linear delay line**. The switched linear delay line consists of two groups of switches and two transmission lines of unequal lengths. The schematic diagram of a single-section (1-bit delay) switched linear delay line is shown in Fig. 4.22. A multi-bit digitally controlled delay line is formed by cascading several similar single-section switch lines. The switched linear delay line uses a pair of PIN diodes or MESFET switches to select different lengths transmission lines, thereby obtaining the time delay difference between the two signals. In the case of transmission line impedance matching, the time delay difference is proportional to the difference between the lengths of the two transmission lines.

If the two transmission lines have the same phase propagation constant β and their length difference is Δl, a time delay or phase shift $\Delta\theta$ will occur after the signal is transmitted through the two paths, that is

$$\Delta\theta = \Delta l(\text{length of time delay}) = \beta\Delta l(\text{phase shift}) \tag{4.29}$$

The switch delay line is simple in principle and easy to implement in structure and is the most commonly used. However, the attention should be paid to avoid the resonance caused when the switch transmission line length is half the wavelength, leading to increased insertion loss.

A wider frequency bandwidth can be obtained by adopting a multi-level cascade to realize a multi-bit digitally controlled delay line method. Still, the size of the switch-line delay line depends on the minimum operating frequency and the number of delay lines. Furthermore, the digitally controlled delay line parameters depend on whether the Single Pole Double Throw (SPDT) switch works satisfactorily. The single-pole double-throw switch in the switch-line delay line usually has two forms: Field Effect Transistor (FET) switch and Micro-Electromechanical Systems (MEMS) switch. MEMS switches have good application potential in microwave control circuits due to their low insertion loss, low power consumption, and easy planar integration. The disadvantage is that the working life and reliability are not high, so the application in high-performance microwave systems is often restricted to a certain extent. The drain and source terminals of the FET switch are usually parallel resonant inductors, which can

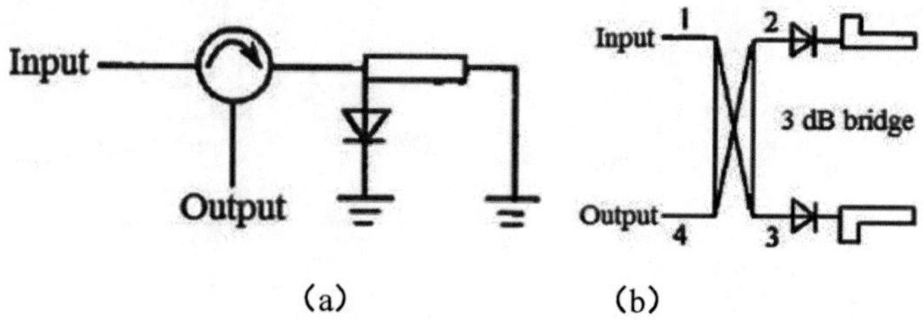

Fig. 4.23 Schematic diagram of reflective delay line, **a** circulator reflective type, **b** coupler reflective type

form a resonant circuit with the off-state drain-source capacitance. Using this feature and Ohmic Electrode Sharing Technology (OEST), broadband and low insertion loss performance of the delay line can be achieved [15].

(2) **Reflective delay line**. The digitally controlled reflection delay line has two deformation structures: the circulator reflection type and the coupler reflection type. The schematic diagram of the single-section delay line is shown in Fig. 4.23.

The shorted position (switch position) of the reflective delay line controls the insertion delay through the signal, and the shorted position is controlled by at least one of the switches shown in Fig. 4.23.

The circulator reflection delay line comprises a circulator and a reflection susceptance network. The reflection susceptance network can be in the form of a short-circuit delay line composed of a parallel switch (susceptance element) and a short-circuit transmission line with a terminal, as shown in Fig. 4.24. After the input signal transmitting to the reflection-type terminal transmission line through the circulator, in the two states of the parallel switch, the signal has reflected the circulator through different short-circuit terminal paths. Finally, it is output by the output port of the circulator.

In the two states of the switch, since the signal transmits and reflects through the reflection terminals of two different paths, for a delay line with a given delay, the length of the terminal transmission line is only half of the delay line. As shown in Fig. 4.24, the length of the transmission line of the reflection terminal is θ, then in the two states of on and off, the delay difference of the output signal is 2θ.

Fig. 4.24 Delay line type reflective termination

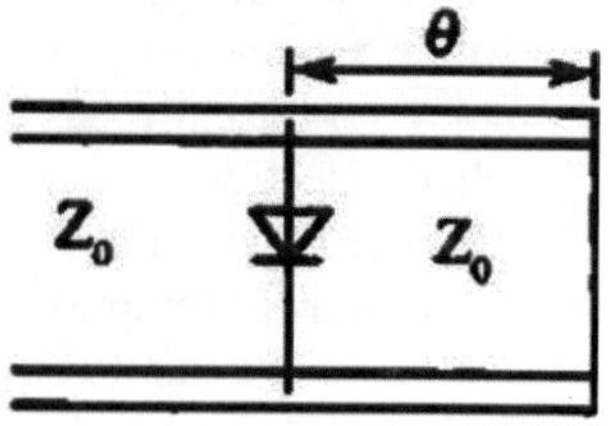

Coupler reflective delay lines consist of a 3 dB coupler (or 3 dB bridge) and two identical switched transmission line reflective terminations, which consist of a series switch and a terminal shorted transmission line. The 3 dB coupler generally adopts the quadrature directional coupler of the branch line or the coupled line structure, and the signal is output at the same amplitude in its straight-through and the coupling arm, and the phase difference is 90°. As shown in Fig. 4.23b, port 1 is the input port, port 2 is the straight-through port, port 3 is the coupling port, port 4 is the isolation port, and ports 2 and 3 are connected to the reflective terminal of the switch transmission line. When the signal is input from port 1, it is equally divided into two signals of equal amplitude and quadrature (phase difference of 90°) after passing through the 3 dB coupler, output from ports 2 and 3, respectively, and then reflected by the short-circuit terminals of ports 2 and 3. The reflected signal of each channel is divided into two equal-amplitude quadrature signals and output from ports 1 and 4, respectively. Because the two-port output signals of 1 and 4 result from the superposition and synthesis of two reflected signals, according to the orthogonality of the 3 dB coupler. If the path lengths of the reflection-type terminal transmission lines connecting ports 2 and 3 are equal, the two reflected signals arriving at port 1 are equal in amplitude and out of phase and therefore cancel each other. The two reflected signals arriving at port 4 are equal in amplitude and phase and superimposed on each other. Finally, there is no signal output at port 1 and complete signal output at port 4. The input signal entering port 1 is all output from port 4 after being reflected by a 3 dB coupler and a reflective terminal.

The function of the switch reflective terminal of the coupler reflective delay line is that when the switch reflective terminal is on, the signal will pass through the switching circuit and be reflected the 3 dB coupler output by the short circuit located far away from the transmission line, In the off state, the signal cannot pass through the switching circuit and is reflected the 3 dB coupler output there. When the switch reflective terminal is turned on and off, a time delay will be generated between the two signals output by the 3 dB coupler. Similar to the circulator reflection delay line, the time delay between the two states of the coupler reflection delay line is also twice the length of the reflection terminal transmission line, 2θ.

4.3.3 Delay Line Component Parameters

Whether it is the element-level delay compensation of the antenna array or the sub-array-level delay compensation, to realize the gain compensation of the transceiver link and the amplitude balance of each delay state, the delay line and the transceiver amplifier circuit are often integrated to form a so-called delay amplifier component. The following mainly discusses its functions and indicator parameters for the delay amplification component. Usually, it has the following functions:

(1) **Real-time delay**

The real-time delay is the most basic function of the delay amplifying component. Its function is to make the RF signal passing through the transmission path have a delay in the time domain through the change of the transmission path length.

(2) **RF transceiver conversion**

Most of the delay amplification components used in active array antenna systems need to have the function of RF transceiver conversion, that is, to meet the mode switching requirements of radar system transceiver time-sharing work. In practical applications, the delay amplifying component is a particular T/R module. Therefore, in addition to the transceiver and amplifying functions of the T/R module, it also has time delay capability. Furthermore, like the T/R module, the delay amplifying component should have a calibration mode for system calibration in addition to the RF signal receiving and transmitting modes of the radar system. Therefore, it needs to have the function of RF signal transmitting, receiving, and correcting.

(3) **RF signal amplification compensation**

The delay amplifying component realizes the time delay of the RF signal through the path length change. For most delay amplifying components, the delay amount reaches more than a dozen wavelengths, the transmission path lengths of different delay states are different, and the total static insertion loss is huge. For example, an X-band 5-bit delay line in the form of a switching line using a Low-Temperature Co-fired Ceramic (LTCC) substrate transmission line has a static insertion loss of about 20 dB. On the other hand, the total insertion loss of the integrated delay amplifying component using the delay line module of the MMIC chip is more significant. Therefore, it is necessary to amplify and compensate the RF signal of the delay line in practical applications. The specific measures are to integrate the delay line and the T/R module circuit to realize the gain compensation of the delay line through the amplification function of the T/R module transceiver link. In addition, the insertion loss of each ground state of the delay amplifying component is very different, resulting in a significant inconsistency in amplitude between different delay states. Therefore, measures should be taken inside each state of the delay line to perform corresponding amplitude equalization compensation.

(4) **Monitoring and protection**

The delay amplifying component is essential in the active array antenna system. For an antenna system configured with a sub-array level delay line, the performance and reliability of the delay amplifying component greatly influence the performance of the antenna system. Therefore, it is necessary to monitor and protect the delay amplifier components accordingly. The main monitoring contents include the current, voltage, temperature, etc., of the corresponding modules in the transceiver link. In addition, measures are taken to protect the delay amplifier components from abnormal states such as overcurrent, overvoltage, and overtemperature.

Typical parameters of delay amplifier components include operating frequency band, delay amount, delay error, amplitude and phase stability, spurious out-of-band suppression, component power efficiency, power capacity, port VSWR, etc. These are just some typical parameters of delay amplifier components but many parameters are not listed. Such as output power, transmit waveform drop, receive noise figure, transmit and receive gain, amplitude (gain) and phase consistency, amplitude (gain) and phase in-band fluctuation, transceiver conversion time, transceiver isolation, phase nonlinearity, amplitude intermodulation, interface requirements, reliability requirements, etc. Depending on the application system requirements, the specific parameters of the delay amplifier components will vary. However, the basic parameters of the delay amplifier components are introduced as follows:

(1) **Operating frequency band**

The operating frequency band is the frequency range within which the delay amplifying components meet specific technical specifications. Usually, the operating frequency band of the delay amplifying component is determined by the operating frequency band of the active array antenna system, and the bandwidth of the delay amplifying component is slightly larger than that of the antenna system.

(2) **The amount of time delay**

Usually, the delay amount of the delay line is represented by the number of wavelengths at the center frequency. The total delay value of the delay line should include three parameters: delay bits, minimum delay amount, and delay step. For example, the delay amount of a delay line is $[1, 2, 4, 8, 16]\lambda$. It shows that the delay line is a 5-bit numerically controlled delay, the minimum delay is λ, the delay step, that is, the minimum quantization bit is λ, and the maximum extension is 31λ.

(3) **Time delay error**

The delay error is also called the equivalent phase accuracy, and there are usually two representation methods: the maximum delay error and the relative delay error. The maximum delay error refers to the maximum value of the difference between the actual delay value and the theoretical value at each frequency point of the delay line in the operating frequency band, the relative delay error refers to the ratio of the maximum delay error to the theoretical value. Since each frequency point of the delay line corresponds to multiple delay states in the frequency band and there are many corresponding delay error test data, the standard deviation or the mean square error is also often used to express the delay error.

(4) **Amplitude and phase stability**

Amplitude and phase stability of delay amplifier components, including time stability, mechanical stability, and thermal stability. Time stability refers to the variation range of the amplitude and phase (delay) of the delay line components within a certain time range. Mechanical stability refers to the range of amplitude and phase (delay) variation of components receiving and transmitting under certain vibration conditions.

Finally, thermal stability refers to the variation range of the transmission and reception amplitude and phase (time delay) of the delay line components within a certain temperature variation range.

(5) **Out-of-band spurious suppression**

Out-of-band spurious suppression refers to the suppression capability of the delay line component in a certain frequency range outside the operating frequency band. It is usually expressed as the ratio of spurious spectral power to carrier power, and the unit is dBc. Spurs generation is often caused by the instability of components in the delay amplifying component or the weak parasitic oscillation and small signal distortion caused by poor circuit design. In addition, due to external interference such as power supply ripple and mechanical vibration, parasitic modulation of signal amplitude and phase may cause spurious out-of-band spectrum.

4.3.4 Real-Time Delay Line Design

In the high-resolution microwave imaging radar system, to realize the wide-angle scanning of the wide-bandwidth array antenna beam, the time delay line (delay amplifying component) is used for delay compensation, which is an effective measure to eliminate the aperture effect of the wideband scanning array antenna. However, large active array antennas usually consist of many components and devices, and weight, power consumption, and cost are other challenges in developing high-performance antenna systems. Therefore, microwave devices of complex array antenna systems, especially critical active devices such as delay amplifier components, mainly adopt integrated design technology to achieve miniaturization, light mass, and high-efficiency performance of delay amplifier components.

The requirements of delay amplifying components vary with the requirements of the application system, and their circuit composition and complexity are also quite different, but their basic components are similar. The basic composition of common delay amplifier components is shown in Fig. 4.25.

As shown in Fig. 4.25, a delay amplifying component is mainly composed of three basic circuits: a delay line circuit, a transceiver compensation circuit, and a power supply and control circuit. Among them, the delay line circuit realizes the multi-bit numerical control time delay of the RF signal, and the transceiver compensation circuit realizes the amplification and gains compensation of the RF transceiver signal. The power supply and control circuit realize the power supply of the active elements of the delay amplifying component, the transceiver conversion, and the amplitude and phase control management through the wave control conversion and the power supply modulation.

RF delay line circuits are mainly switch lines, and multi-digit numerical control delay is achieved by cascading multi-level switch lines. A typical 4-bit delay line circuit is shown in Fig. 4.26.

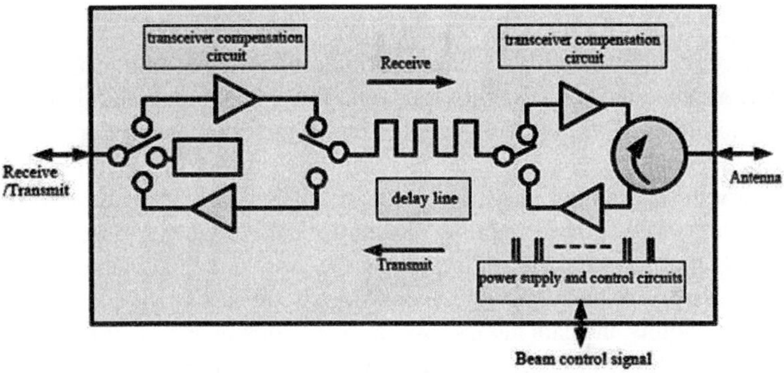

Fig. 4.25 Schematic diagram of the basic compositions of delay amplifier components

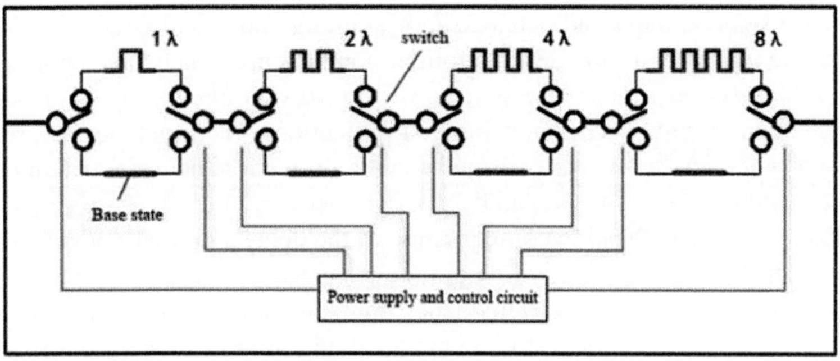

Fig. 4.26 Schematic diagram of the delay line circuit

Delay line circuits based on microwave substrate integration usually have two specific implementation forms: one is based on a multi-level cascade connection of microwave transmission lines and switches. There are various forms of microwave transmission lines, commonly used are microstrip line, multilayer microwave substrate stripline, LTCC stripline, Coplanar Waveguide (CPW), and so on. The other is to use a single-chip microwave integrated circuit time delay line chip, a gain compensation chip, and peripheral power supply and control circuits to achieve planar integration through a microwave substrate or an LTCC substrate. The former is cascaded through the discrete delay transmission lines of the microwave substrate, which can obtain lower insertion loss, but the overall size of the components is more significant, the latter uses a single microwave integrated circuit chip for assembly and integration, and the circuit loss is more significant, but it can achieve miniaturization. Using LTCC transmission substrates to make time delay transmission lines can realize the miniaturization of considerable time delays through multilayer interconnection. Still, LTCC transmission substrates have large microwave losses in high-frequency bands. Therefore, in practical applications, different implementations should be selected according to the overall requirements of the system performance.

The requirements for the delay amplifying components are higher for the multi-functional active array antenna system, and the circuit is more complicated. To obtain better performance and higher stability and reliability, other functional circuits should be added to the basic circuit of the delay amplifying component. For example, to obtain higher transmit channel gain, increasing the power drive amplifier circuit is necessary. In order to protect power devices, it is necessary to add isolators at the output port of the power amplifier. To improve the frequency characteristics of the in-band amplitude and phase, increasing the amplitude and phase equalization circuit is necessary. A temperature compensation circuit needs to be added to improve temperature stability and gain flatness over a wide temperature range. To improve the amplitude consistency of the components and the weighting ability of the receiving amplitude, it is necessary to add an electronically controlled attenuator. In addition, it is necessary to increase the polarization switch to realize the function of changing the polarization and improving its efficiency. Furthermore, filter circuits must be added to improve signal quality and enhance anti-jamming and out-of-band rejection. To improve the reliability of the delay amplifier components, it may also be necessary to add a high-power limiter circuit, a power conversion circuit, and a monitoring and protection circuit. The monitoring and protection circuit includes a variety of fault feedback circuits, such as over-pulse width protection, power signal abnormal protection, power-on self-reset, and data parity check.

The amplitude equalization compensation of the delay amplifier component uses the complementary principle of the amplitude-frequency characteristics of the equalization circuit and the amplitude-frequency characteristics of the delay line circuit to realize the amplitude equalization in the band. Figure 4.27 shows the principle of amplitude equalization compensation of the delay amplifying component.

Figure 4.28 shows the equalization compensation circuit of the single-section delay line element.

Since the attenuation characteristic of the equalizer circuit is that the low-frequency attenuation is large and the high-frequency attenuation is small, which is opposite to the slope of the insertion loss characteristic of the delay line, adding an appropriate equalizing circuit, the delay amplifying component can obtain better in-band amplitude flatness.

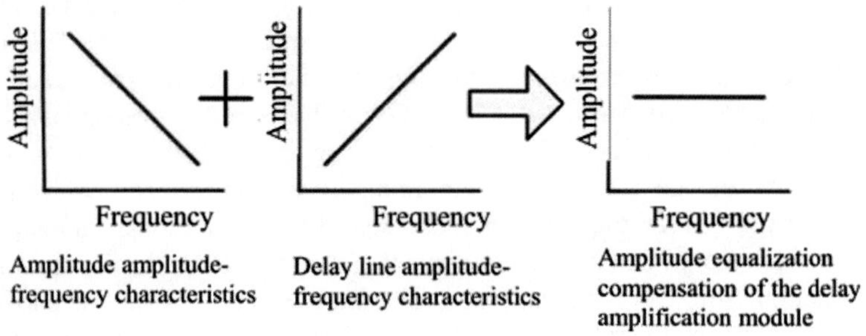

Fig. 4.27 Schematic diagram of the equalizer compensation

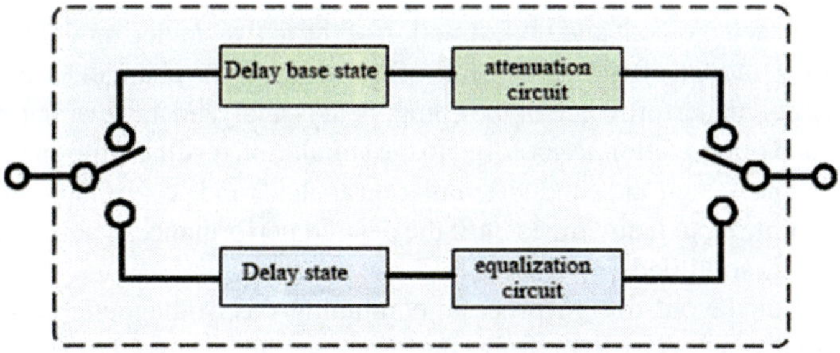

Fig. 4.28 Schematic diagram of the single delay line element

The temperature compensation of the delay amplifier component is to insert a temperature compensation attenuator in the receiving channel of the component. Using the complementary principle of the amplitude temperature characteristic of the temperature compensation attenuator circuit and the temperature characteristic of the delay line receive link gain, the flatness of the in-band gain of the component receiving the channel in a wide temperature range is achieved.

Due to the attenuation characteristics of the temperature compensation attenuator circuit, the attenuation is large at low temperatures and small at high temperatures. Therefore, contrary to the slope of the temperature characteristic of the gain of the receiving channel, after adding the temperature compensation attenuator, the receiving channel of the delay amplifying component can achieve better in-band gain flatness.

During the gain-temperature compensation of the delay line circuit, the theoretical value of the gain-temperature characteristics of all amplifiers and microwave devices in the link is first evaluated, and the gain-temperature characteristic curve of the entire channel is synthesized. Based on this, the attenuation temperature characteristic of the corresponding temperature-compensated attenuator is selected. If necessary, the gain temperature characteristic curve of the component should be obtained experimentally. The circuit design of the delay amplifying component is as follows:

(1) **Circuit layout and simulation**

According to the function and performance requirements of the delay amplifying component, and synthesizing the main component parameters, assembly and integration process methods, heat dissipation measures, interface requirements, and other factors, the principle block diagram of the delay amplifying component is designed, determine the components, microwave substrate materials, assembly configurations, and external interfaces.

According to the principle block diagram, the preliminary layout design of the delay amplifier components is carried out, including the delay line element circuit, the active amplifier circuit, the switch circuit, the compensation circuit, etc., to form the basic design layout. On this basis, the electromagnetic simulation analysis of the delay amplifier circuit is carried out. Through the electromagnetic simulation of the delay line circuit of each element and the extraction of the active device model parameters, the simulation model of the active link of the transceiver channel is established, and the transceiver performance of the component is analyzed and evaluated through simulation and optimization. According to the simulation results, adjust and optimize the circuit layout design and conduct simulation analysis and performance evaluation. Iteratively optimize so many times until the desired performance is achieved and the circuit layout is solidified.

In the circuit layout design process, combining electromagnetic compatibility design is usually necessary to optimize the circuit layout and avoid too many discontinuities and solid mutual coupling in the link. At the same time, the electromagnetic simulation of the cavity should also be carried out to prevent the cavity resonance frequency from falling within the operating frequency range.

The critical point of the circuit layout design of the delay amplifying component is the stability design of the circuit. Unfortunately, in the process of component design, many designers often only pay attention to critical parameters such as the delay accuracy of the component, the gain of the transceiver channel, the consistency of the transceiver amplitude and phase, and the receive noise figure. Sometimes the stability of the component transceiver link also should be considered.

As we all know, the delay amplifier component is like the T/R module, and its internal transceiver channel forms a closed loop through a switch or a circulator. The existence of this closed loop may lead to potential instability. Due to the limited RF isolation performance of switches and circulators, even if the transceiver time-sharing mode is used in most radar system applications, system noise will be generated as long as the component circuits are in operation. Suppose the isolation of the transceiver channel of the component is not high enough. In that case, these system noises may form positive feedback through this internal loop, resulting in unstable problems such as circuit oscillation and self-excitation. Such instability will not only cause instability and deterioration of the amplitude and phase performance of the module but even lead to damage to the module. The overall loop gain must be less than unity to avoid this circuit oscillation instability. Therefore, the sum of the isolation degrees of all switches, delay lines, isolators, etc., in the loop is greater than the sum of the gains of all amplifiers (including power amplifiers and low-noise amplifiers) in the loop.

Another situation that can easily lead to the instability of the delay amplification component is the unreasonable design of the component cavity and layout inside the component. This leads to spatial electromagnetic coupling between the circuits and the existence of a hidden path in the substrate circuit, forming an invisible closed loop, resulting in the instability of the component.

In the circuit layout design of the delay amplifier component, in order to avoid or eliminate the instability of its circuit, the main solutions are as follows:

(1) Timing design. Reasonable design of the pulse modulation timing of the transceiver channel to avoid the intersection of the transceiver timing sequence completely shut down the receive channel during the transmission period and improve the transceiver isolation.

(2) Link gain distribution. Through link analysis, the loss and gain of each component or circuit inside the transceiver link are reasonably allocated, the total gain of the link is controlled, and the impedance matching and isolation performance at all levels are improved. Reduce the gain compression degree of the high-power amplifier in the transmit channel as much as possible to avoid circuit instability or even self-excitation problems caused by increased link gain or too deep gain compression.
(3) Optimize the cavity design. Through the electromagnetic simulation of the cavity, the cavity structure is optimized. Removing barriers or absorbing materials is required to avoid or eliminate cavity resonance, spatial coupling of RF signals inside components, and signal crosstalk between boards.
(4) Power supply filtering. Take necessary filtering and isolation measures for the power supply and control lines to eliminate the mutual coupling between the power supply circuit and the RF and control circuits.
(5) Anti-static design. Take necessary anti-static measures to strengthen the protection of the active RF chip circuit inside the module and improve the overall stability and reliability of the module system.

(2) **Power supply and control circuit**

The auxiliary circuit of the delay amplifying component mainly includes the power supply and control circuit and monitoring and protection circuit. The main function of the power supply and control circuit of the delay amplifying component is to receive the timing signals and power signals from the upper-level wave control unit and control and manage them. For a 4-bit delay amplifier, as shown in Figs. 4.25 and 4.26. When the control signal is sent by the differential timing signal, inside the delay amplifying component, after being converted by the multi-channel differential transceiver, the TTL signal is provided to the PIN switch driver, FET driver, and power modulator, respectively. The timing control of the 4-bit delay line and the sending and receiving compensation T/R are realized through the management of the TTL timing signal and the power signal. Figure 4.29 shows a power supply and control circuit of a 4-bit delay amplifying component using a beam control chip.

As shown in Fig. 4.29, the power supply and control circuit module mainly include a differential interface circuit chip, wave control chip, PIN switch driver, FET driver, power modulation chip, PMOS tube, and other devices. In addition to controlling and managing power supply and timing signals, the circuit can also easily monitor and protect various fault modes, such as fault monitoring and protection of signal pulse width, overvoltage, and abnormal timing combination.

The power supply and control circuit, as the auxiliary circuit of the delay amplifying component, are usually designed and integrated with the delay line circuit and the transceiver compensation T/R circuit, which facilitates the integration, miniaturization and light quality of the entire assembly.

(3) **Link design and analysis**

The link design and analysis of the delay amplifier component is to construct the transceiver link model of the delay amplifier component. Coefficients, excitation,

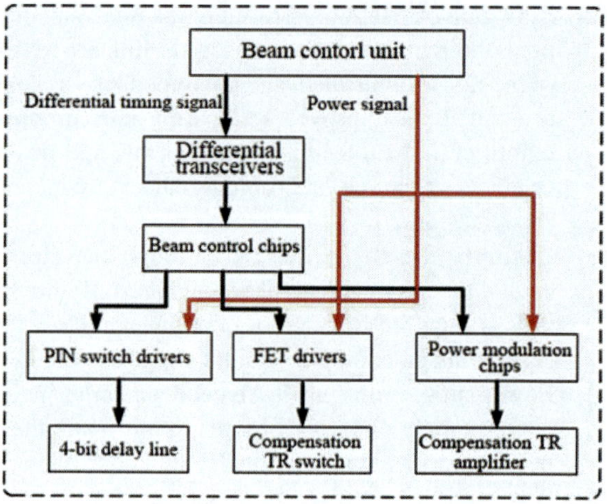

Fig. 4.29 Schematic diagram of the power supply and control circuit of the delay amplifying component

output signal levels, 1 dB input compression point, input signal power level of the power amplifier device, etc., are analyzed, allocated, and evaluated to determine the configuration of the transceiver link and estimate the main performance of the circuit.

The receiving and transmitting working modes of the circuit of the delay amplifying component is modeled. First, the receiving and transmitting links are equivalent to serial link models and are modeled independently. Then, the gain distribution, signal analysis, and estimation are carried out according to the link model.

For a 4-bit delay amplifying component shown in Figs. 4.25 and 4.26, combined with an engineering design example, the signal analysis model of the receiving and transmitting links is shown in Fig. 4.30.

Figure 4.30 shows the index parameters of each module or device in the transceiver link. According to the above link model, combined with the system sensitivity and signal dynamic index requirements, through index allocation and signal analysis, the performance estimation of the delay amplifier component can be obtained: receive noise figure, receive link gain, receive signal dynamics, receive signal 1 dB input compression point input signal power level, transmit excitation signal level, transmit saturated output power, all levels of power amplifier 1 dB input compression point input signal power level, high power amplifier gain compression and so on.

4.4 Real-Time Delay Line Example

A schematic diagram of a C-band five-position switch linear delay amplifier is shown in Fig. 4.31 [16]. The main function of the delay amplifying component is to complete the delay control of transmitting and receiving radio frequency signals, the medium power amplification of transmitting RF excitation signals, the low-noise amplification of receiving RF signals, and the transceiver conversion. In order to improve the system

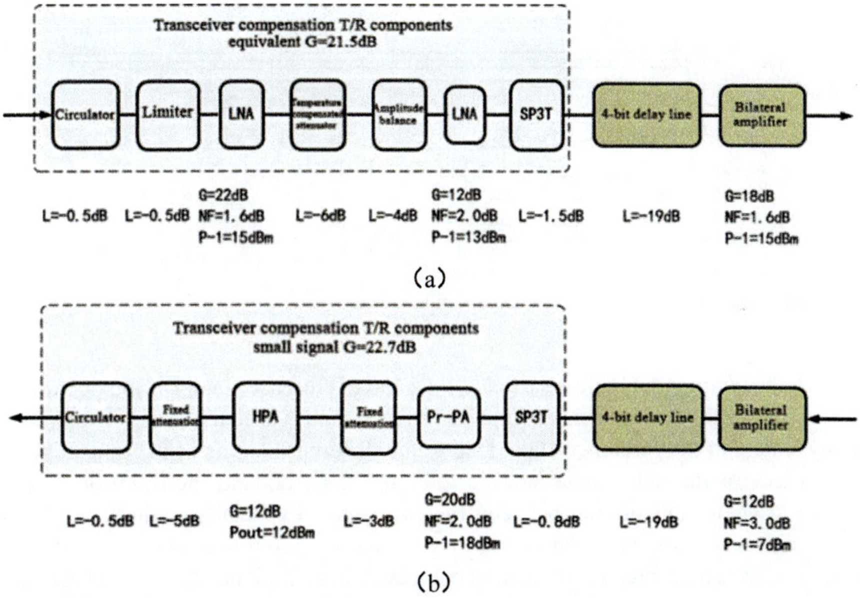

Fig. 4.30 Signal analysis model of the transceiver link of the delay amplification component, **a** receiving link, **b** transmitting link

integration, a 1:2 power division synthesizer is integrated inside the delay amplifying component. As a result, a delay of a 5-bit total length of 23λ is achieved, and the value of the step is λ.

The delay circuit is formed based on the switch microstrip transmission line. Therefore, the design needs to consider the single-pole double-throw switch, the microstrip delay unit circuit, and the various insertion loss compensation. Each bit

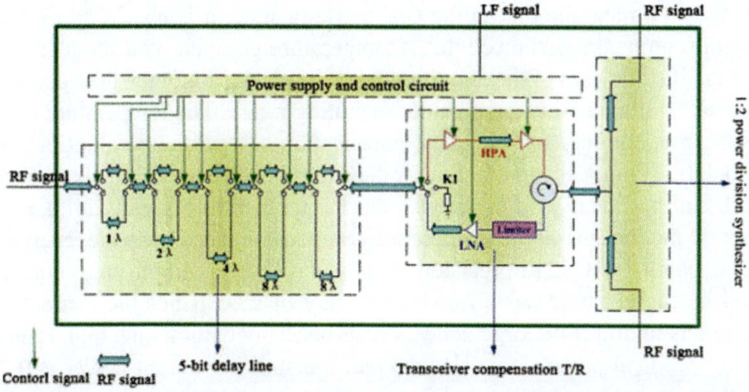

Fig. 4.31 Schematic diagram of the delay amplifier component

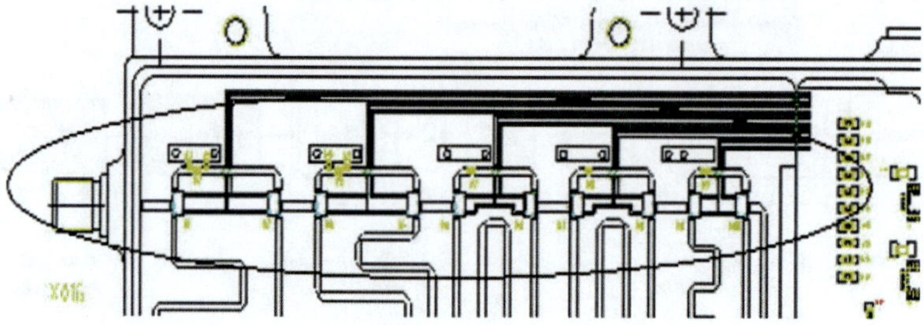

Fig. 4.32 Schematic diagram of the compensation network

state of the component includes a delay state and a reference state. During operation, according to the required delay compensation, a switch is used to select between each delay state and the reference state. Due to the different lengths of the transmission lines used for the delay state and the reference state per bit, the insertion loss is different when transitioning between the two states. Especially in the large delay bit, the insertion loss between the delay state and the reference state may differ by several 1 dB. Therefore, an attenuation compensation circuit needs to be added to the reference state to keep the insertion loss between the delay state and the reference state consistent. In this example, a resistance attenuation network is added to the ground state of each bit for amplitude compensation. The compensation circuit is shown in Fig. 4.32.

The primary function of the transceiver compensation T/R in the delay amplifying component is to complete the compensation and amplification of the delay signal. The difference from the T/R module is that it does not need to provide digital phase shifting and attenuation functions but provides accurate delay signals. The multi-chip assembly circuit is used in this example to realize the transceiver compensation T/R. The multi-chip assembly technology has the advantages of simple and reliable circuits and high circuit integration.

In order to achieve the requirements on the gain and its flatness in a wide temperature range and make the link in the best working state, it is necessary to design a temperature compensation unit circuit. A temperature-compensated attenuator circuit is used to adjust the gain of the amplifiers at all levels to satisfy the gain and flatness requirements within a wide temperature operating range. The temperature compensation circuit is not only for the gain compensation of a particular amplifier but for the temperature compensation design for the entire link. The receiver channel amplifier works in the linear working mode, and the temperature greatly influences the amplifier, so the focus is on the temperature compensation design in the receiver link.

The C-band 5-bit high-precision large-delay delay amplifying component designed in this example has a maximum delay of 23λ. It has the characteristics of high-precision multi-bit large delay, low in-band fluctuation, and high reliability. The main technical parameters achieved are: transmitting output power is 0.25 W, transmitting power consistency and in-band fluctuation is better than ±0.7 dB (all delay states, including parasitic amplitude modulation and consistency), receiving gain fluctuation is better than 1.0 dB (all delay states, including parasitic amplitude modulation), receiving and transmitting delay phase errors are better than ±20° (all delay states, of which nonlinearity ≤5°), temperature stability is better than ≤0.6 dB (Inside operating temperature range).

References

1. Lu J (2019) Design techniques of synthetic aperture radar. Wiley, Hoboken, NJ
2. Wei C, Jin M, He C (2012) Analysis of TTD for wideband phased array. Radar Sci Technol 10(6):668–670
3. Jian W, Xian-rong S, Jian-xin L (2006) Beam-pointing excursion versus frequency and application of realtime delayers for the wideband phased array. J Microw 22(1):23–26
4. Li M, Jude J, Zhu L (2006) Analysis of effect of real-time delay line performance on SAR imaging quality. J China Acad Electron Inf Technol 1(6):536–539
5. Skolnik MI (2008) Radar handbook. 3rd edn. The McGraw-Hill Companies
6. Wang W, Zhang H, Zhang Z et al (2014) Broadband antenna array for SAR applications. In: IEEE antennas and propagation society international symposium (APSURSI). IEEE, Memphis, TN, pp 138–139
7. Huifeng S, Yunkai D, Hong L, Junjun J, Li S (2012) Design and implementation of a C-band wide-band 2-dimentional scanning solid-state active phased array antenna. J Grad Univ Chin Acad Sci 29(2):282–287
8. Zhang J, Li J, Sun H (2010) A study on layered scheme of real-time delayers for the wideband phased array. Modern Radar 32(7):75–78
9. Jin M, Guan W, Guo J, Qi M, Guo Y (2006) Experimental system of a wideband optically controlled phased array antenna. Acta Electron Sinica 34(6):1127–1129
10. Cui N, Kou J, Zhao H et al (2016) Optical true time-delay lines of photonic crystal slow light waveguides used in phased array radar. Acta Opt Sinica 36(6):1–7
11. Wu M, Jin X, Tan J (2001) Digital T/R modules for phased array radar. Modern Radar 23(2):57–60
12. Jiaguo L, Manqin W, Xueming J, Zhengxing F (2003) Active phased-array antenna based on DDS. Acta Electron Sinica 31(2):199–202
13. Wu M (2006) The development of digital array radar. J China Acad Electron Inf Technol 1(1):11–16
14. Robertson I, Lucyszyn S (2001) RFIC and MMIC designand technology. Inst Eng Technol
15. Morrison RA (1982) Improved avionics reliability through phase change conductive cooling. IN: IEEE National telesystems conference. IEEE, Galveston, TX, pp B5.6.1–B5.6.5
16. Haiqing G, Liu Y, He J (2017) Amplitude compensation of C band time delay module. Radar and ECM 37(4):15–19

Chapter 5
Active Array Module Integration

5.1 Introduction

In order to improve the target detection precision and anti-jamming capability of the radar system, in addition to the two key requirements of large aperture and ultra-wideband, the high-resolution microwave imaging active array antenna also requires the antenna system to have multi-function and multimode working capabilities. Therefore, the active array antenna used for high-resolution microwave imaging is a highly complex microwave system. It often contains thousands or even tens of thousands of radiating elements, T/R modules, active modules, RF feeding networks, cable assemblies, frontal power modules and wave control modules, and other hardware devices.

Due to the huge amount of large-scale active array antenna equipment, complex system configuration, weight, power consumption, and development cost, the application of large-scale array antenna systems is seriously limited. Therefore, the weight, power consumption, and cost of the antenna system are the issues that engineers need to focus on and solve in engineering design. In order to solve the limitations of these array applications, in addition to fully optimizing the layout of the array in the telecommunication design, engineers should reduce the size of the array as much as possible and reduce the number of components. The antenna array should also be reasonably divided into some blocks while designing the antenna array architecture and structure. Thus, integration and modularization technologies are used as far as possible to achieve front-end lightweight and low-power design and reduce the costs.

With the development of microelectronics technology, monolithic microwave integrated circuit technology, multi-layer heterogeneous substrate high-density manufacturing technology, and microsystem integration technology. The tens of thousands of radiating elements of the active array antenna are divided into a certain number of standard active array modules. Therefore, using a certain number of standardized active array modules to assemble a larger-scale active array antenna is a more reasonable and feasible implementation scheme for the integration architecture of the array. The active array module integrates the passive antenna array, T/R modules, RF

J. Lu et al., *Active Array Antennas for High Resolution Microwave Imaging Radar*,
https://doi.org/10.1007/978-981-99-1475-3_5

feed network, cable components, wave control module, and power module in high density. The modular design of high-density integration of active array antennas is one of the technical development trends of large active array antennas. The modular integrated architecture can not only realize the larger scale, light mass, and low-power performance of the antenna array but also realize the flexible construction of the array aperture through the 2D building block expansion of the module, and it is easy to debug, repair, and replace. Furthermore, the maintainability and reliability of the active array antenna can effectively improve its performance of the active array antenna and reduce the cost in the life cycle.

5.2 Array Feed Configuration

Active array antennas often contain many active modules (such as T/R and delay amplifier components) and RF feed or beamforming networks. The feed configuration of the active array antenna includes a feed network, an active module, and its control circuit. The feed configuration is critical to the realization of the functions and performance of the active array antenna, such as amplitude and phase control, beamforming, and array control.

Active array antennas mainly adopt a distributed array configuration based on T/R modules and delay amplification components. Each antenna element or several is configured with a T/R module, and multi-level delay amplification components are configured on a single antenna element and antenna sub-array. The distributed array antenna can realize the flexible shaping and scanning control of the antenna beam and has good fault redundancy simultaneously. The influence of a few antenna elements or T/R failure on the antenna performance is very slight. Therefore, this distributed array has higher reliability than the array antenna of the lumped transmission source.

An active array antenna needs to use an RF transmission power distribution network between the microwave signal excitation source and the T/R module. Or use an RF receiving power combination network between the T/R module and the receiver to realize the excitation of the transmitting channel of the T/R module or realize the receiving and combining of echo signals from the target.

In the active array system, in the transmitting mode, the RF signal of the excitation source is distributed and transmitted to each antenna element (or T/R module) of the antenna array, or in the receiving mode, the RF signal from each channel of the array is collected and synthesized and transmitted to the receiver, this process is often called "feeding". Instead, the phase distribution provided to the individual antenna element channels required for beam scanning or beamforming of the array antenna is called "phasing". Array phase feed and power feed are the main functions of the array feed configuration. Among them, there are mainly two feed methods for active array antennas: forced feed and space feed.

Forced feed, also known as constrained feed, mainly includes three forms: series feed, parallel feed, and hybrid feed. Forced feed systems often use microwave transmission lines in the form of waveguides, coaxial lines, strip lines, and microstrip lines so that RF signals are distributed or synthesized in a closed transmission channel according to predetermined amplitude and phase weights. With the development of optoelectronic and optical fiber technology [1], optical divider/combiner also use optical fibers as array transmission lines and build optically controlled phased array systems. Waveguides, coaxial lines, and strip lines are suitable for power division/combination in high-power antenna arrays, while microstrip lines and optical fibers are often used in low-power antenna arrays. Commonly used power divider/combiner include isolated and non-isolated, equal power division, and unequal power division.

The forced feed configuration of an active array antenna mainly consists of RF power divider, phase shifter, and, sometimes, attenuator for amplitude control. For a wideband array antenna, a delay amplification component is usually applied at the sub-array level for delay compensation and amplitude adjustment when the antenna array beam scans at a large angle.

5.2.1 Series Feed

Figure 5.1 shows five common series feed configurations. The most important feature of the series feeding network is that the path lengths from the main network port to the radiation element may not be equal, resulting in the phase of each channel of the antenna array being a highly correlated function of frequency. Therefore, array antennas fed in series are mostly narrowband. A waveguide traveling wave antenna array is a typical series-fed antenna whose beam direction changes with frequency.

In Fig. 5.1, the other four feed configurations except Fig. 5.1e, when the phase shift of the antenna phase shifter changes. Since the electrical length of the path from the main network port to the antenna element is not equal, it must be calculated or considered as a frequency function when designing the antenna array.

Figure 5.1a, b are end-fed array configurations. The antenna radiating elements are cascaded in series along the main feeder, and the coupling feed and amplitude control are performed through directional couplers. To control the beam, a phase shifter is added to the branch line feeding each radiating element, or a phase shifter is added between the adjacent couplers of the main feed line. The amplitude weighting of the antenna array can be achieved by appropriately controlling the coupling of the series coupler. For example, if all cascaded couplers are equally coupled, the amplitude distribution of the antenna array is approximately exponential tapering. Both array configurations are frequency-sensitive and have tighter bandwidth limitations than most feed systems. During the beam control of the series feed array, for the specified beam pointing angle, the phase shift amount of each phase shifter is the same. Thus, only the same phase control is required. This is an advantage of this feed configuration. However, since the serial channel insertion loss increases with the increase of the antenna element sequence, and the phase shift amount of the phase adjustment also increases with the rise of the element sequence, this type of fed antenna array is rarely used.

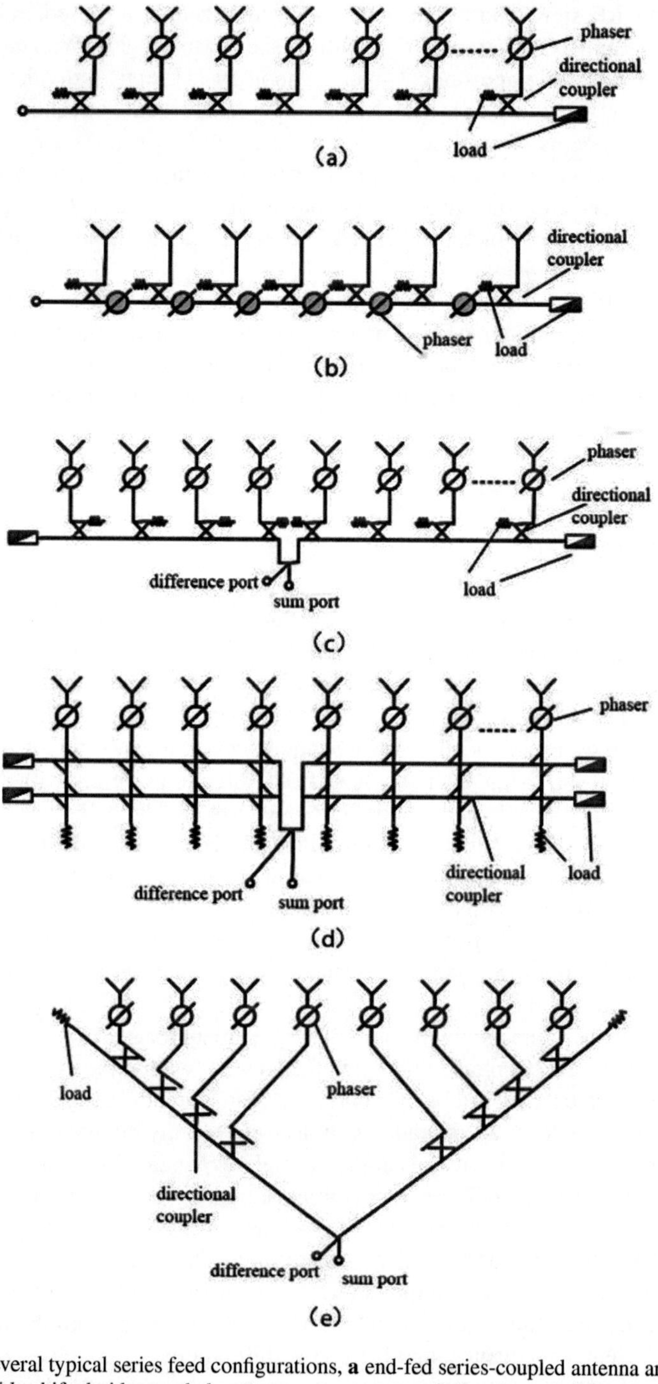

Fig. 5.1 Several typical series feed configurations, **a** end-fed series-coupled antenna array, **b** end-fed series side-shifted side-coupled antenna array, **c** center-fed series-coupled antenna array, **d** center-fed series hybrid network coupled antenna array, **e** center-fed series-coupled antenna array with equal-length path

The feed configuration shown in Fig. 5.1a has lower requirements on the power capacity of the phase shifter, and for a given phase shifter loss, its system loss is also lower. However, the power capacity of the phase shifter with the feed configuration shown in Fig. 5.1b is higher, and the total system loss is higher than that shown in Fig. 5.1a.

Figure 5.1c, d are center-fed array configurations with basically the same bandwidth as parallel-fed network configurations. By feeding through the center of the array, sum-difference beamforming can be achieved. However, the feed array shown in Fig. 5.1c cannot simultaneously achieve optimal sum-difference lobes because this feed configuration has contradictory requirements for the optimal amplitude distribution of the sum-difference beam and cannot be satisfied simultaneously. That is to say, a reasonable sum lobe can be obtained, or a good difference lobe can be obtained. Still, it is almost impossible to get a reasonable sum and difference lobe simultaneously. Adopting the feed configuration shown in Fig. 5.1d can solve the problem that good sum and difference lobes cannot be obtained simultaneously in Fig. 5.1c at the cost of increasing the feeding complexity. The feed shown in Fig. 5.1d adopts two separable center feeders and a hybrid network. Better sum-difference lobes can be obtained simultaneously through independent control of the amplitude distribution of the two cross-feed configurations. The distribution of the two feeders in the hybrid network is required to be orthogonal, i.e., the peak of the pattern of one feeder corresponds to the null of the pattern of the other feeder.

The feed array configuration shown in Fig. 5.1e is a serial feed system with an equal path length. Since each channel path is of equal length, it has a wider frequency band. However, suppose the bandwidth characteristics are already limited by phase shifters, couplers, or phase sweeps. This feed provides slight bandwidth improvement but it adds the size, mass, and cost. Thus, it is often not desirable in the application.

The advantage of the series feed system is that the feed configuration is simple, easy to assemble and integrate, and the cost is low, the disadvantage is that the bandwidth is narrow, and the beam direction of the traveling wave series feed array antenna changes with frequency. Due to the different path lengths, the amplitude and phase of each channel are unbalanced. If there are many nodes in series, a large port VSWR may appear at some frequencies in the band due to periodic reflections.

A typical example of using a series feed configuration is the frequency scanning array antenna [2], as shown in Fig. 5.2, it is a 1D frequency scanning array antenna with a serpentine slow-wave configuration.

In Fig. 5.2, each antenna radiating element couples the RF signal from the serpentine line through a coupler. The feed between two adjacent elements of the array has the same phase value. The excitation phase of the antenna radiating elements delays

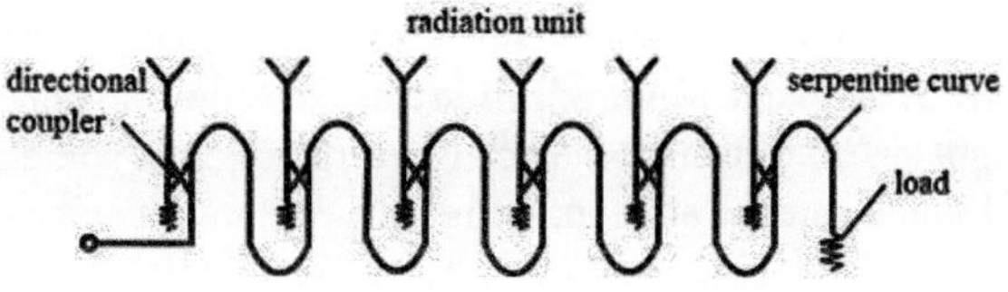

Fig. 5.2 Serpentine one-dimensional frequency scanning array

behind one by one, so it can be considered that each antenna radiating element is sequentially excited by an equivalent slow-wave transmission line. This is also why this feed configuration is often referred to as a slow-wave transmission line structure.

In order to avoid grating lobes in the design of the slow-wave line of the 1D frequency scanning array antenna, the element spacing d should meet the following conditions, namely

$$d \leq \frac{\lambda_1}{1+\sin|\theta_1|} \tag{5.1}$$

where λ_1 is the working wavelength corresponding to the frequency f_1 and θ_1 is the beam direction corresponding to the frequency f_1. Equation (5.1) shows that when the signal frequency changes from f_0 to f_1 and the antenna beam scans from θ_0 to θ_1, the element spacing limitation requirement of the array does not appear in grating lobes.

The relationship between the beam pointing angle θ of a 1D frequency scanning array antenna and the signal frequency or wavelength λ is

$$\theta = \arcsin\left[\frac{\lambda}{\lambda_g}\frac{l}{d} - \frac{\lambda M}{d}\right] \tag{5.2}$$

where λ_g is the guide wavelength in the slow wave line, l is the length of the slow wave line between adjacent elements, and M is a positive integer. Based on the array scanning signal bandwidth, feed loss, and antenna transient effects, in practical applications, M is usually 4 or 5. The slow-wave line structure is widely used in frequency scanning array antennas because of its simple configuration, high power capacity, low loss, and easy manufacturing and engineering implementation.

5.2.2 *Parallel Feed*

The parallel feed structure is more complicated than the series feed configuration, but it can significantly reduce the frequency dependence of the series feed. Therefore, if each antenna radiating element channel uses a delay line instead of a phase shifter in the broadband range, the beam pointing is independent of the frequency when the beam scans.

Figure 5.3 shows two parallel feed array configurations, including phase shifters, power division networks, and sub-array-level delay lines. Figure 5.3a shows the feed configuration of the reactive branch. This kind of feed network configuration is relatively simple. Still, due to the unmatched branch configuration, it cannot absorb the electromagnetic wave reflection caused by the imbalance of the antenna element. This unbalanced reflection may cause re-radiation of some antenna elements, resulting in elevated antenna sidelobes. Figure 5.3b shows the feed configuration of the matched

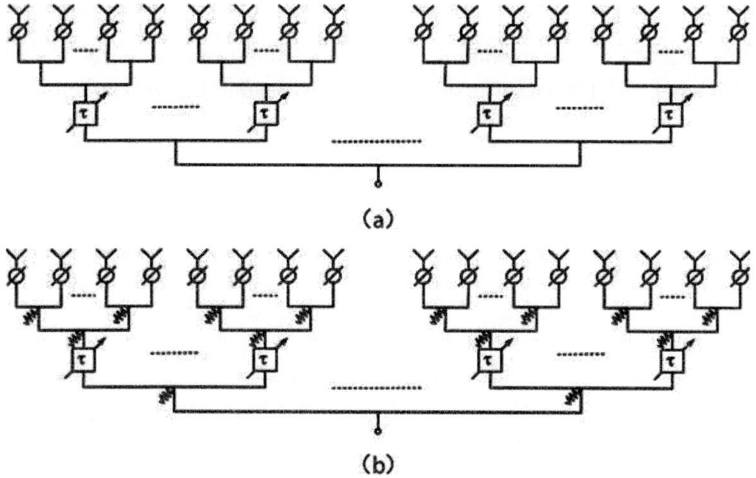

Fig. 5.3 Parallel feed array configuration, **a** reactive branch feed, **b** matching branch feed

branch, which is usually a combination of some matched hybrid networks. The non-in-phase components caused by mismatched reflections of the antenna array and other unbalanced reflections are absorbed by the terminal load, and the in-phase and balanced components return to the input end. As a result, the reflected signal from the aperture will not be radiated again. Therefore, this antenna array with the parallel feed structure can obtain good radiation characteristics. In order to break the periodicity and reduce the maximum quantization sidelobe, a small amount of additional fixed phase shift can be added on the individual paths and compensated by the corresponding adjustment of the phase shifter.

Suppose the phase shifter and delay line in Fig. 5.3 are replaced by T/R modules and delay amplification components. In that case, the feeding configuration will suit the phased array antenna system. The feed configuration includes circuits such as a power divider, T/R modules, delay lines, etc. This is the commonly used feeding configuration for most active array antenna systems. For example, for the shared-transceiver array antenna, two working modes of receiving and transmitting can be realized using the same set of parallel feed structures through the transceiver switch in the T/R module and the delay amplifying component.

The parallel feed structure often divides the antenna array into multiple sub-arrays, and the sum-difference beam is formed by combining sub-arrays in series or splicing.

The advantage of the parallel-fed antenna array configuration is that it can achieve a wider frequency band, better radiation performance, and flexible beam control. Still, the disadvantage is that the system is relatively complex, and the cost is high. Parallel feed structure can realize arbitrary multi-channel power division/combination through the combination of 1:2 or 1:3 primary power divider/combiner. In addition, it is easier to realize the amplitude and phase control of every branch. Therefore, compared with the series feed configuration, the parallel feed structure is more widely used in active array antennas, and sometimes a combination of series and parallel is used [3].

5.2.3 Space Feed

Compared with forced feed, space feed is through the space radiation of the feed source to form a feeding system for the required aperture irradiation at the array feeding port. This feed system can also form the required signal amplitude and phase distribution on the aperture of the antenna array. Space feed is also sometimes called optical feed. the spatial distribution of RF signals of the space-feeding array is analyzed by optical methods.

The advantage of the space feed configuration is that it is simpler than forced feed and does not require complex power dividing or combining. The disadvantage is that it is difficult to control the amplitude taper accurately, the geometric space necessary is large, the spatial signal leakage or loss is large, and there is also potential external signal interference.

There are two basic forms of space feed: transmission and reflection.

(1) **Transmission-type space feed array**

The transmission-type space feed array is also called the lens-type space feed array, and its basic configuration is shown in Fig. 5.4.

The lens-type space feed array comprises a collection front, a phase shifter, and a radiation front. The collection front is also called the inner front, composed of many antenna elements, which are also called collection elements, and they can be arranged on a plane or a curved surface. When the lens array is transmitting, the

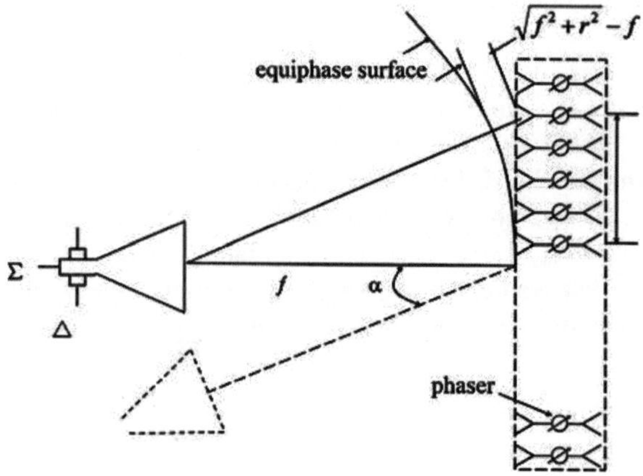

Fig. 5.4 Lens-type space feed array configuration

transmitting signal is irradiated by the illuminating antenna (primary feed) to the collecting antenna element on the inner surface of the array. After receiving the irradiation signal by the collecting antenna element, it is transmitted to the radiation front element (radiating element) through a phase shifter and then radiated to space. For some active array antennas, the signal passing through the phase shifter sometimes needs to be amplified by a power amplifier and then sent to the radiation element.

When the lens array is receiving, the radiation front receives echo signals from space targets. These signals are sent to the phase shifter and then transmitted to the receiving antenna (primary feed) by the collection front antenna element. Similarly, for some active array antennas, the echo signal received by the radiation element may be amplified by low noise before being sent to the phase shifter. That is to say, T/R modules often replace active array antenna applications, the phase shifter in the lens array to achieve flexible control of the amplitude and phase of the transmitting and receiving signals.

The optical feed system (primary feed) provides proper aperture illumination for the lens input, and through the phase (or amplitude) weighting and beam scanning drive management of the phase shifter (or T/R module) in the lens, the lens array antenna can achieve the specific shaped-beam and beam scanning. The radiating elements on both lens sides need to be matched for optimum array performance.

The collection front of the lens is illuminated through the primary feed at the focal point of the lens. The irradiation of the primary feed needs to be carefully designed to obtain the best aperture amplitude distribution with a slight leakage energy loss. The performance of the primary feed significantly influences the performance of the lens array. If it is necessary to realize the sum and difference beam, the structure of the primary feed is usually more complicated.

In the design, special attention should be paid to the matching between the primary feed source and the collection array. The input standing wave of the antenna element of the collection array should be reduced as much as possible to reduce the space feed loss. At the same time, attention should also be paid to the input standing wave of the phase shifter and the radiating element and the mutual coupling effect between the radiating and collecting array antenna elements.

If the space feed array needs to have two functions of transmitting and receiving, two separate feed sources for transmitting and receiving can be used. There is an angular interval α between the transmitting and the receiving feed, as shown by the dotted line in Fig. 5.4. At this time, in the two modes of transmitting and receiving, the phase shifter needs to be independently controlled and adjusted so that the beam points in the same direction in the two modes. The space feed array provides flexibility in optimizing the transmitting aperture distribution, maximizing the radiation power reaching the target, optimizing the receiving sum and difference lobes, and realizing lowside lobes level.

Because the change in feed position corresponds to the time delay required for beam scanning, additional feeds can be added to provide the delay compensation required for beam scanning for a corresponding increase in bandwidth.

Due to the spherical phase front formed by the primary feed, the array phase must be corrected. The required phase calibration is

$$\Delta\varphi = \frac{2\pi}{\lambda}\left(\sqrt{f^2 + r^2} - f\right) = \frac{\pi}{\lambda}\frac{r^2}{f}\left[1 - \frac{1}{4}\left(\frac{r}{f}\right)^2 + \cdots\right] \tag{5.3}$$

When the focal length is large enough, the spherical phase wavefront can be approximated as two orthogonal cylindrical phase wavefronts, which allows the phase difference to be corrected only by row and column control commands. Furthermore, the maximum phase quantization sidelobe level of the array can be reduced by correcting the spherical phase wavefront with the phase shifter.

In the lens-type space feed array, multiple primary feed sources can be used to realize multi-beam. In contrast, a single pulse and the differential beam can be recognized by a similar magic T circuit and multimode horn feed. In addition, multiple beams can be obtained using various feeds, such as a multi-beam Butler matrix feed instead of the primary feed to obtain multiple beams. The advantage is that the Butler matrix can precisely control the irradiation of the array aperture.

In a space feed array system, the illumination pattern of the primary feed covers the entire collection front, providing corresponding amplitude weighting for the radiation front. Therefore, in order to make full use of the energy of the primary feed source and reduce the energy leakage loss at the edge of the array, the number of antenna elements on the collection array can be appropriately increased. As a result, the number of elements in the collection front is greater than that in the radiation front, which is the so-called density-weighted space feed array, as shown in Fig. 5.5.

Near the edge of the collection front, the signals received by several collection elements can be combined and then sent to the radiation element of the emission front through a phase shifter. To sum up, when it is still difficult to obtain low sidelobe weighting of the array only by changing the illumination function of the primary feed to reduce the total number of front elements, it is an excellent choice to adopt the density-weighted space-fed array configuration.

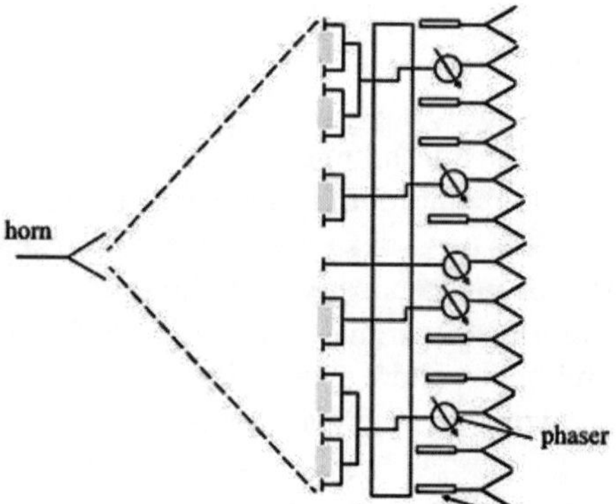

Fig. 5.5 Density-weighted lens space feed array structure

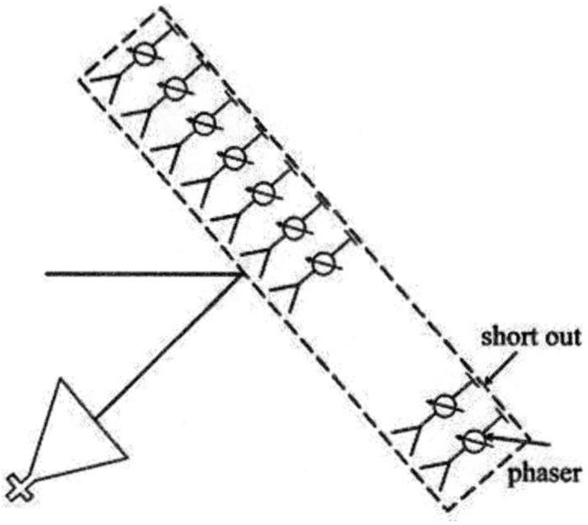

Fig. 5.6 Reflective space feed array configuration

(2) **Reflective space-fed array**

Unlike the lens-type space feed array, the collection front and the radiation front of the reflective space feed array are the same. This is because the phased array aperture acts as a reflector, and the array element acts as both a collecting element and a radiating element, as shown in Fig. 5.6.

For reflective space feed arrays, whether in the transmitting state or receiving state, the signal received by each antenna element is reflected by the short circuit and radiated out from the front after passing through the phase shifter. Since the signal has to pass through the phase shifter twice, the phase shift value is only half that of the lens feed array, but the signal attenuation is doubled.

Such reflection feed array phase shifters are all reciprocal so that the signal has a consistent and controllable phase shift after passing through the phase shifter to and from. There is ample space behind the reflector for the control circuit of the phase shifter.

Due to the feedforward mode, the primary feed has a specific shielding effect on the antenna array, which has an adverse impact on the array aperture gain and the antenna SLL. In order to avoid aperture occlusion, the primary feed can be biased appropriately, as shown in Fig. 5.6. Like the lens feed array, the reflective feed array can also realize the function of transmitting and receiving by separating the receiving and transmitting feeds. For this reason, the phases of the two transmitting and receiving states should be set and controlled separately. Multiple beams can also be obtained using multiple feed combinations.

5.2.4 Multi-beamforming Network

The multi-beamforming network is another feeding network in an array antenna to obtain simultaneous multiple beams. The multi-beam feeding network can simultaneously generate multiple beams covering a large fan-shaped space. For the same beam pointing, each beam of a multi-beam array has essentially the same gain as a single-beam array of the same size and illumination, except for possible increased RF loss in a single-beam array. In addition, each beam of a multi-beamforming network provides an independent beam port.

Multi-beamforming networks are generally divided into two types: series feed and parallel feed. There is also a Rotman lens space feed array that can also form multiple beams.

(1) **Brass multi-beamforming matrix**

The Brass matrix is a series feed multi-beamforming array configuration, as shown in Fig. 5.7. The Brass multi-beamforming array configuration forms multiple beams in only one plane. In order to generate multiple beams, several traveling wave transmission lines terminated with matching loads are coupled to the branch lines of the radiation element through directional couplers, one end of the branch line is connected to the radiation element, and the other end is connected to the matching load. The number of multi-beam is the same as the number of Brass matrix traveling wave transmission lines. The top transmission line produces lateral beams to the center of the array, the second transmission line has beams that are offset from the lateral direction, and so on for the remaining beams.

In order to work effectively in the Brass multi-beamforming array configuration, each radiation beam should meet the beam orthogonality criterion, i.e., the maximum direction of one beam coincides with the null direction of the other beams, and the intersection level between the beams is within 4 dB place. In the Brass feed array configuration, the Brass matrix is lossy. Due to the limited isolation between the traveling wave transmission lines and the limited directivity of the coupler, the feed signal of the transmission line behind the array will be coupled with the upper

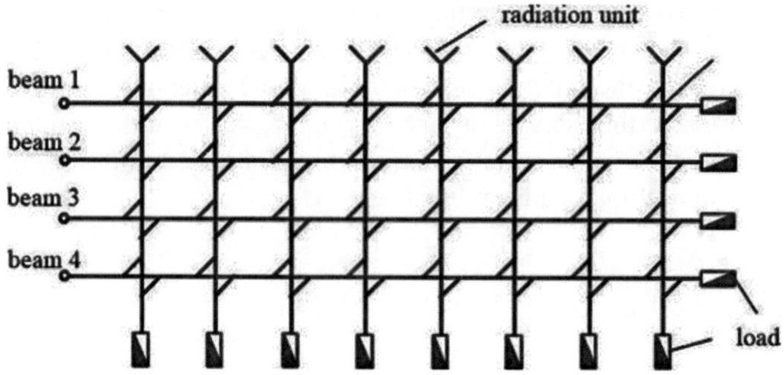

Fig. 5.7 Brass multi-beamforming array configuration

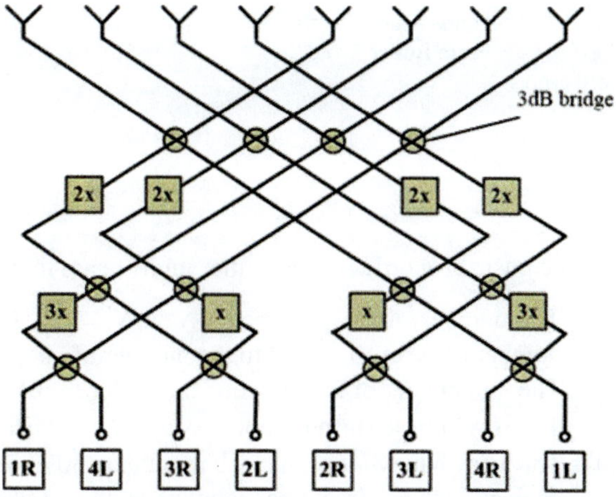

Fig. 5.8 Butler multi-beamforming array configuration

transmission line, i.e., there is cross-coupling between the beam transmission lines. Due to the cross-coupling between beam transmission lines, the inconsistency of couplers, and the existence of amplitude and phase errors between beam channels, it is challenging to obtain ideal orthogonal beams.

Generally, this cross-coupling effect is minimal when the beams corresponding to each beam transmission line are separated by a beam width or more. This is because the impact of the additional feeder does not significantly degrade the pattern produced by each beam transmission line. Thus, the sum of the signals leaked to other feeders is very close to zero, and the power reaching the terminal is minimal.

(2) **Butler multi-beamforming matrix**

The Butler matrix is a parallel feed multi-beamforming array configuration, as shown in Fig. 5.8. The basic composition of the Butler multi-beamforming network is a 3 dB bridge with a 90° phase shift and a fixed phase shifter.

In Fig. 5.8, x is a phase shift of 22.5°. The basic principle of the Butler matrix is to use a circuit to realize the fast Fourier transform and use uniform aperture illumination to form a hardware implementation of a group of orthogonal beams. Each beam obtained with a Butler multi-beamforming network has the same antenna gain as the entire array, so its beamforming network is lossless. The multiple beams obtained by Butler multi-beamforming network are strictly orthogonal to each other, which benefits the synthesis of other complex-shaped antenna beam patterns. Butler multi-beamforming arrays can be used in simultaneous transmitting and receiving modes.

Eight orthogonal beams formed by a Butler multi-beamforming array configuration are shown in Fig. 5.9.

The beams formed by the Butler matrix array intersect at a relative amplitude of $2/\pi$ (~4 dB). Because the beams are orthogonal, there is no cross-coupling loss between the beams. Parallel multi-beam configurations are less intuitive than series feed configurations. However, if a lot of beams are required, a parallel structure is ideal.

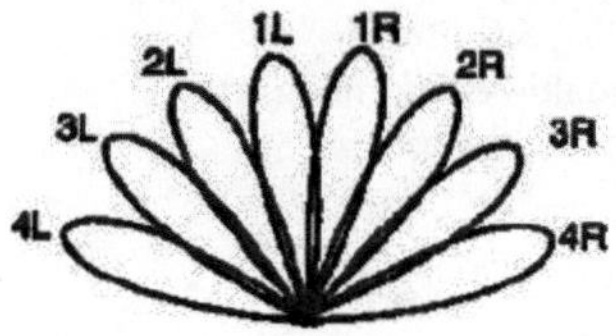

Fig. 5.9 Radiation beams formed by 8-beam Butler feed arrays

The basic properties of the Butler multi-beam matrix are as follows:

(1) The number of array elements N is $N = 2^k$, k is a positive integer, and the number of beams is equal to the number of antenna elements.
(2) The number N_c of the directional coupler (3 dB bridge) is $N_c = \frac{N}{2}\log_2 N$.
(3) The fixed phase shifter number N_φ is $N_\varphi = \frac{N}{2}(\log_2 N - 1)$.
(4) The array bandwidth can reach more than 30%, depending on the bandwidth of devices such as directional couplers and fixed phase shifters.
(5) The insertion loss mainly depends on the loss of the directional coupler and the phase shifter, and each beam shares an antenna aperture.
(6) Antenna aperture weighting: the matrix is uniformly distributed, but $\cos^n x$ weighting can be achieved.

(3) **Rotman lens**

The Rotman lens is a multi-beam forming network structure based on the principle of the equal optical path. A Rotman lens multi-beamforming network has multiple input and output ports. For example, an eight-beam Rotman lens multi-beamforming array is shown in Fig. 5.10.

A Rotman lens multi-beamforming array comprises a beam port profile, lens cavity, array port profile, RF transmission line, and radiation front. The RF signal fed from any beam port of the Rotman lens passes through the focal length path of the positive axis of the lens and then passes through the RF transmission line. The total path length is the same, it has the same delay. This is the equal optical path principle of the Rotman lens to realize multi-beam. As a result, the Rotman lens has the characteristic of forming three perfect focal points in one scanning plane.

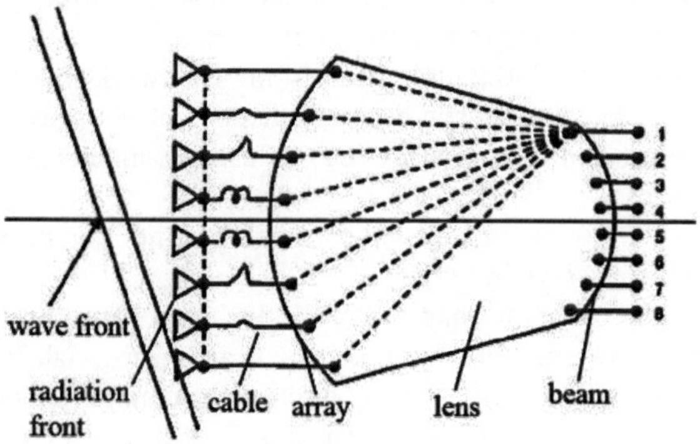

Fig. 5.10 Rotman lens multi-beamforming array

After the RF signals of different beam ports pass through the Rotman lens and the RF transmission line, different phase tapers are formed on the radiation front to form beams in different directions. A Rotman lens can achieve a multi-beam array with the desired intersection level with careful design. Figure 5.10 shows eight symmetrically distributed multi-beam Rotman lens arrays. Using digitally controlled matrix switches, the Rotman lens multi-beam forming array can realize simultaneous multi-beam and single-beam scanning functions. Rotman lenses can be used for transmitting and receiving shared antennas.

Since the Rotman lens is essentially a lens structure deduced based on optical principles, its beamforming network has the characteristics of actual delay, and its beam pointing is theoretically independent of frequency, i.e., its beam pointing does not change with frequency. Thus, Rotman lens is a broadband system. Because of this, Rotman lenses are often used in ultra-wideband microwave networks, such as broadband power dividers, multiplexers, and frequency dividers.

In addition to its ultra-broadband performance, the Rotman lens enables good wide-angle beam scanning over a range of 45°. Furthermore, the Rotman lens has the advantages of a simple circuit, compact size, and flexible design. In the multi-beam array system, the Rotman lens is used to replace the complex multi-beam network of the conventional phased array, which can effectively reduce the amount of equipment and complexity, improve the system reliability, and reduce the cost of the system.

5.3 Modular Integration Architecture

The so-called modular integrated architecture divides the antenna array into multiple antenna sub-blocks or antenna sub-arrays and integrates certain unit channels and active components into an integrated module as the most basic structural unit of the antenna array, called a scalable array module (SAM). First, antenna sub-blocks (antenna sub-arrays) are constructed by expanding the basic structural modules of SAM, and then larger sized antenna arrays are constructed by splicing antenna sub-blocks.

5.3.1 Module Architecture Classification

Based on the characteristics of the modular integrated architecture, the active array antenna can generally be constructed in two structural forms: "brick" and "tile". Typical "brick" and "tile" array architectures are usually shown in Fig. 5.11.

Both the "brick" and "tile" array configurations in Fig. 5.11 include active array antenna elements, active components (T/R modules, delay line components), RF power combining/dividing networks, control drive circuits, power supply circuits, etc. Figure 5.11a is a "brick" array configuration based on the integration of dipole

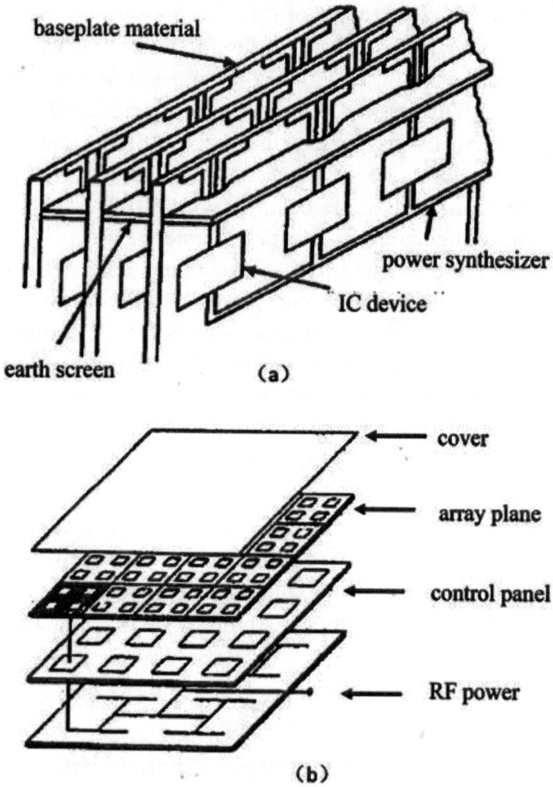

Fig. 5.11 Basic array configuration, **a** brick configuration array, **b** tile configuration array

radiating elements. Each brick can contain several columns or rows of radiating elements, power division networks, and active components. In extreme cases, bricks can be reduced to multi-pass or even single-pass components. Figure 5.11b is a "tile" array structure based on microstrip patch radiating elements. Each tile can contain one or more antenna radiating elements, active components, and excitation and control circuits.

5.3.2 *Brick SAM Module*

The brick SAM module structure is shown in Fig. 5.12. The brick-type SAM module is mainly composed of several columns (or rows) of radiation elements, active components (T/R modules, delay line components), integrated feed network, installation, thermal control, and interface circuits. The integrated feed network includes a radio frequency excitation network, RF calibration network, module power supply circuit, beam control circuit, etc.

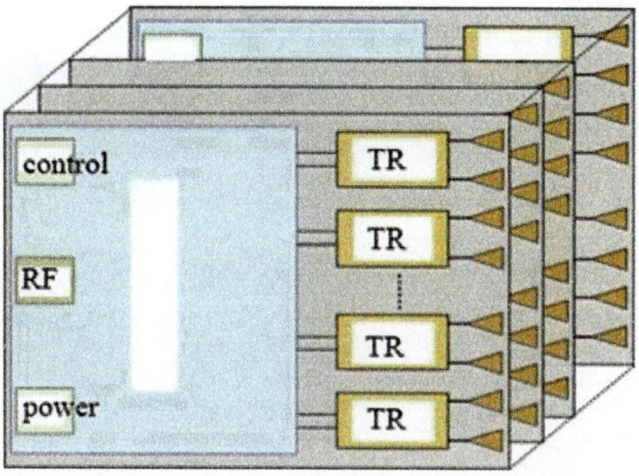

Fig. 5.12 Brick SAM architecture

The brick SAM module is characterized by vertical integration and horizontal assembly, i.e., the active/passive chips and components of each channel are first integrated into a component, and the radiating element, multi-channel component, and comprehensive feed network are vertically integrated into an integrated module. Then, multiple integrated modules are assembled horizontally and constructed as antenna sub-blocks or antenna sub-arrays.

The advantage of the brick-type SAM module is that the internal circuit interconnection is simple, it is easy to debug and maintain, and the development cost is relatively low, the disadvantage is that the integration is low, the volume is large, and the array section is large.

Depending on the conditions and requirements of the active array antenna loading platform, installation space, and supporting equipment layout, the assembly method of the brick SAM module and the antenna array frame may be different. However, there are two main ways to assemble the active antenna array based on the brick SAM module integration: front insertion type and rear insertion type.

(1) **Brick SAM front insertion architecture**

The so-called forward insertion architecture means that the brick-type expandable array module is pushed back into the front frame from the front of the front. It is interconnected with systems such as receivers and excitation sources behind the front through blind-mating interfaces (including RF control, power supply, thermal control interfaces, etc.), as shown in Fig. 5.13.

The forward insertion structure is suitable for occasions where the space behind the antenna array is limited, and it is easy to disassemble and maintain through the front of the array.

The forward insertion architecture integrates the antenna radiation element, T/R modules, integrated feed network, and various signal interfaces into an integrated brick module. Then, the brick-type module is used as the basic structural part, connected to the transceiver system at the rear of the array through a blind-mating interface. Next, a larger active array antenna sub-block or the entire antenna array is constructed through building block expansion.

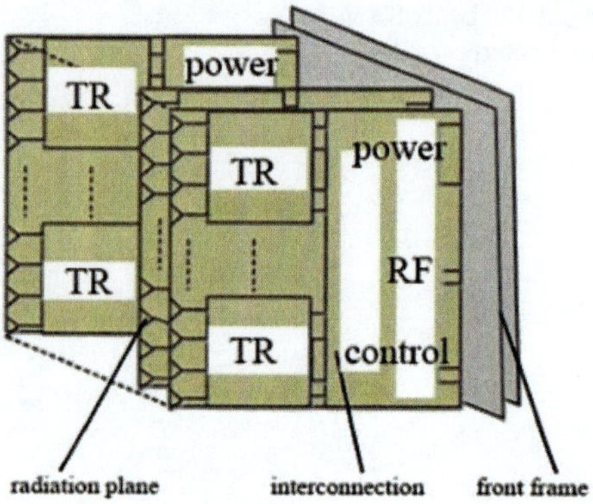

Fig. 5.13 Brick module (SAM) front insertion architecture

(2) Brick SAM backward insertion architecture. The so-called backward insertion architecture refers to the brick module being pushed into the front frame from the rear of the front. It is interconnected with the antenna radiation front through a blind-match interface. At the same time, the control, power supply, and thermal control interfaces of the brick SAM are connected to the corresponding equipment at the rear through blind matching or cable assemblies, as shown in Fig. 5.14.

The backward insertion structure is suitable for occasions where the space in front of the antenna array is limited, and it is convenient to disassemble and maintain through the rear of the array.

The backward insertion architecture integrates T/R modules, an integrated feed network, and various signal interfaces into an integrated brick SAM module. Then, the brick module is used as the basic structural part, connected to the radiation front of the array through a blind-match interface, and a larger active array antenna sub-block or the entire antenna array is constructed through building block expansion. In the backward insertion architecture, the antenna radiation element and the SAM module can be separate devices, and the radiation front and the front frame are installed separately.

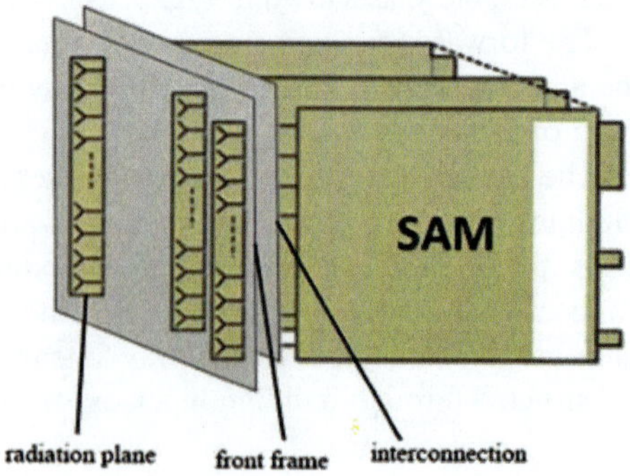

Fig. 5.14 Brick module (SAM) backward insertion architecture

5.3.3 Tile SAM Architecture

Tile scalable array module (T-SAM) comprises several antenna radiation elements, multi-channel T/R modules, delay line components, RF power calibration and calibration network, module wave control, module power supply, interface circuit, thermal control, and support structure. Furthermore, according to the performance and functional requirements of the mission system, sometimes the tile SAM module can also expand the integrated receiver RF front-end circuit, digital transceiver circuit, and even microwave photonic circuit to form a more complex mission system [4].

The T-SAM module adopts board-level integration and chip circuit stacking assembly architecture. Suppose it is layered according to the circuit. In that case, the tile SAM module can typically be divided into two layers of circuits [5], the integrated feed layer and the antenna radiation layer. Among them, the antenna radiation layer mostly adopts low-profile antenna elements such as microstrip patches and cavity-backed antennas, and the integrated feed network integrates RF, control, power supply circuits, and active modules into a multifunctional substrate circuit [6]. The typical structure of a tile SAM module is shown in Fig. 5.15.

The tile SAM module is characterized by horizontal integration and vertical assembly, active/passive chips and components with the same function of multiple channels are integrated on a single-layer circuit, and multi-layer circuits are stacked and integrated through vertical interconnection. Antenna sub-blocks or antenna sub-arrays are constructed through T-SAM module building block expansion, and larger active array antenna sub-arrays or the entire antenna array are constructed through the splicing of antenna sub-blocks.

Compared with the brick antenna structure, the chip array module adopts advanced single-chip microwave integrated circuit technology, micro-assembly technology, micro-system integration technology, etc., to carry out the layered assembly and high-density integration of the array. As a result, it has the advantages of high integration, low profile, light mass, modularization, easy expansion, and easy conformal shape. Therefore, it is very suitable for applications with limited space size, unique loading platforms, and large-aperture, multifunctional microwave imaging radar systems, such as active array antenna systems for airborne and spaceborne platforms.

Because the tile SAM module adopts high-density integration technology and includes complex circuit structures such as RF and low-frequency circuits, high current and small signal circuits, analog and digital signal circuits, etc. Therefore, the technical problems faced in the research of tile SAM modules mainly include mechanical, electrical, and thermal integration design, electromagnetic compatibility

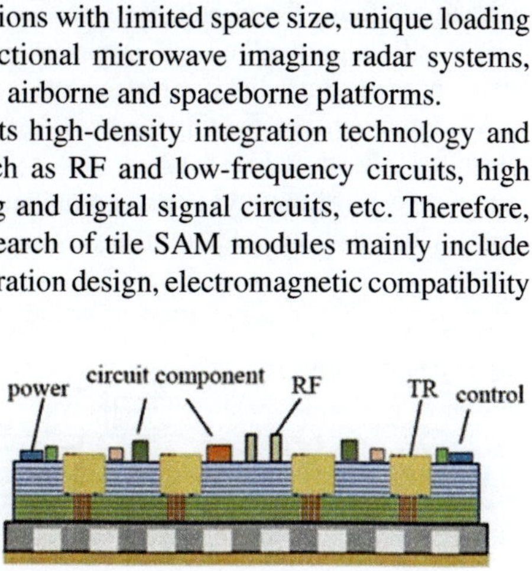

Fig. 5.15 Typical structure of T-SAM

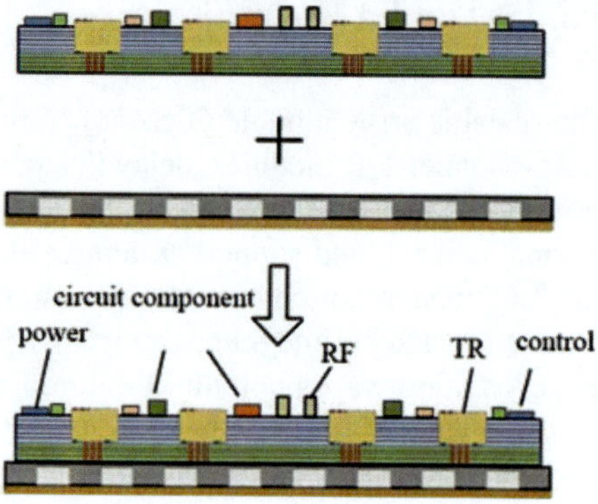

Fig. 5.16 Two-layer circuit integrated tile scalable array module

design of multi-signal circuits, chip stacking integration, vertical interconnection, microsystem packaging and integration, micro-chip T/R modules, heterogeneously integrated feed network design, etc. [7].

The tile SAM module can be stacked and integrated with a multi-layer circuit structure. For example, a tile SAM module integrated with a two-layer circuit architecture is shown in Fig. 5.16.

As shown in Fig. 5.16, the physical structure of the tile SAM module is divided into two layers: the antenna radiation layer and the multifunctional integrated circuit layer. Among them, the multifunctional integrated circuit layer integrates high/low-frequency feeding structures, and the antenna radiation layer has a low-profile radiation element structure and a structural support function. In addition, power heat dissipation devices and circuits are connected to the antenna backplane through the substrate heat sink to achieve efficient heat conduction.

In the tile SAM module shown in Fig. 5.16, the multifunctional substrate circuit (i.e., the integrated feeding network) is the core circuit of the T-SAM module, which is mainly composed of beam control circuit, power supply circuit, monitoring, protection circuit, RF circuit, the interface driving circuit, and other components. It is usually manufactured by composite multilayer PCB substrate and multilayer microwave substrate, and it can also be manufactured by multilayer LTCC substrate.

A comprehensive feeding network made of multilayer PCB and microwave substrate is shown in Fig. 5.17.

Figure 5.17 shows that the comprehensive feeding network adopts a multi-layer PCB substrate and multi-layer microwave substrate integrated laminated

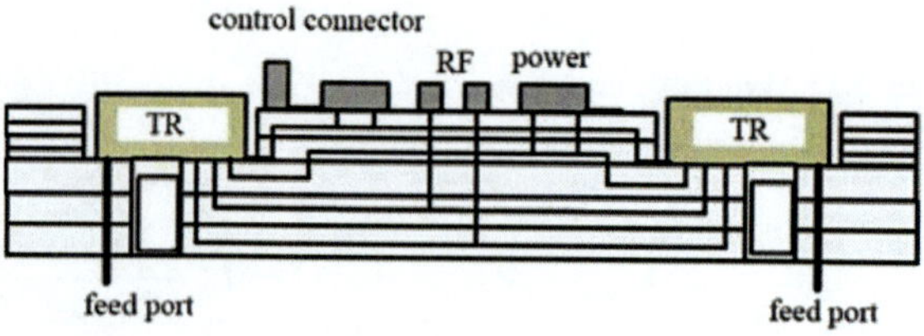

Fig. 5.17 Tile module integrated feeding network

composite process for high-density integration. As a result, the number of circuit layers can reach dozens of layers. The main technical difficulties of the integrated feeding network are the electromagnetic compatibility of multiple circuits, multi-substrate heterogeneous composite, high-density wiring, large-area metalized holes, microstrip embedded resistance, metalized vias and buried holes, fine interlayer vertical interconnection, etc.

The above-mentioned tile SAM module is stacked and assembled with a 2–3 layer circuit structure. According to the task requirements of the antenna system, manufacturing and assembly process capabilities, assembly and maintenance requirements, etc., the division of circuit layers of the tile SAM module is not unique [8].

During the design process of the tiled SAM module, several issues need to be paid special attention to

(1) **Antenna element**

The selection of the antenna element should focus on the adaptability of the element feed mode and the connection between the feed structure and the T/R module or the integrated feed structure. In addition, if there are special requirements for the thickness of the tile SAM module, a radiation element with a low profile should be selected.

(2) **T/R module**

The interconnection mode and connection interface of T/R modules, antenna element, and integrated feed network should be given priority, followed by the encapsulation of T/R modules and their layout in the SAM module. According to needs, in order to facilitate debugging and replacement, the T/R assembly can be arranged at the rear of the SAM module, i.e., on the surface of the integrated feed structure or between the antenna radiation plate and the integrated feed structure. Sometimes, T/R modules do not necessarily have to be modularly packaged. For example, the T/R transceiver channels or main functional circuits can be packaged separately, or the chip circuit can be directly used for distributed layout and integration with other circuits.

(3) **Integrated feeding network**

The first is material selection. According to the requirements of the module working frequency band, transceiver link performance, size, weight, and environmental adaptability, the substrate material of the integrated feed network can be chosen from a variety of materials such as LTCC substrate, high-temperature co-fired ceramic (HTCC) substrate, PCB substrate or microwave substrate. A single substrate material can be used, or multiple material substrates can be used for composite manufacturing. In addition, electromagnetic compatibility, heat conduction and thermal matching, manufacturing yield, etc., are also factors that need to be considered in the design of the integrated feed network.

5.4 RF Link Signal Analysis

The purpose of active array antenna RF link signal analysis is to analyze and evaluate the signal level, link gain, system dynamics, noise figure, etc., of the active array antenna transceiver link through the construction of the active array RF transceiver link model, to determine the performance of the RF link in various working modes of the active array antenna.

According to different antenna working modes, active array antenna link models can be divided into five types [9]: transmitting signal model, receiving calibration and calibration signal model and transmitting calibration and calibration signal model, receiving noise figure analysis model, and receiving system dynamic signal model.

5.4.1 RF Link Model

Active array antennas are mostly serial-parallel combination multi-channel transceiver systems. In order to analyze the signals of various modes of the active array transceiver link, the multi-channel RF transceiver system needs to be equivalent to a serial link model and then use the serial link signal model to analyze and calculate the link signal characteristics.

A typical active array antenna system is shown in Fig. 5.18.

As shown in Fig. 5.18, the active array antenna system mainly includes a radiation element, a T/R module, a primary network (sub-array-level RF power division network), a delay amplification component, and a secondary network (array RF power division network) level 5 circuit. The main port of the secondary network is connected with the receiver and the excitation source (receiving front-end). Therefore, in order to analyze the RF signal link, the active array antenna system can be equivalent to a five-level serial link model, as shown in Fig. 5.19.

In Fig. 5.19, the delay amplification component and the T/R module are the core devices of the active array antenna. The main components of T/R modules

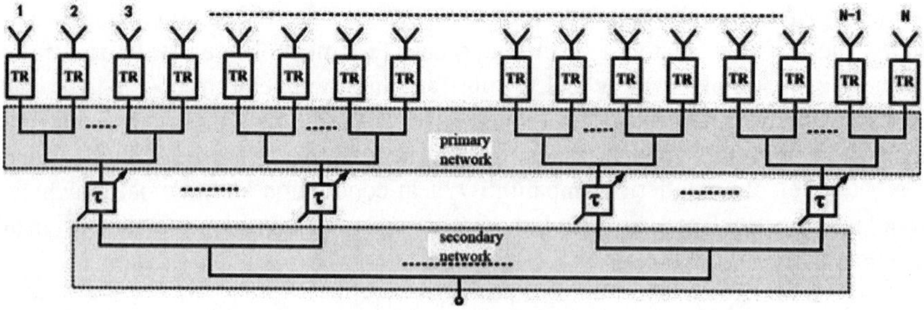

Fig. 5.18 Typical composition diagram of active array antenna system

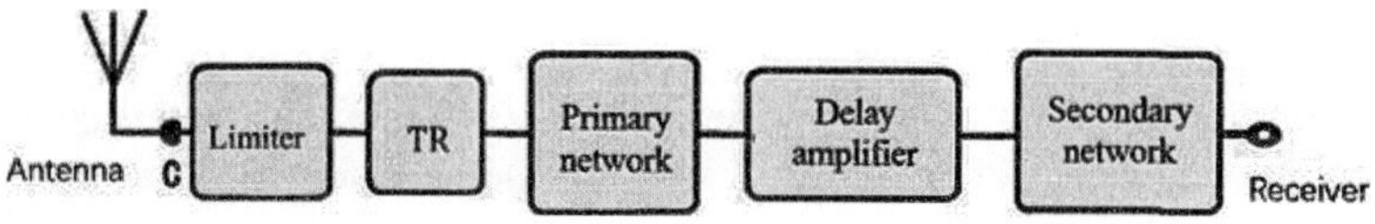

Fig. 5.19 Equivalent model of serial link in active array antenna system

include transceiver switches, power amplifier links, receiver limiters, low noise amplifier circuits, common phase shifters, attenuators, drive amplifiers, and switches for transceivers.

For link signal analysis, in the equivalent series link model, it is necessary to determine the gain and noise figure equivalents of the various stages of the circuit. For the active array module, it is also necessary to specify the excitation level of the power amplifier circuit, the saturation amplification power level, and the signal level of the 1 dB compression point of the received signal.

According to different working modes, the equivalent parameters of related circuits in the RF link are also different.

(1) **The equivalence of the active array antenna transceiver link**

The purpose of the active array antenna transceiver link equivalent is to calculate the antenna system receiving gain, noise figure, and receiving dynamics through equivalent link signal analysis. At the same time, calculate the system gain in the transmitting mode, the active module excitation signal level, etc. The multi-channel active module (T/R module and delay amplifier component shown in Fig. 5.18) needs to be equivalent to a single-channel active module. The multi-channel power divider (the first-level and second-level networks shown in Fig. 5.18) is equivalent to a single-way two-port network.

For the equivalence of two active modules (T/R module and delay amplifier component), since most of their transmitting states work in saturation amplification mode, the gain, output power, and excitation signal level of each channel are almost the same. Therefore, in the transmitting state, the multi-channel active module (T/R module and delay amplifier component) can be equivalent to a single-channel active module with a two-port network.

In the receiving mode, when the system has amplitude and phase weighting, as a reasonable simplification, the weights of the multi-channel active module for array amplitude weighting can be classified into the first-level network and the second-level network for equivalent. In this way, in the receiving state, it can be approximately considered that the receiving parameters (gain and noise figure) of the two active modules of the T/R module and the delay amplifier component of each channel are the same, so the parameters of a single active module can be equivalent to a single-channel active module for two-port networks.

For the equivalent of the primary network and the secondary network, since these two networks are passive power division networks, they can be equivalent to a single-channel two-port network with equal gain (or loss).

For the receiving noise figure analysis model, corresponding to the equivalent gain G_e of the one-way two-port network in Fig. 5.19, the first- and second-level networks are equivalent to the power loss L_r of the network. It can be defined as the ratio of the output power S_{out} of the corresponding multi-channel power division network, including the weighting coefficient to the total input power, that is

$$G_e = L_r = \frac{S_{\text{out}}}{\left|\sum_i \sqrt{S_i} a_i e^{j\theta_i}\right|^2} \tag{5.4}$$

where S_i is the power level received by each network port, a_i and θ_i are the amplitude and phase weighting coefficients corresponding to each channel, including the weighting of active components and the network itself. For equal power division networks, if the array is uniformly weighted, the equivalent gain of the network is equal to the insertion loss of the network.

For the dynamic analysis model of the receiving system, the equivalent gain G_e of the first-level network and the second-level network is equal to the network synthesis gain and then deducts the network synthesis loss, that is,

$$G_e = 10\lg(N) - L_r(\text{dB}) \tag{5.5}$$

where N is the number of input ports of the network and L_r is the decibel number of the combined loss.

In the transmitting state, the equivalent gain G_e corresponding to the first-level network and the second-level network in Fig. 5.19 can be defined as the ratio of the output power sum of each output port of the corresponding multi-channel power division network to the input power S_{in}, that is,

$$G_e = \frac{\sum_i S_{\text{out},i}}{S_{\text{in}}} \tag{5.6}$$

where $S_{\text{out},i}$ is the output power level of each output port of the network, including the network-weighted calibration. If the network is equal in the power division, the total output power is equal to NP_0, where N is the number of network output ports and P_0 is the output power of a single channel.

Of course, the more practical method in engineering is to analyze and calculate the power division level of each port of the primary and secondary networks one by one and take the output power of each network to meet the driving requirements of active components without burning the active components as the design principle.

(2) **Active array antenna calibration is equivalent to calibrating the transceiver link**

Most active arrays often adopt a one-by-one internal calibration method, i.e., through the on–off control of the transceiver channel of the T/R module and the channel-by-channel coupling calibration network, the channel-by-channel calibration of the transmission and reception of the multi-channel active array is realized. In this state, for the equivalence of two active modules (T/R module and delay amplifier component), use the parameters of the specific calibration channel active module to be equivalent to a two-port single-channel active module, For the equivalence of the first-level network and the second-level network, it can also be equal to a single-way two-port network with an equivalent gain of G_e. However, the equivalent gain at this time is only the ratio of the single output power of the power division network to the input power, and its equal gain includes not only the insertion loss but also the power division loss.

5.4.2 RF Link Signal Analysis

According to the above equivalent model, the RF signals of the active array antenna system under different working modes can be analyzed and evaluated. Taking the array antenna system architecture shown in Fig. 5.18 as an example, the analysis and calculation of RF signals in each mode are introduced.

In Fig. 5.18, the array antenna includes a radiating element, a T/R module, a first-level network, a delay amplifier component, and a five-level circuit of a second-level network, which is similar to most active array antenna system architectures, so it has a good representative. For the convenience of quantitative discussion, it is assumed that the number of array elements is 256. The division and combination of signals are realized by two-level RF power division networks of 1:32 and 1:8, respectively. The data given in the following analysis link model is only for the convenience of quantitative analysis and is for reference only.

(1) **Noise figure analysis**

The noise figure calculation link of a 256-element five-level active array antenna with Fig. 5.18 as the reference architecture is shown in Fig. 5.20.

In Fig. 5.20, the effect of the feed loss between the T/R assembly and the antenna element is considered. The primary and secondary networks are 1:32 and 1:8 RF power division networks, respectively, and their equivalent gains are calculated according to the method in the previous section. Here, only the noise figure calculation in the case of uniform weighting is considered. If there is amplitude weighting, the weighted loss can be equivalent to the equivalent gain of the first-level network or the second-level network.

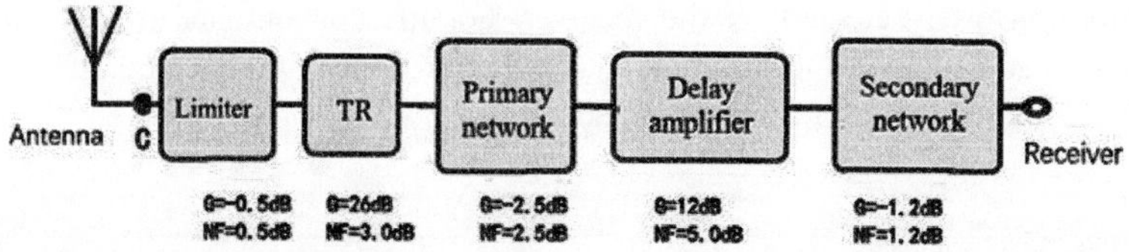

Fig. 5.20 Noise figure analysis model

In Fig. 5.20, with C as the reference point, the system noise figure model looking from C to the receiver is a series model, and the noise figure NF of a series model is

$$\mathrm{NF} = \mathrm{NF}_1 + \frac{(\mathrm{NF}_2 - 1)}{G_1} + \frac{(\mathrm{NF}_3 - 1)}{G_1 G_2} + \cdots + \frac{(\mathrm{NF}_n - 1)}{\prod_{i=1}^{n-1} G_i} \tag{5.7}$$

The equivalent model in Fig. 5.20 and the equivalent parameters at all levels can be analyzed and calculated by Eq. (5.7). The receiving noise figure of the array antenna is about 3.53 dB.

(2) **Receive system gain and dynamic analysis**

The dynamic range of an active array antenna receiving system refers to the maximum and minimum signal level range that the system can receive. For active array antennas, there are usually two concepts of dynamic range. One is the overall dynamics of the antenna system, which is the range of maximum and minimum signal levels that the antenna system can receive. It may be more appropriate to refer to this overall dynamic as the acceptable RF signal level input range of the antenna system. The second is the instantaneous dynamics, which usually refers to the instantaneous dynamics of the receiver, i.e., the maximum and minimum signal level ranges that the receiver can receive in a transient state.

The minimum acceptable signal level in the overall dynamics of the antenna system is the sensitivity $P_{s\min}$ of the system, that is,

$$P_{s\min}(\mathrm{dBm}) = -114 + 10\lg(\mathrm{BW}) + \mathrm{NF} - 10\lg(N) \tag{5.8}$$

where BW is the IF bandwidth of the receiver or the instantaneous signal bandwidth of the antenna system in MHz, NF is the system noise figure, and N is the number of antenna elements. For example, an array antenna with instantaneous signal bandwidth BW = 1600 MHz, noise figure of 3.53, and 256 elements can be analyzed and calculated by Eq. (5.8), and the system sensitivity is about −102.5 dBm.

The maximum receivable signal level in the overall dynamics of the antenna system is the maximum receivable signal level of the active channel of the antenna. The active module (such as the T/R module) receives the 1 dB compression point input signal level, which is the maximum acceptable signal level for the array antenna active module to work linearly.

The instantaneous dynamics of the active array antenna, that is, the instantaneous dynamics of the receiver is mainly related to the effective number of bits of the analog-to-digital converter selected by the receiver. For the convenience of quantitative analysis, as an example, the instantaneous dynamic approximate calculation of the analog-to-digital converter receiver is used

$$D_r(\mathrm{dB}) = 6 \times N \tag{5.9}$$

where N is the adequate number of bits of the analog-to-digital converter, for example, in a 12-bit analog-to-digital converter, whose noise occupies 4 bits and

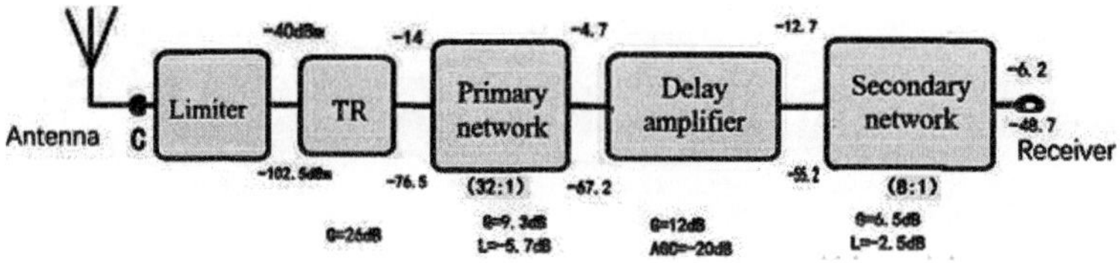

Fig. 5.21 Receiver system dynamic analysis model

effective bits is 8 bits, the receiving dynamic is about 48 dB. The maximum signal level the receiver can accept is approximately the full-scale input power of the analog-to-digital converter. Considering the multi-tone intermodulation and spurious free dynamic range (SFDR) parameters, the maximum acceptable signal power should retain a certain margin.

The gain and dynamic calculation link of a 256-element 5-level active array antenna receiving system with Fig. 5.18 as a reference architecture are shown in Fig. 5.21.

In Fig. 5.21, the primary and secondary networks are 1:32 and 1:8 RF power division networks, respectively. Considering the amplitude-weighted loss, its equivalent gain is calculated according to the previous section method, which equals 9.3 dB and 6.5 dB, respectively. It can be seen from Fig. 5.21 that the input range of the signal level that the system can receive, that is, the total dynamic of the system, is about 62.5 dB, including the synthesis gain of the combination network, the equivalent gain of the receiving link is about 33.8 dB. The total dynamics of the system are much larger than the instantaneous dynamics of the receiver.

In order to ensure the system dynamic range requirements so that the input signal does not enter the nonlinear state under the maximum input dynamic range, a dynamic compression circuit needs to be added to the system link. Dynamic compression methods commonly include Automatic Gain Control (AGC) and Sensitivity Time Control (STC). Here, a 20 dB AGC is added to the delay amplifier component to adjust the overall dynamics of the system so that the dynamic range entering the receiver is about 42.5 dB, which meets the instantaneous dynamic requirements of the receiver.

(3) **Transmit signal analysis**

A 256-element 5-level active array antenna transmission signal analysis link is shown in Fig. 5.22. The purpose of transmitting signal analysis is to determine the output power of the excitation source and the excitation power level of the emission state of each active module through the equivalent link model.

In the transmitting state of the active array antenna, in order to obtain high system efficiency, each active module (T/R module, etc.) often works in a saturated amplification state, i.e., a uniformly weighted mode. Therefore, at this time, each power division network in the antenna system is equally divided in power. As shown in Fig. 5.22, both the primary and secondary networks have equal power division, and the equivalent gain or loss of the network is included in the division loss of the

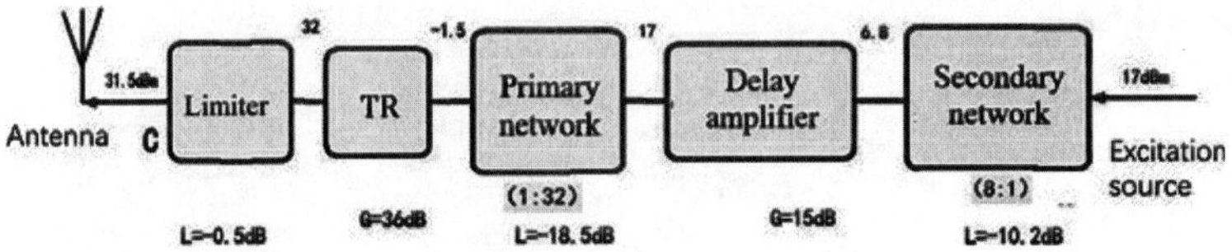

Fig. 5.22 Transmit signal analysis model

network. Since both the T/R module and the delay amplifier component work in a saturated amplification state, according to the gain design of the two components, the excitation signal level of the T/R module is required to be -1.5 ± 1 dBm, and the excitation signal level of the delay amplifier component is 6.5 ± 1 dBm. Therefore, the excitation source needs to provide an excitation signal level of about 17 dBm. According to the signal division of the transmission signal chain shown in Fig. 5.22, after considering the feeding loss, the transmission power fed to the antenna element is about 31.5 dBm (about 1.4 W).

The active array antenna calibration and calibration signal analysis is similar to the transmitted signal analysis. For most active array systems that use channel-by-channel calibration, when the gains or losses of the primary network and the secondary network are equivalent, the power division loss of the power division network should be included. By correcting the link model, the excitation signal level of the calibration system and the signal strength of each node of the link are determined. It is ensured that when the active array antenna is corrected one by one, the calibration receiver and each node in the middle of the link have a sufficiently high signal-to-noise ratio to obtain high calibration accuracy. For the specific link analysis of the calibration signal, please refer to the analysis model of the transmitted signal.

5.5 Miniaturized Transceiver Components

Miniaturized transceiver components (T/R modules) refer to T/R modules designed and manufactured based on MMIC multifunctional chips and adopt microsystem integration technology.

The T/R module is the basis of the active array antenna and the core equipment of the active array antenna. The functions of beam scanning, beam forming, and transmitting and receiving switching of the active array antenna are all realized by T/R modules.

For microwave imaging radar, with the improvement of imaging resolution, the aperture of the antenna increases continuously, and the RF bandwidth also increases constantly. As a result, active array antennas for high-resolution microwave imaging radars often have tens of thousands or even hundreds of thousands of active channels and correspondingly contain many T/R modules. Therefore, the weight, power consumption, and cost of the T/R modules greatly influence the antenna system. In addition, as the operating frequency increases and the frequency bandwidth increases, the element spacing of the array becomes smaller. This requires T/R modules to have a higher degree of integration, so the requirements for the miniaturization and miniaturization of T/R modules are also getting higher and higher.

With the development of microelectronics technology, MMIC technology, microsystem integration technology, and advanced packaging technology, T/R module technology, especially the T/R module technology based on MMIC chip integration, has made significant progress. Mainly manifested in (1) chip integration is getting higher and higher, from a single-function circuit chip to a multifunction circuit single chip, and even the chip of the entire T/R module, (2) continuous breakthroughs in various assembly and packaging technologies, such as LTCC, HTCC, aluminum nitride (AlN), polyimide film and other multi-layer substrate circuit technologies, heterogeneous circuits interconnection and welding technology, microsystem integration technology, such as 3D multi-chip module (3D-MCM), system-in-package (SIP), through-silicon via (TSV), micro-electromechanical system (MEMS), micro-coaxial, integrated passive device (IPD), etc., have been widely used in the design of high-density T/R modules. T/R modules have a variety of functions. In addition to the transceiver function, the current T/R modules can also realize functions such as polarization change, clutter suppression, monitoring and calibration, and digitization.

5.5.1 *Basic Composition*

Although there may be some differences in the composition and function of the T/R modules of different systems according to the different requirements of the application system, the basic composition is roughly similar. For example, the basic composition block diagram of a micro-T/R module based on MMIC multifunctional core chip technology and microsystem integration technology is shown in Fig. 5.23. It is mainly composed of a multi-function chip, a transceiver switch, a limiter, a low-noise amplifier (LNA), a high-power amplifier (HPA), and a power supply and control circuit. Among them, the multi-function chip contains multiple functional circuits, such as a digitally controlled phase shifter, digitally controlled attenuator, amplifier, and control switch.

For the multifunctional active array antenna system, the requirements for the function, performance, and environmental adaptability of the T/R module are relatively high in practical application and engineering, and its circuit composition is relatively complicated. In order to obtain better performance and relatively high stability and reliability, T/R modules often add some other auxiliary circuits to their basic constituent circuits, such as power drive amplifier circuits, power amplifier stage isolation, amplitude-phase equalization circuits, temperature compensation circuit, variable polarization circuit, clutter suppression circuit, etc. In order to improve reliability, it is also possible to add a high-power limiting circuit, power conversion circuit, and monitoring protection circuit. It can realize protection circuits such as

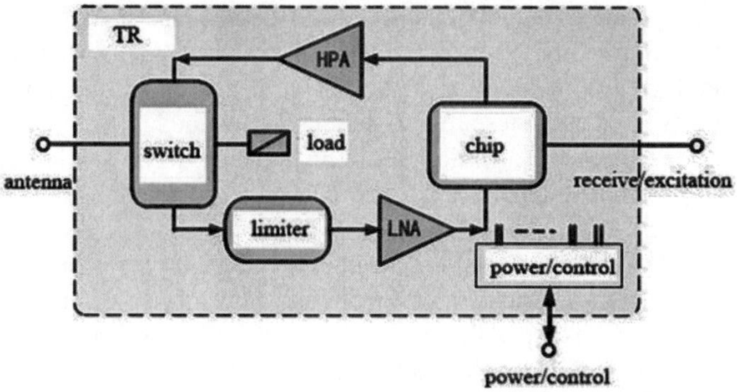

Fig. 5.23 Basic composition block diagram of T/R modules based on multifunctional chips

over-temperature and over-current protection, over-pulse width protection, power signal abnormality protection, power-on self-reset, and data parity check.

For micro-T/R modules, to realize the miniaturization and lightweight of the components, the composition and functions of the components should be simplified appropriately, to avoid the complexity of T/R function and composition, which makes the circuit design and system integration of T/R modules too difficult, resulting in poor reliability and working stability of micro-T/R modules.

5.5.2 *Principle*

The basic function of the T/R module is to realize the transmission and reception of RF signals and the amplitude and phase control of transmitting and receiving signals. The basic principle is that in the transmission mode, according to the control command of the beam control system, power on the relevant components of the transmission link (switches, phase shifters, delay lines, drive amplifiers, high-power amplification) according to a certain sequence, and turn on the transmission link. The RF excitation signal is sent to the transmit channel of the component and is amplified by phase shifting, preamplification, time delay, and high-power amplifier. Then, according to the preset phase and amplitude, it is fed to the antenna element by the transceiver switch (or circulator). In the receiving mode, according to the control command of the beam control system, power on the receiving link-related components (switches, low noise amplifiers, phase shifters, delay lines, attenuators, etc.) according to a certain sequence and connect the receiving link. The RF echo signal received from the antenna element is sent by the transceiver switch (circulator) and then passes through the limiter, low noise amplifier, phase shifter, attenuator, and other circuits. Finally, according to the preset amplitude and phase weighting, it is sent to the RF front end of the receiver.

In addition to the functions of transceiver conversion, RF signal amplification, and monitoring and protection, T/R modules also have the following functions:

(1) **Amplitude and phase adjustment**

The phase shifter of the T/R module realizes the beam scanning of the active array antenna. The array antenna is controlled and managed by the beam control circuit. The RF signal of each channel passes through the digital phase shifter and delay line of the T/R module to generate the phase and delay distribution required for beam scanning and forms a specific phase cone on the beam scanning plane. By changing the phase taper of the array, the radiation beam pointing can be changed, i.e., the beam scanning of the antenna beam in 3D space can be realized. Through T/R module amplitude, phase weighting, and delay compensation, the antenna beam shape control and SLL control can be realized, and the desired antenna radiation characteristics can be obtained. In the antenna transmitting and receiving mode, through the switching of the phase wave control code of the T/R module, the arbitrary beam position switching of the transmitting and receiving beams of the antenna can be realized.

(2) **Variable polarization control**

When the active array antenna adopts two orthogonally polarized antenna elements to realize the multi-polarization function, the T/R module must have multi-polarization channel feeding and polarization conversion functions. This kind of multi-polarization feeding and polarization switching is usually realized through the dual transceiver channels and multimode switch inside the T/R module.

(3) **Monitoring and calibration**

2D active array antenna systems generally contain a large number of T/R modules. Therefore, monitoring and correcting the characteristics of the T/R modules is an important condition for the reliable operation of the radar system.

The monitoring of T/R modules includes two aspects: one is the monitoring of the working conditions of T/R modules. Including the monitoring of the current, voltage, temperature, etc., of the corresponding modules in the transceiver link of the T/R module and taking measures to protect the T/R module from overcurrent, overvoltage, overtemperature, etc. The second is to monitor the RF amplitude and phase characteristics of the transceiver channel of the T/R module. This monitoring is often combined with the active array calibration and calibration system, and the monitoring and calibration functions are implemented through the array calibration and calibration system. Most of the monitoring and calibration of array antenna active channels and T/R modules adopt internal calibration methods. In addition to the transceiver channels of the active array itself, a monitoring and calibration system needs to be formed through each channel RF signal coupling mechanism, the corresponding power division network, and the calibration receiver and excitation device [10]. Through the monitoring and calibration subsystem of the transceiver channel, the monitoring and fault diagnosis of the amplitude and phase characteristics of the receiving and transmitting channels of the active channel and the T/R module can be realized. Furthermore, through the high-precision pattern modeling and the amplitude-phase control function of the T/R module, the fault diagnosis of the channel, the amplitude-phase calibration compensation, and the performance evaluation of the antenna array beam can be realized.

5.5.3 Basic Parameters

The basic parameters that characterize the performance of T/R modules include operating frequency band, phase shift accuracy, transceiver gain, transmit output power, transceiver conversion time, transceiver isolation, amplitude and phase consistency, amplitude and phase in-band fluctuation, amplitude and phase stability, noise figure, nonlinear phase error, phase shift parasitic amplitude modulation, attenuation parasitic phase modulation, spurious out-of-band suppression, component power efficiency, power capacity, port voltage standing wave ratio (VSWR), etc. In practical applications, usually pay attention to some other basic parameters, such as transmit waveform peak drop, transmit pulse rise/fall time, phase noise, receive 1 dB compression point output (input) power, and shape, weight, environmental conditions, interface requirements, reliability, electromagnetic compatibility requirements, etc. Depending on the requirements of the application system, the index parameters of the T/R modules will be somewhat different. Here are some typical performance parameters in practical applications:

(1) **Phase shift accuracy**

Phase shift accuracy is sometimes called phase shift error, which is the difference between the actual value and the nominal value of the phase shift amount of each state of the digitally controlled phase shifter of the T/R module. Phase shift error usually has two representation methods: maximum phase shift error and relative phase shift error, and its definition is similar to the delay error of delay amplifier components.

(2) **Transmitting and receiving conversion time**

The transmitting and receiving transition time of T/R modules is the time it takes to switch from the transmitting state to receiving state or from receiving state to the transmitting state within one working pulse cycle of the radar system. For T/R modules, other circuits, such as drivers, are often required to participate due to the conversion of transceivers. Therefore, when testing transceiver conversion time of the T/R module, the trigger signal of the transceiver conversion given by the beam control should be used as the reference signal. The time between the leading edge of the reference signal and the leading edge of the test signal is the real switching time of the T/R module.

(3) **Transceiver isolation**

Transceiver isolation characterizes the degree of RF signal isolation between the transmit channel and receive channel of the T/R module. It mainly depends on the isolation of transceiver switches, ring isolator components, transceiver channels, or cavities in T/R modules.

(4) **Noise figure**

The noise figure NF of the T/R module is defined as the ratio of the signal-to-noise ratio S_i/N_i at the input end of the receiving channel of the component to the signal-to-noise ratio S_0/N_0 at the output end when the temperature of the source impedance at the input end is 290 K, that is

$$\mathrm{NF} = \frac{S_i/N_i}{S_0/N_0} \tag{5.10}$$

The noise figure represents the strength of the internal noise of the T/R module and is a measure of the contribution of the T/R receiving channel to the total noise power at its output. Its physical meaning is the multiple of the signal-to-noise ratio deterioration after a signal passing through the T/R module.

The noise figure is a dimensionless ratio, usually expressed in dB. The definition of noise figure does not apply to networks that are nonlinear to noise. In order to make the noise figure unique, it is usually stipulated that the source temperature at the input end of the network is 290 K.

(5) **Nonlinear phase error**

In an ideal situation, the T/R module transceiver channel phase has a linear frequency response characteristic. In fact, due to the influence of devices such as power amplifiers and low-noise amplifiers, the phase-frequency characteristics of the transceiver channels of T/R modules present nonlinear characteristics. The T/R module nonlinear phase error is the maximum value of the difference between the T/R module transceiver channel insertion phase measurements in the frequency band and the linear phase fitted by these measurements, usually expressed in peak-to-peak values.

(6) **Phase shift parasitic amplitude modulation**

In an ideal situation, when the T/R module phase shifter is switched in each phase shifting state, only the signal phase changes, but the signal amplitude does not change. In fact, due to the different circuit structures of each state of the phase shifter of the T/R module, the insertion attenuation will also change accordingly when the phase of each state changes. Phase-shift parasitic amplitude modulation is the signal amplitude variation caused by the different phase-shift states of the digital phase shifter of the T/R module.

(7) **Attenuation of parasitic phase modulation**

Similar to phase-shift parasitic amplitude modulation, attenuation parasitic phase modulation refers to the amount of signal phase change caused by the different attenuation states of the digital attenuator of the T/R module.

(8) **Module power efficiency**

The power efficiency of the T/R module is used to indicate the conversion efficiency of the module to convert the DC power to the T/R RF output power. For T/R modules, there are three commonly used expressions for power efficiency.

Drain efficiency (or collector efficiency) of T/R power amplifier η_{DC-RF} is also called DC-RF efficiency. It is defined as the ratio of the RF output power P_{RFOUT} of the power amplifier to the DC power consumed by the power supply (or the DC power fed to the drain of the power amplifier) P_{DC}, that is,

$$\eta_{DC-RF} = \frac{P_{RFOUT}}{P_{DC}} \tag{5.11}$$

Power-Added Efficiency (PAE) is the most commonly used power efficiency of T/R modules, defined as the ratio of the difference between the T/R RF output average power P_{RFOUT} and input average power P_{RFIN} to the DC power P_{DC} consumed by the power supply, that is

$$\text{PAE} = \frac{P_{RFOUT} - P_{RFIN}}{P_{DC}} \tag{5.12}$$

The total module efficiency is defined as the ratio of T/R RF output average power P_{RFOUT} to the total module supply power (RF input average power P_{RFIN} and total DC power consumed by the power supply P_{DC_ALL}), that is

$$\eta_{total} = \frac{P_{RFOUT}}{P_{RFIN} + P_{DC\_ALL}} \tag{5.13}$$

The overall efficiency of the module reflects the total energy the T/R module needs to expend to obtain the desired RF output power. Any energy that is not converted to RF power becomes heat. In Eq. (5.13), the total DC power P_{DC_ALL} consumed by the T/R module power supply includes the sum of the DC power dissipated by the transmitting and receiving channels. Therefore, for pulse-operated T/R modules, when calculating the total DC power consumed by the component power supply, the duty cycle weight should be included in the power dissipation power of the transmitting and receiving channels.

5.5.4 Component Integration Architecture

Traditional T/R modules mostly use single-function chips and discrete components and form brick-structured T/R modules through surface-mounted assembly and hybrid packaging processes. This brick-type T/R module has low integration, large volume, and heavy mass. Moreover, due to a large number of internal interconnections, the total power consumption of the component is large, and the efficiency is low. Therefore, traditional brick-type T/R modules are difficult to meet the requirements of high-performance and high-resolution microwave imaging radar systems. In order to obtain smaller size and weight, this type of component usually adopts a multi-channel integration method, sharing the box body, joints, and internally integrating splitters and directional couplers, etc. [11–13].

Miniaturized T/R modules are based on MMIC multifunctional core chip technology and microsystem integration technology and adopt chip integration architecture [14] to realize high-density integration of T/R modules. Almost all circuits in miniaturized T/R modules are chip circuits, and the RF transceiver circuit has been functionalized into 2–3 chips or even one chip. Chips also realize the power supply and control circuit. Therefore, miniaturized T/R modules have the characteristics of high integration, light mass, high efficiency, and miniaturization or micro-miniaturization. As the microwave frequency increases, the spacing between array antenna elements becomes smaller, and the active channel density becomes larger. It is difficult for traditional T/R modules to match the size of array elements, so miniaturized T/R modules have more significant advantages in applying high-frequency active array antenna systems.

Miniaturized T/R modules are mainly composed of sub-circuits such as the transmitting link, receiving link, transmitting and receiving conversion, and common link. According to specific applications, the specific circuit configuration of miniaturized T/R modules is more or less different. A typical circuit composition of a miniaturized T/R module based on a multifunctional chip is shown in Fig. 5.24.

The micro-T/R assembly shown in Fig. 5.24 uses multiple RF Potassium Arsenide (GaAs) MMIC chips and a power and control circuit chip. RF MMIC chips are multifunction, power amplifier, and low-noise amplifier chips. The multifunctional chip integrates multiple functional circuits such as a digitally controlled phase shifter, digitally controlled attenuator, drive amplifier, delay line, and switch into one chip.

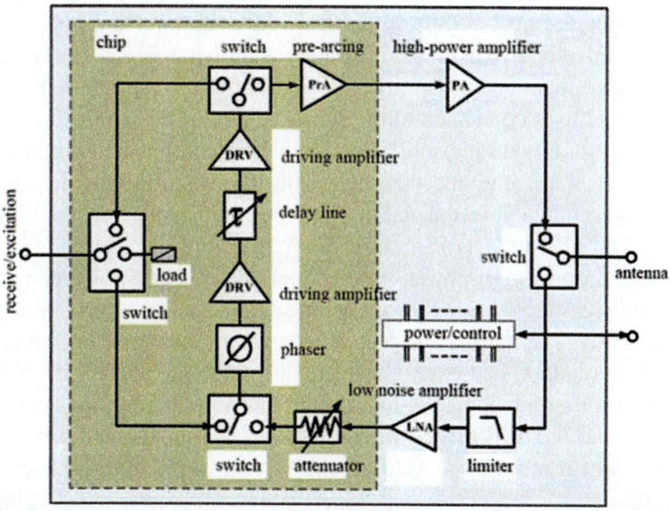

Fig. 5.24 Block diagram of micro-T/R modules based on multifunctional chips

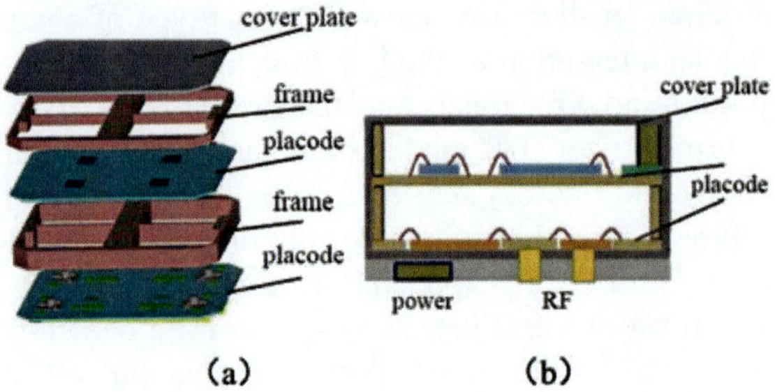

Fig. 5.25 Multi-layer chip integrated architecture micro-T/R module. **a** Basic integrated architecture of T/R module, **b** schematic diagram of T/R module structure

Realize the common circuit of transmitting and receiving of T/R module with a core chip, which greatly improves the integration degree of the module and simplifies the circuit structure. The advantage of this chip architecture is that it has a high degree of integration. At the same time, the power amplifier chip and the low-noise amplifier chip can use different GaAs processes, and each chip circuit can obtain the highest efficiency. Moreover, this chip structure can easily replace GaAs power amplifier chips with potassium nitride (GaN) power amplifier chips for some higher power applications without losing the flexibility of component layout.

Based on the current T/R module integration technology, the miniaturized T/R module mainly adopts chip integration architecture, which has two forms: multi-layer integration and single-layer integration. Furthermore, the packaging scale of T/R modules has various forms, such as single-channel, dual-channel, and multi-channel packaging.

Figure 5.25 shows a multi-layer chip assembly architecture for miniaturized T/R modules. Figure 5.25a is the basic integrated architecture, which mainly includes structures such as substrate circuits, enclosures, and cover plates.

The functional circuits or chips of this multi-layer architecture miniaturized T/R module are mounted on two or more substrates, as shown in Fig. 5.25b is a schematic diagram of the two-layer integrated structure of the micro-T/R module. The T/R functional circuit or chip is connected to the substrate circuit through flip-chip welding or gold wire crimping, and the substrate circuit is interconnected with the external interface through the frame and the shell. The RF isolation between the internal channels of the component is realized through the compartment design of the frame.

The outstanding advantage of multi-layer architecture integration is that RF and low-frequency circuits, high-power circuits, and low-power circuits can be laid out in layers to reduce mutual electromagnetic interference and have better electromagnetic compatibility. The internal space is relatively sufficient, and additional circuits such as filters, couplers, and even intermediate frequency and digital circuits can be arranged. Another advantage is that the circuit is relatively dispersed, the heat density is low, and the difficulty of thermal design is minor. The disadvantages are relatively low integration, large volume and weight, and complex internal signal interconnection.

Figure 5.26 shows a single-layer assembly structure of a miniaturized T/R module, which mainly includes structures such as a substrate, a frame, and a cover.

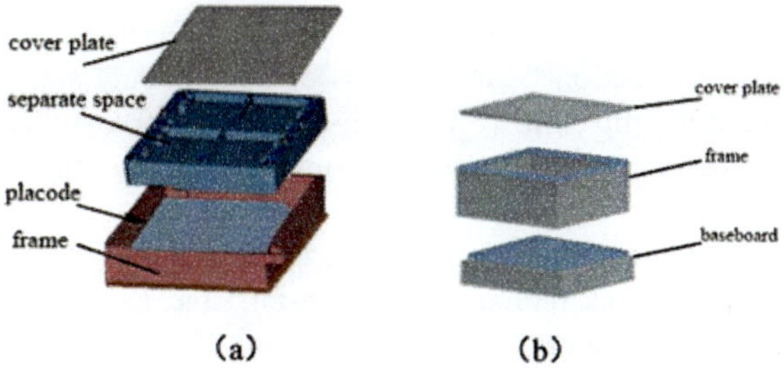

Fig. 5.26 Single-layer chip integrated micro-T/R modules, **a** compartmental structure, **b** single-cavity structure

Fig. 5.27 Schematic diagram of a silicon-based integrated micro-T/R module

This single-layer architecture miniaturizes the functional circuits or dies of the T/R assembly on one or both sides of a substrate. For example, Fig. 5.26a is a single-layer four-channel T/R integrated architecture, and Fig. 5.26b is a single-layer single-channel T/R integrated architecture. Similar to the multi-layer integrated architecture, the functional circuit or chip of the single-layer integrated T/R is also connected to the substrate circuit through flip-chip or gold wire crimping, and the substrate circuit is directly related to the external interface or through a frame. In addition, the single-layer integrated architecture can also use a framed compartment design to achieve RF isolation between the internal channels of the component.

A more straightforward form of single-layer integration architecture is integrated single-layer integration, as shown in Fig. 5.27, which is a schematic diagram of a four-channel miniaturized T/R module packaged with silicon-based materials.

The advantage of single-layer architecture integration is that all circuits and chips are mounted on one or both sides of a single substrate, so it has high integration, small size, and light mass. The disadvantages are high thermal density, complex layout and wiring, and a rugged design for electromagnetic compatibility and thermal design.

For chip-integrated T/R modules, the choice of substrate and circuit design are essential. Substrate materials used in micro-T/R modules mainly include LTCC, AlN, HTCC, silicon substrates, polyimide films, liquid crystal substrates, microwave substrates, etc. Different substrate materials have different mechanical, thermal, and electromagnetic properties. Therefore, factors such as T/R module performance requirements, application environment, assembly process of circuit components and chips, thermal matching between chips and substrates, thermal design measures, and signal interconnection methods should be considered comprehensively.

The frame, shell, and cover materials of miniaturized T/R modules mainly include Kovar, aluminum alloy, high-silicon aluminum, silicon carbide, titanium alloy, and LTCC and HTCC. The choice of frame material should also be considered comprehensively based on factors such as T/R module performance, application environment, manufacturing and packaging process, and thermal design measures.

Commonly used interconnection methods for miniaturized T/R modules are signal interconnection between chips or components and substrates, usually using wire bonding, flip chip, TSV, micro-coaxial, and BGA placement. The signal connection inside or between the substrate and the external interface usually uses wire bonding, metalized via holes, TSV, micro-coaxial, etc.

The main microsystem integration technologies commonly used in miniaturized T/R modules include System-in-Package (SIP), System-on-Chip (SOC), Wafer-Level Packaging (WLP), Multi-chip Modules (MCM), and Integrated Passive Devices (IPD), and etc.

5.5.5 *Circuit Analysis and Design*

Figure 5.24 shows a typical schematic diagram of the miniaturized T/R modules. Based on this architecture analysis, introduce the circuit design of miniaturized T/R modules, optimize the link structure and circuit layout and layout design through component link analysis and circuit simulation. In order to obtain a stable circuit structure and optimal transceiver performance within the frequency band.

Circuit design includes circuit layout design and link analysis, and simulation design. In the actual application process, layout design, link analysis, and simulation design are mutually iterative processes. The electromagnetic compatibility design should usually be combined with the circuit layout design to avoid too many discontinuities and solid mutual coupling in the link. At the same time, the electromagnetic simulation of the cavity should be carried out to avoid the cavity resonance frequency falling within the working frequency band.

(1) **Link analysis and electromagnetic simulation**

Based on the multi-function chip architecture in Fig. 5.24, a miniaturized T/R module transceiver link signal analysis model with an output power of about 3 W is shown in Fig. 5.28.

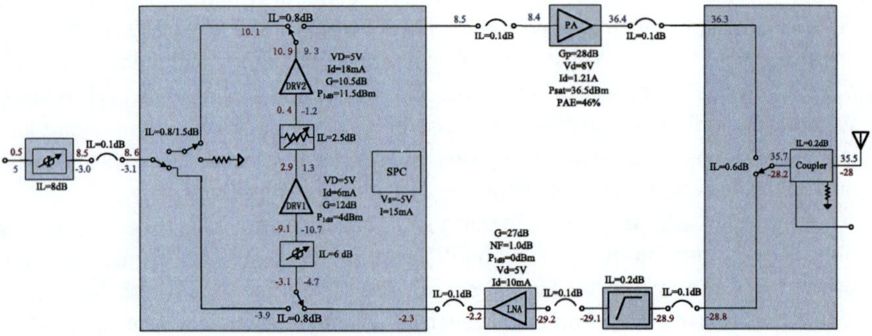

Fig. 5.28 A miniaturized T/R module transceiver link signal analysis model

For the transceiver link model in Fig. 5.28, the electromagnetic simulation design can simulate and optimize the active circuit and field. 3D electromagnetic simulation software is usually used to simulate and extract parameters for the multi-layer transmission circuit of the component substrate, including vertical interconnection, interface lead lap, matching transition, etc. The 2D active circuit simulation software is used to simulate and analyze the receiving and transmitting links of the components, respectively.

Transceiver link analysis uses the basic parameters of each component to extract and establish an equivalent series model to perform approximate calculations and evaluation of link transceiver signals. Its advantages are simple, intuitive, and fast, and its disadvantage is that it cannot consider the influence of electromagnetic fields at discontinuities. The advantage of electromagnetic field simulation is that it can analyze the impact of substrate circuits and connecting circuits on component performance indicators from the perspective of electromagnetic fields. In addition to the signal parameters of the transceiver link, the VSWR of the circuit port and the signal characteristics of each internal node can also be obtained, and the circuit layout can be directly generated simultaneously. The disadvantage is that although it has high simulation accuracy for small-signal active links, it is difficult to obtain accurate circuit parameters for large-signal circuits, so the accuracy of simulation results is low. However, more circuit information can be obtained through co-simulation of 3D field and 2D circuits.

For the T/R module circuit structure shown in Fig. 5.28. Through link analysis and simulation, the excitation signal levels and total output power of power amplifiers can be obtained, as well as performance parameters such as receiving link gain, noise figure, and port VSWR.

(2) **Circuit stability design**

Due to the high-density integration of micro-T/R modules, its circuit layout is dense. It often includes analog/digital circuits, high-power/low-power signals, and other circuit forms. If there is no good circuit layout design. It is easy to cause mutual interference between these circuits, which will affect the stability of the components, and even cause the performance of the components to deteriorate or be damaged.

In order to improve the stability of component operation, proper transceiver link gain design, transceiver link isolation design, cavity separation design, and timing design are essential.

The transceiver link gain (including small signal gain) should not be too high. If the link gain is very high, an isolation circuit should be added between multiple stages. In order to obtain flat power output, large gain compression is often used, which easily leads to excessive link gain and circuit oscillation. In order to make the components work more stably, the gain compression of the transmit chain should not be too significant.

In order to make the component work more stably, in addition to improving the isolation of the transceiver channel as much as possible, timing management should also be used to achieve the isolation of the transceiver channel. The former is the isolation of the circuit in space, and the latter is the isolation in time. In most

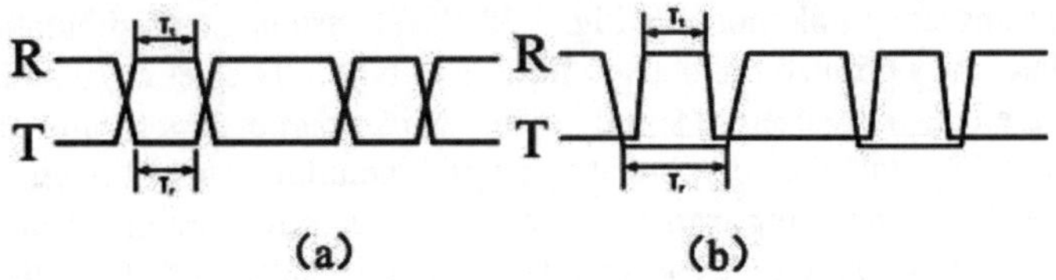

Fig. 5.29 Miniaturized T/R modules transmitting and receiving timing control, **a** wrong timing design, **b** correct timing design

radar systems, the transceiver channel often works in time-sharing mode, and the transceiver channel amplifier realizes time-sharing operation through pulse modulation management. Therefore, to improve the transceiver isolation in the entire pulse cycle, the specific measure should be applied to ensure that the component turns off the transmitting channel during the receiving period and turns off the receiving channel during the transmitting period through timing design. As a result, the control sequence of transmitting and receiving cannot have crossed, i.e., the rising edge of the transmitting control and the falling edge of the receiving control cannot cross, and it is required that the transmitting on-time *Tt* should be less than the receiving cutoff time *Tr*, as shown in Fig. 5.29.

Supposing the timing design with interleaved transceiver control is adopted, as shown in Fig. 5.29a. In that case, it will result in insufficient isolation of components at the moment of transceiver transition, making it difficult for T/R modules to work normally, or there may be potential instability.

(3) **Beam control and timing design**

Most beam control circuits of miniaturized T/R modules use beam control chips to realize timing and power management for high-density integration, miniaturization, and light mass. The primary function of the beam control chip in the T/R module is to receive the serial data sent by the upper-level beam control, convert it into parallel data after judging it is correct, and send it to the corresponding phase shifter, attenuator, or switch. Then, receive the transmitting and receiving control timing is sent by the superior beam control, after judging that it is correct, the level is converted and sent to the corresponding channel switch device, the timing normal/fault indication signal is output.

The block diagram of the beam control chip circuit integrated into a miniaturized T/R module is shown in Fig. 5.30.

In Fig. 5.30, the serial-to-parallel control circuit of the beam control chip converts the external 30-bit serial control signal into a parallel signal. To control the internal digital phase shifter, attenuator, switch, and other circuits of the T/R module to complete functions such as amplitude-phase weighting and mode conversion.

Generally, a beam control chip inside a miniaturized T/R module usually has 6 input control signals. They are serial clock SC, serial data SDin, primary latch END, secondary latch SYN, transmitting timing T/RT, and receiving timing T/RR. For example, a complete write cycle contains 26 data bits, and the corresponding control

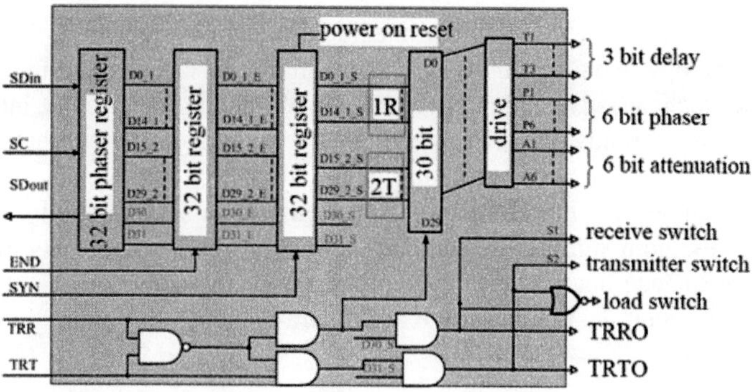

Fig. 5.30 Circuit schematic diagram of miniaturized T/R module beam control chip

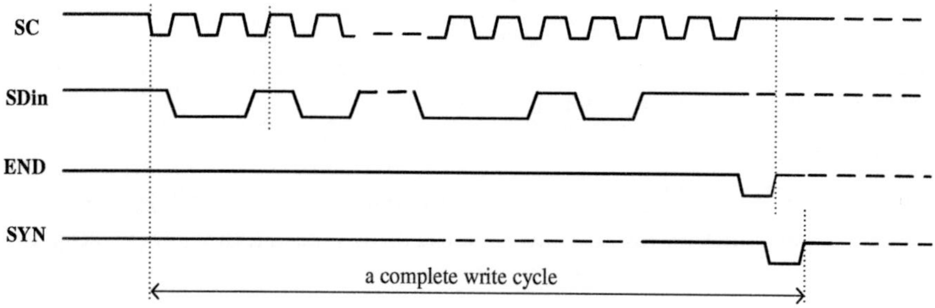

Fig. 5.31 Beam control timing diagram of miniaturized T/R modules

timing is shown in Fig. 5.31. Among them, the rising edge of clock SC shifts data, the rising edge of END latches data, and the rising edge of SYN refreshes data.

5.6 Environmental Adaptive Technology

On the one hand, the active array antenna of spaceborne microwave imaging radar has a straightforward interface relationship with the satellite platform, which is generally determined by the Interface Data Sheet (IDS). Including equipment size and installation size, the center of mass, moment of inertia, power consumption, data transmission rate, and other mechanical interfaces, thermal interfaces, electrical interfaces, etc. On the other hand, it must be able to adapt to the working environment of a rocket launch and on-orbit operation [15].

The active array antenna system often adopts an actively integrated module architecture, which is expanded by multiple active integrated modules. The active integrated module includes several T/R modules, delay amplification components, and antenna radiation elements. In order to meet the requirements of light mass and miniaturization, most of these active integrated modules adopt high-density integration technology, which is high-density integration of circuits and modules into one module, such as active and passive circuits, RF and digital circuits, large signal and small signal circuits, etc. Because the active integrated module has the characteristics of a complex circuit structure, high integration, high power density, and heat dissipation density, therefore, good space environment adaptability, especially electromagnetic compatibility design and thermal design, can not only effectively prevent performance degradation or even damage of the active array antenna, but also be an important guarantee for the active array antenna to obtain high stability and high reliability.

5.6.1 Space Environment Requirements

In addition to the electromagnetic compatibility design and thermal design requirements, the space environment mainly includes:

(1) **Mechanics**

Active phased array antennas are required to adapt to vibration, shock, overload, noise, etc., generated during satellite launch. For this reason, the antenna structure design must have sufficient strength and stiffness, and modal analysis must be carried out to prevent resonance and damage to the active phased array antenna. In addition, to withstand the mechanical environment conditions during the launch phase, the phased array antenna deployment mechanism needs to be locked and unlocked after the satellite enters orbit.

(2) **Weightlessness**

The satellite is in a state of weightlessness in orbit, while the active phased array antenna is in a state of gravity when it is debugged on the ground. Although the support of the air-floating platform is used to simulate the weightlessness state of the active phased array antenna, the plane accuracy and other performance of the antenna are still different in the gravity state and the non-gravity state. Therefore, these need to take appropriate design or ground commissioning measures to ensure that the performance of active phased array antennas in orbit can meet the specified requirements in weightlessness.

(3) **Vacuum**

Due to the extensive use of T/R modules and delay lines, some microwave absorbing materials are sometimes used for active array antennas to overcome the cavity effect or electromagnetic interference of T/R modules and delay lines. T/R modules and delay lines also use composite material glue in manufacturing. These materials will outgas and evaporate in a vacuum state, polluting the chips used in T/R modules, delaying lines, and affecting their lifespan. In addition, vacuum discharge may also cause damage to some circuit components, and the secondary electron multiplication effect may also cause damage to microwave chips. Such issues must be paid attention to in the designing and manufacturing of the active array antennas to ensure compliance with vacuum state requirements.

(4) **Space radiation**

High-energy electrons, protons, and heavy ions will cause damage to the surface of the active array antenna and electronic components, and single-particle flipping and locking will cause CMOS circuits to fail. Therefore, in designing and manufacturing active array antennas, appropriate measures should be taken to adapt to the requirements of the space radiation environment.

Active array antennas are generally the "big energy consumers" of satellites, which include many T/R modules. The efficiency of the final stage amplifier in the T/R modules is the key to the RF power conversion efficiency of phased array antennas. The GaAs monolithic amplifier has an efficiency of about 40%, and the GaN monolithic amplifier has a conversion efficiency of 60–70%. For active array antennas, the energy that has not been turned into RF power will all be turned into heat, which increases the burden of thermal control subsystem.

Due to the non-repairability of satellite equipment in orbit, higher requirements are placed on the high reliability and long life of active array antennas.

5.6.2 Electromagnetic Compatibility Design Technology

The purpose of electromagnetic compatibility design is to ensure that the internal circuits of the active integrated module work in harmony with each other so that the module can work stably and reliably.

The antenna active integrated module is a highly integrated multifunctional microwave module. It includes a variety of active and passive circuits, such as transceiver T/R modules, power supply and control circuits, RF transmission circuits, monitoring and protection circuits, and interface drive circuits. Due to the distributed layout of active integrated modules, the gain of the transceiver link is high, and the RF link is highly sensitive. It is susceptible to internal mutual interference and coupling and is also prone to problems such as oscillation, performance degradation, circuit instability, and short device life. Therefore, electromagnetic compatibility design is one of the critical research points of active integrated modules.

The main content of the electromagnetic compatibility design of the antenna active integrated module is to limit the electromagnetic emission of the interference source, control the propagation of electromagnetic interference, and enhance the anti-interference ability of sensitive equipment.

According to the circuit composition, function, and performance characteristics of the antenna active integrated module, the main measures and methods of electromagnetic compatibility design include:

(1) **Reasonable architecture layout and circuit design**

Make a reasonable layout of the internal circuit layout and wiring, the component configuration, and the relative position of active integrated modules. The basic principle is to keep the sensor and the interference source as far away as possible, the output and input ports are reasonably separated, and the high-level signal and pulse signal leads are arranged separately from the low-level signal lines. As far as possible, electronic circuits (including integrated circuits) with small emissions and powerful anti-interference ability should be selected as the unit circuit of the electronic equipment. For general small-signal amplifiers, the linear dynamic range of the amplifier should be increased as much as possible to improve the overload capability of the circuit and reduce nonlinear distortion. Using a symmetrical circuit structure or filter circuit can effectively suppress the harmonic signal level generated by the power amplifier. In order to minimize the harmonic emission of the amplifier due to nonlinear distortion, feedback, and nonlinear compensation methods can be used to improve the linearity of the amplifier.

(2) **Optimize the cavity design**

Through the 3D electromagnetic field simulation of the cavity, the structure of the cavity is optimized, and the resonance frequency point is adjusted. Take measures such as adding barriers to the cavity design or filling part of the absorbing material to avoid or eliminate cavity resonance, spatial coupling of RF signals inside the module, and signal crosstalk between boards to ensure the stability of the module.

(3) **Optimize timing, filtering, and isolation design**

Reasonably design the pulse modulation timing of the transceiver channel to avoid crossover of the transceiver timing, completely shut down the receiving channel during transmission, improve the isolation of the transceiver, and avoid mutual interference between the transceiver circuits. Filtering is a technology that uses suppression circuits to suppress unwanted signal energy outside the valuable signal spectrum in components, to extract useful signal components from noisy or interfering signals. It is an important measure to suppress and prevent conducted interference. The active integrated module typically employs two main filtering methods. The first involves the use of various filter circuits to suppress the transmission and emission of stray signals in the transmitting link. This method reduces the influence of external interference signals on the sensitive circuits in the receiving link. The second method involves the use of decoupling capacitors to filter out interference signals in the power supply and control lines. This method prevents interstage coupling caused by parasitic feedback, reduces the overshoot effect of logic signals on the power supply, and filters out coupling between the local power supply signal and the control signal. For circuits using multilayer microwave substrates or LTCC substrates, measures such as layering of multilayer boards, metalized vias, and interstage filtering are often used. Realize the isolation of RF, digital and power lines and signals, shielding between signals, and independent grounding design, which can effectively suppress crosstalk between various signals.

(4) **Grounding and shielding design**

The grounding of the active integrated module of the antenna is to suppress the coupling interference generated by the current flowing through the common ground wire and the coupling interference formed by the ground current loop. It is one of the important measures to suppress noise and prevent interference. Good grounding should try to avoid unnecessary ground loops. For active modules with high and low-frequency mixed circuits, the hybrid grounding method should usually be used, high and low-frequency circuits, power and control circuits, high power and low power circuits, etc., should be grounded separately, when a ground loop occurs, use floating ground isolation, The ground wires of low-level circuits with crossovers should be routed vertically, a balanced differential circuit should be used to minimize the crosstalk of the ground potential.

Shielding design encapsulates or isolates the electromagnetic signal of the interference source to suppress or reduce the spatial coupling of the interference source signal, external radiation, mutual interference between circuits, etc. Electromagnetic shielding usually includes electrical shielding, magnetic shielding, and electromagnetic shielding. For active modules, materials such as conductive rubber gaskets and metal diaphragms are often used to shield the holes. The cover plate of the module should be welded and sealed as much as possible. When using coaxial cables to transmit signals, the signal return path should be provided through the shielding layer. In the multilayer circuit of the substrate, different circuits are shielded in layers by metallization. Metalized vias are used for circuit isolation between multilayer circuits. In addition, the separate cavity design or compartment design is adopted to reduce or suppress spatial coupling inside the cavity.

5.6.3 Thermal Design Technology

The purpose of thermal design is to ensure the stable and reliable operation of the device or system by adopting appropriate thermal management measures. For active microwave modules using a large number of RF components, control chips, and RF chips, due to their high integration and high power density, the quality of thermal design greatly impacts the reliable operation of active microwave modules. Poor thermal design often causes severe thermal damage to the lossy microwave module, mainly in performance deterioration or burning. Faults caused by the thermal induction of active integrated modules mainly include functional and physical faults. For active array modules, thermally induced functional failures mainly manifest in performance degradation or even complete failure, such as transceiver gain, noise figure, output power, nonlinear phase, etc. Thermally induced physical failures are mainly manifested in severe physical damage caused by various chips, microwave substrate circuits, and various assembly materials in the assembly. Such as evaporation of organic matter, melting solder joints, thermal stress cracking of leads/joints and welds, cracking of sealants and chips due to fatigue, and deformation due to creep.

Under the condition that the heat flux density dissipated by the active integrated module of the antenna and the allowable temperature rise is clear, it is necessary to study the heat dissipation method and conduction path and the heat dissipation method and heat transfer method. Generally, "natural" air cooling methods relying on free convection and radiation are only effective for heat fluxes below 0.05 W/cm^2 when the temperature difference between the component surface temperature and the ambient temperature is allowed to be below 60 °C. Although forced convection air cooling improves the thermal conductivity by approximately an order of magnitude, the thermal match does not have a heat transfer capability of more than 1 W/cm^2, even with an allowable temperature difference of 100 °C.

The research content of the thermal design of the active integrated module of the antenna includes the following:

(1) **Heat consumption analysis**

According to the composition of the active integrated module of the antenna, determine the main power consumption components or circuits, and calculate the power consumption and heat consumption of each power consumption component according to the operating mode and power supply conditions of the corresponding components. For active integrated modules, its power consumption components mainly include high-power chips, low-noise amplifier (LNA) chips, multi-function chips, power modules, circulators, etc.

(2) **Heat conduction design**

Select the appropriate heat dissipation method according to the thermal distribution of power consumption devices inside the active integrated module. Common heat dissipation methods include conduction heat dissipation (including natural conduction and forced convection, such as liquid cooling), convective heat dissipation (including natural convection and forced convection, such as forced air cooling), radiation heat dissipation, and phase change heat dissipation (using phase change materials).

The final high-power amplifier chip is the main source of heat loss for the antenna active integrated module. Therefore, the selection of the heat dissipation method of the active integrated module should be comprehensively considered in combination with its internal heat consumption distribution, heat consumption density, component assembly and integration structure, environmental conditions, and external application system thermal design conditions. It is the basic requirement of thermal design to achieve good thermal design and high reliability of components through simple and easy heat dissipation measures.

If the heat consumption of the power chip of the active integrated module is not particularly large, it is an economical and effective method to dissipate heat through direct conduction. However, in order to achieve high-efficiency heat dissipation and good thermal stress performance of the power chip, it may be necessary to add a heat sink material with high thermal conductivity and good thermal expansion matching performance between the power chip and the component housing. And, it is assembled by welding to realize an excellent thermal matching design of the power chip.

(3) Thermal model and thermal simulation

Using thermal simulation analysis software for simulation calculation is a practical and effective method for the thermal design of active integrated modules. After completing the heat consumption distribution design and heat consumption analysis of the active integrated module circuit, according to the selected heat conduction mode, combined with the specific circuit composition of the component, assembly and integration process, cavity structure, and other design conditions, as well as the internal and external thermal environment conditions of the module, establish a 3D thermal simulation model of an active integrated module based on the established thermal simulation model and the definition of module materials and external thermal environment conditions. After completing the power consumption conditions and model mesh subdivision parameter settings, a complete 3D thermal simulation analysis can be performed on the module, and the temperature distribution cloud map of each level of the module can be obtained. Using the temperature distribution cloud map obtained by simulation, evaluate the surface temperature distribution and temperature rise data of each heat consumption component of the module, especially the high-power amplifier chip. Then evaluate the feasibility of thermal design measures for active integrated modules according to the maximum operating junction temperature, operating temperature, and reliability index requirements of heat dissipation components or high-power amplifier chips.

In the thermal simulation design, the accuracy of the thermal simulation model is critical, often due to the complexity of the actual circuit and process, and the relatively ideal simulation model, the simulation results have a large deviation. In addition, the mesh division of the simulation model is also significant, and different network division methods greatly impact the calculation efficiency and simulation accuracy. The type of mesh division should be reasonably selected according to the different structural model formats, the simulation scale, complexity, and the calculation precision requirements. The temperature distribution cloud map of the thermal simulation results of a miniature four-channel T/R module is shown in Fig. 5.32.

Due to the error of the model and simulation parameters, the simulation results of the thermal distribution will also have specific errors, so a certain design margin should be considered in the actual heat dissipation design.

(4) Thermal design

According to the heat consumption distribution of the active integrated module circuit system, take appropriate heat control measures and choose a reasonable heat dissipation method. The thermal design measures of active modules need to be considered comprehensively in combination with factors such as radar system operating mode and thermal management measures, environmental conditions, and costs. The heat dissipation methods of active integrated modules usually include conduction heat dissipation, convection heat dissipation, radiation heat dissipation, and phase change heat dissipation. For miniaturized active modules such as T/R modules, integrating high thermal conductivity materials and high-efficiency new heat dissipation technologies (thermal vias, micro-channels) with module circuits and structures is a new

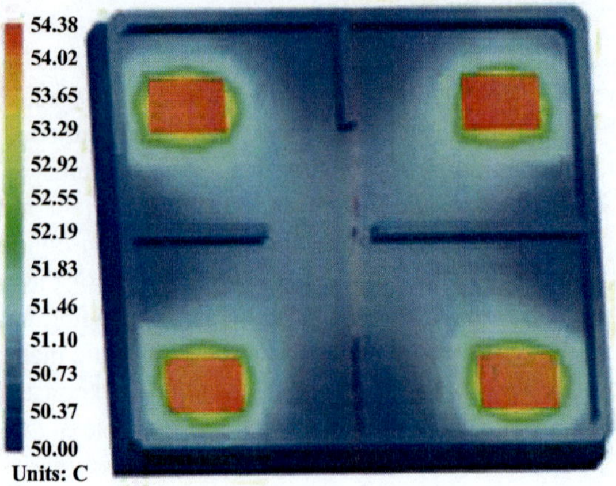

Fig. 5.32 Temperature distribution cloud map of miniaturized four-channel T/R module thermal simulation

approach to thermal management technology. In addition, according to the reliability design requirements of the antenna system, it is also one of the effective measures to improve the operational stability of the module by adopting a derating design to reduce the heat dissipation of the module when necessary, such as voltage and current derating, power derating, working junction temperature derating, shell temperature derating, etc.

Taking the space-borne active array antenna as an example, changes in the external heat flow and internal heat flow of the satellite will change the temperature distribution where the phased array antenna is located. Different radiating surfaces have different heat dissipation capabilities [16], and thermal control measures must be taken to ensure they can meet the requirements of temperature changes. Thermal design is carried out simultaneously with electrical performance and structural design so that the thermal resistance from this part to the heat dissipation surface of the component or satellite is small and meets the requirements. Some parts require insulation from an electrical point of view, but good heat conduction is required from a thermal point of view. For this reason, special measures must be taken, such as using beryllium oxide materials. In addition, some parts do not need or even allow good conductors to lead out heat from an electrical point of view, and special measures must be taken. For example, for the heat conduction problem of the inner conductor (air dielectric plate line network) in the transmission power division network, measures such as blackening the surface of the inner and outer conductors (at the cost of increasing losses) and adding $\lambda/4$ short-circuit posts should be considered.

5.7 Applications

The following is an example of a 64-element active integrated antenna prototype [17] to introduce the tile-integrated architecture and typical performance. Antenna active integrated modules include micro-chip T/R, multi-function chips, RF sub-boards, microelastic connections, etc., which have the advantages of high integration, good assembly process operability, and high interconnection reliability. The antenna active integrated module consists of 64 radiating elements, 64 micro-chip T/R modules, 9 1:8 RF power dividing circuits, 1 multifunctional circuit (including beam control and power supply circuits), and corresponding structural support components and RF and control interfaces. The antenna adopts a layered assembly structure with a total of seven layers of physical structure, including four layers of circuit layers and three layers of structural components. The four-layer circuit includes radiation array, chip T/R modules, RF power combination circuit, power supply, and control circuit. The principle of the chip-integrated array antenna module is shown in Fig. 5.33.

The chip active integrated antenna module includes three external interfaces: RF interface, control, and power interface. The integrated architecture of the chip active integrated array antenna module is shown in Fig. 5.34.

The chip active integrated antenna module locates and assembles the micro-T/R module, radiation element, RF circuit, control, and power circuit through the metal back cavity of the antenna, the structural transition plate, and the supporting frame. The control and RF signals are interconnected through elastic connectors, microblind-mating connectors, and elastic fur buttons respectively. Micro-coaxial connectors and fur buttons realize the vertical interconnection of RF signals between micro-chip T/R modules, radiating elements, and RF boards. In addition, high-density elastic connectors are used to recognize the control/power signal interconnection between chip T/R modules and multi-function boards. The 64-element integrated antenna module prototype is shown in Fig. 5.35, and the test results are shown in Fig. 5.36.

The measurement results of the 64-element active integrated antenna prototype are that within the 10.5% frequency band, the output power of each element is 2.0

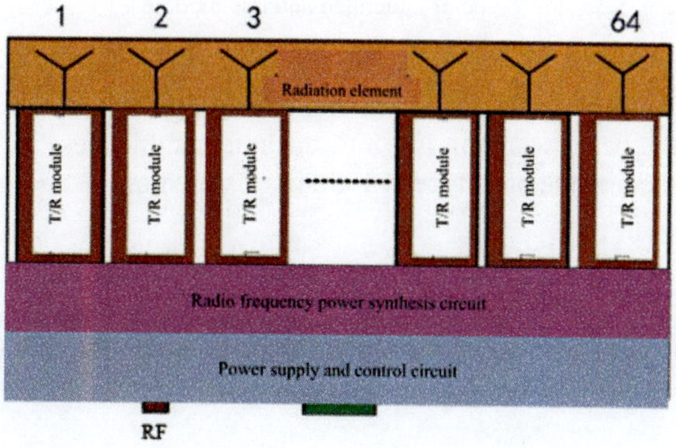

Fig. 5.33 Schematic diagram of chip active integrated antenna module

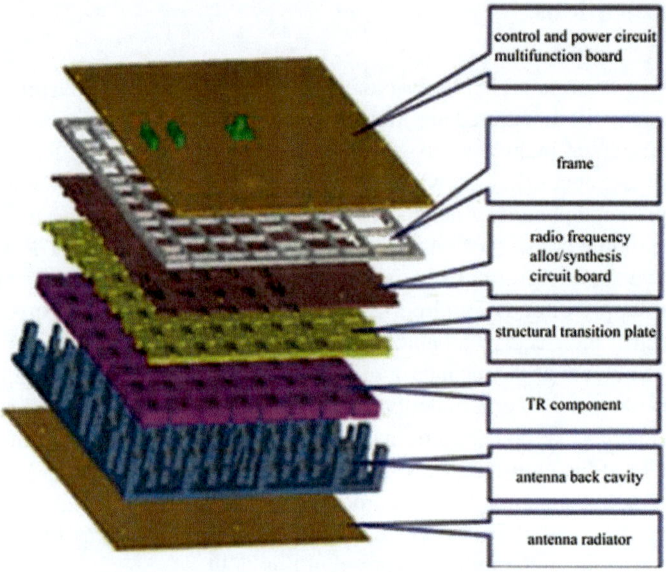

Fig. 5.34 Schematic diagram of chip active integrated antenna module integration architecture

Fig. 5.35 Prototype of integrated antenna module

W (peak value), the receiving gain is 30 dB, the noise figure is better than 2.7 dB, the beam scanning is ≥30°, the antenna thickness is 46 mm, and the dimension of the module prototype is 195 mm × 166 mm.

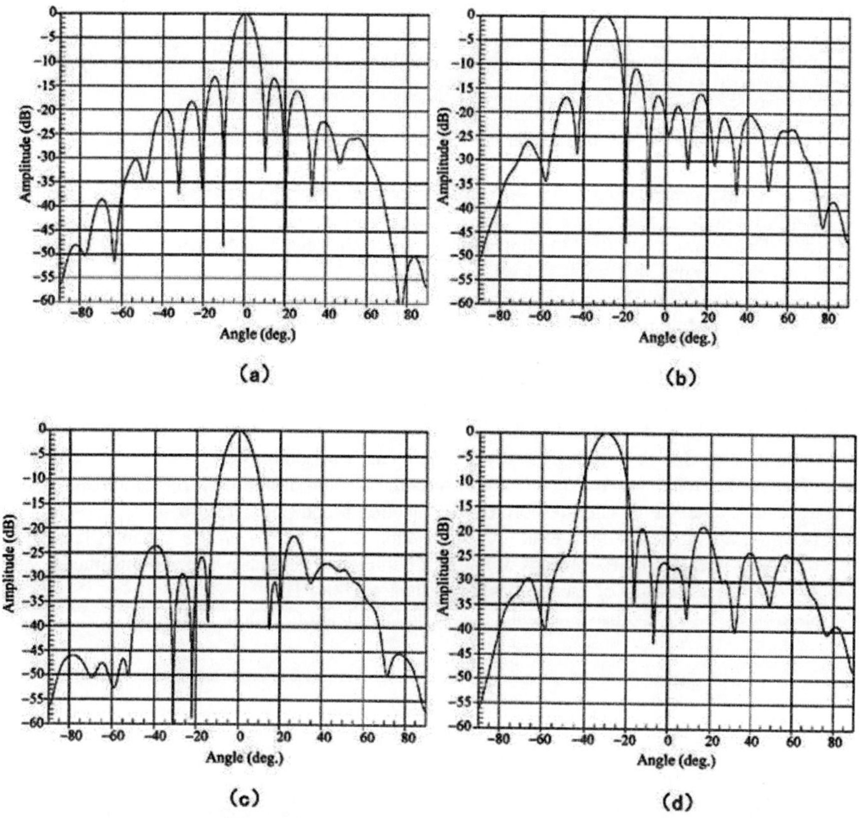

Fig. 5.36 Typical pattern measurement results of the prototype, **a** uniformly weighted normal pattern, **b** uniformly weighted, −30° scan pattern, **c** −25 dB Taylor weighted, normal pattern, **d** −25 dB Taylor weighted, −30° scan pattern

References

1. Wang W, Lu J, Zhang H et al (2016) A brief review of SAR antenna development in China. In: CIE international conference on radar. IEEE, Guangzhou, pp 1–4
2. Wang W (2001) Study on waveguide slow-wave line of frequency scanning antenna. Xidian University
3. Song X, Wang W, Jin M, Wu R (2016) Design of a Ku-band frequency-scanning antenna array fed by hybrid structure. Chin J Radio Sci 31(2):340–345
4. Lu J (2015) The Technique challenges and realization of space-borne digital array SAR. In: IEEE 5th asia-pacific conference on synthetic aperture radar (APSAR). IEEE, Singapore, pp 1–5
5. Wang W, Zhang H, Lu J et al (2015) Dual band dual polarized antenna for SAR. In: IEEE international symposium on antennas and propagation & USNC/URSI national radio science meeting. IEEE, Vancouver, BC, Canada, pp 220–221
6. Wang X, Wang Z, Zou Y (2012) Research on key technologies of active tile array antennas. J Microw (s2):160–162

7. Wang X (2013) A tile antenna based on multi-layer architecture. In: Proceedings of 2013 national conference on antenna technology. Publishing House of Electronics Industry, Beijing, pp 847–849
8. Wang X, Wang Z, Zhang Y (2010) Design and integration of S-band tile array module. In: Proceedings of 11st National Radar conference. National Defense Industry Pree, Beijing, pp 685–687
9. Lu J, Wu M, Chen S, Fang Z (2000) A calibration method of phased array radar based on FFT. Chin J Radio Sci 15(2):221–224
10. Lv C, Dai Y, Wang W (2012) Optimization design of an X-band dual-polarization T/R module. Radar Sci Technol 10(1):108–111
11. Zhang D, Dai Y, Xu J, Liu H (2008) Design and manufacture techniques of A Kind S band T/R module. Modern Radar 30(2):76–78
12. Ge J (2016) Design and implement of ku-band wideband 8-channel T/R module based on LTCC substrate. Shipboard Electron Countermeas 39(2):74–78
13. Jinzhong H, Yu Z, Yang Z (2015) Design of a broadband multi-channel tile-type T/R module. Telecommun Eng 55(1):108–112
14. Imbriale WA, Gao S, Boccia L (2012) Space antenna handbook. Wiley
15. Zhigang W, Tingguang Y, Wei W (2011) Structure design of micro-strip antenna and slotted waveguide antenna for satellite-borne SAR. Electron Mech Eng 27(3):40–43
16. Wang X, Zheng L, Zhang Y et al (2017) Development of a low-profile X-band tile antenna with 64 elements. In: Proceedings of national conference on microwave and millimeter wave 2017. Chinese Institute of Electronics, Hangzhou, pp 809–812
17. Shunshi Z (1991) Microstrip antenna theory and application. Xidain University Press, Xian

Chapter 6
Shared-Aperture Array Antennas

6.1 Introduction

Microwave imaging radar requires the antenna to have sufficient power and gain in order to achieve high-resolution and high-quality imaging. For satellite platforms, microwave imaging radar antennas must meet the constraints of volume, weight, and power consumption. In the case of single band, single polarization, the contradiction between platform capability and microwave imaging radar load is not obvious. With the development of electronic information technology, multi-band and multi-polarization radar image fusion is realized by utilizing the difference in scattering characteristics of electromagnetic waves in different bands and different polarizations of targets. More and more microwave imaging radars use new technologies such as multi-band and multi-polarization to obtain complete information on targets and improve target detection, classification, and recognition capabilities. In the case of multi-band and multi-polarization, the volume and weight of active array antennas integrated with discrete components increase dramatically.

The multi-band, multi-polarization shared-aperture antenna can effectively improve the antenna efficiency, reduce the volume and weight of the microwave imaging system to improve the suitability of the radar to the satellite platform, and reduce the development and launch costs of the platform. However, in the multi-band, multi-polarization shared-aperture antenna structure, due to the nesting and superposition of different antenna elements, electromagnetic field mutual interference and harmful mode spurious radiation are caused, resulting in changes in the isolation inside/outside the working frequency band of the antenna. Therefore, conducting in-depth research on the design of shared-aperture antenna architecture and theoretical simulation analysis is necessary. For example, using mode theory [1] and symmetric method [2] to improve the isolation between polarizations, dual-frequency or multi-frequency, reduce cross-polarization components, improve the radiation efficiency of shared-aperture antennas, etc.

J. Lu et al., *Active Array Antennas for High Resolution Microwave Imaging Radar*,
https://doi.org/10.1007/978-981-99-1475-3_6

6.1.1 Dual-Polarized Antenna Configuration

There are usually two implementations of dual-polarized antennas. One is the simple structural assembly of two antennas with completely independent apertures. The other is the nesting of dual-polarized antennas sharing one physical aperture. The shared antenna physical apertures usually include dual-polarized feed reflector antennas and dual-polarized shared-aperture array antennas. Here we discuss and analyze array antennas.

There are two types of shared-aperture active array antennas, depending on the polarization selection method: one is that the two polarizations of the radiating antenna adopt a shared-aperture design, and the active T/R channel is shared. Each T/R module is connected to a dual-polarization antenna, and the polarization selection is realized by switching. In fact, this architecture is only a dual-polarized antenna array, and its rear end is the same as that of a single-polarized antenna, as shown in Fig. 6.1a. The other is that the antenna array is still dual polarized and shared aperture, and each polarized port of the dual-polarized antenna element is terminated with a T/R module channel. On the RF band, the dual polarizations have completely independent signal chains, while the power supply, control, thermal control, and structural framework are shared, as shown in Fig. 6.1b. The advantage of the former architecture is the low number of active channels and light mass. However, dual polarization at RF cannot realize the simultaneous operation of the entire antenna array. Generally, it works in time division multiplexing, and the RF switch completes the switching between polarizations. The insertion loss of the dual-polarization switch will reduce the radiation efficiency of the entire antenna. The latter architecture avoids the weaknesses of the former but doubles the number of component channels, increasing the complexity, weight, and cost of the antenna system. The choice of the two architectures depends on the radar system requirements, platform suitability, cost requirements, etc.

Many antenna elements can achieve dual-polarization shared apertures, such as the printed dipole, Vivaldi, opened-waveguide, horn, microstrip patch, waveguide slot, and other antennas. However, considering the antenna profile, the antenna elements, such as the printed dipole, Vivaldi, and horn, have a high profile, which is not conducive to low-profile integration, as shown in Fig. 6.2a, b. Therefore, in most applications, waveguide slot antennas and microstrip patch antennas are mainly selected [3], as shown in Fig. 6.2c, d.

6.1.2 Multi-band Dual-Polarization Shared-Aperture Configuration

Similar to dual-polarization ones, multi-band dual-polarization antennas can use independent antenna arrays and assemble them physically. However, this architecture has a large antenna area and is limited by the satellite launch envelope, so its

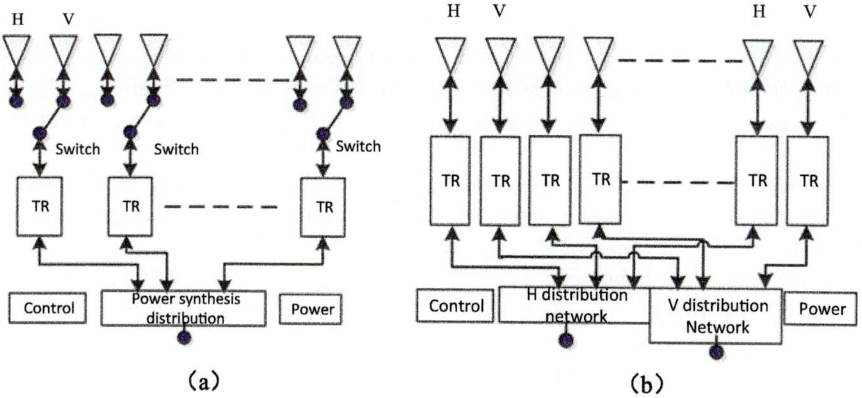

Fig. 6.1 Schematic diagram of dual-polarized antenna architecture. **a** H/V-polarized antenna element sharing T/R modules, **b** H/V-polarized antenna element independent T/R module

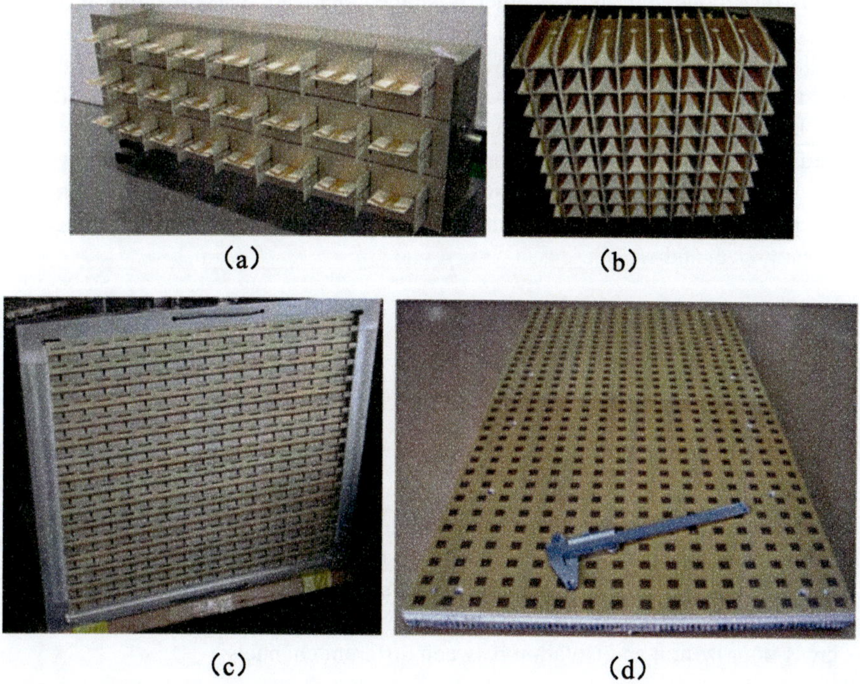

Fig. 6.2 Dual-polarized shared-aperture antenna, **a** printed dipole, **b** vivaldi, **c** waveguide slot, **d** microstrip patch

scope of application is limited. Therefore, a multi-band, dual-polarized feed reflector antenna is a better choice. However, the antenna beam shape of this architecture is fixed, and beam scanning requires electromechanical servo systems, multi-band high-power multi-channel rotary joints, centralized transmitters, etc. and there are multiple single-point failures and the risk of failure is high. In addition, the folded state of the reflector antenna has a large envelope, and the structure of the folded reflector is complex and technically difficult. Usually, only single-band, single-polarization reflector antennas are applied in microwave imaging.

Multi-band dual-polarized active phased array antennas are divided into three categories according to the polarization of each band antenna. The first type is single-polarization work in each band, such as horizontal polarization, vertical polarization, left-handed circular polarization, or right-handed circular polarization. This kind of shared-aperture antenna is relatively easy to implement. The second type is that each band antenna is dual polarized, and this type of antenna design is more complicated. The third category is somewhere in between the first and second types, i.e., one band is dual polarized and the other is single polarized. The operating mode of the dual-polarized antenna in each band has two situations of time division and simultaneous work, as described in Sect. 6.1.1. In addition, according to the scanning range requirements for antenna arrays, there are element-level active arrays and sub-array-level active arrays. Regardless of the classification, the design process of the shared-aperture antenna array is to maximize the sharing of the frame, deployment mechanism, thermal control, power supply, and beam control. Maximize the potential of the shared-aperture antenna architecture and take advantage of its advantages in efficiency, size, and weight.

6.2 Principle

6.2.1 Basic Parameters

Compared to conventional single-band, single-polarized active array antennas, in addition to performance parameters such as bandwidth, gain, beam width, beam shape, sidelobe, efficiency, and port voltage standing wave ratio (VSWR), multi-band, multi-polarization shared-aperture antennas need to focus on parameters such as cross-polarization and isolation between different channels.

(1) Isolation. Isolation parameters come from multi-port networks, such as power divider/combiners, directional couplers, bridges, and circulators. They are used to characterize the degree of crosstalk between channels of different polarizations and frequency bands. The isolation can be expressed as

$$I = 10\ \lg\left(\frac{P_2}{P_1}\right) = 20\ \lg|S_{21}| \tag{6.1}$$

where P_1 is the input power of the main signal channel and P_2 is the output power of the crosstalk signal entering the other channel. This parameter can reflect the amount of signal crosstalk between different channels, and the different channels can be channels of different polarizations or channels of different frequency bands.

(2) Cross polarization. There is a definite electromagnetic field polarization method for a specific electronic information system. For example, horizontal polarization, vertical polarization, left-hand circular polarization, or right-hand circular polarization are usually defined as main polarization, and the polarization orthogonal to it is called cross polarization. Correspondingly, the field component parallel to the field of the reference source is called the co-polarization field or the main polarization field. The field component orthogonal to it is the cross-polarization field. For linear polarization, the main polarization and cross polarization of electromagnetic waves are related to the coordinate system, and the method proposed by A.C. Ludwig is usually used [4]. In applying synthetic aperture radar based on the moving platform, the direction along the flight is called the azimuth direction, and the direction perpendicular to the flight direction is called the range direction. Correspondingly, the antenna whose radiated electric field vector is parallel to the azimuth direction is defined as a horizontally polarized antenna, and the one orthogonal to it is defined as a vertically polarized antenna, as shown in Fig. 6.3. The metric that characterizes antenna cross polarization is usually "cross-polarization discriminant," referred to as polarization purity or cross polarization. Numerically is the power ratio between the electromagnetic field component orthogonal to the main polarization and the main polarization component, and the cross-polarization component can be expressed as

$$\mathrm{XPD} = 10\ \lg\left(\frac{P_X}{P_C}\right) = 20\ \lg\left(\frac{E_X}{E_C}\right)(\mathrm{dB}) \tag{6.2}$$

where P_X is the cross-polarization power, P_C is the main polarization power, E_X is the cross-polarization electric field component, and E_C is the main electric field component. In practical applications, due to the installation or erection error in the darkroom test, the antenna and the probe polarization do not match, and there is a small angle θ, as shown in Fig. 3.9b. This introduces an antenna array cross-polarization measurement error of

$$\mathrm{XPD} = 20\ \lg(\tan\theta) \tag{6.3}$$

Therefore, the accuracy requirements for antenna test erection are usually put forward according to the cross-polarization index requirements.

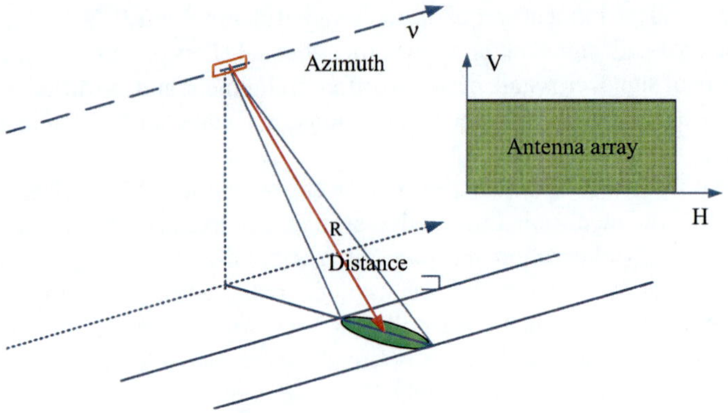

Fig. 6.3 Definition of antenna polarization in the SAR system

6.2.2 Dual-Polarization Shared Aperture

Electronic information systems usually require dual polarization for antennas based on different considerations, such as polarization multiplexing and polarization diversity, anti-jamming of dual polarization/multi-polarization in radar, and polarization information acquisition. The critical problem of the dual-polarization co-aperture antenna is to solve the crosstalk coupling in the channel and improve the isolation, that is to say, it is necessary to solve the problem of the isolation between the two ports of the main polarization and the cross polarization. As a result, the cross-polarization component of two orthogonal main polarizations is suppressed, and the crosstalk problem of two polarization vectors is solved. Furthermore, since most receivers are packaged in metal boxes, the channel isolation can achieve excellent performance if two completely independent RF signal channels are used. In this case, the isolation between the input/output ports of the two polarizations of the antenna element and the cross polarization of the space radiation field is mainly solved.

(1) Homogeneous dual-polarized antenna. Isomorphic dual-polarized antennas are dual-polarized antennas that use the same type of radiation mechanism antenna elements, such as square/circular microstrip patch antennas, square/circular horns, and open waveguides. The dual-polarized shared-aperture antenna excites two orthogonal internal fields on the antenna and radiates two completely orthogonal electromagnetic waves. In order to obtain excellent port isolation and cross-polarization performance, the two polarization modes are usually selected to work in the main mode and ensure their orthogonality.

Taking the microstrip patch antenna as an example, the dual-polarized square microstrip patch works in TM_{01} and TM_{10} modes, respectively, as shown in Fig. 6.4a. J_s^m is the equivalent magnetic current and E_z is the tangential electric field on the surrounding magnetic walls. Excitation at points C and D or between the two can obtain the main mode TM_{01} to work and radiate vertically polarized waves, excite at points A and B to obtain TM_{10} mode and radiate horizontally polarized waves. It can be seen from Fig. 6.4a that the excitation points of the two main modes are mutually zero points of the electric field. Therefore, theoretically, the two polarization isolations are infinite, but considering the perturbation of the excitation structure and the excitation structure is not infinitesimal, the mutual coupling still exists. Therefore, choose differential excitation to eliminate port coupling, that is, points *A* and *B* are excited in the anti-phase at the same time, and points *C* and *D* are excited in the anti-phase. As shown in Fig. 6.4b, when higher order modes are generated, ensuring the isolation between the two polarizations will be challenging. In addition, from the magnetic current distribution in Fig. 6.4b, the radiation fields of higher order modes are basically in the anti-phase state. In the normal direction of the antenna, the electromagnetic fields cancel each other, affecting the antenna gain and SLL.

(2) Heterogeneous dual-polarized antenna. A heterogeneous dual-polarized antenna is composed of antenna elements with different radiation mechanisms, such as slot and dipole nested dual-polarized antennas, as shown in Fig. 6.5.

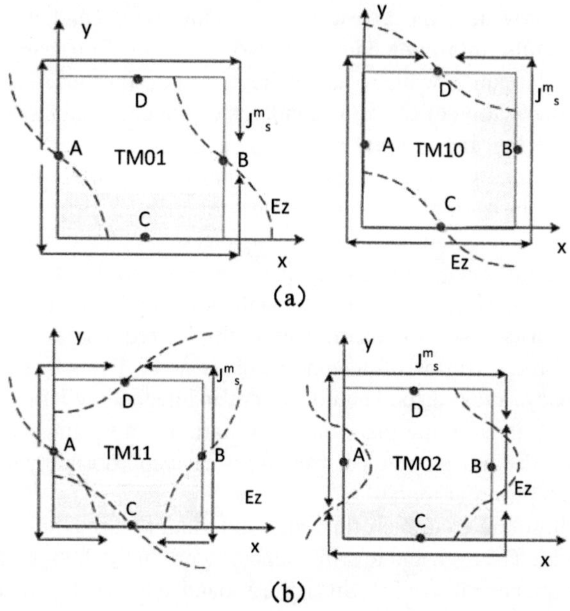

Fig. 6.4 Square microstrip patch dual-polarization mode, **a** main mode, **b** high-order mode

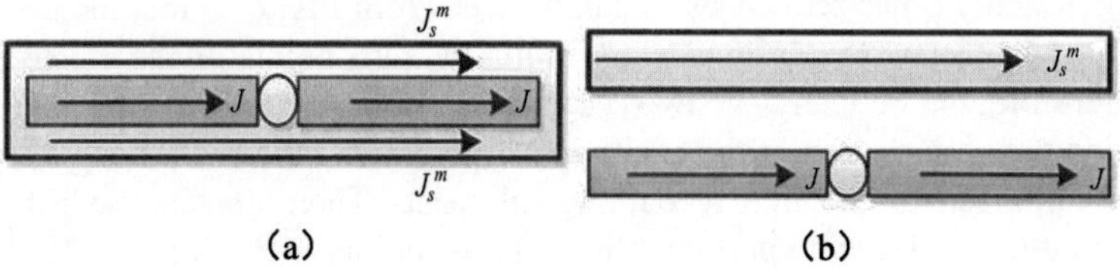

Fig. 6.5 Slot + dipole co-aperture antenna, **a** nested arrangement of dual-polarized antennas, **b** parallel arrangement of dual-polarized antennas

6.2.3 Multi-band, Multi-polarization Shared Aperture

Similar to the single-band and dual-polarization shared-aperture antenna, the multi-band and multi-polarization shared-aperture antenna mainly solves the implementation methods of multiple antennas in overlapping, nesting, and alternating architectures. On the one hand, it is necessary to ensure the effective radiation and impedance matching of each polarization antenna in each band. On the other hand, it is necessary to eliminate or weaken the crosstalk between multi-band and multi-polarization channels. As a result, improving antenna polarization isolation, band isolation, and reducing cross polarization while reducing the impact of spurious radiation on SLLs.

In the dual polarization of the same band, the orthogonal main mode is generally selected to ensure that the internal and external fields are orthogonal, the excitation points are mutually zero to ensure the polarization isolation of the antenna. However, cross polarization and isolation performance are affected in multi-band shared-aperture antennas due to the inevitable electromagnetic coupling and parasitic radiation. Therefore, a symmetrical distribution is usually used to suppress the high-order mode of the high-band antenna on the low-band radiation antenna. Under such conditions, it is easier to ensure even multiple frequencies than various odd frequencies, and the selection of antenna forms is more diverse. However, when a high-low frequency ratio is an odd number or not an integer, it will bring great trouble to the design. In this case, the main mode excitation and symmetrical distribution are still used, and the anti-phase feed structure is used appropriately to eliminate the influence of the high-order mode.

Figure 6.6 shows an example of a dual-band, dual-polarization shared-aperture antenna layout [5]. The high and low frequencies work in the X and S bands, respectively, with a frequency ratio of 3:1. Since the X-band antenna element is a microstrip radiating element, the S-band is a slot emitting element, it is a heterogeneous radiating element on the band. In the X-band, both polarizations are microstrip antenna elements, while in the S-band, both polarizations are slot antenna elements, so in terms of polarization, they are all isomorphic radiating elements.

Combined with the schematic diagram of the antenna array layout shown in Fig. 6.6, it can be known that both high- and low-frequency antennas work with the main mode. In order to suppress the electromagnetic energy coupled from the X-band to the S-band, the X-band antenna is arranged in a mirror-symmetrical

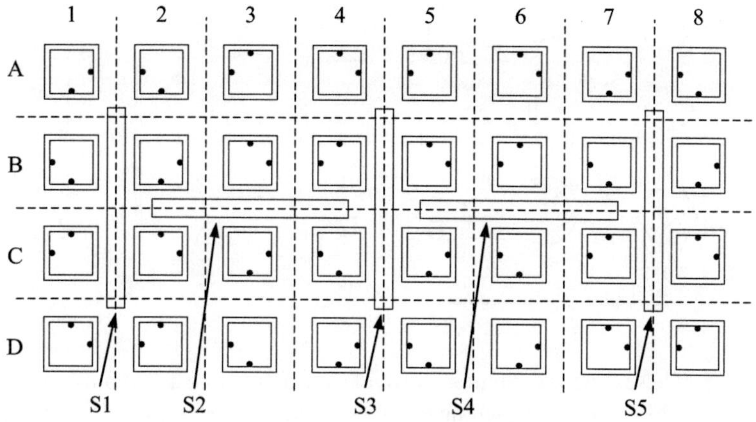

Fig. 6.6 Schematic diagram of dual-band dual-polarization layout and excitation

manner according to the S-band antenna. As shown in the S1 element in Fig. 6.6, the first and second columns around it is mirror-symmetrical to S1, and the upper and lower rows B and C of element S2 are X-band mirror-symmetrical to S2. S3, S4, and S5 are also laid out according to the same principle.

6.3 Antenna Elements

Microwave imaging radar systems usually require active array antennas to scan at a large angle in one dimension but only a few degrees in the other dimension. In view of weight, volume, and cost constraints, active array antennas generally adopt a sub-array structure, and a transceiver module is connected to a sub-array composed of multiple antenna elements in the azimuth direction. For the space-borne platform, due to the limitation of its size, weight, and power consumption by the platform capability, the sub-array architecture is given priority for the active phased array antenna.

6.3.1 Dielectric-Based Antennas

There are many types of antenna elements. From the difficulty of realizing a shared aperture, dielectric-based antennas are flexible, especially the combination of microstrip patches, planar oscillators, and microstrip slot antennas. Microstrip patch antennas can be rectangular, square, circular, cross-shaped, hollow, etc., while planar oscillators and slot antennas are spatially embodied in a 1D structure, as shown

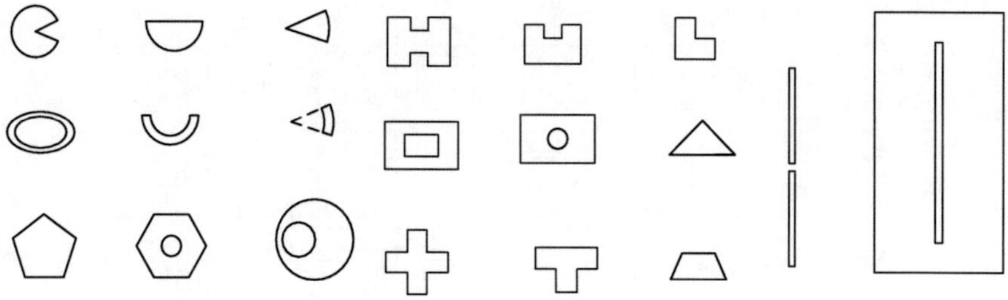

Fig. 6.7 Dielectric-based antenna element

in Fig. 6.7. These antenna elements provide great freedom for multi-antenna co-aperture design. Dielectric-based antennas with different shapes and structures can be nested, adjacent, and stacked to form shared-aperture antennas, so they are widely used in linear polarization, dual-polarization, circular polarization, multi-frequency, and multi-band antennas.

6.3.2 Metal-Based Antennas

Metal-based antennas mainly include slotted waveguide antennas, apertured waveguides, horns, and metal resonant cavity slot antennas, as shown in Fig. 6.8. Its main advantage lies in its high efficiency, which can facilitate the use of the sub-array structure. In addition, the overall design of the sub-array antenna can be used as the installation structure of T/R modules, beam control, and power supply modules. It can even have both heat dissipation and heat sink functions.

6.3.3 Hybrid-Based Antennas

The shared aperture of the hybrid-based antenna usually chooses the metal base antenna to be placed on the bottom layer, which serves as the ground of the structural member and the dielectric base antenna. As shown in Fig. 6.9a, the antenna working in the high-frequency band selects a waveguide slot array, and the low-frequency band uses a printed dipole. The two polarizations are orthogonal to reduce the crosstalk between them. The radiation surface of the waveguide slot array constitutes a plane of the metal ground, and the printed dipole is arranged on the waveguide array. The waveguide slot array is not only the ground reflection surface of the printed dipole but also its structural installation surface.

To sum up, planar dielectric-based antennas have advantages in terms of volume, weight, and integration, but the low efficiency of sub-array antennas is an inherent

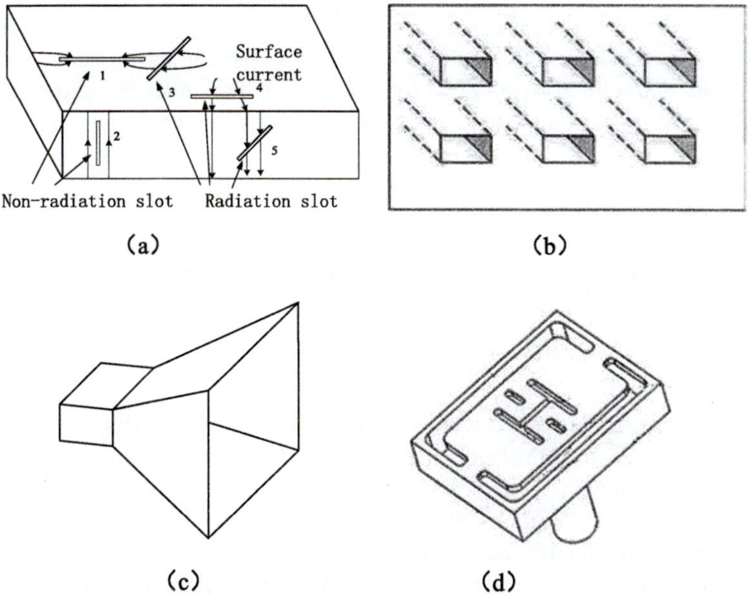

Fig. 6.8 Metal-based antenna element, **a** waveguide slot, **b** open waveguide, **c** rectangular horn, **d** cavity-backed antenna

Fig. 6.9 Hybrid-based antenna, **a** waveguide slot + printed dipole antenna, **b** metal cavity-backed microstrip antenna

defect. On the other hand, metal-based antennas have advantages in terms of efficiency, structural mechanics, and thermal conductivity but are disadvantaged in volume, weight, and integration. Therefore, in the design of shared-aperture antennas, various factors such as operating frequency band, bandwidth, polarization, and sub-array size can be considered comprehensively. Flexible application of the respective advantages of dielectric-based and metal-based antennas choose a hybrid combination. For example, choose a dielectric-based antenna in the low-frequency band and choose a metal-based antenna in the high-frequency band. In the dual-band shared-aperture antenna, generally select a dielectric-based antenna with a small number of sub-array elements, and choose a metal-based antenna with a large number of elements. Hybrid-based antennas are also ideal for achieving shared apertures. For example, in broadband cavity-backed antennas, the resonant cavity adopts a metal frame cavity and a microstrip circuit to realize the radiating surface. As shown in Fig. 6.9b, the grid-shaped metal cavity-backed is used simultaneously as a structure and a heat sink, and other components are installed on its back.

6.4 Dual-Polarized Microstrip Antennas

6.4.1 Microstrip Antenna Elements

The dual-polarized microstrip antennas are mainly classified according to the feeding mode into probe-feed multi-layer microstrip patch antennas, coplanar microstrip line-fed patch antennas, and slot-coupled microstrip antennas. Compared with single-polarized antennas, dual-polarized antennas increase key parameters such as polarization isolation and cross polarization.

The basic condition for the dual-polarization work of the microstrip antenna is to establish two orthogonally polarized electromagnetic fields that are mutually isolated, port-matched, and radiated. Ideally, dual-polarized orthogonal main mode excitation, no mutual coupling at the ports, and no distortion of the electromagnetic field of the radiating element can obtain excellent polarization isolation and cross-polarization performance. Assuming that the thickness of the microstrip antenna is much smaller than the wavelength, the space between the microstrip patch and the ground plate is regarded as a resonant cavity with magnetic walls around and electric walls up and down. According to Maxwell's equations and cavity boundary conditions, we can get

$$\begin{cases} L_r = \sum\limits_{m,n} B_{\text{mn}} \cos\frac{m\pi x}{\alpha}\cos\frac{n\pi y}{\alpha} \\ B_{\text{mn}} = jk_0\eta_0 I_0 \frac{\delta_{0m}\delta_{0n}}{\alpha^2\left(k^2-k_{\text{mn}}^2\right)}\cos\frac{m\pi x_0}{\alpha}\cos\frac{n\pi y_0}{\alpha}\, j_0\left(\frac{m\pi d_0}{2\alpha}\right) \end{cases} \tag{6.4}$$

where δ_{0m} and δ_{0n} are Neumann functions, x_0 and y_0 are the positions of the excitation points, and d_0 is the width of the feed microstrip line.

The dual-polarized antenna element excites a pair of orthogonally polarized working modes, TM_{10} and TM_{01}, through two independent feed points, as shown in Fig. 6.10. Vertical polarization mainly works in TM_{01} mode and horizontal polarization mainly works in TM_{10} mode. In addition to the working modes, the two feed points also excite higher order modes. As shown in Fig. 6.10b–d, the high-order modes of the square patch are usually considered only TM_{11}, TM_{02}, and TM_{20} modes. Equation (6.4) shows that when the feed point is in the middle of the side, the coefficient B_{11} is zero. Therefore, the element will not excite the TM_{11} mode, and the electric fields at port H and port V of TM_{02} and TM_{20} are not zero, thus generating coupling excitation and increasing the coupling between the two ports. If you want to improve the isolation between the ports, you can only reduce the excitation coupling of the high-order mode to the port. Comparing Fig. 6.10b, c with Fig. 6.10d, e, the direction of the electric field that generates TM_{02} and TM_{20} modes when the left side is excited alone is just opposite to that when the right side is excited alone. Therefore, equal-amplitude anti-phase feeding can be used on both sides simultaneously, as shown in Fig. 6.10f. This way, the excitation of TM_{02} and TM_{20} mode to port V can theoretically be eliminated. Since the fields excited by the two modes on the two sides are just in anti-phase, they also cancel each other out for the total port H, thereby further reducing the coupling between the two ports and improving the isolation of the polarized ports. The cross-polarization level is reduced due to the radiative cancellation suppression of the higher order modes.

If the double-point side-feed form is changed to the double-point angle-feed form, not only can the dual-polarization characteristics be obtained, but also the double-point angle-feed square microstrip antenna can improve the isolation [6]. Compared with the two-point side-feed square microstrip patch antenna, its port isolation has an advantage of 10 dB. When forming a dual-polarized microstrip antenna line array, the port isolation of −33 dB can be obtained [7], and the cross polarization is around −20 dB. However, when the angle-feed square microstrip antenna forms an antenna array, its diagonal size increases by two times relative to the side length of the square patch. Therefore, the occupied space of the non-antenna element is correspondingly increased, which is not conducive to the wiring of the broadband feed combination/division network.

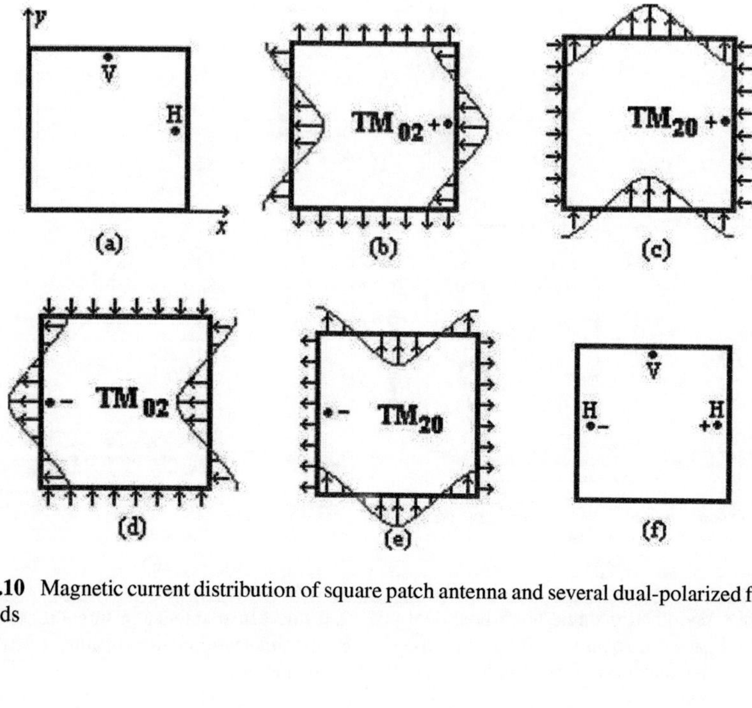

Fig. 6.10 Magnetic current distribution of square patch antenna and several dual-polarized feeding methods

Figure 6.11a is a side-feed structure of a dual-polarized microstrip antenna element [8]. The antenna element uses two independent feed ports. For vertical polarization, the microstrip line on the same layer as the patch antenna is directly connected to the feed. For horizontal polarization, the microstrip line under the floor of the patch antenna is used to couple and feed through the gap opened on the floor. In order to increase the antenna bandwidth, a double-layer microstrip structure is chosen.

The dual-polarization simulation results of this structure applied to the X-band are shown in Fig. 6.11c–e. S_{11} in Fig. 6.11c represents the coplanar microstrip feed port, i.e., the vertically polarized port. S_{22} represents the slot-coupled feed port, i.e.,

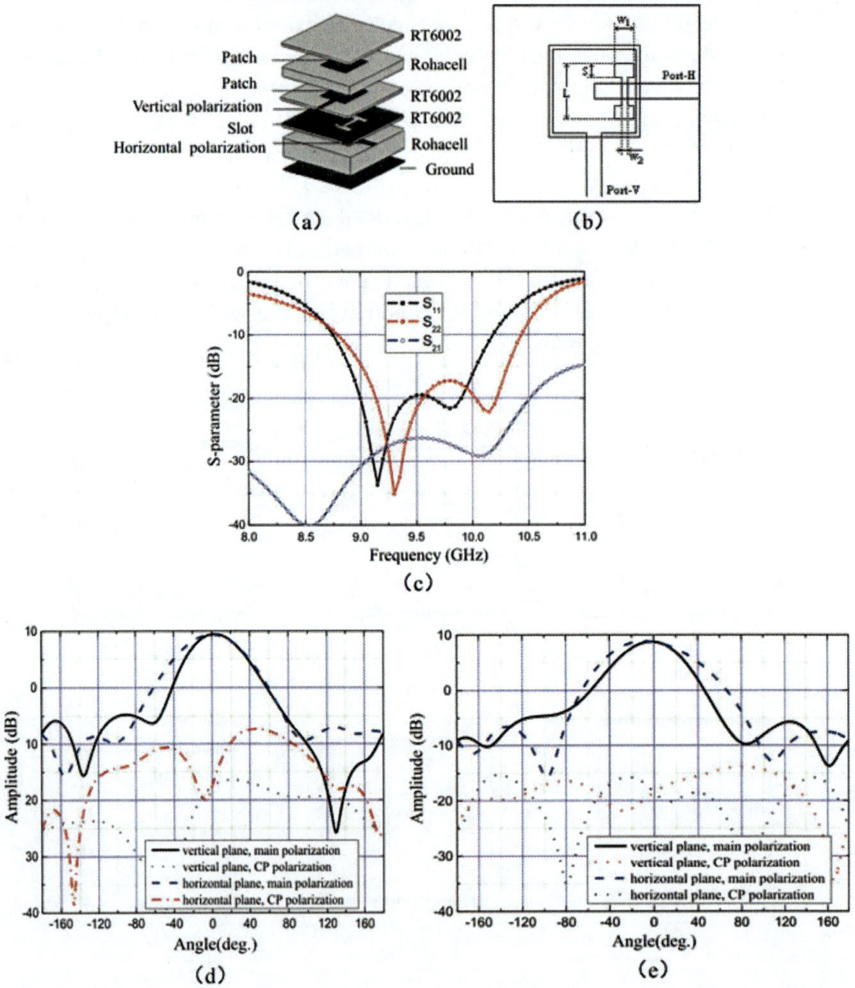

Fig. 6.11 Asymmetric single-point feed dual-polarized microstrip antenna, **a** three-dimensional structure, **b** side-fed square patch antenna structure, **c** S-parameter simulation results, **d** vertically polarized antenna pattern, **e** horizontally polarized antenna pattern

the horizontally polarized port. The dimensions of the antenna element are: parasitic square patch 10 × 10 mm^2; radiation patch 9 × 9 mm^2, $W_1 = 2$ mm, $W_2 = 1$ mm, $L = 7.4$ mm, $S = 2$ mm; radiation slot off center by 3.4 mm; and length of the coupling feed microstrip short circuit is 3.3 mm. The bandwidths of the two ports of the antenna element with reflection loss less than −10 dB reach 15.2% and 17.2%, respectively. The frequency range includes 8.75–10.19 and 8.79–10.44 GHz, and the isolation between the two ports reaches −20 dB.

Figure 6.11d, e shows the radiation patterns of the two main polarizations and their cross polarizations of the antenna. For vertical polarization, cross polarization is below −25 dB on the E-plane and below −17 dB on the H-plane. For the horizontal polarization, the cross-polarization components of the E-plane and H-plane are similar, both about −23 dB lower than the main polarization. The aforementioned poor isolation and cross-polarization performance are caused by the feed structure breaking the 2D symmetry of the antenna.

Based on the above analysis, a single coupling slot on one side of the microstrip radiating patch breaks the symmetry of the radiating antenna. The horizontally polarized ports are symmetrically arranged on both sides of the vertically polarized feeder through two identical slot couplings. The two coupling slots are fed through an equal power splitter with a phase difference of 180°, as shown in Fig. 6.12a. This structure increases the symmetry of the antenna and reduces the influence of the electromagnetic field distortion caused by the asymmetric structure on the antenna performance. Figure 6.12b shows the calculated and experimental values of the isolation of the two polarized ports, and the isolation reaches −40 dB. H-and V-polarized radiation cross polarized below −30 dB and −35 dB, respectively [9].

Figure 6.13 shows a typical symmetrical structure. The two feeds of the antenna element are both in the form of slot coupling, which is a double-layer microstrip antenna structure. In Fig. 6.13a, the coupled excitation of two slots on the side realizes dual polarization, and its port isolation is only −18 dB [10]. Change the position of the coupling slot to make the two have 1D symmetry, as shown in Fig. 6.13b,

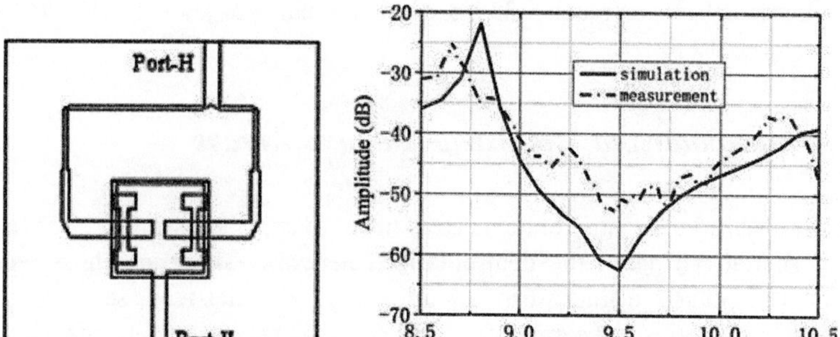

Fig. 6.12 Improved mixed-feed dual-polarized microstrip antenna, **a** antenna element structure, **b** S-parameter test results

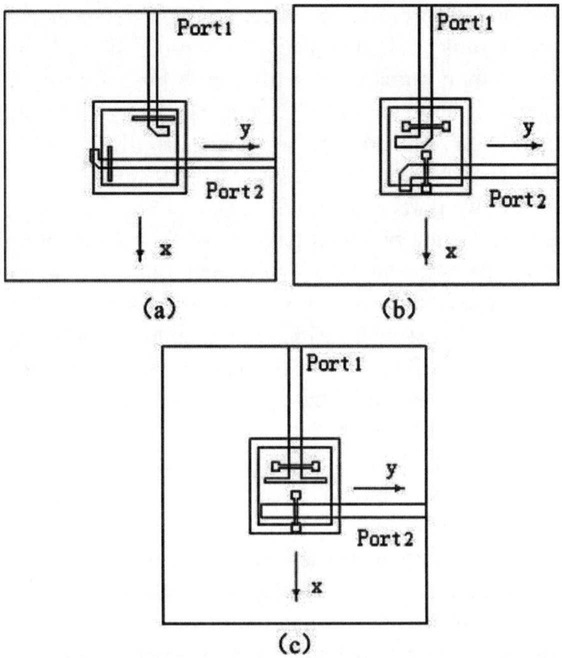

Fig. 6.13 Slot-coupled dual-polarized microstrip antenna structure, **a** dual-polarized asymmetrically excited microstrip antenna structure, **b** dual-polarized symmetrically excited antenna structure (L-shaped short circuit), **c** dual-polarized symmetrically excited microstrip antenna structure (T-shaped short wire)

the antenna port isolation reaches 36 dB, and the cross-polarization is better than −25 dB [11]. Next, change the terminal L-shaped short circuit to a T-shaped structure [12], as shown in Fig. 6.13c. Due to the use of a symmetrical feed structure, the cross polarization of the two polarized ports of the antenna can reach about −28 dB and the port isolation can reach −33 dB within the band, as shown in Fig. 6.14.

6.4.2 Dual-Polarized Microstrip Antenna Arrays

The dual-polarized microstrip antenna array needs to consider issues such as feeding loss and spatial wiring of the dual-polarized network. There are three types of feeds for dual-polarized microstrip antenna arrays: parallel feed, series feed, and a combination of series and parallel feeds. Figure 6.15a is a schematic diagram of an eight-element parallel feed mode. Multi-level 1:2 power splitters constitute a parallel feed network, ensuring equal electrical lengths from the main port to each radiating element, thus having good broadband equal phase performance. However, its network occupies a large space and is limited by the spacing of antenna elements (y-direction).

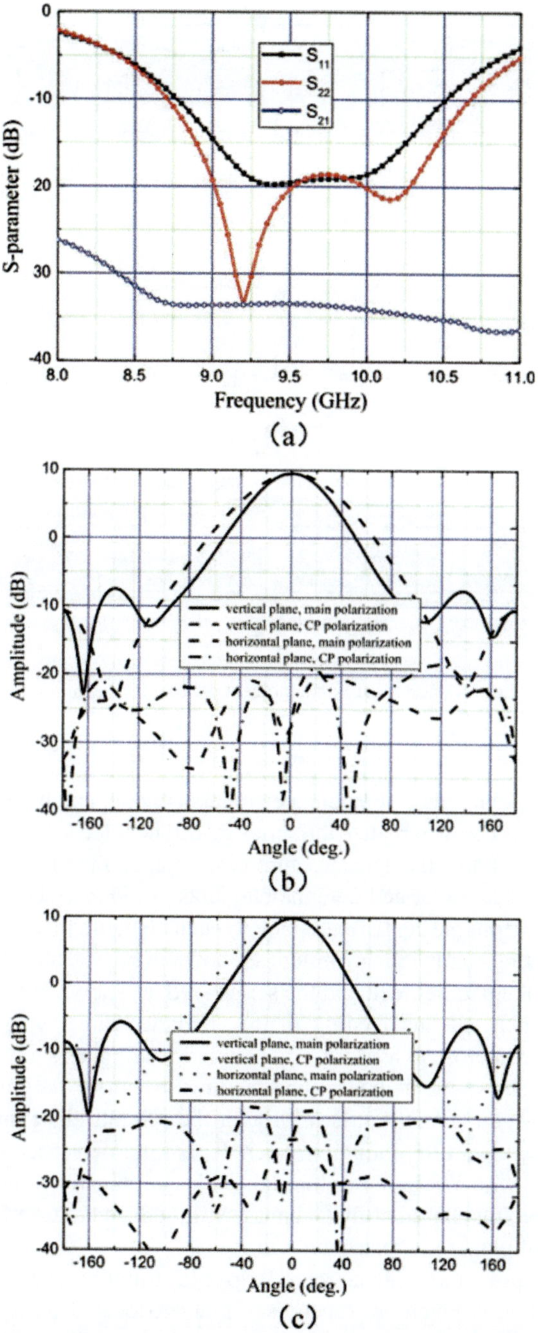

Fig. 6.14 Simulation results of slot-coupled dual-polarized microstrip antenna. **a** S-parameter simulation results, **b** vertically polarized antenna pattern, **c** horizontally polarized antenna pattern

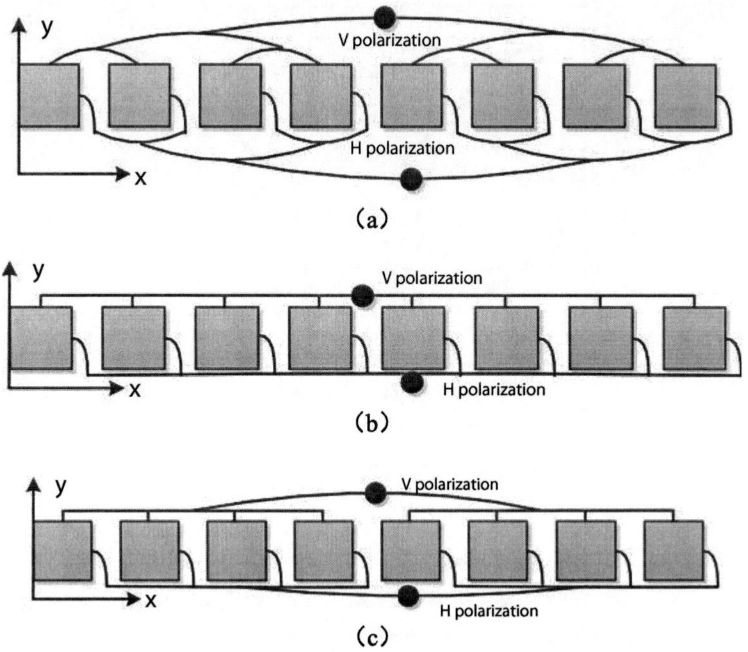

Fig. 6.15 Sub-array network combination methods, **a** parallel feed dual polarization, **b** serial feed dual polarization, **c** series and parallel feed combined dual polarization

Especially when the number of antenna elements is large, and the vertical scanning angle is large, it isn't easy to realize the series feed. Then, feeders are used to connect multiple feed points directly. The structure is simple, and the requirement for large spacing between dual-polarized antenna elements is reduced. Due to the different electrical lengths from the center feed point to each element, there is in-band dispersion, so its operating bandwidth is limited, as shown in Fig. 6.15b. The series-parallel feed is between the parallel feed and the series feed, as shown in Fig. 6.15c.

Microstrip patch antenna feeding mainly includes three types: coaxial probe feeding, coplanar microstrip feeding, and coupled feeding, as shown in Fig. 6.16.

Figure 6.16a shows the coaxial probe feeding structure, the inner conductor of the coaxial connector directly passes through the ground plane and the dielectric layer and connects with the radiation patch. This kind of vertical interconnection can be connected with T/R modules through cables or plug-ins, which is suitable for element-level feeding phased array antennas. Still, it does not take full advantage of the advantages of planar microstrip circuits.

Figure 6.16b shows the coplanar microstrip feed structure. The feeding microstrip line and the radiation patch are on the same layer, located on the surface of the dielectric board. They are designed and processed integrally with the microstrip power divider, suitable for sub-array-level feeding phased array antennas. However, since the feeder and the radiation patch are on the same layer, there is a "grab" for

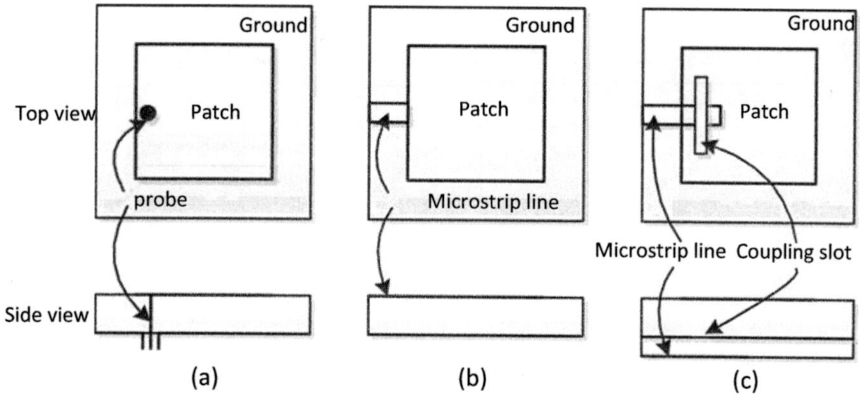

Fig. 6.16 Microstrip typical feed mode

area resources between the feeder network and the patch. In addition, the spurious radiation of the feeder network cannot be ignored under the low cross-polarization requirement.

Figure 6.16c shows the coupled feed structure. The slot-coupled patch antenna requires at least two layers of dielectric plates, and the feed microstrip line is coupled and excited through the slender slots on the ground plate of the antenna. The feeder and the antenna are isolated by the ground plane, so the area used in the sub-array network design is increased, and the parasitic radiation is eliminated simultaneously. The disadvantage is that there are many dielectric layers.

In order to further improve the dual-polarization isolation and cross-polarization performance, a symmetrical design is also added to the dual-polarization antenna design to suppress the influence of the high-order mode of the element. Figure 6.17a, b shows two-element microstrip linear arrays. Each patch contains two excitation points and produces two linearly polarized waves. "H" indicates the port that excites the horizontally polarized wave and "V" indicates the port that excites the vertically polarized wave. The symbol "+" indicates that the ports corresponding to the two elements are fed equal amplitude and phase, and "−" indicates that the ports corresponding to the two elements are fed equal amplitude and anti-phase. The direction diagram of the electric field in the far area of the antenna is

$$E(\theta, \varphi) = \begin{cases} E^h(\theta, \varphi) \\ E^v(\theta, \varphi) \end{cases} \tag{6.5}$$

where E^h is the horizontal polarization component and E^v is the vertical polarization component. The electric field of the two-element array is the superposition of the electric fields of the two elements in the far region, namely,

$$E_{2\times1}(\theta, \varphi) = G_L E_L(\theta, \varphi) + G_R E_R(\theta, \varphi)$$

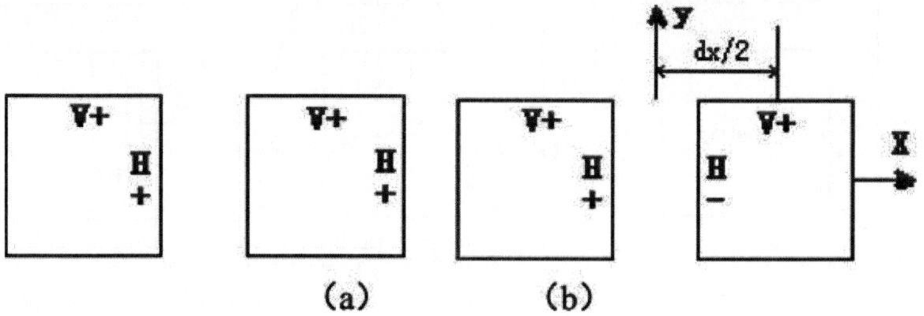

Fig. 6.17 The excitation combination of two-element array layout, **a** two-element in-phase feeding, **b** two-element anti-phase feeding

$$G_L = A_L e^{-jB}, G_R = A_R e^{jB}, B = \frac{\pi dx}{\lambda_0}\cos\varphi\sin\theta \tag{6.6}$$

where $E_L(\theta,\varphi)$ is the radiated electric field of the left antenna element, $E_R(\theta,\varphi)$ is the radiated electric field of the right unit, and λ_0 is the free space wavelength. For the structure shown in Fig. 6.17a, since the two ports adopt equal-amplitude and same-phase feed, take $A_L^H = A_R^H = A_L^V = A_R^V = \frac{1}{\sqrt{2}}$.

Using the principle of odd–even mode symmetry, the electric field of the left element can be expressed as

$$E_L(\theta,\varphi) = \begin{cases} E^h(\theta,\varphi) = E^{he}(\theta,\varphi) + E^{h0}(\theta,\varphi) \\ E^v(\theta,\varphi) = E^{ve}(\theta,\varphi) + E^{v0}(\theta,\varphi) \end{cases} \tag{6.7}$$

The electric field in the right cell can be expressed as

$$E_R(\theta,\varphi) = \begin{cases} E^h(\theta,\pi-\varphi) = E^{he}(\theta,\varphi) - E^{h0}(\theta,\varphi) \\ E^v(\theta,\pi-\varphi) = E^{ve}(\theta,\varphi) - E^{v0}(\theta,\varphi) \end{cases} \tag{6.8}$$

Substituting Eqs. (6.7) and (6.8) into Eq. (6.6), we can get

$$E^H_{A_{2\times1}}(\theta,\varphi) = \sqrt{2}\begin{cases} \cos \mathrm{B}\mathrm{E}^{Hhe}(\theta,\varphi) - j\sin \mathrm{B}\mathrm{E}^{Hh0}(\theta,\varphi) \\ \cos \mathrm{B}\mathrm{E}^{Hve}(\theta,\varphi) - j\sin \mathrm{B}\mathrm{E}^{Hv0}(\theta,\varphi) \end{cases} \tag{6.9}$$

$$E^V_{A_{2\times1}}(\theta,\varphi) = \sqrt{2}\begin{cases} \cos \mathrm{B}\mathrm{E}^{Vhe}(\theta,\varphi) - j\sin \mathrm{B}\mathrm{E}^{Vh0}(\theta,\varphi) \\ \cos \mathrm{B}\mathrm{E}^{Vve}(\theta,\varphi) - j\sin \mathrm{B}\mathrm{E}^{Vv0}(\theta,\varphi) \end{cases} \tag{6.10}$$

For the structure shown in Fig. 6.17b, the horizontal polarization port adopts equal-amplitude anti-phase feed and $A_L^H = -A_R^H = \frac{1}{\sqrt{2}}$, the vertical polarization port adopts equal-amplitude in-phase feed and $A_L^V = A_R^V = \frac{1}{\sqrt{2}}$. The electric field of the left unit is still Eq. (6.7), and the electric field of the right element becomes

$$E_R(\theta,\varphi)=\begin{cases}-E^h(\theta,\pi-\varphi)=-E^{he}(\theta,\varphi)+E^{h0}(\theta,\varphi)\\ E^v(\theta,\pi-\varphi)=E^{ve}(\theta,\varphi)-E^{v0}(\theta,\varphi)\end{cases} \tag{6.11}$$

Therefore, the electric field of the structure shown in Fig. 6.17b can be obtained as

$$E^H_{B_{2\times1}}(\theta,\varphi)=\sqrt{2}\begin{cases}\cos \mathrm{B}E^{Hhe}(\theta,\varphi)-j\sin \mathrm{B}E^{Hh0}(\theta,\varphi)\\ -j\sin \mathrm{B}E^{Hve}(\theta,\varphi)+\cos \mathrm{B}E^{Hv0}(\theta,\varphi)\end{cases} \tag{6.12}$$

$$E^V_{B_{2\times1}}(\theta,\varphi)=\sqrt{2}\begin{cases}-j\sin \mathrm{B}E^{Vhe}(\theta,\varphi)+\cos \mathrm{B}E^{Vh0}(\theta,\varphi)\\ \cos \mathrm{B}E^{Vve}(\theta,\varphi)-j\sin \mathrm{B}E^{Vv0}(\theta,\varphi)\end{cases} \tag{6.13}$$

In addition, according to the odd–even mode principle, there are

$$\begin{cases}E^e(\theta,\varphi)=E^e(\theta,-\varphi)\\ E^0(\theta,\varphi)=-E^0(\theta,-\varphi)\end{cases} \tag{6.14}$$

Setting $\varphi=0$, we can get

$$\begin{cases}E^{h0}(\theta,0)=0\\ E^{v0}(\theta,0)=0\end{cases} \tag{6.15}$$

Then, when $\varphi=0$, Eqs. (6.9) and (6.10) can be simplified as

$$E^H_{A_{2\times1}}(\theta,0)=\sqrt{2}\begin{cases}\cos \mathrm{B}E^{Hhe}(\theta,0)\\ \cos \mathrm{B}E^{Hve}(\theta,0)\end{cases} \tag{6.16}$$

$$E^V_{A_{2\times1}}(\theta,0)=\sqrt{2}\begin{cases}\cos \mathrm{B}E^{Hhe}(\theta,0)\\ \cos \mathrm{B}E^{Hve}(\theta,0)\end{cases} \tag{6.17}$$

Similarly, Eqs. (6.12) and (6.13) can also be simplified as

$$E^H_{B_{2\times1}}(\theta,0)=\sqrt{2}\begin{cases}\cos \mathrm{B}E^{Hhe}(\theta,0)\\ -j\sin \mathrm{B}E^{Hve}(\theta,0)\end{cases} \tag{6.18}$$

$$E^V_{B_{2\times1}}(\theta,0)=\sqrt{2}\begin{cases}-j\sin \mathrm{B}E^{Vhe}(\theta,0)\\ \cos \mathrm{B}E^{Vve}(\theta,0)\end{cases} \tag{6.19}$$

Define the cross polarization of port H as $\left|E^{Hv}(\theta,\varphi)\right|/\left|E^H(\theta,\varphi)\right|$ and the cross polarization of port V as $\left|E^{Vh}(\theta,\varphi)\right|/\left|E^V(\theta,\varphi)\right|$. It compares the two structures in Fig. 6.17a, b, on the $\varphi=0°$ plane, the cos B in the cross-polarization component changes to sin B. In order to suppress the grating lobe level, $d_x\le\lambda_0/2$ is usually taken, and the range of B is $0\sim\pi/2$. It can be seen that the cross-polarization component of the structure in Fig. 6.17a is the largest when $\theta=0°$. However, the cross-polarization component of the structure in Fig. 6.17b is the smallest when

$\theta = 0°$, thus suppressing the cross polarization in the main lobe but raising the cross-polarization level outside the main lobe.

The equal-amplitude anti-phase feeding of paired elements can better suppress the cross polarization in the main lobe but raise the cross-polarization level outside the main lobe. Therefore, optimizing the symmetry between multiple pairs of elements and sub-arrays in the antenna array to suppress cross polarization in a large range is necessary. Figure 6.18a–d shows the different combinations of four elements [13, 14], and Fig. 6.19 shows the four combinations of main polarization and their cross-polarization patterns in turn. The cross-polarization levels are −17.5 dB and −23.8 dB, −22.2 dB and −22.5 dB, −24.4 dB and −20.5 dB, −27.4 dB and −30.2 dB, respectively. It can be seen that the cross-polarization levels of the four structures are relatively low in the main lobe but relatively high outside the main lobe because the paired elements are fed with equal amplitude and anti-phase. However, compared with the other three structures, the cross-polarization level outside the main lobe of the structure in Fig. 6.18d has decreased significantly (about 6–10 dB). The main reason is that the feeding structure of this array suppresses not only the radiation of TM_{02} mode but also the radiation of TM_{20} mode, while the other three forms only suppress the radiation of TM_{02} mode.

The dual-polarized microstrip antenna array comprises antenna elements, feed structures, and power combination/distributing networks. The following will be analyzed and discussed in conjunction with practical examples. Figure 6.20 shows several typical structures of series-fed and series-parallel combinations. The antenna sub-array in Fig. 6.20a is center-fed six-elements, coplanar microstrip line series feed [15]. Figure 6.20b shows an eight-elements series feed structure [16]. Both polarizations choose the open-circuit microstrip line coupling feeding method. Due to the reverse feeding and the symmetrical distribution between the line arrays, the cross polarization is well suppressed. Figure 6.20c shows a 16-elements antenna series-parallel feed structure, which adopts the hybrid feeding method of coplanar microstrip and slot coupling [17]. Due to a large number of elements and narrow bandwidth, the horizontal polarization is fed by coplanar microstrip lines, and its cross-polarization performance is poor. The vertical polarization is fed by the slot coupling method located on the central axis of the patch, and the cross-polarization component is relatively low.

With the improvement of the antenna operating bandwidth and cross-polarization suppression requirements, it is sometimes necessary to select a parallel feed combination/division network, and there is a problem of "competition" for area resources between the parallel feed network and the radiation patch in coplanar microstrip line feeding. Therefore, it is chosen to feed the combination/division network below the metal ground and stimulate the microstrip antenna through the slot coupling to provide ample space for network design and layout. A C-band eight-elements antenna array adopts the form of a single-layer patch antenna [18], as shown in Fig. 6.21a. Since the dual-polarized feeding in the antenna array adopts slot coupling, and both slots deviate from the square antenna axes, it greatly influences the electromagnetic field distribution of the dual-polarized main mode of the antenna.

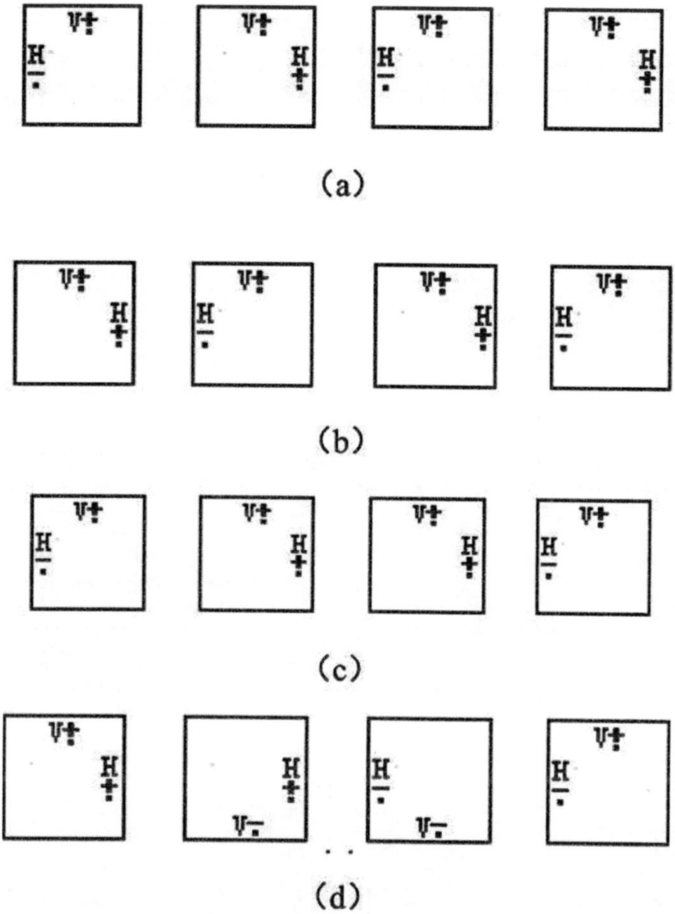

Fig. 6.18 Multi-array layout incentive combination, **a** method one, **b** method two, **c** method three, **d** method four

Figure 6.21b is an X-band eight-elements parallel-fed dual-polarization array [12] designed for symmetry improvement, and its performance is shown in Fig. 6.22. The reflection loss of the two polarization ports of the antenna is less than −10 dB, the bandwidth reaches 20%, and the port isolation is better than −35 dB in the whole band. Furthermore, radiation patterns excited by two ports show that the antenna cross-polarization level is below −37 dB in the main lobe.

If the antenna requires broadband work, the number of elements in the antenna sub-array is large. Although the above-mentioned double-slit coupling feeding method is adopted, the layout of the two parallel feeding networks is still very crowded, and the hybrid feeding method will be an appropriate choice. Taking the four-elements feed distribution as a basic combination, the radiating elements are connected into a 16-elements antenna array. Based on the same purpose, in the vertical direction, the

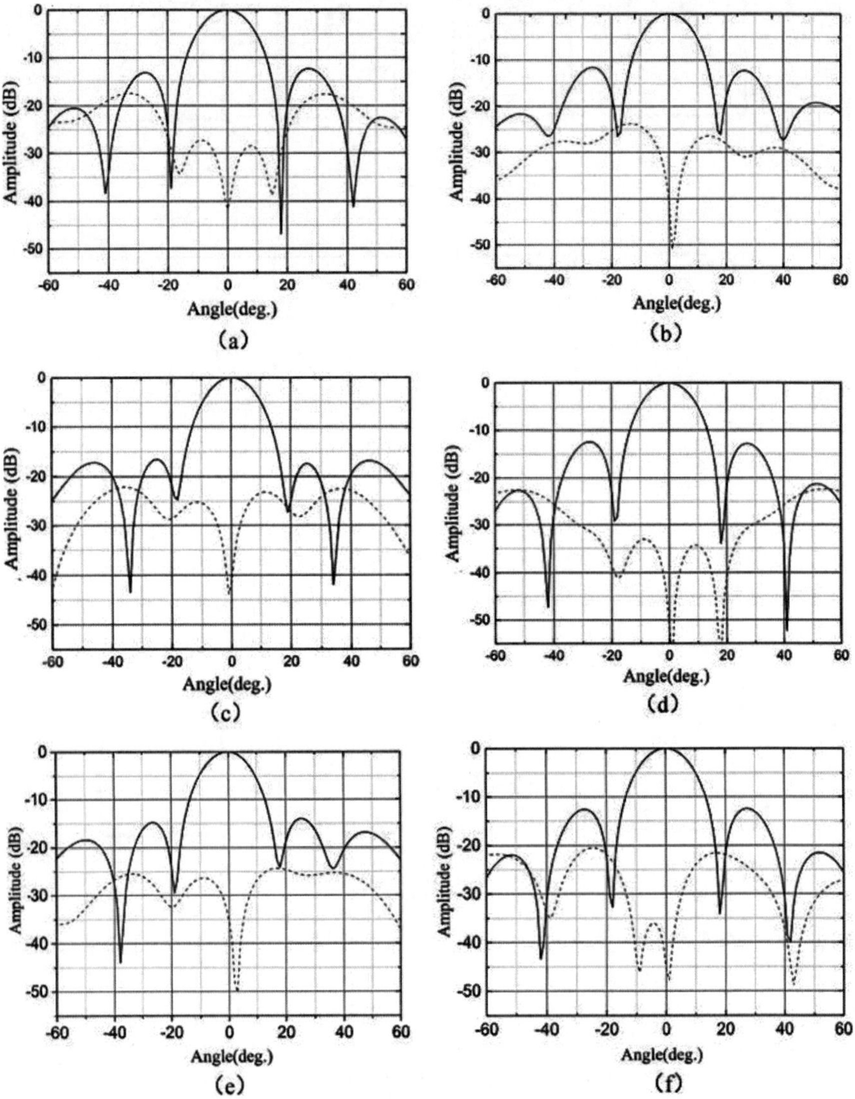

Fig. 6.19 Four-elements combined dual-polarization array calculation and measured pattern, **a** combination mode-horizontal polarization, **b** combination mode-vertical polarization, **c** combination mode-two horizontal polarization, **d** combination mode-two vertical polarization, **e** combination of three horizontal polarizations, **f** combination of three vertical polarizations, **g** combination of four horizontal polarizations, **h** combination of four vertical polarizations

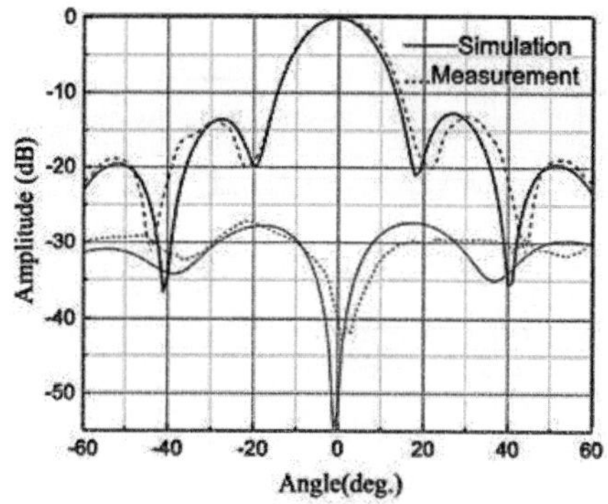

(g)

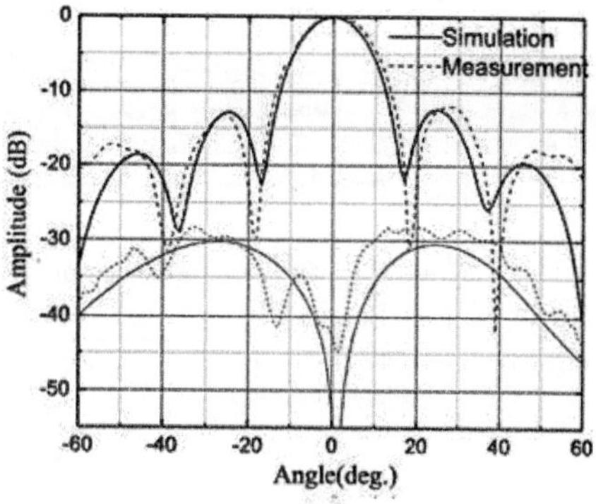

(h)

Fig. 6.19 (continued)

mirroring method is adopted between the linear arrays so that the antenna array is also symmetrical in the vertical direction. The horizontally polarized ports between adjacent linear arrays are fed equal amplitude and in-phase, while the vertically polarized ports are fed equal amplitude and anti-phase. Figure 6.23a shows the basic feed combination of arrays. The 16×16 element dual-polarization array designed and processed in this way is shown in Fig. 6.23b, and the back support is a honeycomb sandwich lightweight structural plate, and the antenna works in the X-band [19]. The test results are shown in Figs. 6.24, 6.25 and 6.26. The dual-polarization impedance bandwidth with a VSWR of less than 1.5 reaches 18.5%, and the polarization isolation is better than 45 dB. The cross polarization in the main lobe of the two main planes of the horizontal polarized array is lower than −44 dB and lower than −38 dB in the wide-angle range. The horizontal plane is slightly worse in the vertical polarized array, lower than −35 dB in the main lobe and lower than −34 dB in the wide-angle range. The vertical and horizontal polarized arrays are equivalent. The antenna gain of the horizontal polarized array is slightly lower than that of the vertical polarized array, and the antenna efficiency is higher than 47.6%.

6.4.3 Dual Circular Polarized Antennas

The electric field vector of any electromagnetic wave in space can be expressed as

$$E(\theta, \varphi) = \theta E_\theta(\theta, \varphi)e^{j\phi_1} + \varphi E_\varphi(\theta, \varphi)e^{j\phi_2} \tag{6.20}$$

where θ, φ are the far-field orthogonal bases, $E_\theta(\theta, \varphi)$ and $E_\varphi(\theta, \varphi)$ are the far-field orthogonal electric field components radiated by the antenna, and ϕ_1 and ϕ_2 are the phases of the two orthogonal components. When $E_\theta(\theta, \varphi) = E_\varphi(\theta, \varphi)$, and $\phi_1 - \phi_2 = \pm 90°$, the terminal of the radiated electric field vector is in the plane

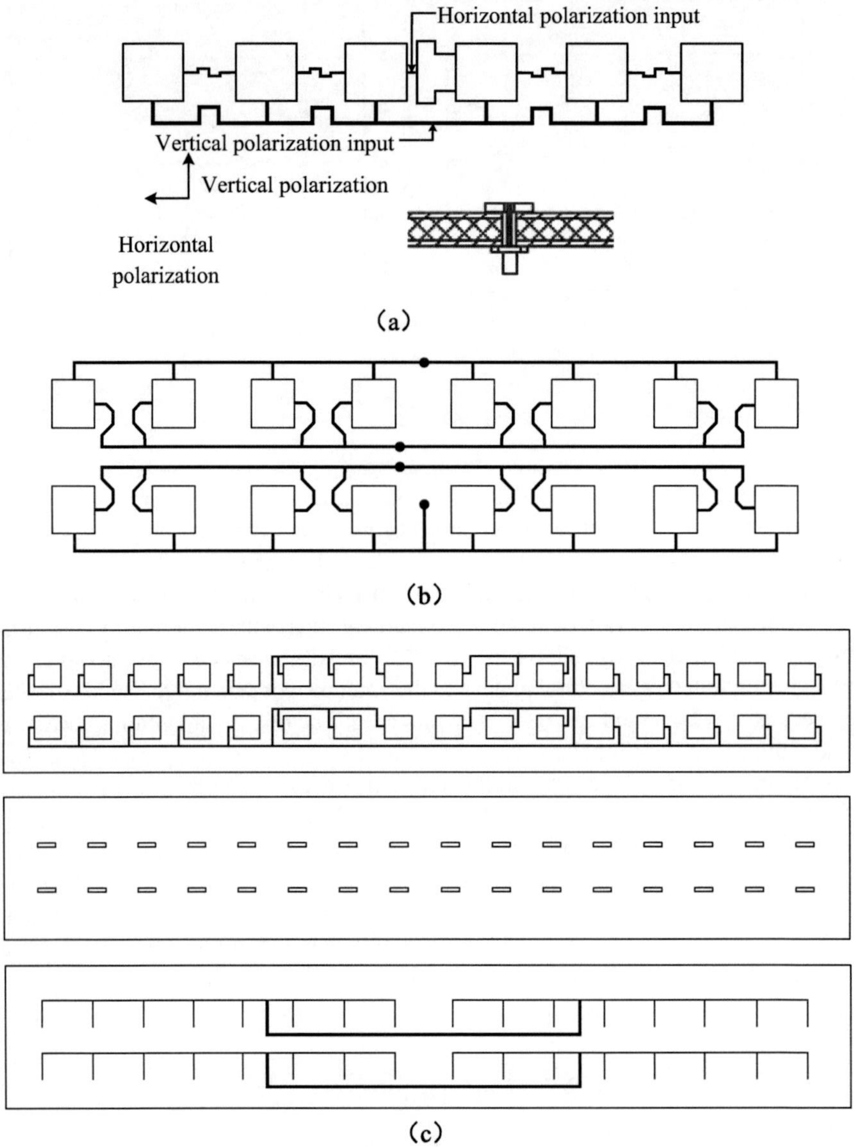

Fig. 6.20 Serial/parallel antenna array, **a** coplanar microstrip series feed, **b** coupled series feed, **c** series-parallel combined hybrid feed

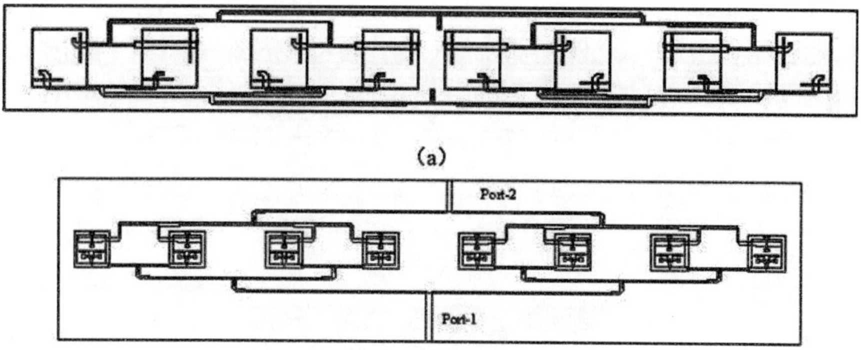

Fig. 6.21 Parallel feed slot-coupled antenna array, **a** eight-elements single-layer patch antenna, **b** eight-elements parallel feed dual-polarization array

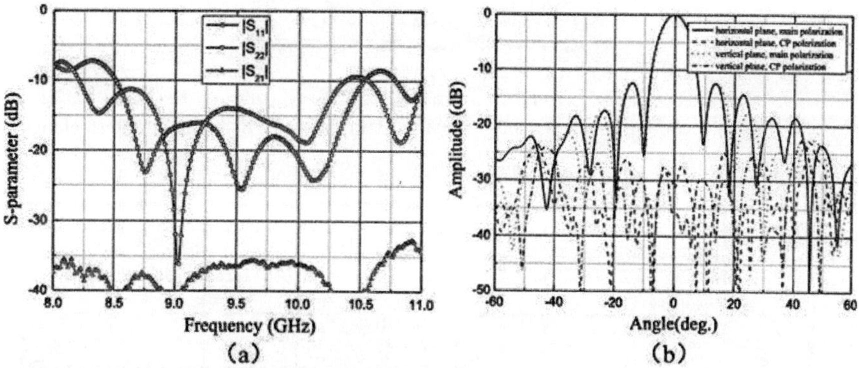

Fig. 6.22 Eight-elements parallel feed dual-polarization array test results, **a** port reflection loss and isolation, **b** pattern and cross polarization

perpendicular to the wave vector, and the trajectory with time appears as a circle, that is, circular polarization, as shown in Fig. 6.27. The circularly polarized antenna is very easy to understand and design from the essence of its electromagnetic field, i.e., the excitation radiation amplitude is equal, the space is orthogonal, and the two phases are 90° electromagnetic waves. A circularly polarized antenna can be obtained if the excitation and radiation of any radiator meet these three elements. Single circularly polarized antennas mainly include single-point, double-point, and four-point excitation methods [20–22].

Starting from the basic concept, if dual-linear polarization is realized, it is easy to obtain dual circularly polarized antennas. Based on dual linearly polarized antennas, a leading or lagging 90° phase difference is added to the dual-polarized channels. This phase difference can be realized at the system level in active array antennas or the element level. At the system level, in the dual-channel linearly polarized

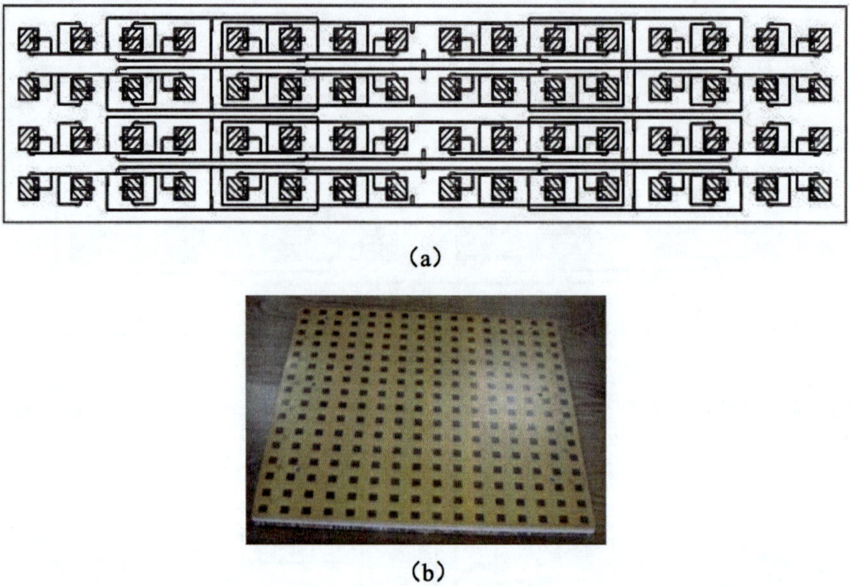

Fig. 6.23 Mixed feeding 16-elements sub-array, **a** sub-array structure diagram, **b** 16 × 16-elements antenna array prototype

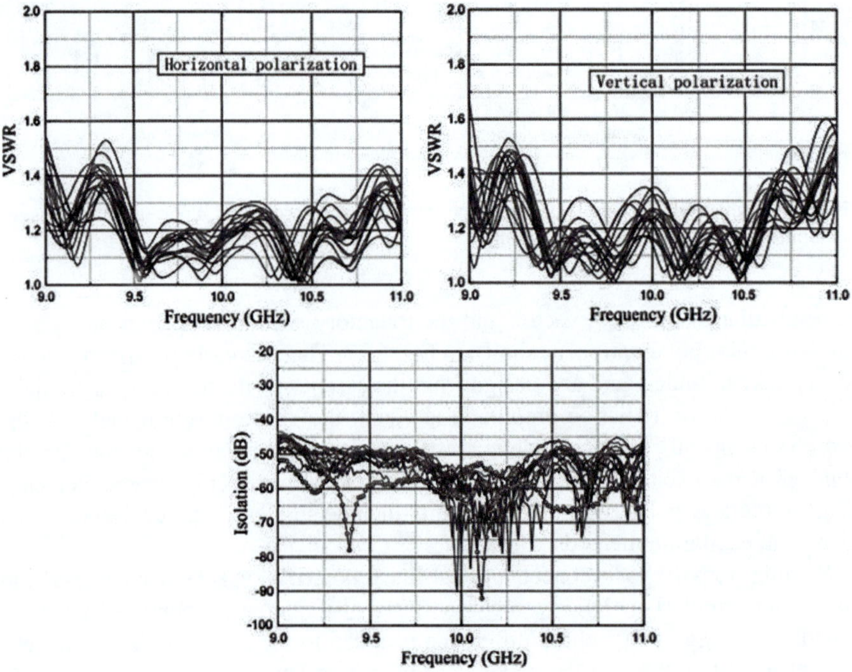

Fig. 6.24 Antenna port isolation test results

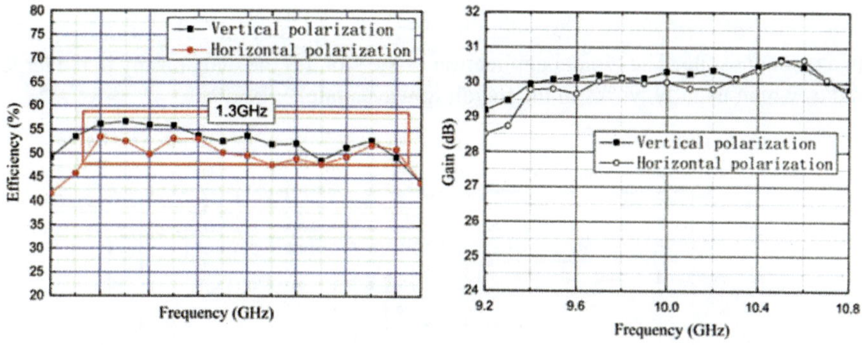

Fig. 6.25 Antenna gain and efficiency test results

active array antenna, the phase difference between the two polarized channels is obtained through unified control of the antenna array phase shifter. This method is very flexible and can obtain dual linear and circular polarization working modes in the active array antenna. At the element level, dual circular polarization is realized on the antenna element. Its advantage is that it can provide any linear polarization working mode under antenna transmission saturation amplification [23]. This working mode is usually used in satellite communication but also provides a new technical route for future microwave imaging technology.

The direct and effective method of the element-level dual circularly polarized antenna is to add a 90° electric bridge at the two excitation terminals based on the dual-polarized antenna, as shown in Fig. 6.28. First, to improve the bandwidth and axial ratio (AR) performance, each linear polarization is synthesized by dual-point anti-phase feed. Then a broadband 90° electric bridge is added.

6.5 Dual-Polarization Waveguide Slot Array Antennas

Despite the rapid development of microstrip antennas in microwave imaging radar applications, waveguide slot array antennas still have advantages. It is mainly manifested in the operational characteristics of broadband, dual polarization, high efficiency, low cross polarization, good electrical conductivity, thermal conductivity, and structural strength. In addition, waveguide slot array antennas can also achieve lighter weight under advanced manufacturing conditions.

Based on the slot radiation mechanism of the waveguide slot, as shown in Fig. 6.29, there are five types of slots in the rectangular cross-sectional metal waveguide. Among them slots 1 and 2 are parallel to the surface current, so they have no radiation. Slots 3, 4, and 5 cut the metal surface current, forming a discontinuous electromagnetic field and generating electromagnetic radiation, which can be used as the radiation element of the antenna. Traditionally, slots 4 and 5 are selected as the parallel radiation element in the waveguide. The standard application of slot 4 is a concentratedly fed planar waveguide slot array, and the slot line array usually works in a resonant state. The typical application of the slot 5 is the planar array in the large radar, which usually works in the traveling wave state.

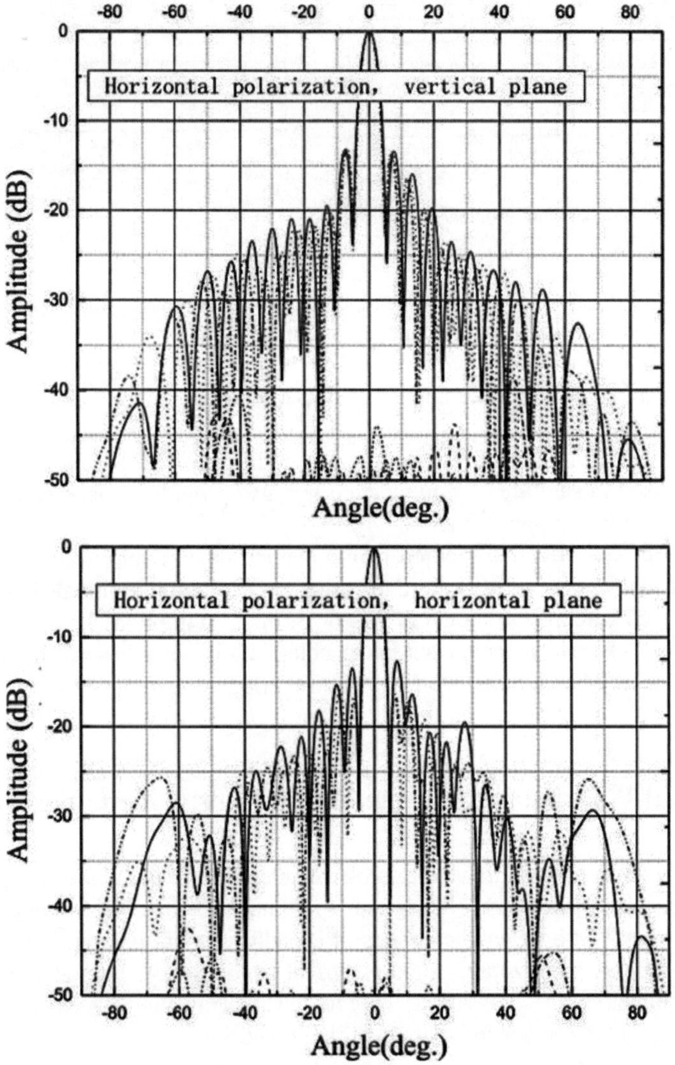

Fig. 6.26 Antenna array pattern and cross-polarization test results

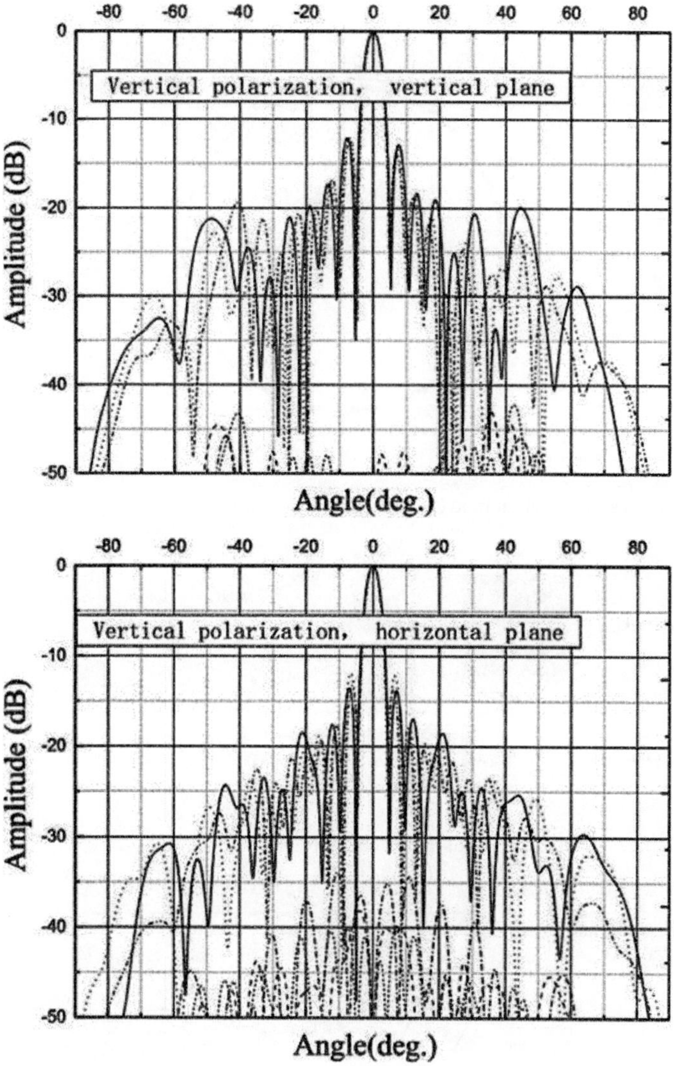

Fig. 6.26 (continued)

6.5.1 *Waveguide Slot Configuration*

From the perspective of the electromagnetic wave radiation mechanism, any guided electromagnetic field will produce radiation as long as it meets the two basic conditions of discontinuity and being in free space. Therefore, the slender slot on the metal wall of the closed cavity is closest to the ideal linearly polarized antenna. As shown in Fig. 6.30a, the smaller the gap between the upper and lower metal edges of the slit, the smaller the distortion of the electric field, and the electric field is completely

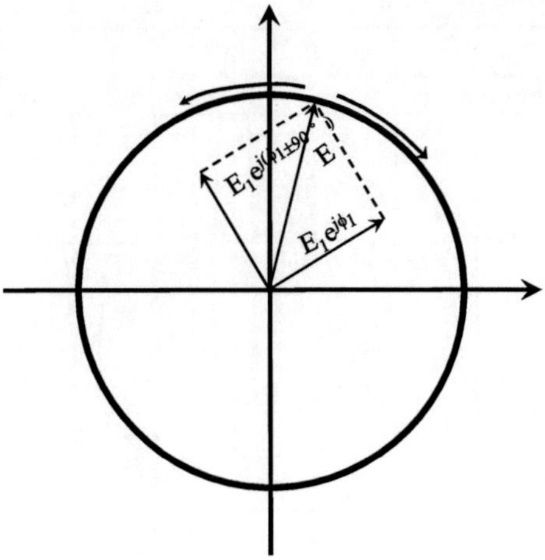

Fig. 6.27 Circular polarization field electric field locus

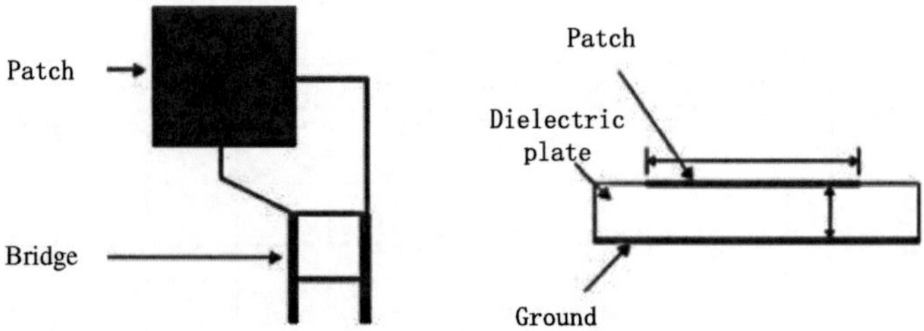

Fig. 6.28 Typical structure of a dual-polarized microstrip patch

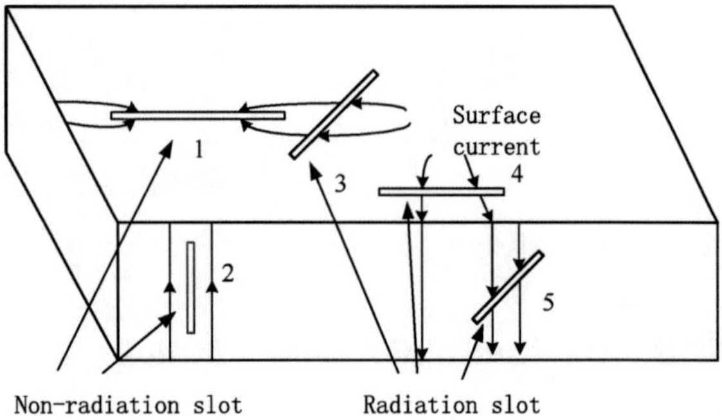

Fig. 6.29 Rectangular waveguide wall slot

perpendicular to the slit. In addition, the feeding structure is completely closed to avoid the parasitic radiation of other structures, so the cross-polarization component of the radiated electromagnetic wave is close to zero. Considering the influence of

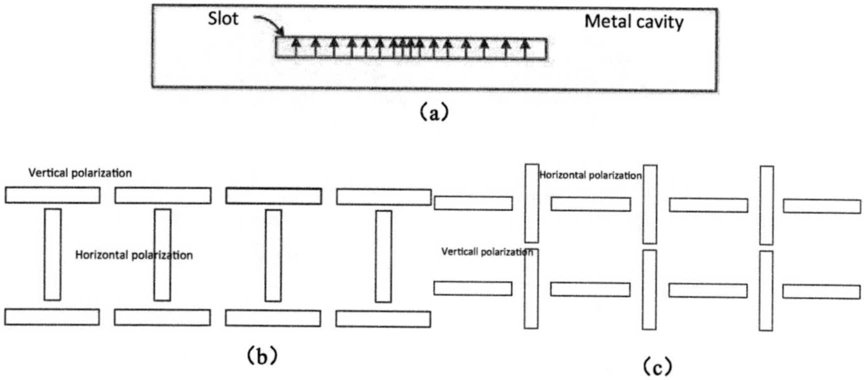

Fig. 6.30 Slot waveguide antenna, **a** closed cavity slender radiation slot, **b** dual-polarization slot arrangement I, **c** dual-polarization slot arrangement II

parasitic radiation caused by the mutual coupling of radiation slots on cross polarization and deterioration of port isolation, a symmetrical array structure is selected, as shown in Fig. 6.30b, c. In arrangement I, the two kinds of polarization radiation slits are distributed continuously in the horizontal direction while arranged alternately in the vertical direction. This arrangement is suitable for horizontal linear arrays based on the structural characteristics of rectangular metal cavities. Arrangement II is suitable for vertical linear arrays.

The waveguide slot array antenna is based on a closed metal waveguide, and the only source of radiated electromagnetic waves is the long slot that cuts the current on the metal wall, such as slots 3, 4, and 5 in Fig. 6.29. Therefore, good dual-polarized antenna performance can be obtained by properly designing the radiation slot and its arrangement.

In microwave imaging radar, the sub-array-level active array antenna has a smaller scanning angle in the horizontal direction. Still, a larger scanning angle is in the range direction, so most of the antenna element sub-arrays are arranged in parallel in the horizontal direction. In a dual-polarized array, the vertical polarization is realized by the waveguide broad-wall slot antenna, and the waveguide narrow-wall slot realizes the horizontal polarization. Based on this working state, the dual-polarized radiation slots in Fig. 6.30c are arranged alternately in the horizontal direction, which is difficult to realize, and the arrangement in Fig. 6.30b can only be selected.

Figure 6.31a shows a rectangular waveguide broadside slot antenna and its array arrangement. Slot No. 4 in Fig. 6.29 is selected as the radiation slot, which meets the most basic conditions as an antenna. Usually, its length is about half a wavelength, and the radiation intensity of the slot is determined by the distance away from the central axis. The current distribution in Fig. 6.29 shows that the polarity of the radial slit deviates from the symmetrical position of the opposite center line. Therefore, although slot 3 can also obtain electromagnetic radiation, compared with slot 4, its cross-polarization problem is more prominent, so slot 3 is rarely selected as the waveguide broadside radiation slot in practice.

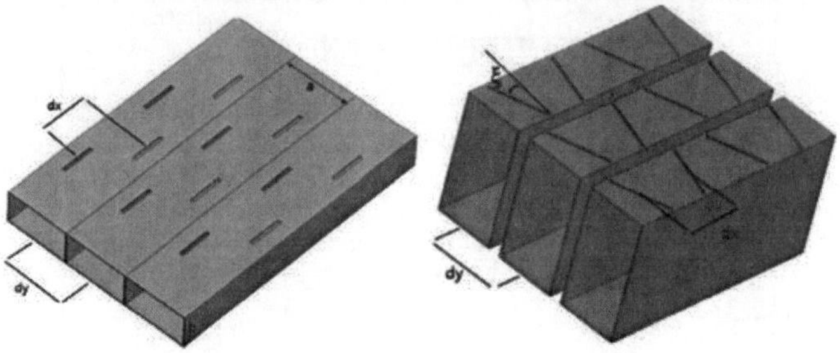

Fig. 6.31 Waveguide slot array antenna, **a** rectangular waveguide broad-wall slot array antenna, **b** rectangular waveguide narrow-wall inclined slot array antenna

Figure 6.31b shows a rectangular waveguide narrow-wall inclined slot array antenna and its array arrangement. Slot No. 5 in Fig. 6.29 is selected as the radiation slot, which constitutes the basic antenna form. The amplitude depends on the inclination angle ξ of the slot, and its left or right inclination determines the polarity of the radiated electromagnetic field. Since the width of the narrow wall of the rectangular waveguide cannot meet the half-wavelength resonance requirement of the radiation slot, it is usually necessary to cut the broad wall of the waveguide in a "Π" shape in practical applications.

Based on the above analysis, two orthogonally polarized antennas can be formed by slots 4 and 5 in Fig. 6.29. The size and direction of the bias or tilt angle control their radiation intensity and polarity. When the gap d_x of the slots on the same waveguide is selected as half of the waveguide wavelength, the slots are alternately offset or reversed, and one end is input. The other end is short-circuited, and a standing wave array antenna can be constructed. When $0 < d_x < \lambda_g/2$ or $\lambda_g/2 < d_x < \lambda_g$ is selected, and the other end of the waveguide is connected with a matching load, a traveling wave array antenna is formed.

Due to the large size of the waveguide antenna, there is a problem of structural interference when selecting a standard waveguide to realize a dual-polarized antenna array, so the method of compressing the size of the waveguide is often used. The compression of the dimensions of the rectangular waveguide should ensure single-mode transmission. Since the size of the waveguide is related to the working frequency range, considering the high-order mode interference antenna radiation, the ratio of the highest frequency to the lowest frequency in the working frequency range should not be too large, and $f_{\max}/f_{\min} \leq 1.5$ is generally taken. In addition, the working frequency band should keep a certain distance from the cutoff frequency of the main mode. Generally, $f_{\min}/f_c \leq 1.25$, and the cutoff wavelength of the main mode is $\lambda_c = 2a$. Therefore, the condition for only transmitting TE_{10} single mode is

$$\frac{\lambda}{2} < a < \lambda \tag{6.21}$$

In order to cut off the TE_{01} mode, it is required

$$b < \frac{\lambda}{2} \tag{6.22}$$

In addition, the choice of waveguide size also affects the transmission loss, a factor that needs to be considered in high-efficiency antennas. For the TE_{m0} transmission mode, the decay constant is

$$a_{m0} = \frac{1}{b} \frac{R_f}{\sqrt{1 - \left(\frac{\lambda}{\lambda_c}\right)^2}} \left[1 + \frac{2b}{a}\left(\frac{\lambda}{\lambda_c}\right)^2\right] \text{(Np/m)} \tag{6.23}$$

where R_f is the normalized surface resistance, a is the waveguide width, and b is the waveguide height. Ideally, when $a/b = 0.85$, the waveguide loss is the smallest, but it does not meet the single-mode transmission constraints, usually choose $a/b = 2$. But for compact sizes, half-height waveguides are often chosen, and sometimes quarter-height waveguides are chosen.

For the compression of the waveguide width, ridge waveguide structures are usually used, including single-ridge waveguides, double-ridge waveguides, etc.

6.5.2 Bandwidth Widening Technology

The slot waveguide traveling wave antenna has a broader bandwidth than the resonant antenna array, but the beam pointing of this type of antenna scans with the frequency change. Therefore, in the application of high-resolution microwave imaging radar, the beam dispersion effect of the antenna will significantly affect the image resolution. The basic principle is that high resolution is obtained through the instantaneous bandwidth in the range direction. However, simultaneously, the beam dispersion of the traveling wave array narrows the effective bandwidth of the illuminated target, resulting in a deterioration of the resolution.

The bandwidth of the waveguide resonant array antenna is limited, as shown in Fig. 6.32. One end of the rectangular waveguide is short-circuited, and the other is used as an input/output port. There are four radiation slots on the broad side of the waveguide. The four-elements resonant array is designed with a center frequency of F_0, and the element spacing is $\lambda_{g_0}/2$. Each radiation slot is located at the peak point of the electric field intensity. On the one hand, as the operating frequency shifts from the center frequency to both sides of high F_H and low F_L, its guide wavelength λ_{g_0} changes, and the radiation slot gradually deviates from the peak point of the electric field intensity corresponding to this frequency. As the bandwidth increases, the greater the change of the guide wavelength is, the more the peak points of the electric field intensity corresponding to the high and low frequencies F_H and F_L deviate, and the radiation pattern deteriorates more seriously. On the other hand,

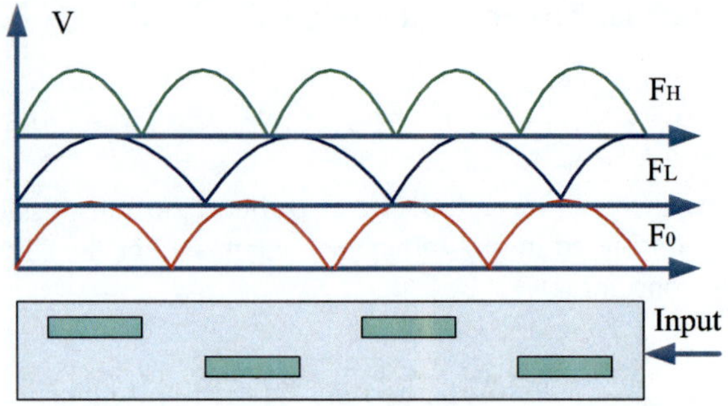

Fig. 6.32 Waveguide slot resonator array bandwidth limitation

when the bandwidth is constant, the more elements there are, the farther the radiation gap position is from the short-circuit surface of the terminal. When the operating frequency deviates from the intermediate frequency, the greater the accumulation of radiation slot peak deviation, the faster the radiation pattern will deteriorate.

Theoretically, by changing the susceptance curve of the slot, the pattern bandwidth of the waveguide resonant antenna array can be improved, and the impedance bandwidth can be expanded by adding an impedance-matching structure. However, the most direct and effective way to broaden the operating bandwidth of waveguide resonant arrays is still to reduce the number of elements in the resonant array, divide the resonant array with more elements into multiple sub-arrays, and use the parallel feed network to feed and excite the sub-arrays. For example, as shown in Fig. 6.33, 16 units are fed in the middle, and each resonant array has 8 elements. Adding a layer of power dividers reduces the number of elements in each resonant array to four. It reduces the number of resonant array elements by half, so the bandwidth of the antenna pattern is greatly expanded.

In space-borne microwave imaging applications, 1D resonant arrays are usually selected to meet the requirements of large-angle scanning in the range and only

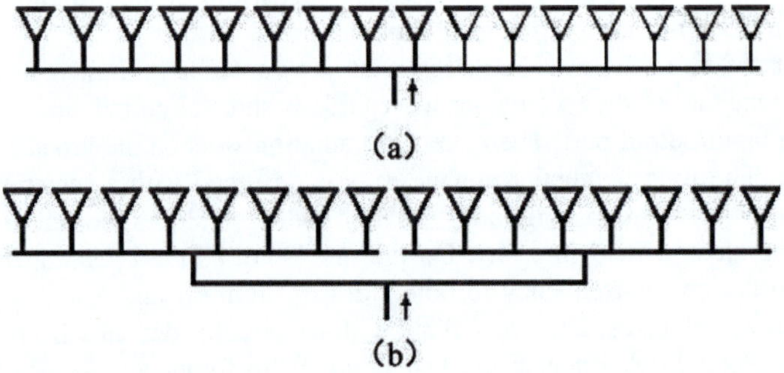

Fig. 6.33 Segmented power division excited resonant antenna, **a** mid-fed 16-elements array, **b** primary power division 16-elements array

a few degrees in the azimuth. A waveguide power divider is designed below the radiation waveguide, and the two realize microwave power division and excitation through slot coupling or T-joint to widen the bandwidth. This approach applies to slot resonant arrays with rectangular waveguides. Figure 6.34 shows two structures of ridge waveguide power splitters. Figure 6.34a is a parallel arrangement of radiating ridge waveguides, and feeding ridge waveguides up and down [24], and Fig. 6.34b is a back-to-back ridge waveguide structure [25]. The radiation slot is opened on the broad wall of the upper ridge waveguide, the feed ridge waveguide and the radiation waveguide are coupled through the "I"-shaped slot, and the port uses a coaxial waveguide "T" joint.

Since this broadband waveguide slot antenna array is added with a waveguide power divider and integrated with an interference suppression filter, it provides anti-out-of-band interference characteristics for the microwave imaging radar system, as shown in Fig. 6.35. This combined design has no additional structure and minimizes insertion loss. In some lower resolution applications, the antenna bandwidth is moderate, and no additional power divider is needed. At this time, other means are needed to realize the filtering function of the antenna, such as a waveguide slot antenna with a metamaterial surface structure [26].

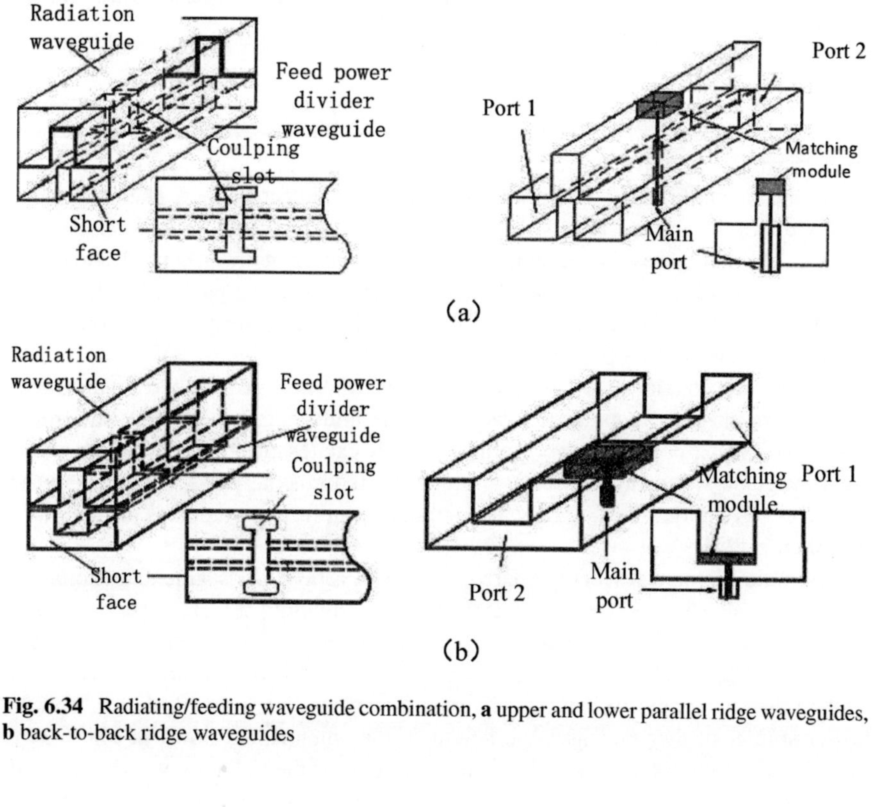

Fig. 6.34 Radiating/feeding waveguide combination, **a** upper and lower parallel ridge waveguides, **b** back-to-back ridge waveguides

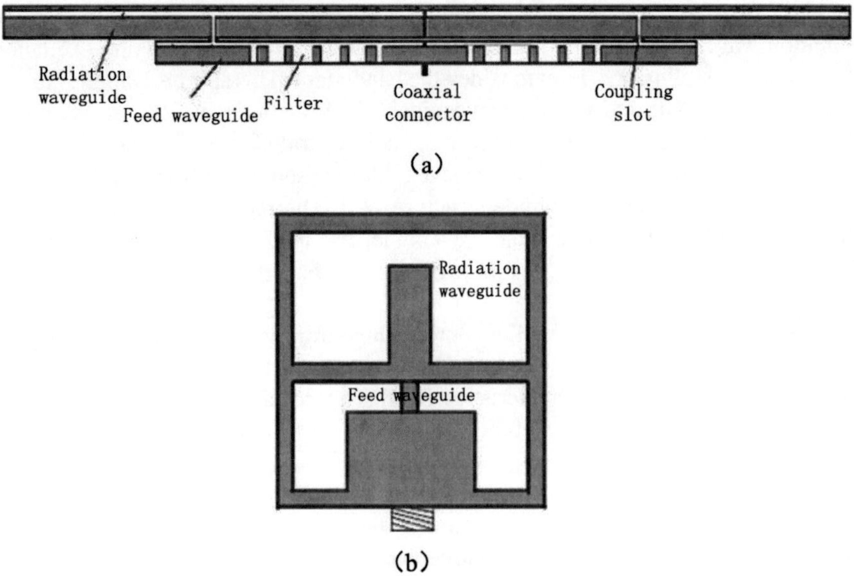

Fig. 6.35 Integrated filter anti-jamming waveguide slot antenna, **a** longitudinal section, **b** cross section

Figure 6.36a is a part of an X-band 16-elements section-feed standing wave array. The antenna layout in Fig. 6.34a is selected. The thickness of the antenna is only 13 mm, which is equivalent to the height of the standard waveguide. Figure 6.36b shows the calculation and test results of the antenna reflection loss. The impedance bandwidth of the antenna port with a reflection loss of less than −15 dB can be seen to reach 14.9%. Figure 6.36c–e shows the antenna pattern of lower, central, and upper frequency points and their cross-polarization test results. The pattern bandwidth of the maximum sidelobe below −11.5 dB reaches 10%, and the cross polarization in the main lobe is below −37 dB.

Suppose the coaxial structure is combined with the waveguide slot resonant array. In that case, an inner conductor is added in the multi-antenna element waveguide cavity to form a coaxial power splitter, which excites the radiation elements in the waveguide slot array in sections. Since the fed coaxial power splitter and the radiation waveguide share a cavity, this structure can further compress the height of the antenna [27].

For multi-element waveguide narrow-wall slot arrays, the segmented excitation method can also expand the antenna operating bandwidth [28]. However, for structural reasons, the slotted antenna with a narrow wall of the rectangular waveguide has a relatively high thickness, and the thickness of a single-layer line array is the broad wall of the waveguide. When the segmented waveguide is excited, the waveguide power splitter further increases the height of the antenna, as shown in Fig. 6.37a. Slits are made on the narrow wall of the upper part of the radiation waveguide, and the main port of the lower part of the feed power sub-waveguide is excited by the

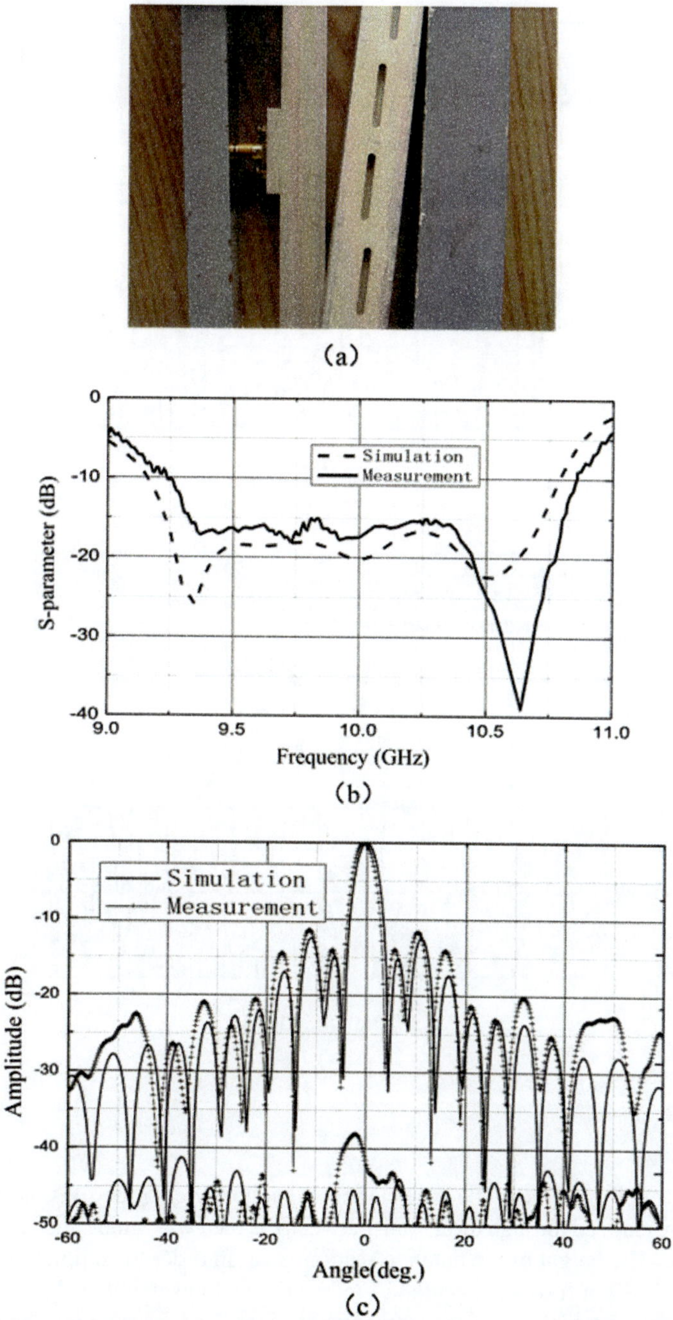

Fig. 6.36 16 Element resonant array antenna, **a** antenna photo, **b** antenna port reflection loss, **c** 9.5 GHz direction and cross polarization, **d** 10 GHz direction and cross polarization, **e** 10.5 GHz direction and cross polarization

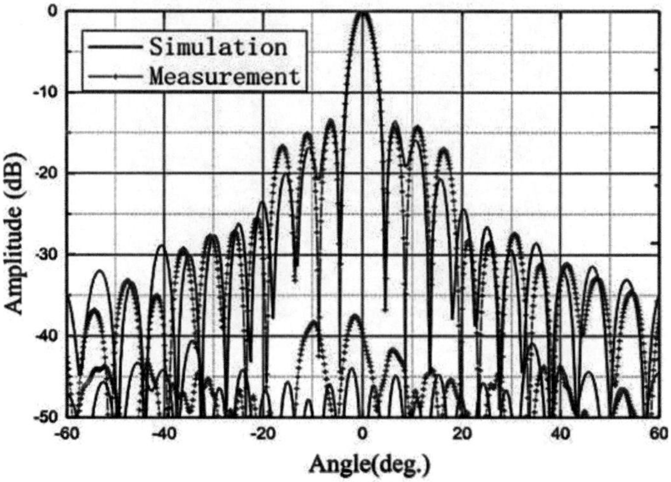

(d)

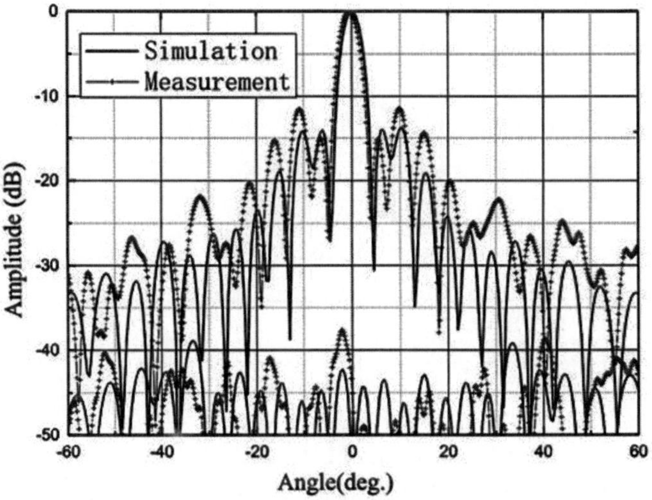

(e)

Fig. 6.36 (continued)

magnetic coupling of the probe. The electromagnetic wave power division and transmission are realized through the T-joint between the feed waveguide and the radiation waveguide. The height of the antenna section is 2a. In order to compress the height of the antenna, the feed waveguide in Fig. 6.37b, c adopts double-ridge and single-ridge structures, which compresses the height of the feed waveguide. The radiation waveguide in Fig. 6.37d, e also adopt symmetrical double-ridge waveguide structures, compressing the radiation waveguide and the feed waveguide. Figure 6.37f shows

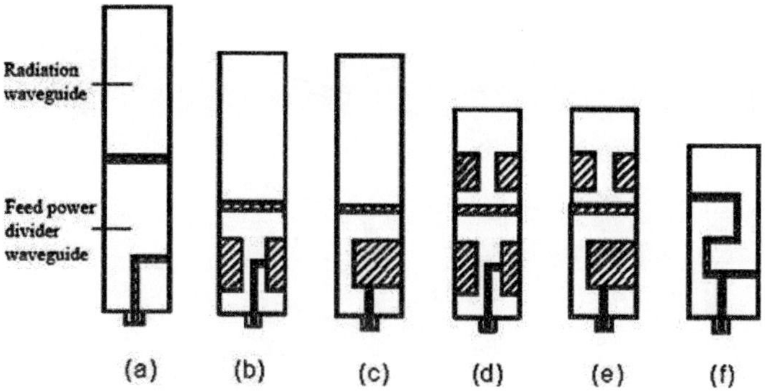

Fig. 6.37 Relationship between waveguide narrow-wall slot waveguide and power divider waveguide

a complementary structure of radiation asymmetric ridge waveguide and feeding asymmetric ridge waveguide, further compressing the antenna height [12].

6.5.3 Cross-Polarization Suppression

In Fig. 6.29, the longitudinal slot 4 deviates from the central axis of the waveguide, and the coupling between the slots causes the radiated electromagnetic field to distort and raise the cross-polarization component. In addition, when the ridge waveguide broadside slot array beam scans, parasitic sidelobes will appear. The anti-phase feed is used to eliminate the normal cross-polarization component. When this method deviates from the normal airspace, the anti-phase cancellation effect is weakened due to the increase of the spatial phase difference, resulting in worsened cross polarization. Cross polarization is further aggravated when the antenna array scans, especially when the spatial phase difference approaches 180°. In Fig. 6.29, the waveguide narrow-wall inclined slot array constructed by slot 5 inevitably has cross-polarization components, and the traditional solution is to eliminate the cross-polarization components by feeding in pairs with opposite phases between adjacent linear arrays. Also, there is the problem of weakening cross-polarized pairs when deviating from normal. The cross-polarization suppression of these two polarized antennas is analyzed below.

For waveguide broad-wall longitudinal radiating slot arrays, the structures of asymmetric single-ridge waveguides [28] and twisted ridge waveguides [29] trade the asymmetry inside the waveguide cavity for the symmetry of the radiating slot. As shown in Fig. 6.38, since the slender radiation slots are located on a straight line, oblique current radiation caused by mutual coupling between bias slots is excluded. As a result, the cross polarization inside the main lobe of this type of antenna is usually lower than −60 dB, and the cross polarization outside the main lobe can also reach about −50 dB.

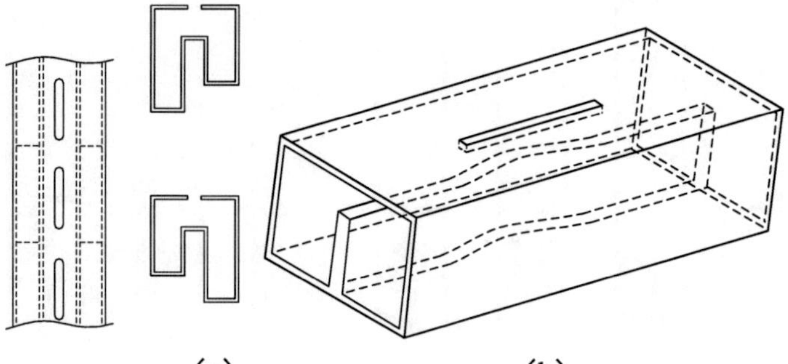

Fig. 6.38 Waveguide broadside non-offset slot array, **a** asymmetrical single-ridge waveguide slot array antenna, **b** twisted ridge waveguide slot array antenna

Similarly, the waveguide narrow-wall inclined slot array antenna naturally radiates obliquely polarized electromagnetic waves, which are traditionally arranged in pairs of linear array mirrors, and the symmetry of the array offsets the cross-polarized components. In order to reduce the amount of equipment and increase the gain of the antenna array, the linear array spacing in the planar array should be selected as large as possible, usually under the condition that no grating lobes appear when scanning the maximum angle, that is,

$$d_x \leq \frac{\lambda_H}{1 + |\sin\theta_{\max}|}\left(1 - \frac{1}{N}\right) \tag{6.24}$$

where λ_H is the wavelength in free space corresponding to the highest frequency point in the working frequency band, $\theta_{\max}$ is the maximum scanning angle, and N is the number of elements. The element spacing usually chooses the maximum value in Eq. (6.24), as shown in Fig. 6.39a. At this cell spacing, the spatial phase difference from the normal θ angle is

$$\varphi = \frac{2\pi}{\lambda} d_x \sin\theta \tag{6.25}$$

In the case of selected antenna element spacing, the array half-space coverage θ is within the range of $\pm 90°$, and the spatial phase difference is proportional to d_x, so the larger d_x is, the worse the cross-polarization cancellation effect is from the normal direction.

The inclined slot of the symmetrical single-ridge waveguide presents a natural anti-phase characteristic to the two narrow side surface currents of the antenna, as shown in Fig. 6.39. The components of the radiation field on both sides of the "V"-shaped slit parallel to the waveguide axis are superimposed in phase [30], while the field components perpendicular to the axis are anti-phase and canceled. Since the pairs of inclined slots are distributed on a single waveguide, the element spacing is reduced exponentially, as shown in Fig. 6.39b. The spatial phase difference away

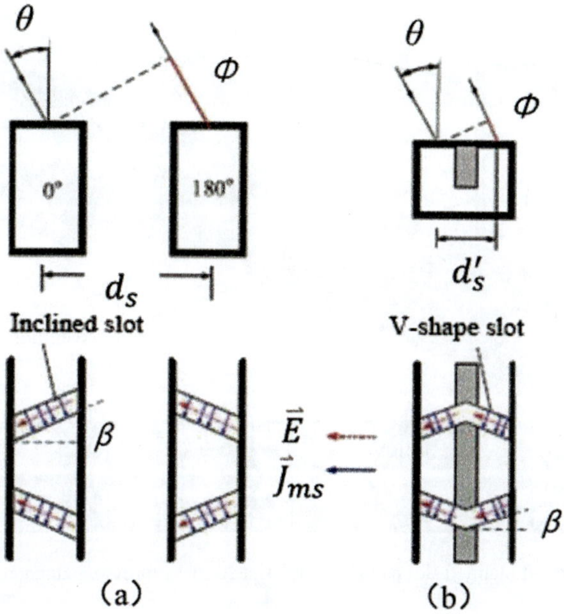

Fig. 6.39 Principles of cross-polarization suppression. **a** Two linear arrays excited in opposite phases, **b** single ridge waveguide pair of inclined slots

from the normal direction is small. It has little influence on the cancellation of cross-polarization components, and good polarization purity can be obtained in a wide spatial domain, as shown in Fig. 6.40.

Figure 6.41 shows several typical cross-polarization suppression structures, such as inclined metal rods or dielectric sheets with inclined metal films attached to the surface to disturb the electromagnetic field distribution. The structure in Fig. 6.41a needs oblique perforation, installation, and fixation of oblique metal rods on the narrow and broad walls [31], which adds great difficulty to machining. In the structure of Fig. 6.41b, a thin dielectric sheet [32] is inserted in the metal waveguide. The engineering performance is poor, considering factors such as machining accuracy, positioning and fixing reliability, and instability under hot and cold conditions due to different material expansion coefficients. Based on machining and welding, it is more manufacturable to choose an integrated metal sheet to excite non-inclined gaps [33], as shown in Fig. 6.41c. If the rectangular metal diaphragm is used instead of the right-angled trapezoidal structure, the same effect can be obtained, and the manufacturability can be further improved. The structure shown in Fig. 6.41d can also obtain the cross-polarization suppression effect using the twisted double-ridge waveguide narrow side non-inclined slots [34]. This kind of structure is also relatively simple to machine. Compared with the structure shown in Fig. 6.41c, one layer is added to the machined parts. For the above types of antennas with non-inclined

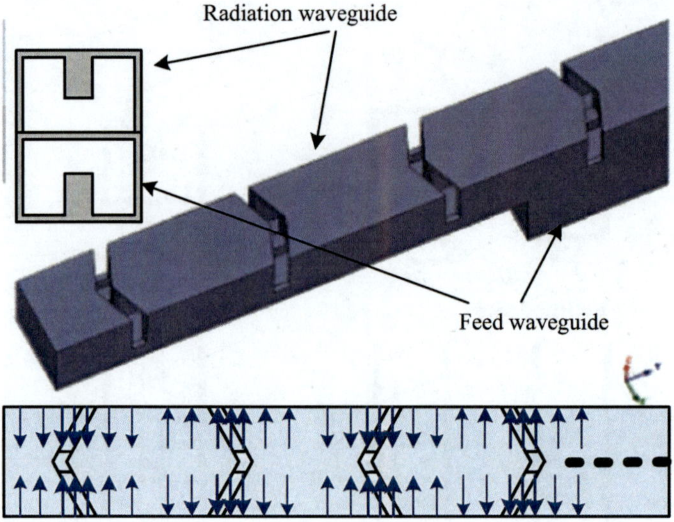

Fig. 6.40 Broadband inclined slot pair cross-polarized self-suppressing antenna

slot structures, due to the orthogonality of the external structure, the cross polarization of the antenna can reach about −60 dB if the slot with a narrower width is selected. Considering the manufacturability of the segmentally excited waveguide power splitter for the broadband resonant antenna, it is possible to change the oblique slots in the ridge waveguide inclined slot resonator array to non-inclined slots excited by metal diaphragms. As shown in Fig. 6.41e, broadband and low cross-polarization characteristics are obtained.

6.5.4 Dual-Polarization Waveguide Slot Array Antennas

The most commonly used dual-polarization waveguide slot array antenna arranges the waveguide broad-wall slot array and narrow-wall slot array, as shown in Fig. 6.42a. However, this structure has apparent defects, such as large volumes and heavy weight. Therefore, a compact alternately arranged dual-polarization structure had been developed to improve the antenna polarization purity and port isolation, as shown in Fig. 6.42b. Its horizontally polarized linear array adopts non-slanted slots on the narrow wall of the rectangular waveguide excited by metal rods. The vertical slots on the broad wall of the single-ridge waveguide are used for vertical polarization so that the waveguides of the dual-polarized antenna are arranged alternately in the same layer. Compared with the layout in Fig. 6.42a, the height of the antenna is reduced, so the cross-sectional width of the dual-polarized waveguides arranged alternately can be effectively compressed.

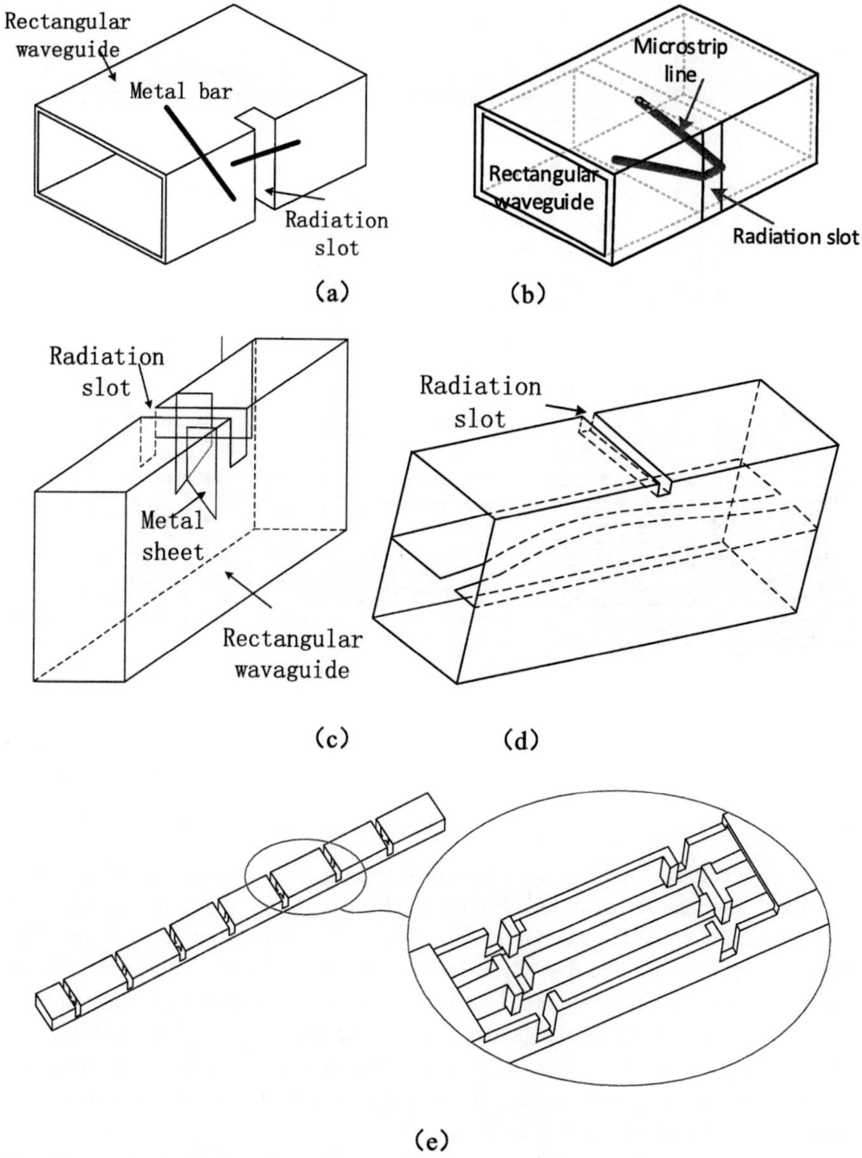

Fig. 6.41 Waveguide narrow-wall radiation slot cross-polarization suppression structure, **a** perturbation of inclined metal rod, **b** perturbation of dielectric metal film, **c** perturbation of metal sheet, **d** twisted double-ridge waveguide, **e** symmetrical single-ridge waveguide metal film sheet disturbance straight seam

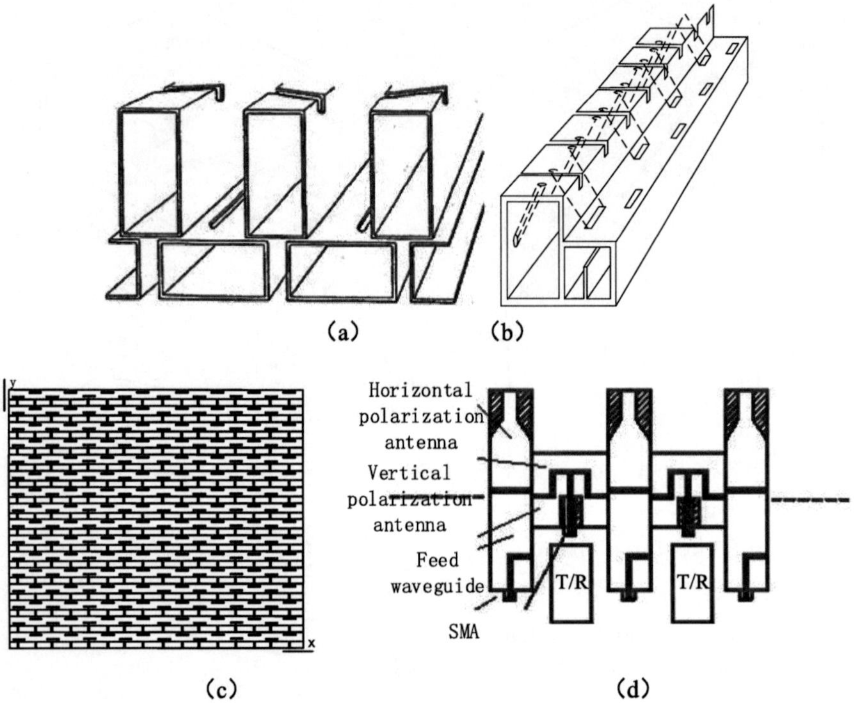

Fig. 6.42 Basic layout of dual-polarized waveguide slot array antenna, **a** simple assembly, **b** alternate arrangement, **c** broadband dual-polarization top view, **d** broadband dual-polarization cross section

In high resolution, multi-polarized microwave imaging systems, broadband dual-polarized antennas are required. Figure 6.42 only uses a single-layer waveguide cavity, which is limited by the bandwidth of the waveguide standing wave array. The more elements in a single sub-array, the narrower the bandwidth. There are two ways to expand the bandwidth: one is to increase the number of active channels in the sub-array-level phased array antenna and reduce the number of elements in the sub-array. This approach is very effective and can increase the horizontal scanning range, the second is sub-array segmentation and multi-point excitation, thereby expanding the bandwidth of the sub-array [24–27].

For the active array antenna in the microwave imaging radar system, due to the particularity of the application platform, the volume and weight of the antenna have been raised to the same level of importance as the electrical performance. When designing a dual-polarized antenna array, it's important to optimize the waveguide cavity structure to achieve a light, low profile, high-strength, and easy-to-process antenna array. This involves considering different waveguide structure combinations of two broadband linear polarized antennas, as well as the mechanical processing characteristics, in order to ensure the dual-polarized antenna is suitable for integrated processing. Additionally, it's important to pursue superior performance, such as bandwidth, isolation, and cross-polarization.

Based on the above basic analysis, in the dual-polarized waveguide slot array antenna, the waveguides of the vertically polarized and horizontally polarized arrays adopt common lateral walls and horizontal metal walls on the same layer. Figure 6.43 shows an integrated dual-polarized antenna structure [35] consisting of two linear arrays arranged in parallel, each linear array having 16 radiation slots. To simplify the structure and reduce the difficulty of manufacturing, the vertical polarization is realized by the longitudinally offset slot array of the symmetrical single-ridge waveguide, and the horizontal polarization is realized by the narrow side "V"-shaped slot array of the symmetrical ridge waveguide. Since the antenna array does not adopt a completely symmetrical element structure, it can be optimized on the antenna array to cancel the cross-polarization component. When the frequency is low, and machining conditions permit, the "V"-shaped gap can be replaced by the straight seam in Fig. 6.41e [36]. The dual-polarized 16-elements linear array is equally divided into four 4-elements end-fed resonant arrays, which are excited by a 1:2 waveguide power splitter with equal amplitude and same phase mid-fed to expand the working bandwidth of the waveguide resonant array.

According to the processing and mechanical properties, the scale of the overall processing of the integrated dual-polarization array is determined. Figure 6.44 shows a planar array of 16 pairs of dual-polarized linear arrays. The method of suppressing cross polarization by paired anti-phase excitation is selected for the horizontally polarized array. The antenna works in the X-band. Since the straight waveguide section of the waveguide power splitter provides space for the waveguide filter, the antenna array integrates a filtering structure. As a result, compared with the vertical polarization array, the horizontal polarization array is shifted upward by one waveguide cavity height, and the height of the waveguide cavity in the same layer is the same, effectively reducing the number of machining parts and reducing the difficulty of welding.

Figure 6.45 shows the test results of the antenna port reflection loss of the prototype, and the bandwidth of the dual-polarized antenna is greater than 10%. Figure 6.46 shows the test results of the main polarization and cross-polarization radiation patterns within the working frequency band of the prototype. It can be seen that the upper, central, and lower frequency points are approximately ideal uniform distribution characteristics. The first side lobe is about −13 dB, the cross polarization in the main lobe is lower than −43 dB, and the far zone is lower than −30 dB. The antenna efficiency is greater than 82% compared with the directivity coefficient. Since the antenna integrates a waveguide filter, its out-of-band suppression reaches 25 dB and more than 30 dB outside the operating frequency band side frequency of 500 MHz.

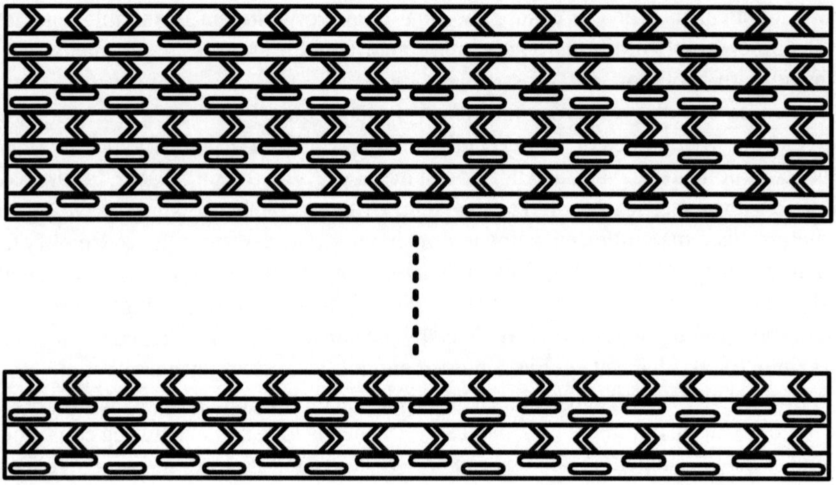

(a)

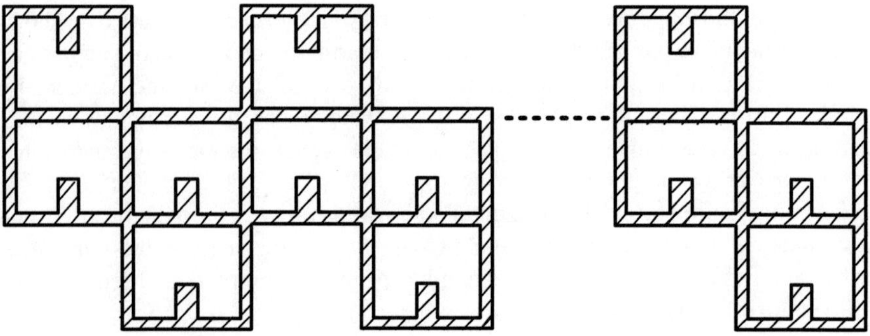

(b)

Fig. 6.43 Schematic diagram of the integrated broadband dual-polarized waveguide slot array antenna, **a** top view, **b** cross-sectional view

6.5.5 Dual Circularly Polarized Slot Waveguide Antennas

There have been many studies on single circularly polarized waveguide slot antennas, mainly focusing on traveling wave array antennas. However, these circularly polarized structures have large lateral dimensions, making it difficult to implement dual-polarized array antennas, especially when the array requires 1D large-angle scanning.

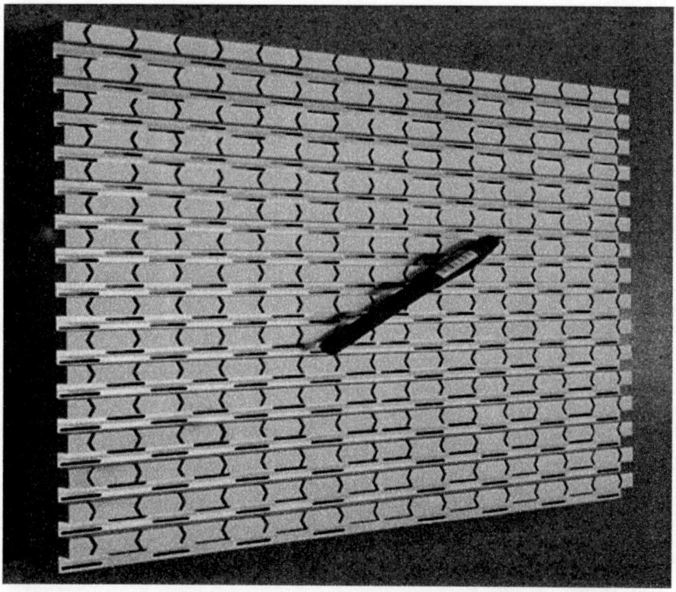

Fig. 6.44 Photograph of integrated broadband dual-polarization slot array antenna

By placing single circularly polarized waveguide antennas side by side, a dual circularly polarized waveguide array can be realized. In a single circularly polarized antenna array, a slender slit is opened on the concave metal wall in the middle of the symmetrical single-ridge waveguide to obtain vertical polarization, and the symmetrical "eight" slits on the two protruding ridges realize horizontal polarization. The depth of the depression in the ridge waveguide is chosen to be a quarter wavelength. The heights of the two polarization radiation slots make the dual-polarization radiation electromagnetic wave obtain a 90° phase difference. By changing the offset position of the slit, left-handed or right-handed polarized waves can be obtained. Figure 6.47a–c shows a single circularly polarized antenna structure [37]. Based on single circular polarization, left and right circular polarization antennas are arranged in parallel to form double circular polarization [38]. Figure 6.47d shows a top-view layout structure of dual circularly polarized antennas. The transformation of left and right circular polarization only changes the offset of the slender longitudinal slot.

6.5.6 Dual-Polarized Aperture Waveguide Antennas

The square aperture waveguide or the four-ridge waveguide has a natural 2D symmetry, and the technology for exciting two orthogonal internal fields is relatively mature. The analysis mainly focuses on the design of the power combination/division network in sub-array-level applications. Dual-polarized aperture waveguide antennas

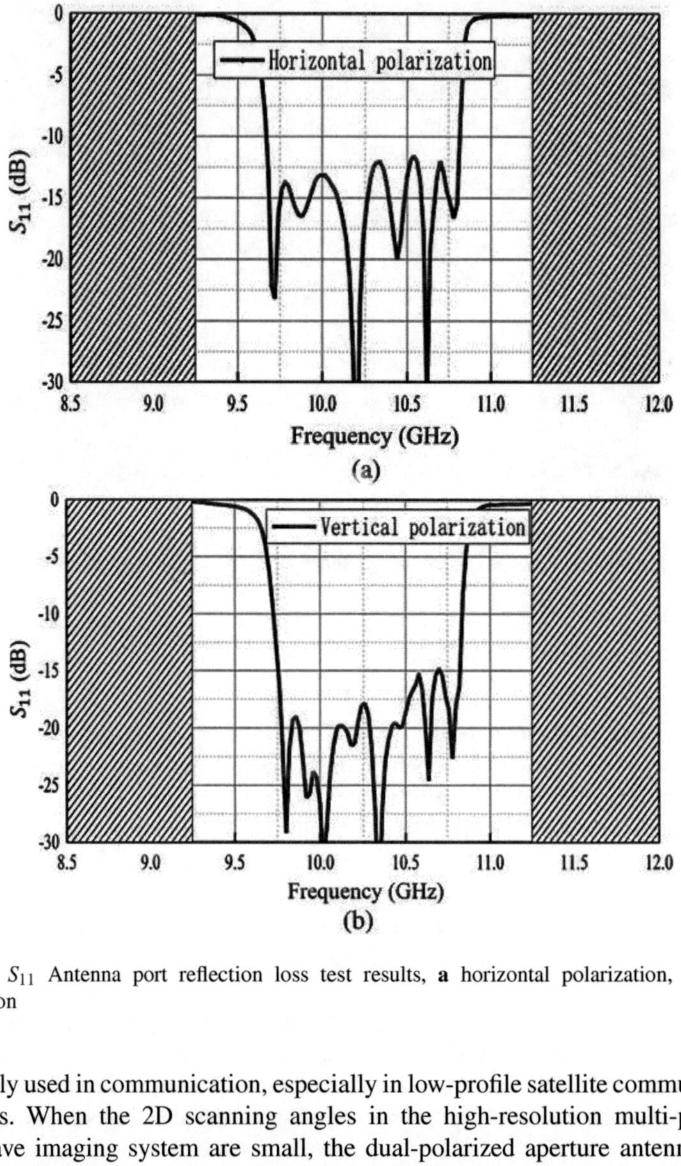

Fig. 6.45 S_{11} Antenna port reflection loss test results, **a** horizontal polarization, **b** vertical polarization

are widely used in communication, especially in low-profile satellite communication terminals. When the 2D scanning angles in the high-resolution multi-polarized microwave imaging system are small, the dual-polarized aperture antenna shows better practicability. In order to simplify the feeding network of this antenna, the radiation opening is usually a large square resonant cavity. In order to improve the electromagnetic field distribution of the aperture antenna array in the high-frequency band, a square grid is added to the opening of the resonant cavity, which is approximately a 2×2 unit, as shown in Fig. 6.48a. The bottom is excited by a dual-polarized combined waveguide to form a dual linearly polarized antenna array, which can also be combined with a broadband waveguide bridge to form a dual circular polarization [39].

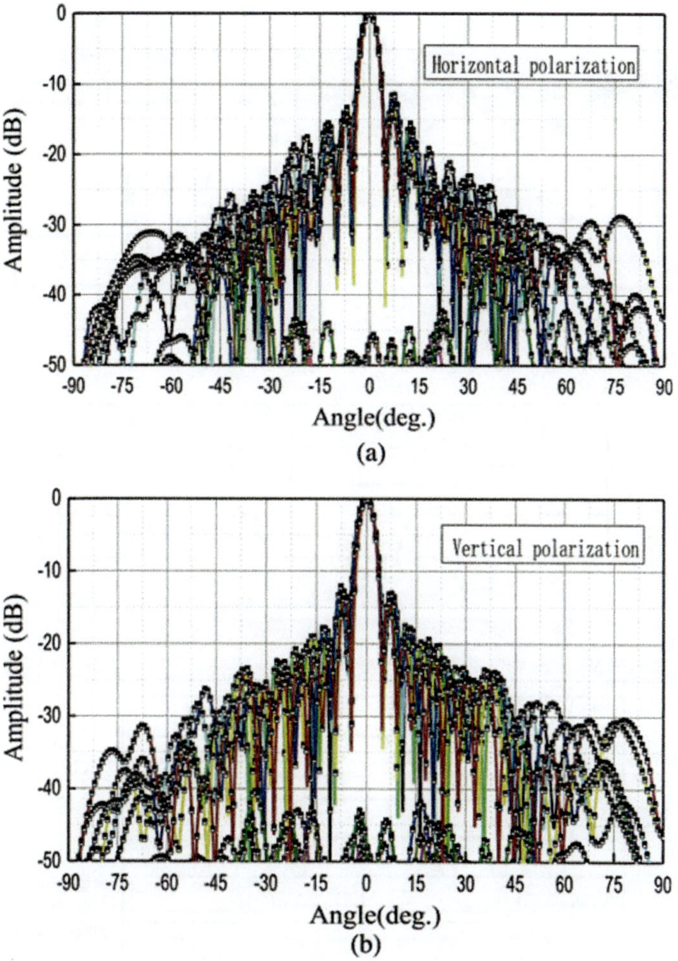

Fig. 6.46 Antenna port reflection loss test results, **a** horizontal polarization, **b** vertical polarization

Figure 6.49 shows the test results of a Ku-band dual-polarized aperture waveguide antenna. The VSWR of the vertically polarized port is less than 1.8 in the receiving frequency band (14.0–14.5 GHz) and transmitting frequency band (12.25–12.75 GHz). The VSWR of the horizontally polarized port is less than 1.9 in the receiving and transmitting frequency bands. As a result, the impedance bandwidth is more significant than 16.8%. By comparing the test gain and directivity coefficient, the antenna efficiency is greater than 80%.

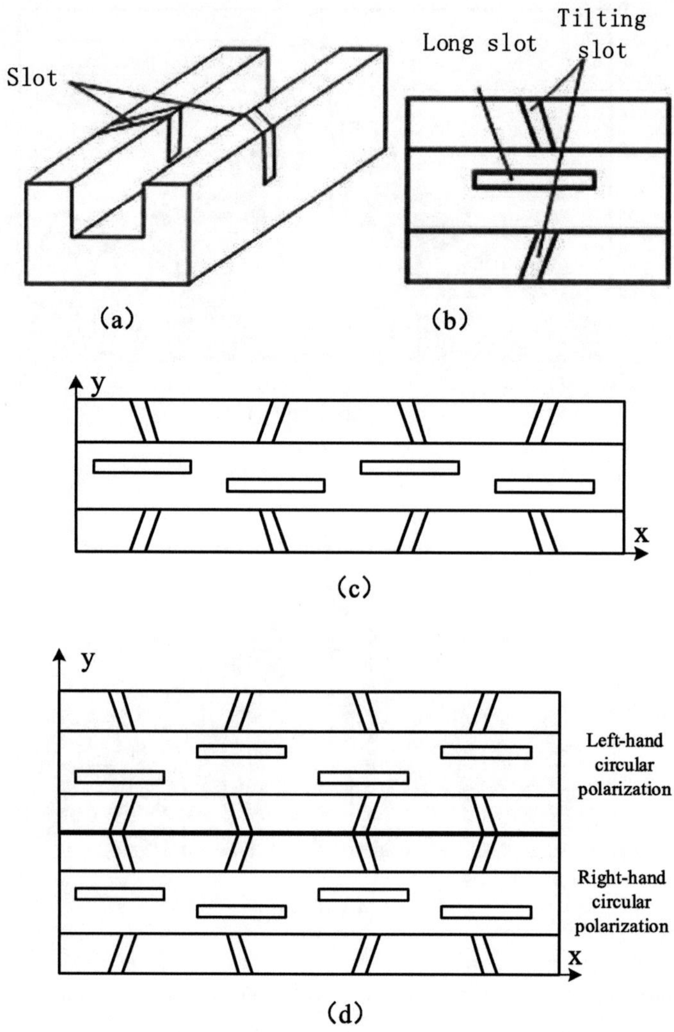

Fig. 6.47 Waveguide circularly polarized antenna, **a** 3D structure of the element, **b** top view of the element, **c** top view of a single circularly polarized array, **d** top view of a dual circularly polarized array

6.6 Multi-band and Multi-polarized Shared Aperture

Multi-band and multi-polarized shared-aperture antennas are multiple antennas sharing one physical antenna aperture. Research and analysis focus on arranging multiple antennas in a limited space to ensure that grating lobes do not appear or control grating lobes within a certain range. The dense radiating antenna structure

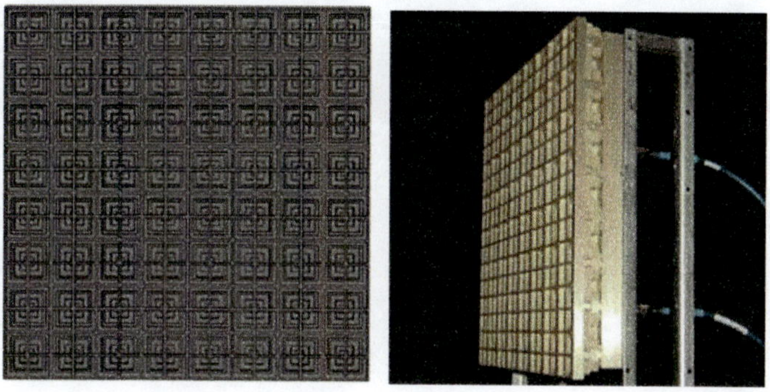

Fig. 6.48 Dual-polarized aperture waveguide slot array antenna, **a** top view of the array layout, **b** photograph

causes mutual coupling and interference, which affects the cross-polarization performance of every single polarization, the isolation between dual polarizations, and the isolation between different bands. Multi-band and multi-polarized shared-aperture antennas mainly include dual-band, triple-band, or even quad-band according to the working frequency band and single polarization and dual polarization according to the number of polarizations. Different numbers of frequency bands and polarization combinations constitute a multi-band and multi-polarized shared-aperture antenna series. Antenna implementation forms cover dielectric, metal, and hybrid-based antennas according to material selection. Dielectric-based microstrip antennas have a high degree of freedom in realizing multiple antenna nesting and superposition due to the diversity of forms and the small space occupied by the feeder transmission line. Therefore, it has great advantages in the design of multi-band and multi-polarized shared-aperture antennas. However, metal-based waveguide antennas occupy a large 3D space due to the limitation of cutoff frequency and the characteristics of the cavity structure. The interference of multi-antenna shared-aperture structures is serious, making the design extremely difficult. But its advantages are also very obvious: high efficiency and light electromechanical integration [40]. Discussions are made below based on different combinations and practical examples.

6.6.1 Dual-Band Single Polarization

In some applications, the antenna array operates in dual bands with only one polarization per band, effectively two antennas sharing the same aperture. In order to distinguish it from a shared-aperture antenna whose two frequency bands are dual polarized, it is defined here as a dual-band single-polarized shared-aperture antenna. Dielectric-based or metal-based antennas can realize this situation. Dielectric-based antennas are microstrip patch antenna arrays, as shown in Fig. 6.50a. The low-band

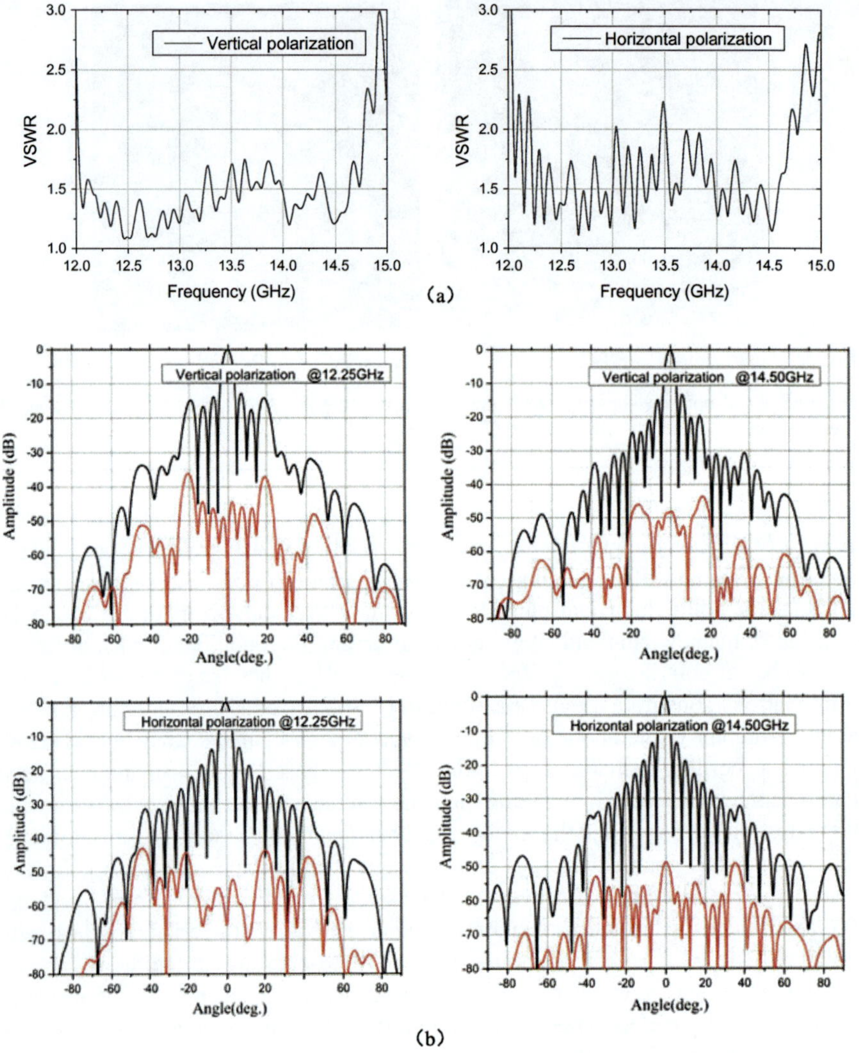

Fig. 6.49 Test results of dual-polarized aperture waveguide. **a** Measured results of the VSWR, **b** measured results of radiation pattern

antenna corresponds to the long side of the rectangular patch, and the high-frequency band corresponds to the short side. This antenna is limited by the frequency ratio of the two bands, that is, the ratio between the high-frequency and low-frequency bands. In the case of a relatively large frequency, the spacing between elements parallel to this dimension is relatively large due to the size constraint of the long side of the rectangular patch antenna. Compared with the high-frequency band, this spacing does not meet the constraint conditions for suppressing lobes. As a result, grating lobes will appear in the high-frequency band antenna array in this dimension. Compared with the low-frequency antenna, the element spacing is too small for the short side of the high-frequency patch, resulting in too dense active channels of the active array antenna.

Figure 6.50b, c shows two combinations of a dual-band waveguide slot antenna array. In combination mode I, the low-frequency band is a waveguide broad-wall longitudinal slot array, and the high-frequency band is a narrow-wall inclined slot array. The broad wall of the waveguide side is larger than the its narrow one. Therefore, the frequency ratio in this combination is small. In combination with method II, the waveguide broad-wall slot array is selected for the high-frequency band, and the waveguide narrow-wall slot array is selected for the low-frequency band. Since the ratio of the waveguide width to the narrow side is about 2:1 or even greater, the frequency of this combination is relatively large.

Figure 6.50d shows a photo of a dual-band single-polarization array [41]. The waveguide narrow-wall slanted slot array constitutes a 35 GHz horizontally polarized array, while the ridge waveguide broad-wall slot array constitutes a 30 GHz vertically polarized array, and the antenna cross-polarization is less than −25 dB. From the perspective of symmetry, the vertical polarization chooses the offset slot. Although the horizontal polarization adopts a paired anti-phase feed structure of adjacent linear arrays, the inclined slots in the horizontal direction are not placed in the middle of the vertical slots. As a result, the vertical and horizontal polarization slots are coupled to each other, and the parasitic radiation cannot be canceled, which increases the cross-polarization component.

Usually, in practical applications, the scanning angles of the antennas in the two frequency bands are the same, and the element spacing of the antennas in the two frequency bands is selected to be equal to the electric length. However, when the frequency ratio of the two frequency bands is not even, it is necessary to compromise the spacing of the antenna elements and try to make multiple high-band antenna elements symmetrically distributed around the low-frequency antenna element to reduce the coupling between the dual band and the influence of cross polarization.

6.6.2 *Dual-Band and Dual-Polarization Shared-Aperture Antennas*

There are many ways to realize the dual-band and dual-polarized shared-aperture antenna.

(1) Dual-band single antenna, i.e., one antenna element, has multiple resonant frequencies. Usually, this kind of antenna has the disadvantages of narrow bandwidth, high cross polarization, relatively small frequency ratio, difficulty in achieving dual polarization, and high active channel density in active array antennas. Therefore, it is mainly used in the fields of communication and navigation.

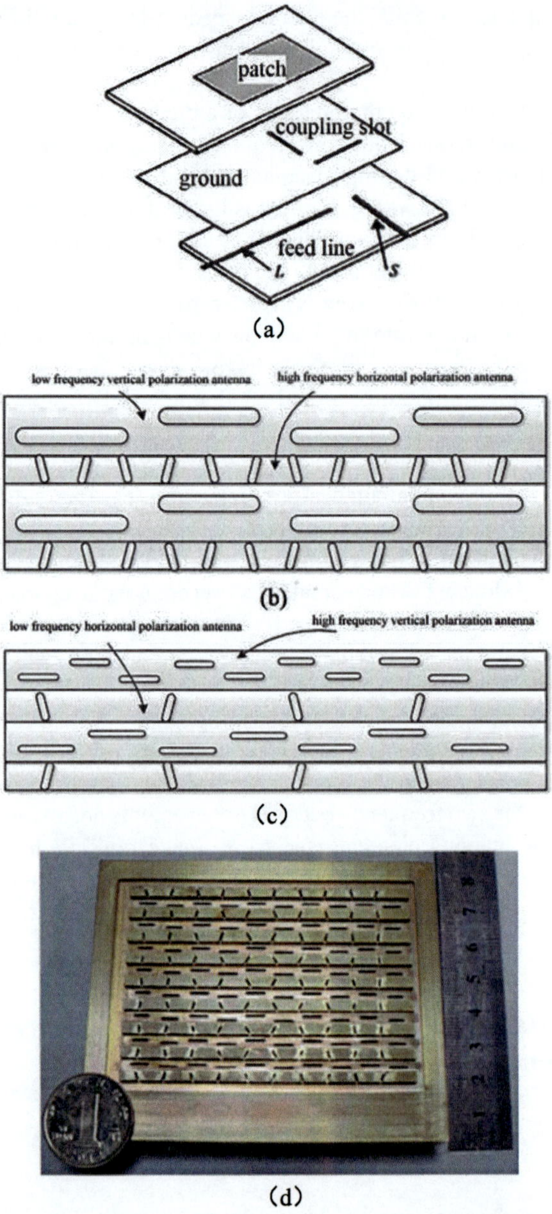

Fig. 6.50 Dual-band single-polarized antenna. **a** Rectangular microstrip patch dual-band antenna combination, **b** dual-band dual-polarization waveguide antenna combination method I, **c** dual-band dual-polarization waveguide antenna combination method II, **d** photo of the Ka-band dual-band antenna

(2) The low-frequency antenna array "dugs out" space to embed the high-frequency antenna array. This shared-aperture embedding method with the array as an element sacrifices the performance of the low-frequency antenna array and is suitable for application scenarios that do not require high antenna sidelobes.
(3) Dual-band element interweaving and nesting. This element has many forms and a large degree of freedom in combination. Usually, the high-frequency antenna element chooses a microstrip patch and forms an array according to the antenna scanning requirements.

In the interweaving and nesting combination of dual-band elements, the high-frequency dual-polarized antenna element chooses a square or circular patch with good symmetry. Low-frequency elements can be classified into two types. The first type is a dual-polarized independent patch antenna or slot antenna. The second type is a dual-polarized patch antenna in a "+" shape or a circular/square patch with a central opening.

(1) **S- and X-band dual-polarized shared-aperture antenna**

For slender-shaped patches, you can choose a planar printed dipole antenna or a narrow microstrip patch with a large aspect ratio. Figure 6.51 shows a typical dual-polarization shared-aperture antenna configuration [5]. The low frequency works in the S-band, the high frequency operates in the X-band, and the frequency ratio is 1:3. The X-band dual-polarized antenna is a square patch in which the vertical polarization is excited by coplanar microstrip feeding. The horizontal polarization is excited by slot coupling. The two feed structures guarantee 1D symmetry, thus improving port isolation. The S-band is dual polarized by two independent elongated patches, also fed by coaxial probes. In order to expand the working bandwidth, the antennas of the two frequency bands adopt a double-layer patch structure, and the S-band excitation patch also adds a stub line to adjust the port matching so that the entire feed microstrip line is in the shape of a "cross."

As shown in Fig. 6.51b, according to circuit layering, the lowest layer is the X-band horizontally polarized feeding microstrip line. The second layer is the metal ground, and the X-band coupling slot is opened on this layer. The third layer is the X-band square patch antenna, and the coplanar microstrip line fed by vertical polarization is also on this layer. The S-band "+"-shaped feed microstrip is located between the X-band patch antenna. The fourth layer has X-band and slender strip patches, the fifth layer only has S-band parasitic patches. Between the various circuit layers is a dielectric board.

(2) **L-band slot C-band microstrip dual-band dual-polarized shared-aperture antenna**

Replacing the slender patch with the slender slot constitutes a dual-band dual-polarization shared-aperture combination of the high-band square patch and the low-band slot antenna [42]. Through the layered design of the feed network, the dual-polarized work of the two frequency band antennas is realized, as shown in Fig. 6.52.

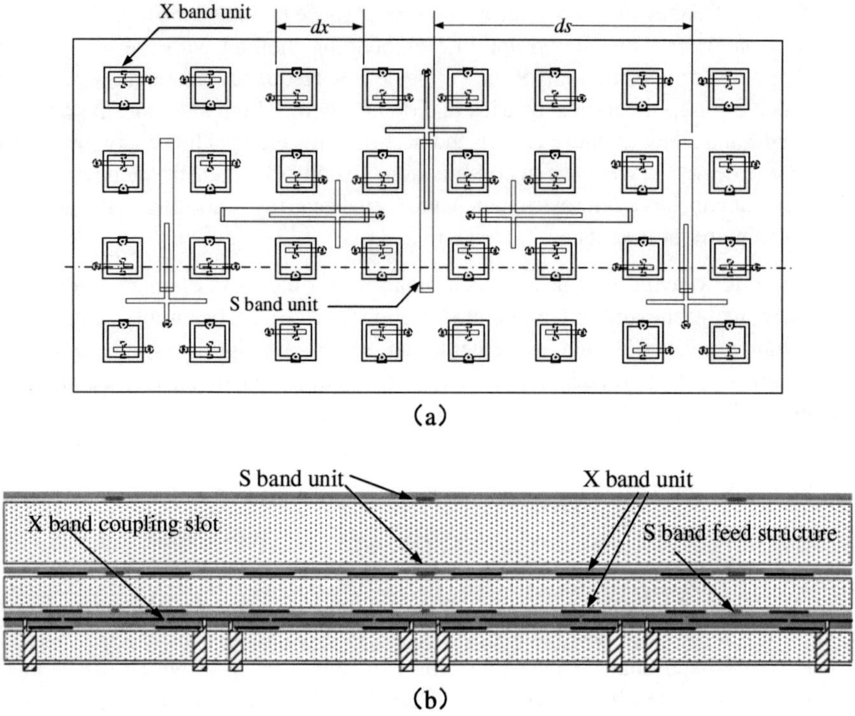

Fig. 6.51 Narrow patch/square patch combination dual-band dual-polarized shared-aperture antenna. **a** Antenna array layout, **b** antenna cross-sectional structure diagram

The dual-polarized antenna is two completely independent antenna elements when choosing a slender patch or a slit as the low-frequency antenna. However, in some applications, the phase centers of dual-polarized antennas coincide, or circular polarization needs to be formed by dual-polarized antennas. At this time, dual-polarized antennas need to share a radiator, such as a square microstrip patch. The square microstrip patch antenna is selected for the low-frequency band. It is necessary to solve the problem of interference between the high and low-frequency antenna structures when the high-frequency band antennas are arranged at equal intervals. The low-frequency square patch antenna occupies a large area. The high-frequency antenna is only arranged around the low-frequency antenna, resulting in too large spacing between high-frequency antenna elements. As a result, the antenna array is prone to grating lobes.

(3) **L/S dual-band circularly polarized shared-aperture antenna**

Figure 6.53a shows a dual-band dual circular polarization shared-aperture antenna structure. The S-band radiating elements all choose circular patches, and a single L-band radiating element is placed in the gap between the four S-band elements and is shaped like a "cross" [43]. The L-band patch uses a four-point feed to ensure

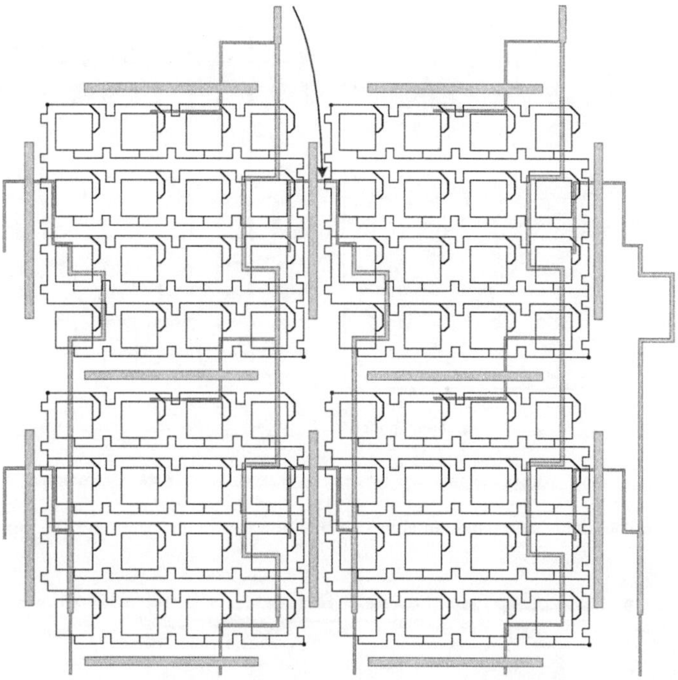

Fig. 6.52 L-slot C-patch dual-band dual-polarized shared-aperture antenna

the orthogonality of the two main mode electromagnetic fields. The high-frequency element in Fig. 6.53b chooses a square patch, and the low-frequency antenna uses a "+"-shaped patch to realize the complementary nesting of high- and low-frequency antenna structures. In addition, the low-frequency antenna can also have multiple holes in its patch, and the high-frequency antenna can be placed in the hole to achieve structurally inclusive nesting. Figure 6.54 shows that the element can be round or square.

(4) **L/C dual-band dual-polarized shared-aperture antenna**

By placing waveguide slot antennas in two frequency bands side by side, a dual-band dual-polarization shared-aperture waveguide slot antenna can be realized. The high-band dual-polarized antenna chooses the form of a dual-polarized slot waveguide antenna, leaving a gap between the dual-polarized waveguides. A low-band radiation slot is cut at the slot, and a metal wall is attached to the back of the high-band dual-polarized antenna to form a low-band ridge waveguide cavity. This is an L/C-band dual-polarized shared-aperture antenna configuration. The C-band dual-polarized radiation slots choose the slender longitudinal slots on the wide side of the ridge waveguide and the "V"-shaped slots on the narrow side of the ridge waveguide. Since the offset position of the L-band slender slots is not adjustable, the gap between the C-band dual-polarization waveguides is increased by 1 or 2 slots. Instead of a

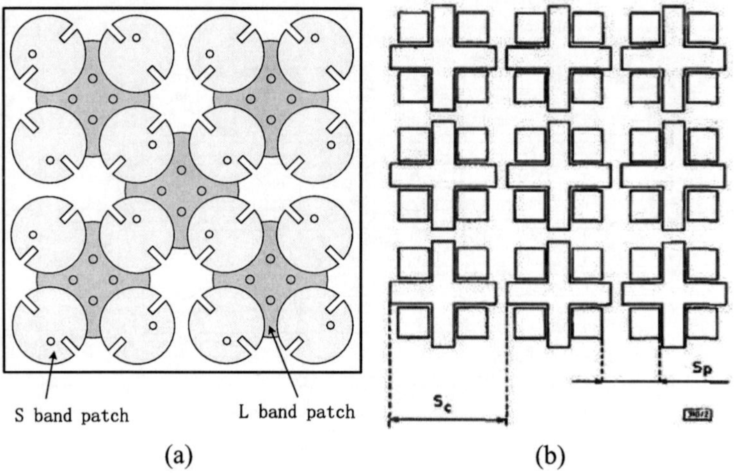

Fig. 6.53 Alternate dual-band patch antenna combination, **a** circular patch, **b** square patch

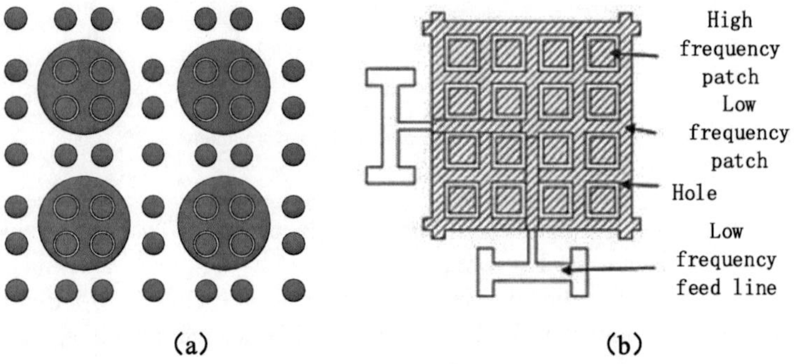

Fig. 6.54 The high-frequency antenna combination nested in the middle hole of the low-frequency patch, **a** circular patch nesting, **b** square patch nesting

single slot in the traditional waveguide broadside slot array, it adjusts the radiation impedance matching. This structure can achieve the dual-band dual-polarization high-efficiency antenna shared-aperture target. Still, because the L-band chooses a waveguide structure, it is large in size and heavy in weight. Replacing the L-band waveguide structure with a resonant cavity structure solved this problem [44]. As shown in Figs. 6.55 and 6.56, the radiating slot is still inserted between the C-band dual-polarized waveguides, and the antenna achieves 100 MHz bandwidth in both bands with an efficiency greater than 85%. Figure 6.57 shows the VSWR results of the dual-band antenna port, and the bandwidth reaches 3.7% and 8%, respectively. The pattern and cross polarization in Fig. 6.58 show that the cross-polarization component in the main lobe is more than 45 dB below the main polarization.

Fig. 6.55 L/C dual-band dual-polarized waveguide slot array shared-aperture antenna

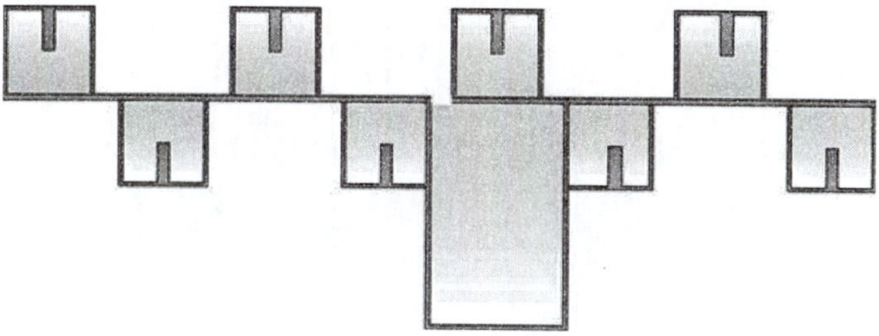

Fig. 6.56 Schematic diagram of the L/C dual-band dual-polarized waveguide slot array shared-aperture antenna

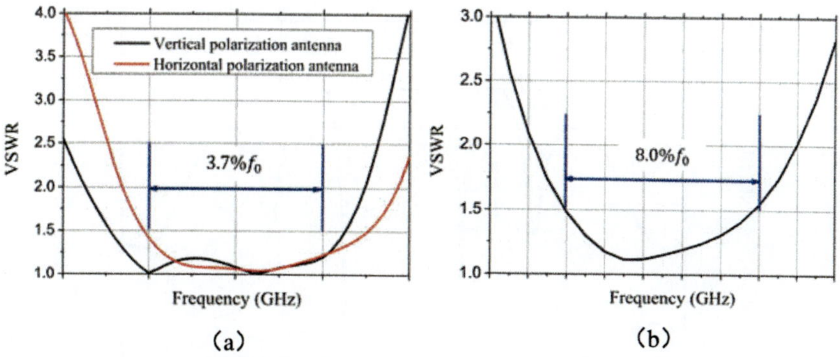

Fig. 6.57 Port VSWR. **a** C-band, **b** L-band

(5) **Ka/Ku dual-band single-polarized shared-aperture antenna**

In some applications, the number of elements in the antenna sub-array is small. In this case, a flexible dielectric-based antenna and a low-loss high-strength metal-based antenna can be selected. Figure 6.59 shows a typical configuration. The Ka-band single-polarization selection high-efficiency waveguide slot array antenna is located on the bottom layer, which also functions as a structural support plate, metal ground, and active component mounting plate. The Ku-band dual-polarized microstrip patch antenna is placed on the surface of the waveguide, and the radiation patch is located

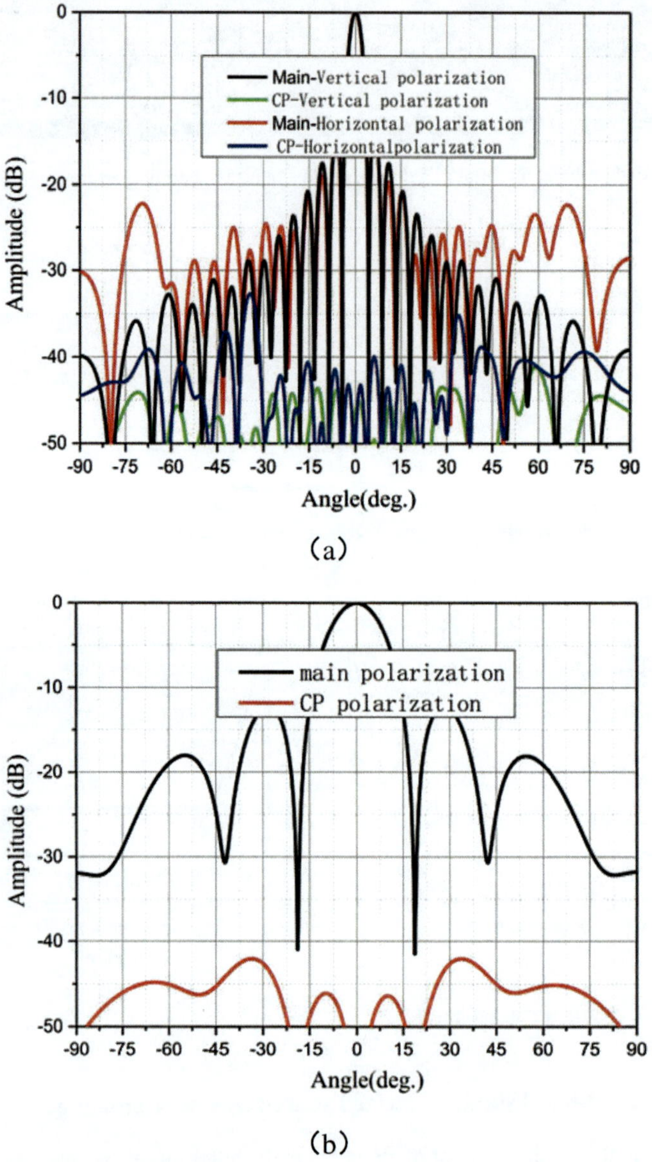

Fig. 6.58 Antenna linear array cut pattern and cross polarization. **a** C-band, **b** L-band

Fig. 6.59 Ka waveguide slot Ku dual-polarized microstrip shared-aperture antenna

in the gap between the radiation slots of the waveguide. A broadband shared-aperture antenna can be obtained when the dielectric base planar microstrip patch antenna is replaced by an element placed vertically on the dielectric base. This kind of application usually has two bands, each band antenna only works with a single polarization, and the two polarizations are orthogonal to minimize mutual interference.

6.6.3 Three-Band Dual-Polarized Shared-Aperture Antennas

Multi-band dual-polarized shared-aperture antennas are a huge challenge. In the case of three bands, each band is dual polarized, equivalent to six antennas sharing one physical aperture. So many elements are nested, mixed, and overlapped in one physical aperture, which is very difficult to realize structurally. In addition, in the element-level active array antenna, the back of the antenna will be covered with output/input connectors of six antennas. The six sets of active channels (T/R modules) cannot be densely distributed and tiled on the back of the antenna, and cables are required to connect the antenna element and the T/R modules when cascaded installation is used, which will result in greater insertion loss. It is also difficult to realize six sets of power division excitation networks in sub-array-level active array antennas.

In microwave imaging radar, the purpose of using multi-band and multi-polarization is to obtain multi-dimensional images of the same target, which requires multi-band and multi-polarization beams to cover the same area. Therefore, when there is a large difference in frequency ratio, the size of the antenna arrays in each frequency band is the same, but the size of the physical aperture is different. Therefore, it is proposed that two high-band antennas share the same aperture as the low-band antenna. Thus, two dual-band dual-polarization shared-aperture antennas are spliced into a physical array [45]. The antenna array layout is shown in Fig. 6.60. The antenna operates in the L-, S-, and X-bands with a frequency ratio of 1:3:8. The whole front is divided into L/S dual-band dual-polarization shared-aperture area, L-band dual-polarization area, and L/X dual-band dual-polarization shared-aperture area. Among them, the L/S and L/X sub-arrays are, respectively, located on both sides of the full array, and the L-band dual-polarization sub-array is used in the middle to supplement the aperture length of the required gain.

The S- and X-band elements are arranged in a square lattice, and the adjacent elements are fed in pairs to suppress the cross-polarization level in the main lobe and improve the polarization isolation of the array. The element structure is similar to the X-band dual polarization, while the L-band uses a planar dipole. The L-band elements in the single L, L/S, and L/X dual-polarization arrays adopt the same structure to

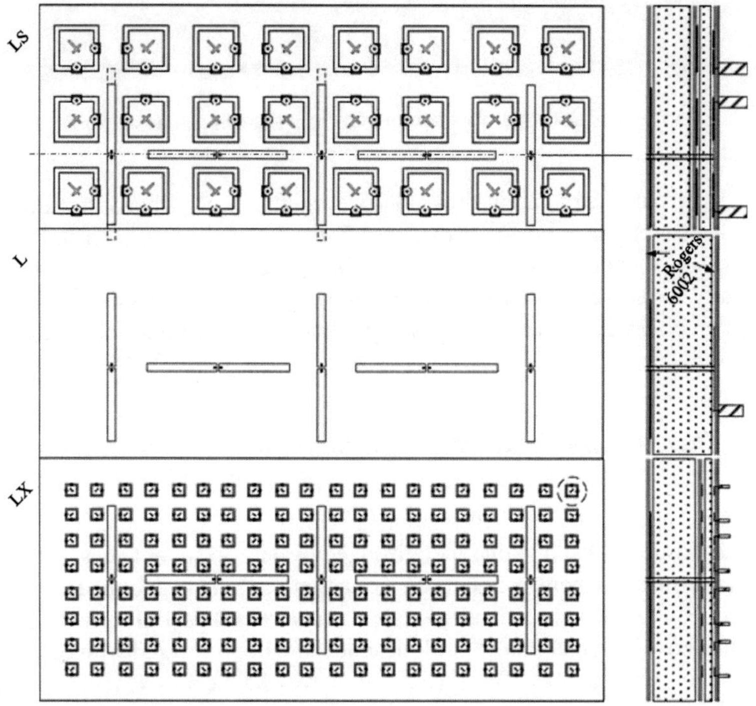

Fig. 6.60 Three-band dual-polarized shared-aperture antenna layout

ensure the consistency of radiation characteristics and S parameters of the port, which is beneficial to splicing L-band arrays. The L-band elements are interleaved in the X (or S)-band slots in a "T" shape.

This scheme decomposes the design work of the three-band dual-polarized shared-aperture antenna array into two dual-band dual-polarization shared-aperture antenna arrays and one single-band dual-polarization antenna array. Therefore, the design experience and processing technology of the previous dual-band dual-polarized shared-aperture antenna array can be used for reference, which greatly simplifies the design of the three-band dual polarization and reduces the risk of development.

To ensure the consistency of the characteristics of the L-band antenna in the amplitude/phase band of the whole array, in the three area designs, the L-band antenna remains unchanged, and planar dipole elements are used. Both the S-band and X-band elements use a square stacked patch structure. In addition, an asymmetric isolation slot was etched on the drive patch to improve the polarization isolation of the dual-probe feed. After testing the experimental antenna, the L-band bandwidth reaches 13.4%, the port polarization isolation is better than 37 dB, and the cross-polarization level is lower than −30 dB. The measured bandwidth of the S-band is 14.8%, the in-band array isolation is better than 45 dB, and the cross-polarization level is lower than −30 dB. The X-band bandwidth is 16.8%, the in-band array isolation is better than 43 dB, and the cross-polarization level is lower than −35 dB.

References

1. Wei W, Shunshi Z, Xianling L (2007) Improved antenna design based on aesthetics in electromagnetic engineering. J Appl Sci 25(4):424–427
2. Ludwig AC (1973) The definition for cross polarization. IEEE Trans Antennas Propag 21(1):16–19
3. Wang W, Lu J, Zhang H et al (2016) A brief review of SAR antenna development in China. In: CIE international conference on radar. IEEE, Guangzhou, pp 1–4
4. Zhu S, Shunshi Z, Lingbing K et al (2012) Dual-band dual-polarized microstrip array with fractional frequency ratio. Electron Lett 12:674–676
5. Shunshi Z, Xuexia Y, Shichang G et al (2002) Corner-fed microstrip antenna element and arrays for dual-polarization operation. IEEE Trans Antennas Propag 50(10):1474–1480
6. Xianling L, Shunshi Z, Wei W (2005) Dual-polarized corner-fed patch antenna array with high isolation. Microw Opt Technol Lett 47(6):520–522
7. Wang W, Zhong S, Liang X (2004) A dual-polarized stacked microstrip antenna subarray for X-band SAR application. In: IEEE antennas and propagation society symposium. IEEE, Monterey, CA, pp 1604–1606
8. Xianling L, Shunshi Z, Wei W (2005) Design of dual-polarized microstrip patch antennas with excellent polarization purity. Microw Opt Technol Lett 44(4):329–331
9. Tso HH, Wang YM, Kilburg F et al (1988) Aperture-coupled patch antenna with wide-bandwidth and dual-polarization capabilities. In: IEEE antennas and propagation society international symposium, antennas and propagation. IEEE, Syracuse, NY, pp 936–939
10. Gao S, Li L, Leong MS et al (2003) Dual-polarized slot-coupled planar antenna with wide bandwidth. IEEE Trans Antennas Propag 51(3):441–448
11. Wei W (2005) Broadband printed antennas and dual-polarized microstrip and waveguide slot antenna arrays. Shanghai University, Shanghai
12. Xianling L, Shunshi Z, Wei W (2005) The cross-polarization suppression of dual-polarized linear microstrip antenna arrays. J Microw 21(1):22–25
13. Xianling L, Shunshi Z, Wei W (2004) Cross-polarization suppression of dual-polarization linear microstrip antenna arrays. Microw Opt Technol Lett 42(6):448–451
14. Zawadzki M, Huang J (1999) A dual-polarized microstrip subarray antenna for an inflatable L-band synthetic aperture radar. In: IEEE antennas and propagation society international symposium. IEEE, Orlando, FL, pp 276–279
15. Patel PD (1996) A dual polarization microstrip antenna with low cross-polarization for SAR application. In: IEEE antennas and propagation society international symposium. IEEE, Baltimore, MD, pp 1536–1539
16. Kabacik P, Bialkowski M (1998) Microstrip patch antenna design considerations for airborne and spaceborne applications. In: IEEE antennas and propagation society international symposium. IEEE, Atlanta, GA, pp 2120–2123
17. Rostan F, Wiesbeck W (1995) Dual polarized microstrip patch arrays for the next generation of spaceborne synthetic aperture radars. In: IEEE antennas and propagation society international symposium. IEEE, Firenze, pp 2277–2279
18. Lu J, Cao J, Liu H (2003) A circular polarization phased array antenna system for satellite communication system. Radar Sci Technol 1(1):54–59
19. Wei W, Xianling L, Yumei Z et al (2007) Experimental characterization of a broadband dual-polarized microstrip antenna for X-band SAR applications. Microw Opt Technol Lett 49(3):649–652
20. Targonsji SD, Pozar DM (1993) Design of wideband circularly polarized aperture-coupled microstrip antennas. IEEE Trans Antennas Propag 41(2):214–220
21. Lu J (2001) Research on a rectangular cavity crossed slot antenna. J Microw 17(1):1–6
22. Wu R, Zou Y, Wang W (2012) Study of the polarization tracking for the antenna of mobile satellite terminal. J Microw 28(3):48–50
23. Gao S, Luo Q, Zhu F (2014) Circularly polarized antennas. Wiley, West Sussex

24. Wei W, Shunshi Z, Yumei Z et al (2006) A broad bandwidth slotted ridged-waveguide antenna array. IEEE Trans Antennas Propag 54(8):2416–2420
25. Wei W, Zhi Z, Xiaochuan F et al (2019) A waveguide slot filtering antenna with an embedded metamaterial structure. IEEE Trans Antennas Propag 67(5):2954–2960
26. Zhang H, Wang W, Jin M et al (2016) A novel broadband slotted waveguide antenna array in the single waveguide cavity. In: 9th international conference on microwave and millimeter wave technology. IEEE, Beijing, pp 1–3
27. Lu J, Wu S, Chen S (1998) Analysis for slot array antennas fed by asymmetric single-ridge waveguide. Chin J Radio Sci 13(4):388–392. (in Chinese)
28. Moradian M, Tayarani M, Khalaj AM (2011) Planar slotted array antenna fed by single wiggly-ridge waveguide. IEEE Antenna Wirel Propag Lett 10:764–767
29. Wang W, Zhang H, Qi M (2012) Ridged waveguide slot antenna array with low cross-polarization. In: International conference of microwave and millimeter technology (ICMMT). IEEE, Shenzhen, pp 1–3
30. Hashemi YS, Elliott RS (1990) Analysis of untilted edge slot excited by tilted wires. IEEE Trans Antennas Propag 38(11):1737–1745
31. Hirokawa J, Kildal PS (1997) Excition of an untilted narrow-wall slot in a rectangular waveguide by using etched strips on a dielectric plate. IEEE Trans Antennas Propag 45(6):1032–1037
32. Wei W, Shunshi Z, Jian J et al (2005) An untilted edge-slotted waveguide antenna array with very low cross-polarization. Microw Opt Technol Lett 44(1):91–93
33. Mallahzadeh AR, Mohammad Ali Nezhad S (2012) An ultralow cross-polarization slot array antenna in narrow wall of angled ridge waveguide. J Commun Eng 1(1):46–59
34. Lu J, Zhang H, Wang W et al (2019) Broadband dual-polarized waveguide slot filtenna array with low cross-polarization and high-efficiency. IEEE Trans Antennas Propag 67(1):151–159
35. Zhang H, Wang W, Zhang Z et al (2012) A novel dual-polarized waveguide antenna with low cross-polarization for SAR applications. In: International conference on microwave and millimeter wave technology, pp 1–4
36. Wang W, Zhang H, Jin M et al (2016) A circularly polarized waveguide slot antenna. In: IEEE Antennas and propagation society international symposium. IEEE, pp 1709–1710
37. Zhang H, Wang W, Zheng Z et al (2018) A novel dual circularly-polarized waveguide antenna array. In: 2018 IEEE-APS symposium on antennas and propagation and URSI CNC/USNC join meeting (AP-S/URSI 2018). IEEE, Boston, Massachusetts, USA, pp 509–510
38. Zhang H, Wang W, Liang X, Jin M, Lu X et al (2017) Design of a dual circular polarization waveguide antenna array with wide band and low axial ratio. Radar Sci Technol 15(1):50–54
39. Li T, Meng H, Dou W (2014) Design and implementation of dual-frequency dual-polarization slotted waveguide antenna array for Ka-band application. IEEE Antenna Wirel Propag Lett 13:1317–1320
40. Lu J, Wang W, Lu X, Zhang H et al (2020) Research on three matching problems in waveguide slot antenna. Radar Sci Technol 18(2):115–123
41. Pozar DM, Schauhert DH, Targonski SD et al (1998) A dual-band dual-polarized array for spaceborne SAR. In: IEEE antennas and propagation society international symposium. IEEE, Atlanta, Georgia, pp 2112–2115
42. Chen M, Jiang Z, Wang W, Li L (2017) Design of L/S dual-band circularly polarized shared-aperture microstrip antenna array. In: Proceedings of 2017 National conference on antenna technology. Chinese Institute of Electronics, Xian, pp 282–284
43. Shunshi Z, Zhu S, Lingbing K et al (2012) Tri-band dual-polarization shared-aperture microstrip array for SAR applications. IEEE Trans Antennas Propag 60(9):4157–4165
44. Wang W, Zhang H, Lu J et al (2015) Dual band dual polarized antenna for SAR. In: IEEE international symposium on antennas and propagation & USNC/URSI National radio science meeting. IEEE, Vancouver, BC, Canada, pp 220–221
45. Kumar G, Ray KP (2003) Broadband microstrip antennas. Artech House, Norwood, MA, pp 12–13

Chapter 7
Active Antenna-in-Package Arrays

7.1 Introduction

Active Antenna-in-Package (AiP) technology integrates antenna elements, various chips, components, etc. into one package to realize system-level antenna functions based on packaging materials and integrated process technologies. An array of active AiP units or array composed of multiple AiP units is called an active AiP array.

Integrated circuit chips were invented in the 1960s [1]. Since the advent of the first integrated circuit chip, the integration level of chips has increased rapidly at the speed predicted by Moore's law. Higher integration means that the functions of the integrated circuit chips are becoming more and more diversified, such as microwave signal generation, amplification, phase shifting, and switching. In order to realize such a variety of functions, integrated circuit chips need to be hybrid integrated. Chips are integrated with other heterogeneous chips or components through interconnection. At the same time, it must ensure that various chips can work in harmony without interfering with each other, let alone being affected by the external environment.

From a functional point of view, an active AiP is an integrated package of the antenna and the RF transceiver system to form a package as a whole. This enables the entire system to obtain more stable electrical characteristics and higher integration level while improving the signal integrity and reliability of the system.

From the design point of view, an active AiP refers to the layout of antenna elements, chips, components, and other functional modules which matches the design requirements of the electromechanical and thermal multi-physical quantities. Then micro-assembly techniques such as bonding and welding are carried out, making it an overall system to meet the specific performance and functional requirements of the antenna system.

J. Lu et al., *Active Array Antennas for High Resolution Microwave Imaging Radar*,
https://doi.org/10.1007/978-981-99-1475-3_7

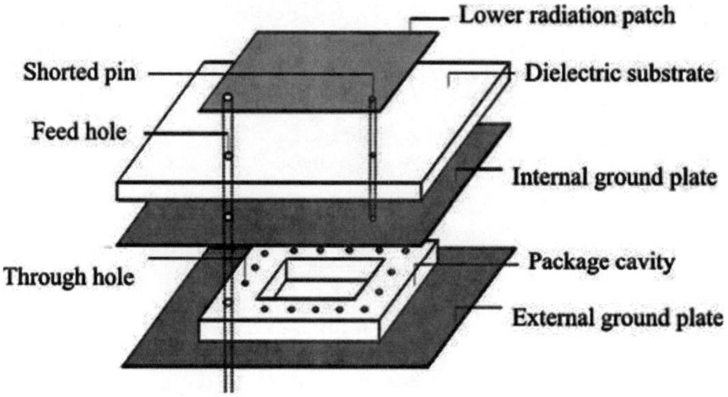

Fig. 7.1 Basic configuration of AiP

7.1.1 AiP Configuration

AiP is a new type of active antenna proposed in the early twentieth century [2], which inherits and develops the integration concept of microstrip antenna, multi-chip circuit module, and tile phased array antenna structure. The development of integrated circuit technology has promoted the emergence and development of active AiP. The AiP technology extends antenna tentacles to the fields of integrated circuits, packaging, materials, and processes. It advocates multidisciplinary collaborative design and system-level optimization while considering antenna performance, cost, and volume.

The basic configuration of the AiP is shown in Fig. 7.1, comprising three parts: the ordinary antenna, the package cavity, and the package ground. The package cavity is arranged between the antenna ground and the package ground. Chips and circuit components are placed in this cavity and connected to external circuits through bonding wires and via holes. At the same time, the antenna ground and the package ground will be connected through metal vias. Finally, the AiP must be installed on a system board for testing and use.

The antenna structure shown in Fig. 7.1 integrates the antenna and the RF integrated circuit chip into one integrated package, making the antenna element and the chips to be a whole system, which can effectively realize the miniaturization and integration of the RF system. The distance between devices is getting shorter with the improvement of radio frequency and front-end integration. As the distance between the device and the radiation unit decreases significantly, the micro-scale interconnection's parasitic effects become more remarkable.

7.1.2 AiP Active Arrays

An active AiP unit or a plurality of antenna elements packaged into an array is called an AiP active array. The main purposes of arranging the antenna elements into an antenna array are as follows: firstly, it is to combine multiple unit antenna beams to increase the gain of the antenna system, and secondly, it is to carry out beam scanning by phase shifter in the antenna element channel, changing the original mechanical scanning method into electronic scanning way.

There are two main types of active array AiP. One type is that the active antenna is packaged in a single channel, and the antenna elements are arranged in an array according to the design requirements. Figure 7.2 shows a single-channel AiP, in which No. 3 is the antenna substrate, which is usually microstrip-fed or cavity-backed, and the materials are commonly used in low-temperature co-fired ceramics or aluminum nitride ceramics. No. 8 and No. 9 are multi-function chips, such as transceiver switches, high-power amplifiers, low-noise amplifiers, switches, phase shifters, attenuators, and pre-power amplifiers. The chips are interconnected by gold wire (No. 6), the output of the high-power amplifier is connected to the antenna through the transceiver switch, and the receiving end of the transceiver switch is connected to the low-noise amplifier. A single antenna element active package input and output ball grid array (BGA) (No. 7) is connected to the next-level power combination/distributing network. Through a LTCC feed base plate (No. 2), a certain number of single antenna element packages can be formed into a sub-array active array AiP, and then a larger array of active AiPs can be formed through PCB substrate (No. 1).

Another array type integrates multiple antenna elements and their active circuit chips into an AiP array, as shown in Fig. 7.3.

The difference between Figs. 7.2 and 7.3 is that the multi-antenna element AiP integrates a certain number of antenna elements (such as multiple chips, integrated feed network, heat sink, and multi-function board) into one package. A sub-array-level active AiP is directly formed.

In practical application, the AiP configuration, packaging method, and packaging material are optimized, according to the antenna's operating frequency, frequency bandwidth, and single-channel antenna power.

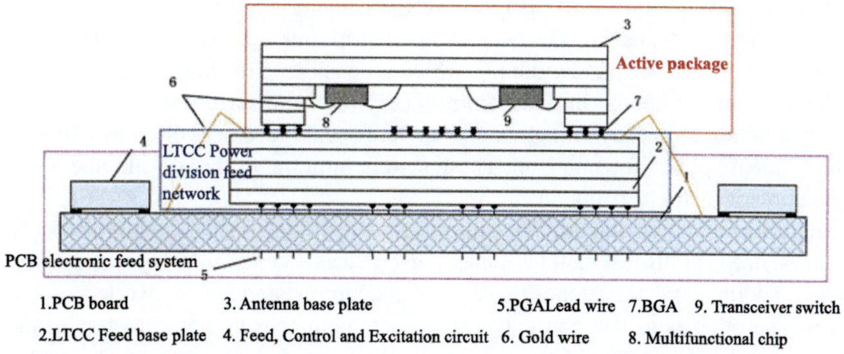

Fig. 7.2 Single-channel AiP

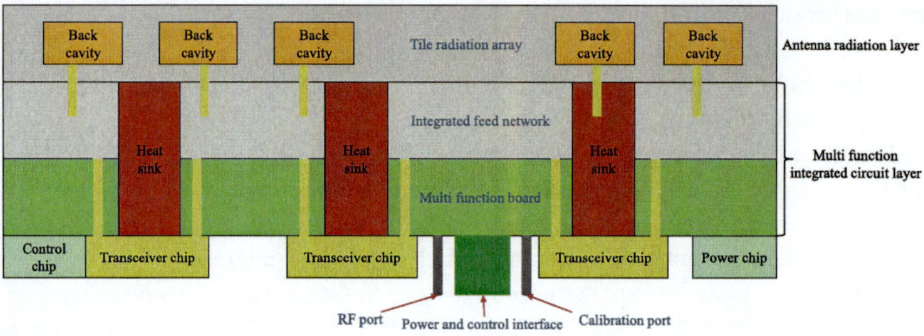

Fig. 7.3 Multi-antenna element AiP

7.2 AiP Elements

High-resolution Earth observation synthetic aperture radars usually use broadband active array antennas. Typical antenna elements commonly used in the design of broadband active array antennas include tapered slot antennas, long slot antennas, and dipole antennas. To reduce the antenna profile, resonant antennas are usually used [3], such as microstrip resonant antennas and cavity-backed antennas for AiPs.

7.2.1 Multi-layer Microstrip Antennas

Microstrip antennas have many advantages: planar structure, small size, light mass, low profile, easy installation, and effective integration with semiconductor integrated circuits. The processing and manufacturing cost is low, and it is easy to be mass-produced. The excitation feed is flexible, making it easy to obtain linear polarization, dual polarization, circular polarization, and other states, and it is also easy to realize dual-band or even multi-band operation of the antenna. However, due to its high-quality factor at resonance, the general microstrip antenna usually has less than 1–5% of the impedance bandwidth [4], which is considered the biggest shortcoming of the microstrip antenna. In addition, it is easy to excite surface waves, resulting in low radiation efficiency and gain, and the large-scale feed network loss will lead to the decrease of the antenna gain. These factors limit the further development and application of microstrip antennas.

There are two types of methods to improve the operating bandwidth of microstrip antennas: one is multi-resonant point methods, such as double/multi-layer microstrip patches, adding parasitic patches to planes, slot coupling, and feed network compensation. The other is to reduce the Q-value, such as increasing the thickness of the substrate and reducing the dielectric constant of the dielectric plate. The multi-layer microstrip patch antenna obtains multi-frequency resonance through patches of different sizes on the upper and lower layers, which can effectively broaden the

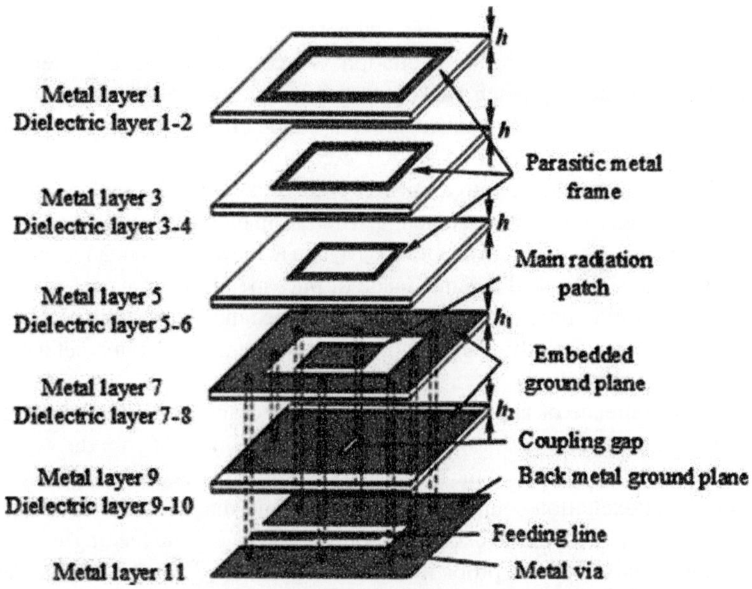

Fig. 7.4 Schematic diagram of the structure of a multi-layer parasitic microstrip antenna

bandwidth. Usually, the double-layer microstrip patch can obtain more than 10% of the working bandwidth [5]. Slot coupling is due to the intervention of multiple variables, such as patches, coupling slots, and short-circuit lines, to obtain a wider operating bandwidth. With increasing the air layer, a bandwidth greater than 20% can be obtained [6]. The coupled microstrip line matching technique can achieve an impedance bandwidth of more than 53% [7].

The sacrifice of some aspects of performance parameters usually accompany the planarization and miniaturization of the antenna. If some new structures can be realized, the space occupied by the antenna can be fully utilized, such as the multi-layer structure of Low Temperature Co-fired Ceramic (LTCC) technology, as shown in Fig. 7.4. In order to optimize the antenna design, it's worth considering making the structure three-dimensional and selecting a suitable broadband microstrip antenna structure. This can help ensure that the performance parameters, such as gain and bandwidth, meet the application requirements while minimizing the overall size of the antenna. In addition, reconfiguring the RF connector and embedding it inside the medium can effectively prevent the RF connector from being damaged in actual use and improve the stability and reliability of the antenna.

7.2.2 Cavity-Backed Antennas

Cavity-backed antennas usually include metal cavity-backed antennas and substrate-integrated waveguide (SIW) cavity-backed antennas.

The research on the metal cavity-backed antenna began in the 1950s [8], which consists of a metal cavity and an excitation source. Cavity-backed antennas can be divided into microstrip patch-backed antennas, microstrip or metal slot cavity-backed antennas, dipole cavity-backed antennas, and helical cavity-backed antennas according to the type of excitation source. According to the polarization form of the antenna, it can be divided into a linearly polarized cavity-backed antenna and a circularly polarized cavity-backed antenna. Microstrip patch cavity-backed antennas composed of patches of different shapes [9, 10] can achieve an impedance bandwidth of 45% [10]. The slot-backed antenna made of pure metal materials can also obtain an extensive operating bandwidth, of which the impedance bandwidth reaches 43% [11]. In addition, the specially designed slot-backed antenna with pure metal material has certain filtering ability [12].

For the convenience of analysis, the cavity-backed antenna is decomposed into several independent parts, as shown in Fig. 7.5. Here J_a and M_a are the equivalent current and magnetic current on the back cavity aperture surface S, V_r is the resonant cavity, E_0 is the excitation source with current J_0 and magnetic current M_0, M_i is the matching component at the excitation source, and W_g is the feeding waveguide. Therefore, the analysis of the problem of the cavity-backed antenna can be decomposed into three basic problems: intra-cavity problem, extra-cavity problem, and input problem. Among them, the intra-cavity problem is the most critical because the electric field distribution of the intra-cavity mode directly determines the electric field amplitude and phase distribution at the radiation aperture. The external problem can be solved according to the electric field distribution at the aperture. On the other hand, the input problem is mainly related to the impedance-matching relationship between the excitation source in the cavity and the feed waveguide. It also has a certain influence on the intra-cavity problem.

For a cavity, especially a cavity filled with medium, the quality factor Q_A of the cavity is relatively high. Therefore, the cavity can be regarded as a resonant cavity, and the electric field distribution in the cavity can be obtained by analyzing the eigenmodes in the cavity. However, compared with the general closed resonant cavity, the cavity antenna has unique properties [13]. The cavity antenna will cause an electric field disturbance, which will lead to the shift of the antenna's resonant frequency and the narrowing of the fundamental mode spectrum.

Fig. 7.5 Schematic diagram of the structure of a cavity-backed antenna

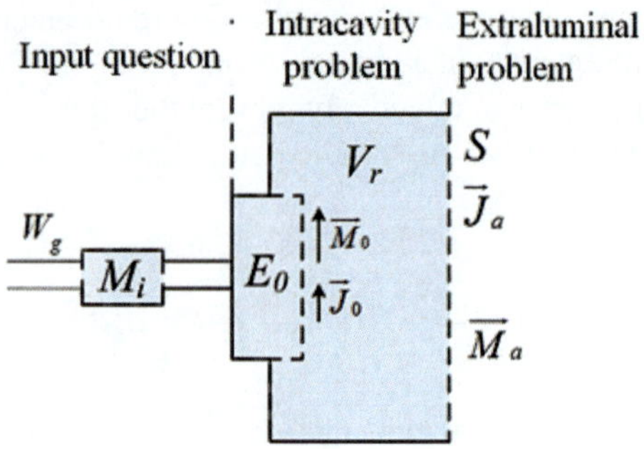

The quality factor of the cavity antenna is

$$\frac{1}{Q_A} = \frac{1}{Q_{\text{wall}}} + \frac{1}{Q_{\text{med}}} + \frac{1}{Q_{\text{rad}}} \tag{7.1}$$

where Q_{wall}, Q_{med}, and Q_{rad} are some quality factors of the antenna, which are, respectively, related to the loss of the metal wall, the loss of the cavity medium, and the radiation loss.

It can be seen from Eq. (7.1) that as the cavity antenna is equivalent to a resonant cavity with one open end, two-dimensional resonance occurs in the cross section of the cavity. This resonant mode is the same as the traditional waveguide resonant mode. Therefore, using the familiar waveguide mode to analyze the cavity-backed antenna mode will greatly simplify the problem.

In the microwave frequency band, the metal-backed cavity antenna has the characteristic that different polarizations are directly generated by the excitation source, which is generally fed by a coaxial line. However, this feeding form is inconvenient to connect with other components in the system, and when forming an antenna array, the form of the coaxial feeding network is relatively complicated. In addition, since the material of the antenna cavity is metal, when the antenna array is formed, the volume of the antenna will be too large, and it will not be easy to integrate.

To meet the needs of modern information systems, the antenna design should be miniaturized and integrated as much as possible to meet the performance indicators. It can be effectively connected with other components in the system. Therefore, substrate-integrated waveguide structures are gradually being used in cavity-backed antennas due to their advantages of low profile, low loss, and high integration. At the same time, the antenna elements, various chips, and feeding network are integrated into the dielectric board using low-temperature co-fired ceramic technology and multi-layer printed circuit board (PCB) technology, which effectively improves the integration and stability of the active AiP.

The SIW concept started from the substrate-integrated circuit (SIC). The typical SIW structure is shown in Fig. 7.6, of which the electric field is similar to the main mode in the ordinary rectangular waveguide. Still, due to the through-hole gap, there will be a certain leakage of the electromagnetic fields. Nevertheless, this structure can effectively realize the integration of passive and active circuits, miniaturize the microwave and millimeter-wave system, and even make the entire microwave and millimeter-wave system in one package. Its propagation characteristics are similar to rectangular metal waveguides. Therefore, microwave, millimeter-wave, or even submillimeter-wave components and systems composed of SIW structures have the advantages of high Q-value, high power capacity, and easy integration with other planar circuits and chips. At the same time, since the whole structure is wholly composed of metalized through-hole arrays on the dielectric substrate, this structure can be accurately realized by ordinary PCB technology, LTCC technology, or even thin film circuit technology.

Figure 7.7 shows a configuration diagram of a commonly used SIW. The upper and lower surfaces of a dielectric substrate with a thickness h are metalized and

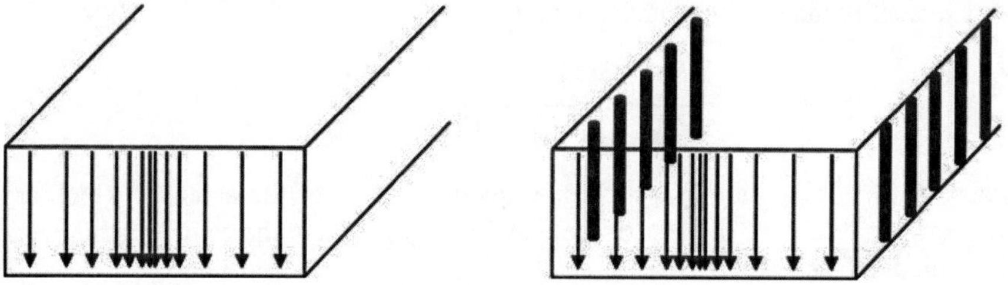

Fig. 7.6 Typical SIW structure and equivalent rectangular cavity

are connected by two rows of metalized through holes arranged periodically in the middle. When the arrangement density of the metalized via holes meets certain requirements, it can effectively prevent electromagnetic waves from leaking between the hole gaps, thereby forming a waveguide structure similar to a rectangular waveguide on the cross section of the substrate. The two rows of metalized holes are equivalent to the side walls of the metal waveguide, and the upper and lower metalized surfaces are the metal waveguide's upper and lower metal walls. Among them, the metalized via hole diameter D, the hole array period p, and the distance W between the centers of two rows of metal holes are the width of the integrated waveguide on the substrate. Numerical analysis and experiments of electromagnetic fields have proved [14] that the guided waves of substrate-integrated waveguides are similar to rectangular waveguides, so when designing microwave circuits with substrate-integrated waveguides, the design principles of metal waveguides can be referred to.

For the substrate material of the substrate-integrated waveguide, if a low dielectric constant substrate is selected, the non-radiative part of the electromagnetic field energy coupled to the medium will be reduced so that the radiation conductance of the antenna equivalent circuit will increase, the antenna Q-value will decrease, and the bandwidth will increase. If a high dielectric constant substrate is used, it is often used in some planar circuits, such as power dividers, couplers, and filters, and its ability to "constrain" field energy is strong. The relationship between the Q-value of the antenna and the dielectric constant of the substrate, and the thickness of the plate can be expressed as

Fig. 7.7 Schematic diagram of SIW structure

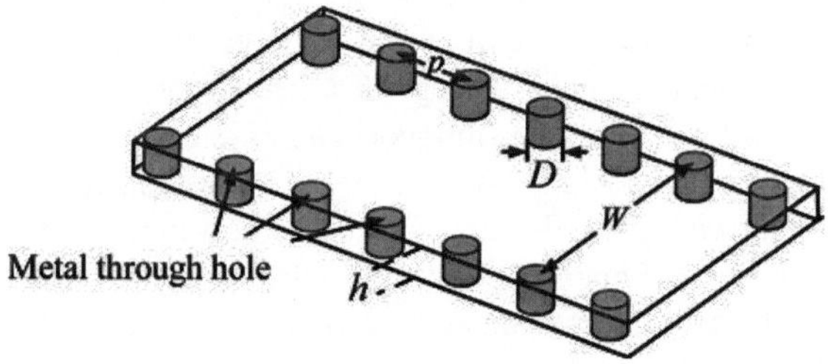

$$Q = \frac{\varepsilon_r}{\mu_0 h} A \tag{7.2}$$

where ε_r is the relative permittivity of the medium, μ_0 is the permeability constant, h is the thickness of the substrate, and A is a constant.

It can be seen from Eq. (7.2) that when the antenna h increases, the Q-value will decrease, and the antenna bandwidth will be significantly improved, but simply increasing the dielectric thickness will often sacrifice the antenna performance. This is because as the thickness of the medium increases, the surface wave of the medium will increase. The surface wave effect may not be very noticeable for low-frequency applications, but as the frequency increases, the antenna efficiency will drop significantly. For example, suppose a coplanar transmission line such as a microstrip or CPW is used to excite the antenna. In that case, the current distribution on the antenna surface will propagate along the surface of the medium, and the far-field radiation pattern will be significantly shifted axially compared with the original pattern.

7.2.3 Bandwidth and Impedance Matching

The antenna bandwidth is closely related to the impedance bandwidth of the antenna. The antenna bandwidth is an important parameter in antenna design, and it is used to describe the amount of radiation capability of the antenna that changes with the operating frequency. For the resonant antenna, the expression of the impedance bandwidth BW is [15]

$$\mathrm{BW} = \frac{\mathrm{VSWR} - 1}{\mathrm{Q}\sqrt{\mathrm{VSWR}}} \times 100\% \tag{7.3}$$

$$\frac{1}{Q} = \frac{1}{Q_r} + \frac{1}{Q_c} + \frac{1}{Q_d} + \frac{1}{Q_{\mathrm{sw}}} = \tan \delta_{\mathrm{eff}} \tag{7.4}$$

where $\tan \delta_{\mathrm{eff}}$ is the equivalent dielectric loss tangent, Q is the quality factor of the antenna, Q_r is the quality factor caused by antenna radiation, Q_c is the quality factor caused by conductor loss, Q_d is the quality factor caused by dielectric loss, and Q_{sw} is the quality factor caused by surface waves. The quality factor is

$$Q_r = \frac{\varepsilon_r W L}{120 \lambda_0 h G_r} \tag{7.5}$$

$$Q_c = \pi h \sqrt{\frac{120 \sigma_c}{\lambda_0}} \tag{7.6}$$

$$Q_d = \frac{1}{\tan \delta} \tag{7.7}$$

$$Q_{sw} = \left(\frac{\lambda_0}{3.4h\sqrt{\varepsilon_r - 1}}\right)Q_r \tag{7.8}$$

where λ_0 is the wavelength of free space, G_r is the equivalent radiation conductance, σ_c is the conductivity of the conductor, ε_r is the relative permittivity of the medium, $\tan\delta$ is the loss tangent of the medium, h is the thickness of the medium, and W and L are the equivalent width and length of the antenna.

(1) **Reduce the quality factor and increase the antenna bandwidth**

Equation (7.3) shows that a higher dielectric constant will make the quality factor Q of the antenna higher so that it has high energy storage and cannot radiate energy well. Therefore, in order to increase the impedance bandwidth of the antenna, it is necessary to reduce the quality factor Q of the antenna, that is, to choose a larger substrate thickness or a lower dielectric constant. When the material of the antenna dielectric substrate is selected, the tangent loss $\tan\delta$ is determined. Generally, an effective way to increase the antenna bandwidth is to change the equivalent width-to-length ratio W/L of the antenna or use other loading methods to expand the bandwidth.

Optimizing the antenna dielectric layer's structure is an effective way to increase the antenna bandwidth. Reducing the dielectric constant of the antenna dielectric material can increase the impedance bandwidth. In addition to using composite materials and new materials, the antenna dielectric layer structure can also be optimized to reduce the equivalent dielectric constant. For example, a plurality of air cavities is arranged in the dielectric layer using cavity technology to reduce the equal dielectric constant without increasing the height of the dielectric layer.

Optimizing the shape of the antenna can also increase the bandwidth of the antenna. For example, changing the shape of the antenna radiating element by introducing meander lines or other special fractal mathematical means can increase the effective path through which its current passes, change its equivalent width-to-length ratio to a certain extent, and increase its impedance bandwidth.

(2) **Impedance matching improves antenna bandwidth**

The input impedance of an antenna is defined as the ratio of the voltage and current at the antenna input. Its value characterizes the matching condition between the antenna and the feeder (the feeder is the connection between the antenna and the transmitter or receiver) and reflects the quality of the energy conversion between the radiated wave and the guided wave. Antenna input impedance is usually expressed as

$$Z_{in} = \frac{U_{in}}{I_{in}} = \frac{P_{in}}{\frac{1}{2}|I_{in}|^2} = R_{in} + jX_{in} \tag{7.9}$$

$$P_{in} = \mathrm{re}(P_{in}) + \mathrm{im}(P_{in}) \tag{7.10}$$

where R_{in} is the input resistance, X_{in} is the input reactance, P_{in} is the power fed to the antenna, composed of real input and imaginary power. Considering the structure loss of the antenna itself and its surroundings (including the losses due to the lossy medium, the support structure, the grounding device, the radio waves absorbed by the surrounding objects, etc.), the real input power is

$$\mathrm{re}(P_{\mathrm{in}}) = P_{\mathrm{rad}} + P_n \tag{7.11}$$

where P_{rad} is the antenna radiation power and P_n is the dissipated power.

It can be seen from Eq. (7.9) that if the ohmic loss of the antenna conductor is considered, the loss caused by the input resistance R_{in} consists of two parts: one is the power caused by radiation loss and the other is the power caused by ohmic loss.

When the input impedance of the antenna does not match the characteristic impedance, it will cause reflection, and the reflection coefficient Γ, VSWR, and return loss L can be used to express the impedance-matching conditions, that is,

$$\Gamma = \frac{Z_{\mathrm{in}} - Z_0}{Z_{\mathrm{in}} + Z_0} \tag{7.12}$$

$$\mathrm{VSWR} = \frac{|V_{\max}|}{|V_{\min}|} = \frac{1+\Gamma}{1-\Gamma} \tag{7.13}$$

$$L = 10\log\left(\frac{P_{\mathrm{in}}}{P_{\mathrm{ref}}}\right) = 20\log(\Gamma) \tag{7.14}$$

where Z_{in} is the input impedance of the antenna, Z_0 is the characteristic impedance of the antenna transmission line, $V_{\max}$ is the maximum voltage on the transmission line, $V_{\min}$ is the minimum voltage on the transmission line, P_{in} is the incident power, and P_{ref} is the reflected power.

It can be seen from Eq. (7.9) that when the input impedance of the antenna is much greater than the characteristic impedance, the reflection coefficient $\Gamma = 1$, and total reflection occurs at this time. On the other hand, when the input impedance is equal to the characteristic impedance, the impedance of the input end is matched, and all the electromagnetic energy will enter the antenna radiation element. At this time, the reflection coefficient $\Gamma = 0$, reaching the ideal situation. However, the antenna usually cannot be completely matched, and voltage standing wave ratio VSWR and return loss L are often used to indicate the antenna's matching degree and working condition.

The input impedance of the antenna is related to factors such as the geometric shape, size, position of the feed point, operating wavelength, and surrounding environment of the antenna. Antenna impedance matching is to eliminate the reactive component in the input impedance of the antenna so that the resistive component approaches the characteristic impedance of the feeder line. In the AiP, the integration level is getting higher and higher, and the length of the feeder is getting shorter and shorter, which may enter a micro-scale. This micro-scale's impedance-matching characteristics will change, producing mismatch effect.

Optimizing the matching circuit network is a method to reduce the micro-scale feeder's parasitic effect. Adding loss components such as capacitor patches or resistors at the feed end of the antenna is equivalent to reducing the quality factor and increasing the impedance bandwidth to a certain extent. However, adding a matching network will degrade performance, such as antenna gain and radiation efficiency.

The multimode resonance technology increases the impedance bandwidth of the antenna by changing the position of the coaxial feeder on the radiation patch so that the antenna generates multiple resonance frequency points. Multiple radiating antenna elements can also be stacked vertically so that different antenna radiating elements work in adjacent frequency bands to expand the antenna bandwidth.

7.3 Multi-layer Vertical Interconnect Technology

The basic structure of AiP can generally be divided into two types: horizontal configuration and vertical configuration, as shown in Fig. 7.8.

As shown in Fig. 7.8a, the AiP with a horizontal structure places the chip and the antenna horizontally in two cavities. The connection between the antenna and the chip is realized by bonding wires or flip-chip welding technology. This kind of antenna structure is easy to realize. However, because of the horizontal arrangement, the cross-sectional area of the entire AiP is too large, and it is difficult to improve the level of integration in the horizontal direction.

Figure 7.8b shows the AiP with a vertical structure. By packaging the antenna vertically above the chip, flip-chip welding and metal interconnection via technology are used to realize the feeding of the antenna end. This structure greatly reduces the size of the microwave RF system and can also introduce a metal shielding layer between the chip and the antenna to reduce mutual interference. The vertical interconnection between the AiP units and the active transceiver chips plays an important role in realizing the low profile and light weight of the active array antenna system. The parasitic effect caused by this vertical feeding interconnection will affect the radiation efficiency of the antenna, the input impedance of the antenna, the insertion loss of the feed line, and the electromagnetic compatibility inside the AiP.

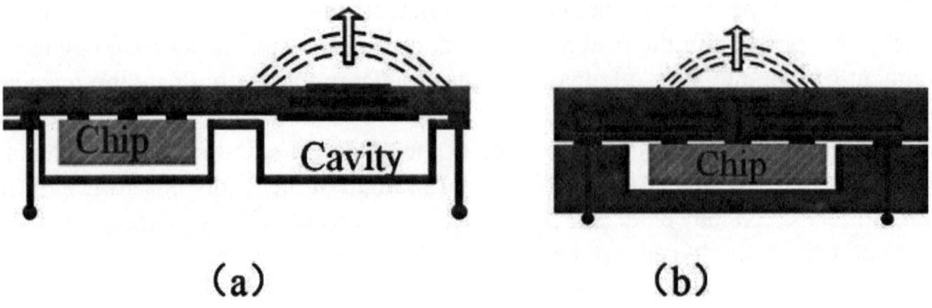

Fig. 7.8 AiP configuration, **a** horizontal configuration, **b** vertical configuration

The vertical interconnection structure mainly includes land Fuzz Button interconnection, land BGA interconnection, land grid array (LGA) interconnection, land interlayer interconnection, through-silicon via (TSV) interconnection, etc.

7.3.1 Land Fuzz Button Interconnection

Fuzz Button is generally a columnar body made of thin metal wires, having a certain degree of elasticity, and is wrapped with a supporting medium on the outer layer, such as polytetrafluoroethylene. The physical photo of Fuzz Buttons is shown in Fig. 7.9. The metal wire used for Fuzz Buttons usually has good electrical properties and a certain degree of elasticity. This is because the Fuzz Button is often pressed between two plates, and the good elasticity makes the Fuzz Button closely in contact with the circuit. The metal materials that can be selected are usually molybdenum (Mo), beryllium copper (BeCu), wolfram (W), and nickel–chromium alloy (NiCr). In general, to further enhance the electrical properties of the button, increase the oxidation resistance, and reduce the contact resistance between the button and the pad, a layer of aurum (Au) is plated on the surface of the wire.

There are usually two types of fur buttons used in the vertical interconnection technology of microwave circuits: one is the coaxial form, as shown in Fig. 7.10a, which uses the Fuzz Button as the central conductor. The other is the three-wire form (three-wire), as shown in Fig. 7.10b. There are three Fuzz Buttons putting side by side. The middle Fuzz Button is used as a signal channel, and the other two are grounded to form a microwave ground.

The Fuzz Button is a high-performance, commonly used, customizable connection device which can be made into a connection size that meets the requirements according to different circuit forms. Fuzz Button can be used in a variety of packaging forms, such as Pin Grid Array (PGA), BGA, Land Grid Array (LGA), Cylinder Grid Array (CGA), Quad Flat No-lead Package (QFN), and Multi-Chip Module (MCM).

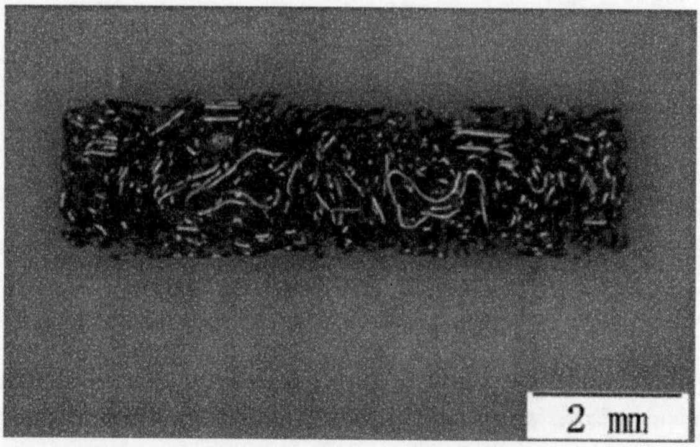

Fig. 7.9 Photograph of Fuzz button

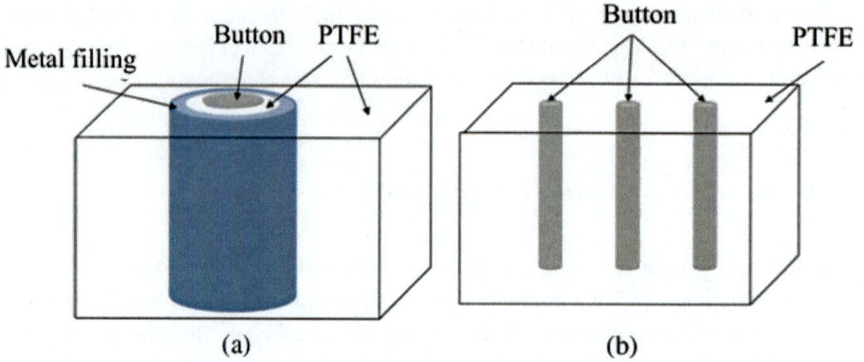

Fig. 7.10 Common forms of Fuzz button, **a** coaxial form, **b** three-wire form

The radius and height of Fuzz Button are important parameters for matching connections. Compared with other connection technologies, such as Pogo Pin and Spring Probes, and various welding technologies, the Fuzz Button has great advantages. Not only can it reduce signal distortion, but it can also transmit signals up to 40 GHz [16], and it has the advantages of high stability, shock resistance, long service life, etc.

Fuzz Button has its drawbacks. Since the Fuzz Button needs to be pressed on the circuit layer, the circuit layer often needs to be designed in a "face-to-face" form or use metalized through holes to transmit the signal to the back layer. The "face-to-face" format generally causes crosstalk of signals between two layers of circuits. And if the distance is large enough, the device size will be increased. If the distance is large enough, the length of the Fuzz Button will also increase, which will increase the loss of the transition structure. Furthermore, reaching the back layer through the metalized hole will bring more processing difficulties, such as the alignment problem between the hole and the Fuzz Button. This defect is not unique to the Fuzz Button but is caused by parasitic effect of the micro-scale interconnection. The main reason is that the electromagnetic field is discontinuous, the continuous transition line becomes shorter, the high-order modes are not sufficiently attenuated, and the "face-to-face" distance between the two adjacent layers of circuits is too short. As a result, the signal crosstalk between the two layers is relatively serious. Considering these factors, the upper and lower circuits should avoid forming parallel circuits as much as possible when designing circuits using the Fuzz Button technology.

7.3.2 Land BGA Interconnection

When analyzing and designing active packaged array antennas, one of the core issues is to solve the three-dimensional transmission of microwave signals. For the traditional two-dimensional packaging module, the transmission of signals is

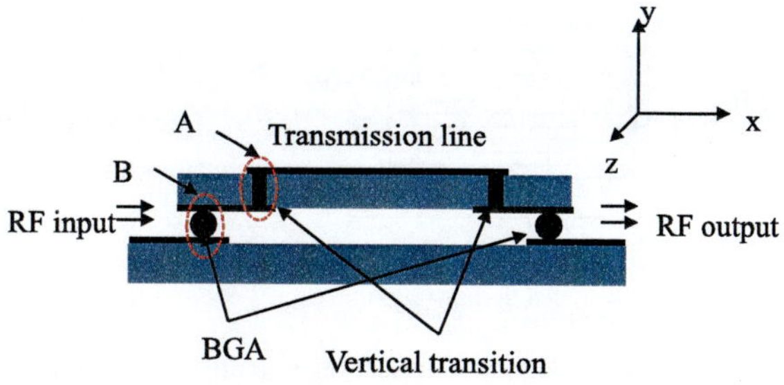

Fig. 7.11 Schematic diagram of the land BGA vertical interconnection

based on the transmission in the *X*–*Y*-plane, as shown in Fig. 7.11. But in the three-dimensional system, in addition to the transmission of the microwave signal in the *X*–*Y*-plane, the transmission in the *Z*-axis direction is realized. In the application of three-dimensional integrated systems, the BGA technology is commonly used for transmitting microwave signals along the *Z*-axis, which uses I/O terminals as cylinders to establish high-density connections with different components in the system. Although this method increases the number of I/O pins, the overall electrothermal performance of the circuit system can be well improved when BGA is used for soldering the pins. As a result, the thickness and parasitic parameters of the entire integrated circuit are greatly reduced, and the working frequency can also be increased.

Figure 7.11 shows a schematic diagram of the vertical interconnection between the upper and lower substrates realized by BGA technology and made of different materials. The structure consists of upper substrates, lower substrates, BGA solder balls in the middle, and several BGA solder balls are produced by reflow soldering technology between the two substrates. BGA solder balls play a supporting role in the transmission channel in this structure, and the transmission of microwave signals is completed between the upper and lower substrates.

7.3.3 Land LGA Interconnection

The Land Grid Array (LGA) technology uses solder between the upper and lower substrates to realize the physical connection between different components. Thus, the transition between the two base layers caused by BGA solder balls can be well avoided. The core of LGA technology is that it can realize seamless connection between two substrates, reduce parasitic effects inside the system, and adopt the transition structure to increase the operating frequency of the entire system effectively. A typical application is to use a Quad Flat No-lead Package (QFN) to realize the heat conduction inside the whole system with a large area of exposure on the substrate.

Traditionally, QFN technology is usually used to realize the packaging of a single MMIC chip. This packaging technology has a single function and less application

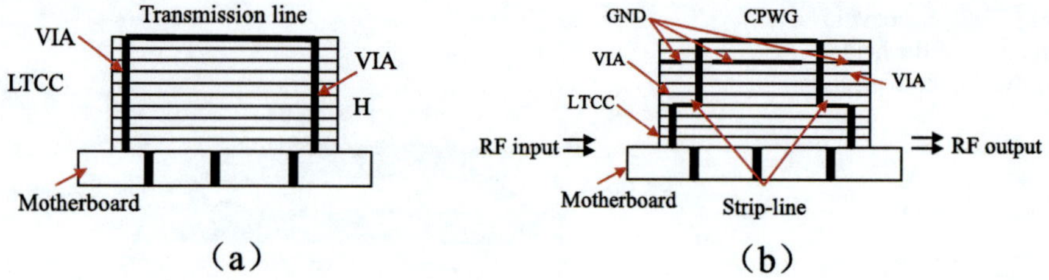

Fig. 7.12 Land LGA interconnection, **a** traditional mode, **b** improved mode

selectivity, which cannot meet the application requirements of multiple chips. For example, in order to improve the performance diversity of the entire package, it is necessary to realize the integrated package of multiple chips in the same substrate, which can only be achieved by using multi-layer substrate materials. Figure 7.12a shows the LGA connection mode diagram of the multi-layer substrate. In this structure, multi-layer ceramic materials are used, and bare chips are used inside to complete the integrated packaging of the multi-layer substrate. In the lower motherboard, PCB material is generally used, which is relatively cheaper, and easier to process. This material can be used as the bottom large-area substrate.

In practical applications, the LGA vertical interconnection structure designed with multi-layer substrates still has certain limitations. For example, to improve the microwave performance in the microwave and millimeter-wave frequency bands, it is necessary to reduce the thickness of the upper substrate as much as possible so that the mechanical performance and reliability can be greatly reduced. However, when the thickness of the upper substrate is reduced, the wiring capacity of the upper substrate is also greatly reduced at the same time.

In order to solve this problem effectively, a new LGA structure is generally adopted, as shown in Fig. 7.12b. This structure inserts a metal ground layer between the upper and lower substrates. The transmission line transition can be realized between the upper and lower substrates within the circuit structure. The essence of this structure is to divide the hole transmission structure into two transition structures with different lengths, which can effectively reduce the inductance effect in the circuit and broaden the working bandwidth of the entire circuit.

7.3.4 Intra-Board Layer-to-Layer Interconnect

The vertical interconnect structure adopts the interlayer interconnection method within the board, which follows the development of LTCC and Al_2O_3-HTCC, AlN-HTCC multi-layer ceramic packaging technology. Compared with inter-board interconnection technologies such as Fuzz Button, micro-connectors, BGA, and LGA, intra-board interlayer interconnection technology has the advantages of the shorter path, smaller microwave transmission loss, and simpler structure. The schematic

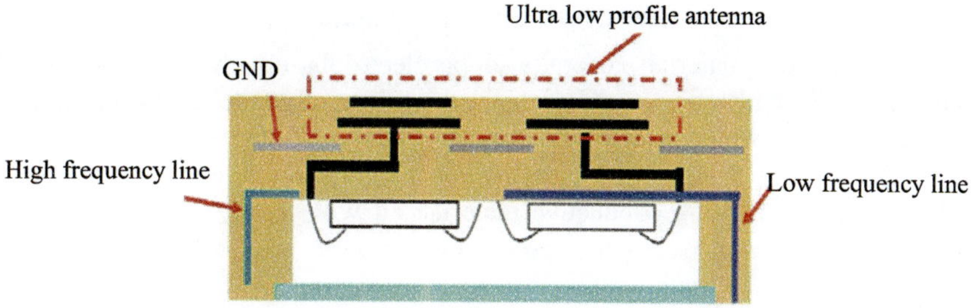

Fig. 7.13 Intra-board layer-to-layer interconnection

diagram of the vertical interconnection between the typical AiP and the inner layer of the active transceiver chipboard is shown in Fig. 7.13.

7.3.5 Through-Silicon Via

Through-Silicon Via technology (TSV), known as through-hole interconnection technology, is the core technology to realize chip-level three-dimensional integration. Unlike the traditional wiring on the chip surface, TSV technology is a vertical connection that penetrates the chip, as shown in Fig. 7.14. TSV technology requires no wires but makes holes in the vertical direction between wafers to realize the transmission of heat and signals between chips. It overcomes the limitations of traditional integration technology, such as many leads and large time delays. With the development of silicon wafer stacking technology, the use of TSV technology applied to 3D integrated chips has become the main development direction of System-on-Chip (SOC) technology.

TSV technology has three core technical points: through holes penetrating the substrate, insulating layer, and conductor filling. Vias are formed by using ion etching, wet etching, mechanical drilling, and laser drilling. The insulating layer is usually

Fig. 7.14 3D interconnect stacking between chips

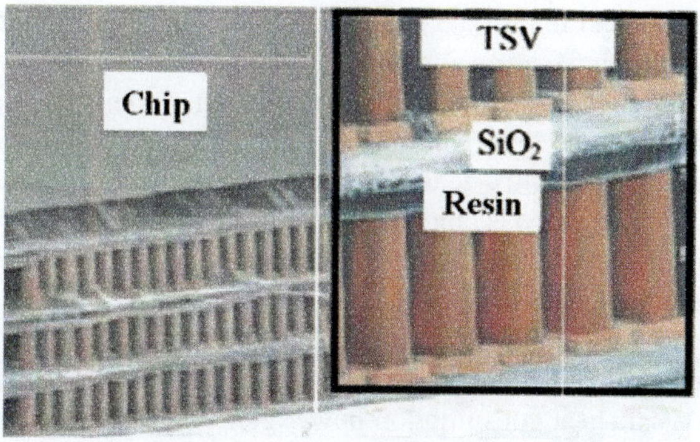

formed by thermal oxidation to form SiO_2 or sputtering insulating materials. The filling of conductive materials generally adopts electroplating, low-resistance silicon filling, or solder ball reflow. Among them, the filling of conductor material is the key and difficulty, which affects the application range, reliability, production cost, and production cycle of TSV technology. For integrated circuits, if unevenness or voids appear when filling conductive materials, it will greatly impact the circuit performance.

Compared with the traditional bonded stacking technology, TSV technology has the advantages of higher interconnection density, shorter transmission path, lower power consumption, and lower cost.

Of course, TSV technology also has some disadvantages. Since the silicon crystal lattice's shape is a regular tetrahedron, the etched transmission line slot's cross section cannot be rectangular and often can only be made into a trapezoid. It has little effect on low-frequency circuits, but the effect cannot be ignored in high-frequency bands.

7.4 Thermal Design and Heat Dissipation Technology

The active packaged array antenna system mainly comprises a passive antenna array, T/R modules, feed networks, wave control modules, a power supply, an antenna frame, and other components, among which T/R modules and the high-density power supply generate a lot of heat. Since many T/R modules are used in a large active array antenna, the phase error of each phase shifter in the T/R module, the phase error of the transmission line, and the phase characteristics of each device are closely related to the operating temperature. Therefore, the difference between the highest and lowest allowable ambient temperature in the antenna array is generally less than 10 °C. Otherwise, the difference between each phase state of the T/R module and the nominal value will be relatively large. As a result, the purpose of the thermal design and analysis of the active array antenna is to ensure that the T/R modules and the power supply work in an appropriate temperature range and the operating temperature of each element is consistent as possible. In addition, the thermal deformation of the antenna array will cause problems such as decreased antenna gain, increased sidelobe level, and inaccurate beam direction, which will reduce the electrical performance of the active array antenna.

Active array antenna package is a high-density integration of an antenna array, active transceiver circuit, feed network, and wave control circuit in one package. Higher heat flux density requires higher heat dissipation technology, and comprehensive analysis and thermal design are required from the System-in-Package (SiP) perspective. At the same time, combination of the knowledge of thermodynamics, materials science, electromagnetics, and other disciplines is necessary to carry out multi-physics field matching analysis on temperature field, stress field, electromagnetic field, etc.

The heat consumption power generated inside the active array package antenna mainly includes two parts: one is the heat consumption power P_1 generated by the active transceiver circuit and the other is the heat consumption power P_2 caused by the feeder loss between the power amplifier output terminal and the antenna element. Their calculation formulas are

$$P_1 = N \times P \times \left(\frac{1}{C_1} - 1\right) \quad (7.15)$$

$$P_2 = N \times P \times C_2 \quad (7.16)$$

where N is the total number of antenna elements; P is the output power of the power amplifier of the unit; C_1 is the efficiency of the power amplifier, which is generally only 15–40%; and C_2 is the loss coefficient, which is generally small and is 1–10%.

The external heat of the active packaged array antenna mainly comes from the temperature change in its working environment. Therefore, it is necessary to design the heat dissipation of the active array antenna package according to different application scenarios of the system. For example, for ground active phased array antennas with low heat flux density, it is feasible to adopt the traditional forced air-cooling method for heat dissipation. This scattering method is easy to install and meets the requirements of heat dissipation, applied in many engineering examples. On the other hand, for the airborne active phased array antenna with high heat flux density, the traditional air-cooling method is difficult to meet the heat dissipation requirements due to the increased thermal power consumption of the active array antenna. Therefore, more and more engineering examples and studies have focused on the forced liquid cooling method. For space-borne active array antennas with high heat flux density and strict requirements on system volume and weight, the thermal control system realizes the heat exchange between the antenna and the space environment inside the antenna. Thermal control consists of thermal control coatings, multi-layer insulation components, temperature control heating loops, phase change heat pipes, thermistors, thermally conductive fillers, and other products [17].

7.4.1 Analysis of Chip Heat Dissipation

In order to reduce the size and difficulty of the active array antenna package, the active circuit is usually implemented by mixing and integrating multiple MMICs. Then, the large-scale antenna and the micro-scale active circuit are packaged and integrated, which forms the active AiP.

High power and miniaturization are the basic requirements of active array antenna systems for MMIC. With the development of integrated circuit technology, the integration level of MMIC is higher and higher, so the power density and the heat flux density increase substantially, leading to the increase of junction temperature in chip.

If the junction temperature of the chips is too high, the reliability of the MMIC will be reduced, and it may even cause serious consequences such as failure and burnout. In addition, the phase shifter in the MMIC is the most sensitive to temperature.

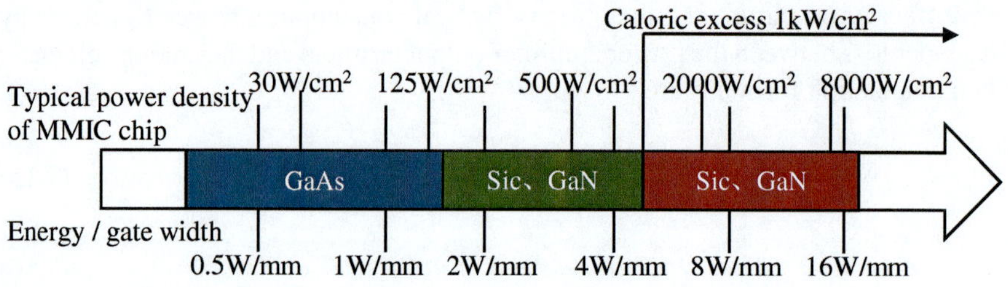

Fig. 7.15 Heat flux of MMIC

Different operating temperatures will cause different phase deviations of the phase shifter, affecting the active array antenna beamforming and scanning performance.

Figure 7.15 shows the development trend of the heat flux of the MMIC on the active phased array antenna array. It can be seen from Fig. 7.15 that with the development of integrated circuits, the power density of the MMIC on the antenna will exceed 1 kW/cm^2, which brings a huge challenge to the heat dissipation of the active array package antenna array.

The traditional convection cooling method often requires additional equipment, such as fans. Although heat dissipation can achieve the effect, it increases the weight and volume of the system, which is unfavorable to the miniaturization and light weight of the phased array antenna system. In addition, this convection method cannot dissipate heat for the active array AiP in the MMIC. The heat radiation method has low heat transfer efficiency and poor heat dissipation effect, which cannot meet the long-time working requirement of the MMIC.

In general, heat conduction is the most direct and effective way to dissipate heat from the MMIC chip. That is, the packaging substrate material with high thermal conductivity is used to quickly and effectively transfer the heat on the chip to the cold end, thereby reducing the chip's junction temperature. Furthermore, the heat conduction method not only effectively dissipates heat from the MMIC chip but also ensures that the overall weight and volume of the system will not increase too much.

In practical applications, the MMIC chip is not closely attached to the substrate or heat sink usually. There is often an air gap between the MMIC chip and substrate due to surface roughness. However, the thermal conductivity of air is very low. The heat flow channel produces a large thermal resistance at the air gap, which is unfavorable for system heat dissipation. In order to solve this problem, it is necessary to fill the air gap with thermally conductive silicone grease, conductive silver glue, or metal solder with relatively high thermal conductivity to reduce thermal resistance and improve heat dissipation efficiency.

7.4.2 Microchannel Cold Plate

The heat dissipation design of a conventional active array antenna is mostly to add a heat sink outside the active circuit module packaging shell. The heat sink mainly includes a finned heat sink and a cold plate. Fins are used when the calorific value is not very high, which belongs to air cooling. Cold plates are usually used in places with high heat generation and can be used in combination with forced air or liquid cooling. The heat dissipation design of active array antennas mostly uses the cold plate method. The cold plate can achieve good results with an efficient heat dissipation design and advanced processing technology. However, an air-cooled cold plate or a liquid-cooled cold plate has the characteristics of large space and heavyweight, which is a significant constraint factor for active array antenna systems with high efficiency and low-profile requirements. Therefore, active array antenna systems need to adopt more efficient and refined thermal design methods.

Since the active array antenna package integrates the antenna and the active transceiver circuit into one module, the integration degree is high. The power density of the packaged active array antenna system is also improved while reducing the system's volume. One of the key issues is how to systematically optimize the thermal design and enhance the heat dissipation capability of the system. The emergence and development of microchannel cold plate technology have strongly supported the solution to the technical problems of thermal design of active array antenna systems.

Microchannel cold plate technology is a new type of heat dissipation technology with high efficiency, small size, low thermal resistance, and easy integration with the packaging shell. Figure 7.16 shows a schematic diagram of the active array package antenna microchannel cold plate. It can be seen that the microchannel cold plate heat dissipation technology has the advantages of good heat dissipation effect as well as great shape which is relatively light and thin. Therefore, it can be easily integrated with an antenna array, active transceiver circuit, wave control circuit, and other integrated packages. In addition, the microchannel cold plate made of metal material with high structural strength can also be used as the system support plate of the active array package antenna, effectively reducing the section height of the system.

The research on microchannel cold plate technology mainly focuses on high strength and high thermal conductivity cold plate materials, wide temperature thermal phase change materials, microchannel circuit topology technology, and

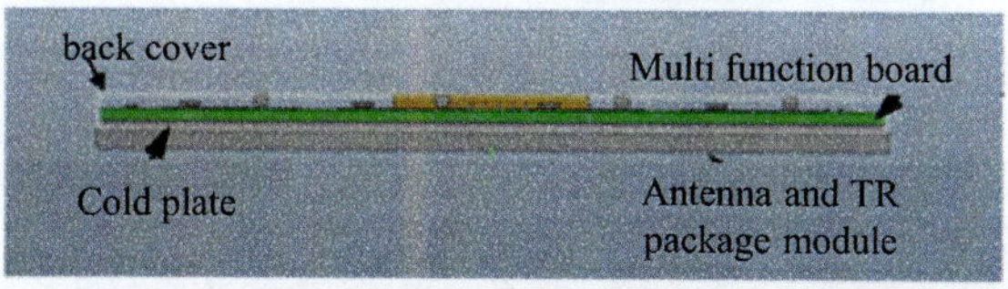

Fig. 7.16 Schematic diagram of the microchannel cold plate structure of the active array AiP

new processing technology of microchannel cold plate. For example, to make a microchannel cold plate using multi-layer ceramic wiring substrate technology such as LTCC or HTCC can integrate the AiP and the liquid cold plate, minimizing the volume of the phased array antenna system.

7.4.3 Thermal Simulation Technology

To carry out the thermal simulation of an active array antenna package, a comprehensive analysis and thermal design approach is required from the perspective of SiP. This involves combining knowledge from multiple disciplines such as thermodynamics, thermal science, materials science, electromagnetics, and others to conduct multi-physics field matching analysis and simulation for stress field, temperature field, and electromagnetic field, among others.

In the following, the thermal deformation of the active AiP unit is analyzed as an example. Since the deformation affects the flatness of the array antenna and the position deviation between the elements, it causes the phase error of the surface field at the antenna port so that the structural deformation error will be introduced into the antenna pattern function as an additional phase factor. Based on this idea, the three-field coupling simulation analysis process of temperature field, displacement field, and electromagnetic field can be established, as shown in Fig. 7.17.

Thermal simulations of active AiPs and multi-physics simulations are challenges.

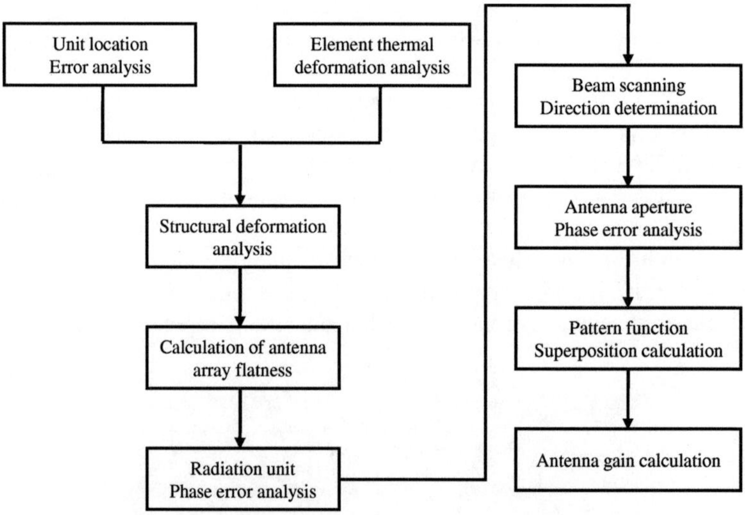

Fig. 7.17 Multiphysics simulation analysis process of thermally induced antenna position change

(1) **Scale matching theory and method**

The large-scale antenna and the micro-scale chip are integrated into the same package. The electromagnetic field radiated by the large-scale antenna interacts with the microscopic electromagnetic signals of different small-scale chips. In addition, the package includes the crosstalk effect of RF signals and the skin effect of analog signals, digital signals, and RF signals at the micro-scale.

(2) **Signal flow layout and optimization technology**

In the active antenna package, it is necessary to study the basic theory of mechanical, electrical, and thermal multi-physics layout and to optimize the flow of RF, analog and digital signals, heat generation, heat dissipation channel flow, and various force conduction flows.

(3) **Research on parasitic effects of micro-scale connections**

In the package body of the active package antenna, the three-dimensional micro-scale connection lead to the discontinuity of the electromagnetic field. The characteristics and changes of the eigenmodes of the electromagnetic field under specific boundary conditions are studied.

7.5 Embedded Microwave Devices

With the rapid development of large-scale integrated circuits, new electronic materials, and high-density packaging and interconnection technologies, active AiPs are developing in the direction of low profile, light mass, and high integration. Monolithic integrated circuits, such as low-noise amplifiers, power amplifiers, mixers, switches, and frequency multipliers, have made great progress in miniaturization. In comparison, it is of great significance to further realize the miniaturization, light mass, and high integration of passive components (such as resistors, capacitors, inductors, couplers, duplexers, filters, and power dividers).

Passive components can be integrated into electronic systems in three primary ways: through separate placement, integrated packaging, or embedded within the circuit board.The separate type is to install the packaged components into the system. Its integrated level is low, but the process is relatively mature. The integrated type is to integrate unpackaged passive components into the system with a certain interconnection method. The embedded type is to fabricate passive components directly on the substrate and connects them internally, realizing the shortest interconnection and the highest packaging efficiency. Embedded passive component integration technology is the basis for realizing active packaged array antennas.

LTCC technology is based on thick-film technology and ceramic multi-layer technology. Mechanical drilling/laser drilling and screen printing are used to fill the conductive paste into the through-hole pores on the green ceramic tape. The required circuit layout is produced by precise conductive paste printing, and multiple passive components are embedded in it. This embedded passive device has the characteristics of high processing precision, good consistency, flexible design, and excellent high-frequency performance. It is convenient for hybrid integration with other integrated circuits and components. LTCC technology provides great convenience and possibility for the design of miniaturization, broadband, and integration with active or passive devices using via holes, open cavities, and flexible metal layer printing. At the same time, LTCC technology can also take advantage of its three-dimensional integration advantages to integrate passive circuits, such as various resistors, high-Q capacitors, inductors, and even filters, minimizing the number and weight of independent components, as well as reducing the complexity and cost of the system.

Figure 7.18 shows a schematic diagram of a three-dimensional integrated package with LTCC as the substrate. On the uppermost LTCC substrate are ICs, active devices, and modules, such as PIN tubes and triodes. There are many built-in passive devices inside the multi-layer substrate, such as various inductors, capacitors, and embedded resistors, which are distributed in the board. It is connected to devices and ICs on the board through holes or through plate capacitors with LTCC as the medium to form a microsystem with certain functions. LTCC can be used as a carrier for functional modules, built-in passive components, surface mount IC, and active devices. When in use, it is connected to other LTCC substrates through BGA welding technology so that a variety of dedicated multi-chip modules (MCM) can be designed. LTCC technology can also be applied to design some independent standard packaged devices, such as filters and duplexers. These devices can be mounted on the LTCC substrate as part of the MCM system and also be used in any other place as a general-purpose device with standard input and output interface conditions.

7.5.1 Inductors, Capacitors and Resistors

(1) **Embedded inductors**

As a widely used and indispensable component in microwave RF systems, the performance of inductors is a key factor that determines the performance of the overall circuit. Inductors occupy the largest area compared to other passive components, such as resistors and capacitors. Therefore, how to reduce the size of inductors without affecting the inductor's performance has become an urgent problem to be solved for miniaturization. LTCC technology is an effective way to realize the miniaturization of embedded inductors. By giving full play to the advantages of LTCC technology's 3D multi-layer and embedded passive components, it not only effectively reduces the area occupied by inductors but also greatly improves the Q-value of inductors.

In general, the three main parameters of the effective inductance L_{eff}, Q-value and self-resonant frequency (SRF) are used to characterize the embedded inductance.

The effective inductance L_{eff} is defined as the total value of the inductance at the inductance input port. The input impedance is

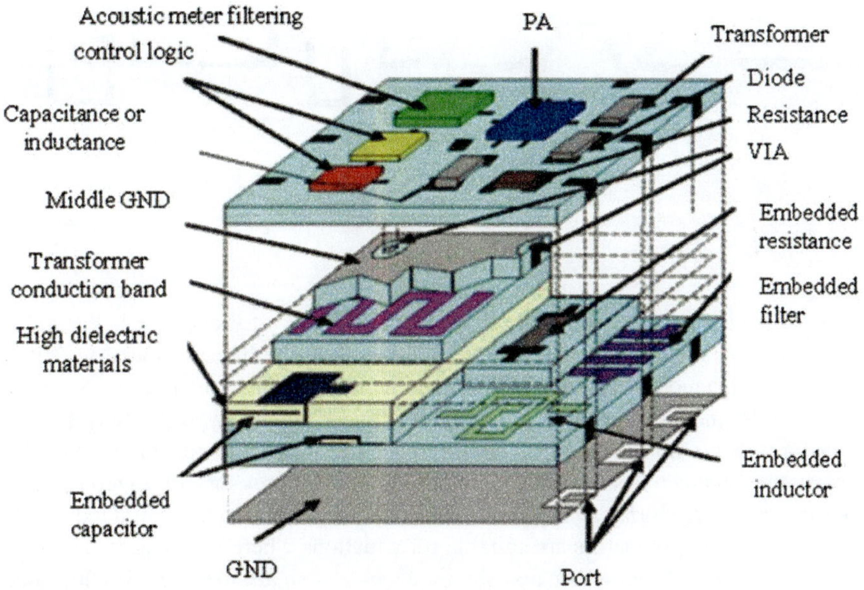

Fig. 7.18 Internal structure diagram of a typical LTCC module

$$Z_{\text{in}} = R + jX \tag{7.17}$$

Then the effective inductance is

$$L_{\text{eff}} = \frac{X}{\omega} = \frac{X}{2\pi f} \tag{7.18}$$

where R is the real part of the input impedance, X is the imaginary part of the input impedance, ω is the angular frequency, and f is the frequency.

There are different definitions for the Q-value. For LTCC-embedded inductors, the Q-value is the ratio of the imaginary part of Zin to the real part.

$$Q = \text{mag}\frac{\text{Im}(Z_{\text{in}})}{\text{Re}(Z_{\text{in}})} \tag{7.19}$$

The main factors affecting the Q-value of the embedded inductor are divided into two aspects: one is the conductor loss caused by the inductance metal conductor and shape factors and the other is the dielectric loss caused by the LTCC materials.

The self-resonant frequency defines the inflection point at which Z_{in} changes from inductive to capacitive due to parasitic effects, which determines the effective frequency range of the inductance. The main factor affecting the self-resonant frequency is the parasitic capacitance effect. In engineering design, the commonly used formula for estimating the parasitic capacitance is

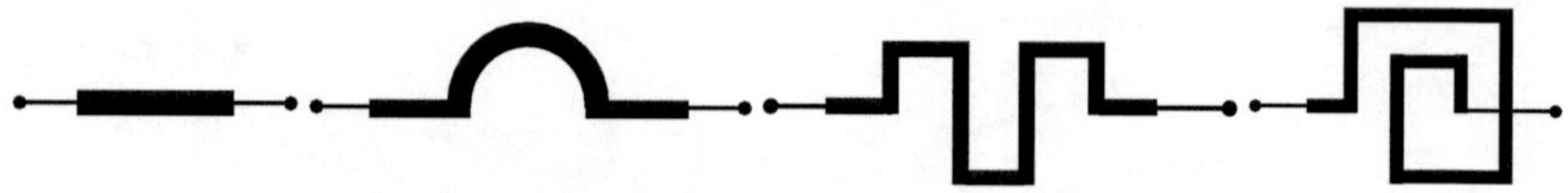

Fig. 7.19 Common single-layer inductors

$$C = \frac{\varepsilon_0 \varepsilon_r w L}{h} \tag{7.20}$$

where ε_r is the relative permittivity, ε_0 is the permittivity of vacuum, w is the width of the inductor metal winding, L is the total length of the inductor metal winding, and h is the distance from the inductor metal winding plane to the ground plane.

Embedded inductors in LTCC are generally divided into two types, namely, single-layer form and multi-layer form. The difference lies in whether the whole is on the same plane. Figure 7.19 shows the common single-layer inductors. From left to right, they are linear inductors, ring inductors, folded inductors, and square spiral inductors.

Linear and ring inductors are suitable for situations where the inductance of the circuit is not large, mainly because the inductance is small, the mutual inductance between line segments is zero, and the Q-value is high. Folded inductors are mostly suitable for circuits with low characteristic impedance or weak coupling, of which the composition includes multiple small broken line segments. Because of this structure, the weak coupling phenomenon with negative mutual inductance occurs between adjacent conductors. Therefore, folded and spiral inductors can achieve larger inductance values. In comparison, the circular spiral inductor is more dominant in the Q-value and self-resonant frequency under the premise of the same size. Still, the inductance per unit area is inferior to that of the square spiral inductor. However, in practical applications, the Q-value is often given up to make the device more compact, and a square spiral inductor with a more favorable size is selected.

Making multi-layer inductors is an advantage of the LTCC structure. This kind of inductor fully uses three-dimensional space to transform the original flat configuration into a three-dimensional vertical form, effectively reducing the occupied area and realizing the goal of miniaturization of inductors. Common multi-layer inductors are shown in Fig. 7.20: Planar, Offset, Stack, and 3D Helical.

It can be seen from Table 7.1 that the 3D spiral inductor is stronger than the other three ones in terms of the required area, self-resonant frequency, and quality factor. However, 3D spiral inductors require the most layers, increasing the difficulties in the design process. The 3D spiral inductor utilizes the characteristics of the multi-layer 3D layout of the LTCC. The layout can be extended in the vertical direction to form a multi-layer spiral inductor. Figure 7.21 shows two typical multi-layer spiral inductor structures. The structure in Fig. 7.21b can effectively reduce the influence of parasitic capacitance and expand the operation frequency range. Using the multi-layer spiral inductor structure, the mutual inductance between the three-dimensional wires can be effectively used to increase the inductance value, and more layers can be used to reduce the area occupied by the inductor.

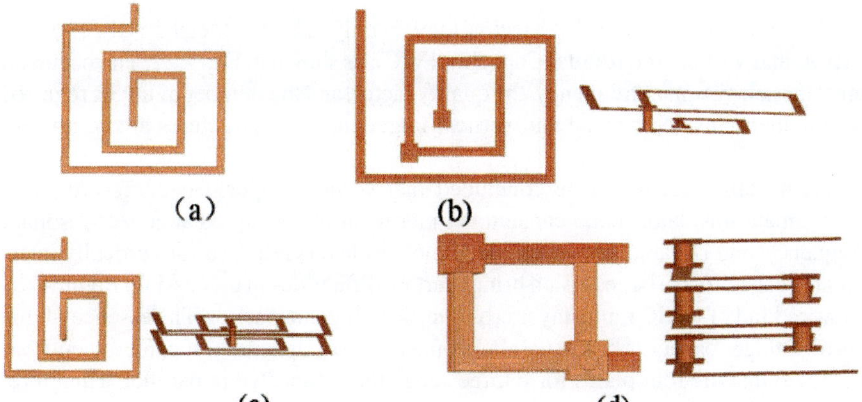

Fig. 7.20 Structures of multi-layer inductors, **a** planar, **b** offset, **c** stack, **d** 3D helical

Table 7.1 Parameters influencing factors

Structure	Planar	Offset	Stack	Helical
Area occupied (at the same effective inductance value)	Maximum	Medium	Small	Minimum
Self-resonant frequency (at the same effective inductance value)	Minimum	Medium	High	Maximum
Q-factor (at the same effective inductance value)	Minimum	Medium	High	Maximum
Required layers	Minimum	Small	Small	Maximum

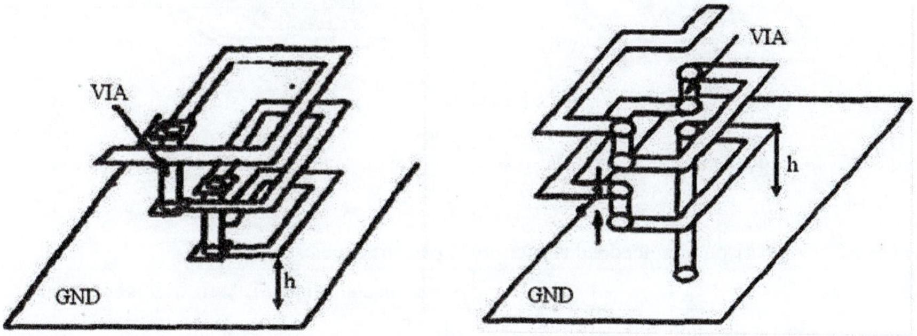

Fig. 7.21 Multi-layer helical inductor structure

(2) **Embedded capacitors**

Similar to the embedded inductance, the embedded capacitance can also be characterized by three physical parameters: effective capacitance C_{eff}, Q-value, and self-resonant frequency SRF.

Capacitors based on LTCC technology mainly include Metal-Insulator-Metal (MIM) and Vertical Interdigital Capacitor (VIC), as shown in Fig. 7.22. The occupied area, the self-resonant frequency, the quality factor, and the number of layers required for the metal-insulator-metal and vertical finger capacitor structures are compared, as listed in Table 7.2.

From Table 7.2, it can be concluded that vertical finger capacitors are superior to metal-insulator-metal capacitors in terms of the occupied area, self-resonant frequency, and Q-value. However, the number of layers required for vertically interpolated capacitors is large, which brings certain difficulties in processing. Figure 7.23 shows a kind of LTCC multi-layer cross-vertical finger capacitor. The essence of this vertical finger capacitor is the parallel connection of capacitors. It can be seen from Fig. 7.23 that the four plates form three capacitors connected in parallel. Therefore, with the same area, the value of the vertical finger capacitance is about three times that of the metal-insulator-metal capacitance. The size of the capacitance is not only related to the area and mutual distance of the plates but to the distance between the ground and the plates. Generally speaking, compared with double-parallel-plate

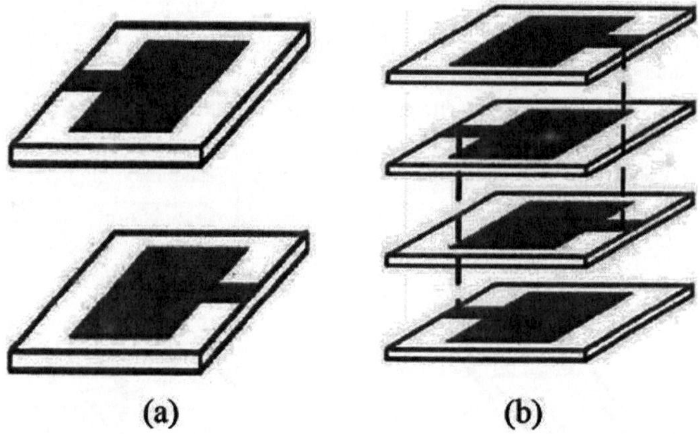

Fig. 7.22 Embedded capacitors, **a** metal-insulator-metal capacitors, **b** vertical finger capacitors

Table 7.2 Comparison of embedded capacitors in two structures

Structure	Metal-insulator-metal	Vertical finger capacitor
Occupied area (at the same effective inductance value)	Big	Small
Self-resonant frequency (at the same effective inductance value)	Low	High
Q-factor (at the same effective inductance value)	Low	High
Required layers	Low	High

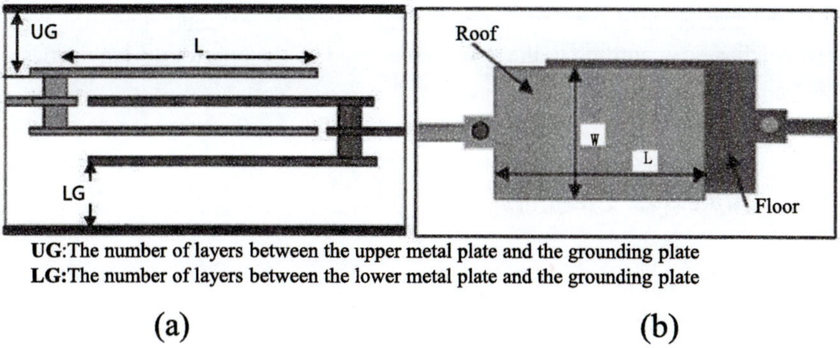

Fig. 7.23 Multi-layer interleaved vertical finger structure capacitor

capacitors, multi-layer crossover capacitors have more advantages in realizing small areas and large capacitance. The capacitance value is constant. The more layers of capacitance, the smaller the area.

(3) Embedded resistors

The use of embedded resistors can reduce the size of electronic products and improve performance. The typical structure of embedded resistors is shown in Fig. 7.24. The formula for calculating the resistance value R of embedded resistors is

$$\begin{aligned} R &= R_S \cdot N \\ R_S &= \rho / d \\ N &= L / W \end{aligned} \tag{7.21}$$

where R_S is the square resistance, ρ is the material resistivity, d is the film thickness of the printed resistor, N is the square number, L is the length of the resistance, and W is the width of the resistance.

When designing an embedded resistor, the size of the resistor is related to the power density of the ceramic substrate and the allowable power of the resistor. The power consumed by the resistor is related to its resistance value. The smaller the resistance

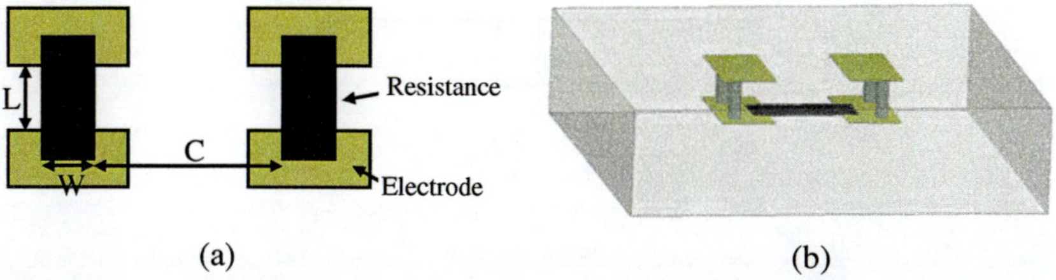

Fig. 7.24 Embedded resistor, **a** schematic diagram of embedded resistor, **b** structure diagram of embedded resistor

value, the greater the power consumption. The size of the embedded resistor can be flexibly adjusted according to the power consumed by the resistor during the design process.

7.5.2 Duplexers, Couplers

(1) **Embedded duplexers**

Generally, a multi-channel frequency synthesis and separation device, a duplexer, is configured at the front end of the electronic information system. A duplexer is usually composed of a transmitter filter, a receiver filter, and a combination circuit. The combination circuit is generally a phase adjustment circuit added to reduce the mutual influence between the transmitter and receiver filters, which consists of transmission lines, resistors, capacitors, and other reactive components.

As shown in Fig. 7.25, it is a typical duplexer. For the classic LC structure, 3D spiral inductors and vertical finger capacitors are generally used, so that larger capacitance and inductance values can be obtained in a smaller space.

The basic parameters of the duplexer include operating frequency, bandwidth, insertion loss, isolation, matching impedance, and maximum power.

The duplexer based on LTCC technology has the characteristics of integration, small size, light mass, high reliability, excellent electrical performance, simple structure, high yield, low cost, and stable performance. Figure 7.26 shows a typical LTCC duplexer, the entire circuit consists of 18 metal and 17 ceramic layers. The resonators are folded vertically and horizontally in the middle section on layers 6, 8, 9, 10, and 12. Metal vias are used to connect circuits on different layers. Etching squares on layers 3, 7, 11, and 15 allows two ceramic plates to be firmly bonded together for increased reliability. In addition, due to the design flexibility of the multi-layer LTCC structure, the bandwidth and the frequency of two passbands can be controlled separately.

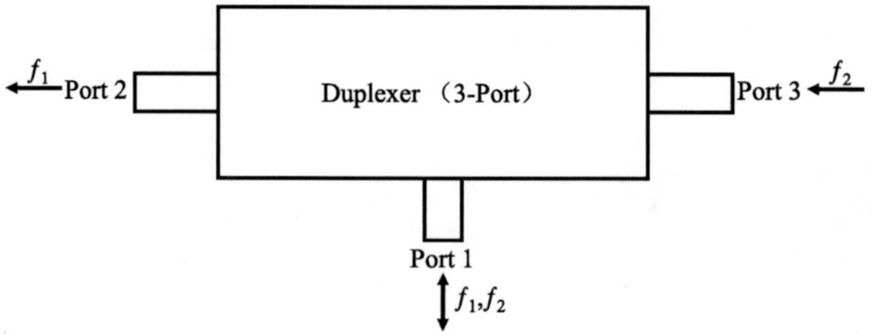

Fig. 7.25 Duplexer model

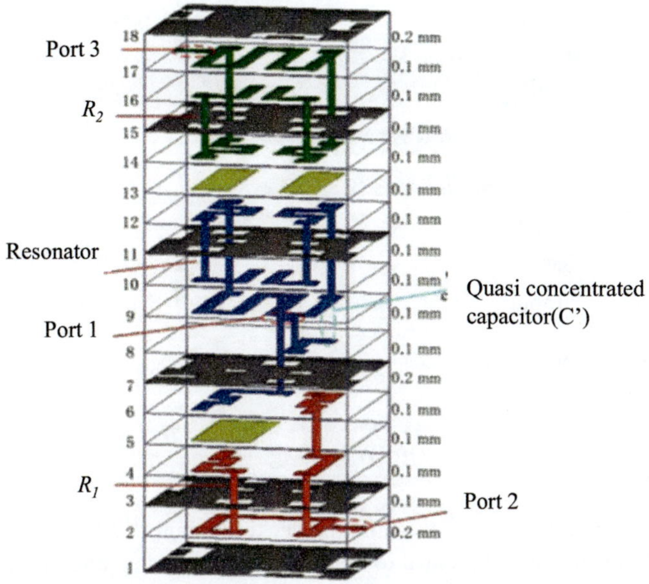

Fig. 7.26 Diagram of the LTCC duplexer

(2) **Embedded couplers**

The directional coupler is a common RF device in microwave circuits, of which the function combines and distributes the signal power according to the set direction and ratio. Figure 7.27 shows a typical four-port directional coupler. Part of the signal power at the input terminal directly reaches an output terminal, and the other part is coupled to the coupling output terminal through coupling. The coupling power is required to be transmitted only to the coupling terminal, and the other port is an isolation port. There is no output power. Therefore, the output power of the two output ports can be equal or unequal. When the power of these two output ports is equal, half of the input power is coupled to the coupling port. At this time, it is called a 3 dB directional coupler.

The basic parameters of a directional coupler include coupling, directivity, isolation, and bandwidth.

The coupling factor is the ratio of the output power at the coupling port to the input power at the input port, i.e.,

$$C(\text{dB}) = 10 \lg \frac{P_3}{P_1} \tag{7.22}$$

The directivity is the relationship between the output power at the coupled port and the output power at the isolated port, i.e.,

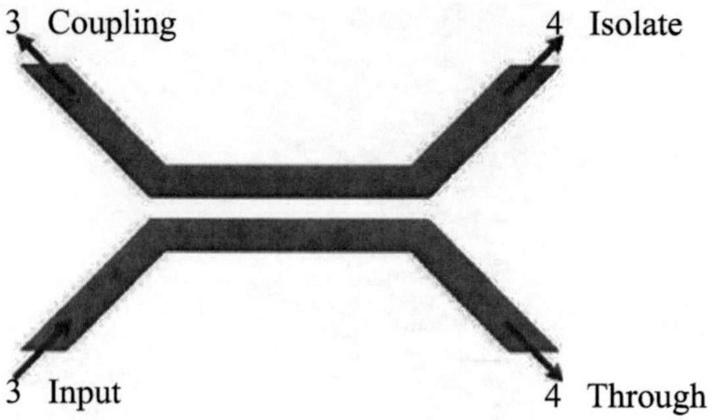

Fig. 7.27 Geometry of a directional coupler

$$D(\text{dB}) = 10\ \lg \frac{P_3}{P_4} \tag{7.23}$$

The isolation is the ratio of the input power to the output power of the isolated port. Isolation is numerically equal to the sum of coupling and directivity, i.e.,

$$I(\text{dB}) = 10\ \lg \frac{P_1}{P_4} = -C(\text{dB}) + D(\text{dB}) \tag{7.24}$$

Embedded couplers usually use coupled striplines, mainly side-coupled striplines and broadside-coupled striplines, as shown in Fig. 7.28. b is the distance between the upper and lower ground planes, t is the thickness of the conductor, w is the width of the conductor, and s is the distance between the conductors. However, the side-coupled stripline is a single-layer structure, which is not conducive to saving areas. In addition, it cannot take advantage of LTCC technology to realize a three-dimensional structure. Therefore, the broadside coupled stripline structure is usually used in the design of LTCC couplers.

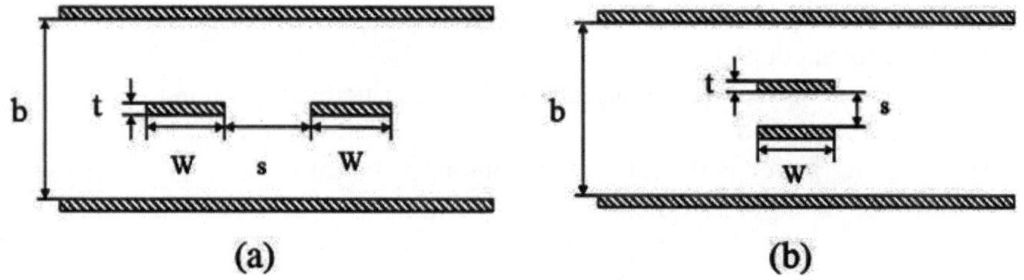

Fig. 7.28 Coupled stripline, **a** side-coupled stripline, **b** broadside-coupled stripline

7.5.3 Filters

A filter is a microwave/millimeter-wave device with frequency-response control properties, allowing frequencies to be selectively passed and suppressing unwanted or harmful frequencies.

The basic components of passive microwave filters are capacitors, inductors, and resistors. Using low-loss LTCC materials and a multi-layer substrate structure, the layout of the filter can be changed from a plane to a three-dimensional structure, and a tiny stacked filter can be produced. The electromagnetic coupling characteristics brought by the three-dimensional structure are used to optimize the filter's performance and improve the filter's integration and reliability.

Figure 7.29 shows the general structure of the LTCC multi-layer microwave passive filter. The microwave circuit is printed on the LTCC substrate. The upper and lower layers of the device serve as shielding layers to protect against electromagnetic interference both inside and outside the device. The middle layer contains the circuit structure responsible for filtering. The specific style and the number of layers of the pattern layer are determined according to the required filter parameters.

When comprehensively analyzing and designing filters, the basic parameters are cutoff frequency, bandwidth, insertion loss in the passband, return loss, out-of-band rejection, ripple and squareness coefficient, etc.

The technical difficulty of LTCC-embedded multi-layer filter design is mainly the influence of the parasitic effects. Since the LTCC filter is a multi-layer microstructure, and the layers are interconnected via holes, electromagnetic field coupling effects will occur when the resonating frequency is greater than 1.0 GHz.

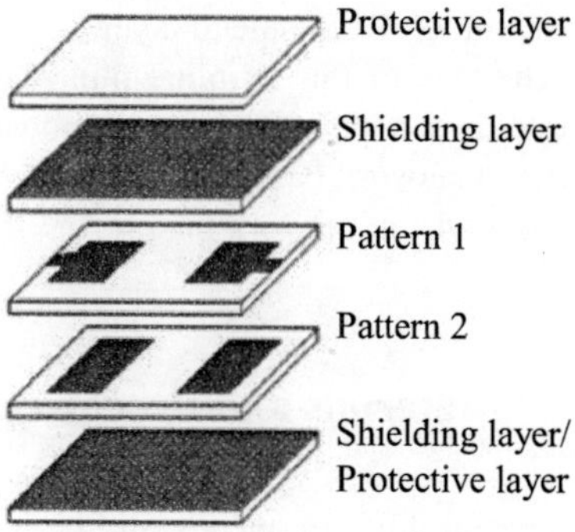

Fig. 7.29 Schematic diagram of the LTCC filter

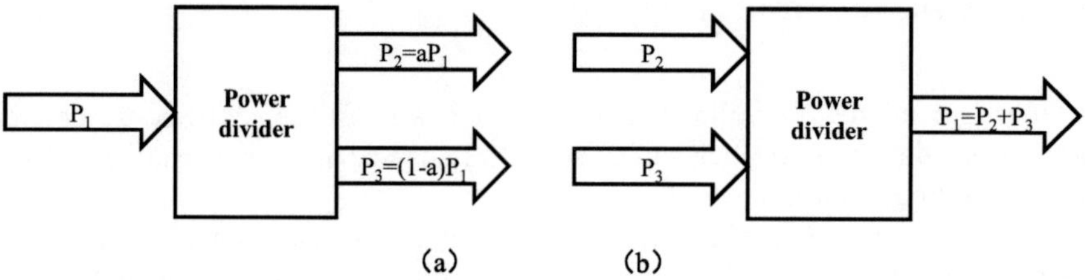

Fig. 7.30 Geometries of power division and combination, **a** power division, **b** power combination

7.5.4 *Power Dividers/Combiners*

The main function of the power divider/combiner is to distribute and combine power, as shown in Fig. 7.30. Usually, it is equal power combiner/divider, but there are also unequal power combinations/divisions.

The Wilkinson power divider network is a simple three-port network which is lossless when the input/output ports are matched. The basic parameters of the power division/combination network are frequency bandwidth, insertion loss, incident power range, isolation, and standing wave ratio.

Figure 7.31 shows a second-order multi-layer stripline Wilkinson power divider network designed based on the LTCC technology. The size of the power division network is 9 mm × 10 mm × 4.5 mm. And there are six metal layers in total. Three of the metal layers are used for the transmission line wiring: the layers where the input and output ports are located, the layers where the two branch lines are located, and the other three metal layers are metal grounds. Adjacent metal layers are connected through metal transition holes. The isolation resistor is a square resistance embedded in the middle of the metal layer.

The area of this stripline multi-layer power splitter is only 10% of that of the single-layer coplanar waveguide power splitter. It can be seen that the stripline power division network based on LTCC technology is superior to the coplanar waveguide power division network in size.

7.6 Materials and Processes of AiP

In general, high-frequency and high-speed circuits mainly depend on the components' performance and integration methods. Also, the impact of the substrate materials and manufacturing processes used to integrate these components cannot be underestimated. For example, the material and structure of the substrate used in the active AiP, the layout of the substrate circuit as well as the length and spacing of the leads will all affect the electrical characteristics.

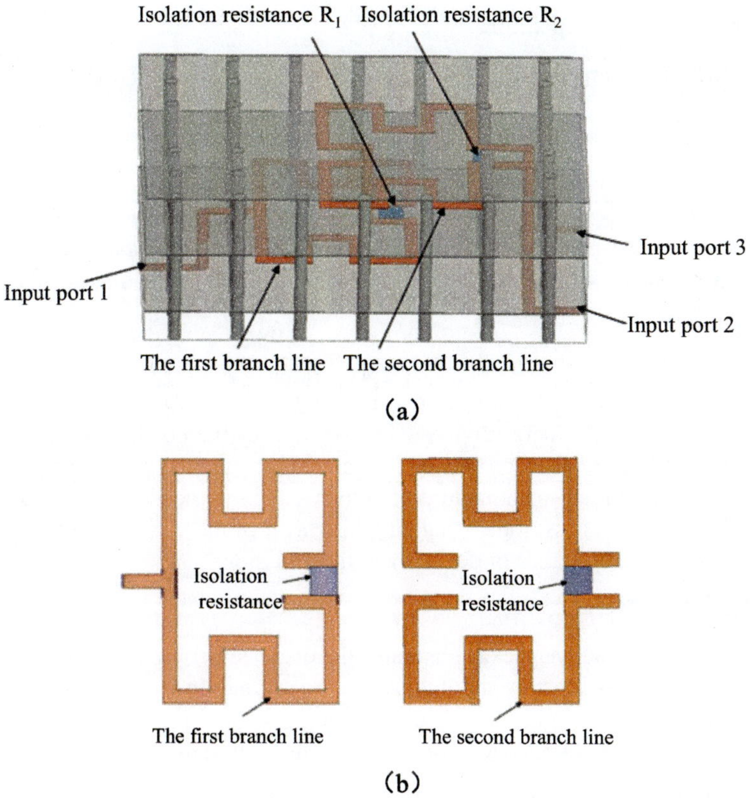

Fig. 7.31 Geometry of the second-order multi-layer stripline power splitter, **a** front view, **b** plan view of two branch lines

With the development of active package antenna technology, in order to maintain the inherent performance of the MMIC chip without degrading signal transmission performance, the issues are usually concerned as follows.

(1) **Signal transmission delay time**

For non-ferromagnetic media, under unit length, the propagation delay time T_{pd} of electromagnetic waves in the media is

$$T_{\text{pd}} = \frac{\sqrt{\varepsilon_r}}{C} \tag{7.25}$$

where ε_r is the relative permittivity and C is the speed of light. In order to perform high-speed and low-latency signal transmission, materials with lower dielectric constants are generally used in active package antennas.

(2) **Characteristic impedance**

The characteristic impedance Z_0 can be expressed as

$$Z_0 = \sqrt{\frac{R + jwL}{G + jwC}} \tag{7.26}$$

where R is the resistance, L is the inductance, G is the conductance, and C is the capacitance. The following empirical formula can be used to calculate Z_0, i.e.,

$$Z_0 = \frac{8}{\sqrt{\varepsilon_r + 1.41}} \ln \frac{5.98h}{0.8w + t} \tag{7.27}$$

where ε_r is the relative permittivity, w is the width of the conductor line, t is the thickness, and h is the distance between the signal line and the ground plane. Obviously, in the design and manufacture of active AiPs, it is necessary to control the shape, size, and thickness of the insulating interlayer of the wiring pattern to match the characteristic impedance of devices and transmission lines.

(3) **Parasitic effects**

When designing an active package antenna, the distance between the leads between devices should be as short as possible. At the same time, low-resistivity conductor and substrate materials with low dielectric constants should be used. The parasitic effects of connecting wire, conductor, and substrate materials should be avoided.

(4) **Intermodulation noise**

When designing the active package antenna, too close and parallel arrangement of the signal lines should be avoided as much as possible.

The above requirements are reflected in the performance requirements of the active package antenna substrate material, mainly reflected in the need to use materials with a lower dielectric constant and conductor materials with high conductivity. Active AiP substrate materials mainly include multi-layer wiring ceramic and multi-layer organics. The ceramic materials mainly include LTCC and High Temperature Co-fired Ceramics (HTCC). Organic materials mainly include PCBs and molding compounds.

7.6.1 LTCC Materials and Processes

LTCC is a multi-layer wiring ceramic technology formed by stacking and sintering ceramic green tapes printed with conductive patterns and interconnected holes. Common LTCC material properties are listed in Tables 7.3 and 7.4.

Table 7.3 Foreign LTCC material performance

Supplier	Model	Dielectric constant	Dielectric loss	TCE ppm/°C	Thermal conductivity w/m K	Flexural strength MPa
Dupont	9k7	7.1	0.001 (10 GHz)	4.4	4.6	230
Dupont	951	7.8	0.014 (10 GHz)	5.8	3.3	320
Ferro	A6M	5.9	<0.002 (1–100 GHz)	7.8	2	170
Kyocera	GL330	7.7	0.0005 (2 GHz)	8.2	4.3	400
NTK	Noc	5.9	0.0006 (3 GHz)	5.2	3	250

Table 7.4 Domestic LTCC substrate material performance

LTCC material system	Dielectric constant	Dielectric loss	TCE ppm/°C	Thermal conductivity w/m K	Flexural strength MPa	Research unit supplier
SiO_2–Al_2O_4–MgO	4.7	0.00046 (1 MHz)	3.05	1.097	177	China Electric 43 CETC 43
CaO–B_2O_4–SiO_2	5.7	0.0024 (10 MHz)	–	–	–	Nanjing Tech University of Technology
CaO–B_2O_4–SiO_2	4.9	0.0014	3.52	–	160	University of Electronic Science and Technology
CaO–B_2O_4–SiO_2	5.8	0.000046 (1 MHz)	–	–	–	
MgO–Al_2O_4–SiO_2	6.13	0.00423 (1 GHz)	–	–	–	Shanghai Institute of Silicon Ceramics
CaO–Al_2O_4–SiO_2	6.13	0.00257 (1 MHz)	–	2.62	110	National Defense University of Science and Technology University

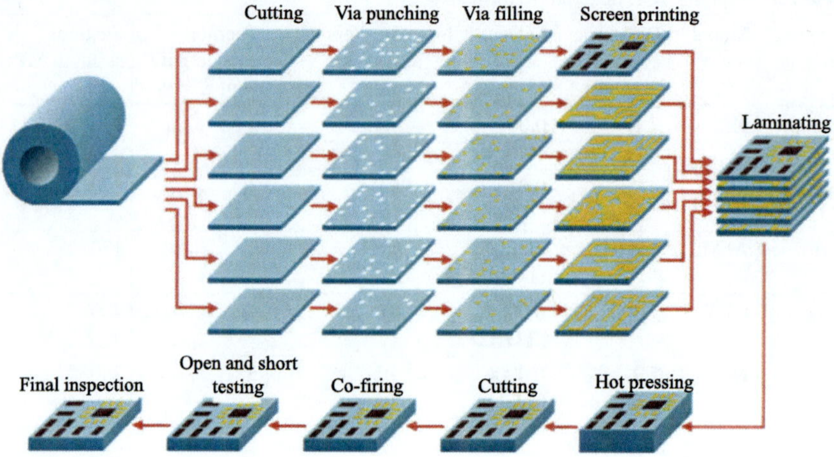

Fig. 7.32 LTCC process flow

LTCC substrate manufacturing technology is a parallel processing technology that processes all layers separately and then jointly fires them into a ceramic substrate for a ceramic package. Firstly, the green ceramic tape is cut to a certain size, punched, printed circuit, laminated, and finally fired to form a single ceramic multi-layer circuit substrate, as shown in Fig. 7.32.

The characteristic of LTCC is that it is easy to realize more wiring layers and increase assembly density. Due to the simultaneous completion of multi-layer conductive strip metallization sintering and multi-layer ceramic green sheet firing, the number of layers of the co-fired ceramic multi-layer substrate can be made, and higher wiring density can be obtained. Due to the low sintering temperature of low-temperature co-firing, usually 850 °C, it can be co-fired with thick-film passive devices such as resistors, capacitors, and inductors. It is easy to embed components, increase assembly density, and achieve multi-function. Before the substrate is fired, the quality inspection of each wiring layer and interconnection holes is conducive to improving the yield and quality of the multi-layer substrate. LTCC can use highly conductive materials (such as Ag–Pd, Au, and Cu) for multi-layer metallization wiring to reduce signal transmission loss, which is more conducive to high-density multi-layer wiring and has good high-frequency characteristics and high-speed transmission characteristics. At the same time, the use of low dielectric constant materials makes it very suitable for applications in the high-frequency field.

7.6.2 HTCC Materials and Processes

Typical representatives of HTCC are two ceramic materials, alumina (Al_2O_3-HTCC) and aluminum nitride (AlN-HTCC). Table 7.5 compares these two HTCC ceramic materials and Ferro A6M-type LTCC ceramic materials.

From Table 7.5, it can be found that for low-power active packaged antennas, it is more appropriate to use Ferro A6M material, for high-power packaged antenna components, it is more appropriate to use Ferro A6M material combined with AlN-HTCC material. In addition, since HTCC can only use tungsten or molybdenum metals with low conductivity as wires, compared with LTCC materials, the conductor resistance loss will be relatively large. Therefore, reducing the resistivity of HTCC conductor materials has become a significant problem to solve.

Except for the different materials used, the manufacturing process of HTCC substrate is roughly the same as that of LTCC, as shown in Fig. 7.33. But the sintering temperature is much higher, most of the sintering temperature of Al_2O_3-HTCC needs to reach 1650 °C, and the sintering temperature of AlN-HTCC often needs to reach above 1800 °C. In addition, since HTCC uses tungsten or molybdenum as the conductor material, it cannot be directly used for micro-assembly processes such as gold wire bonding and reflow soldering. Instead, it needs to be electroplated or electroless plated with nickel and gold on the surface of the tungsten or molybdenum conductor to meet the packaging requirements.

Table 7.5 Performance comparison of the ceramic materials

	Al_2O_{43}-HTCC	AlN-HTCC	Ferro A6M
Dielectric constant	9.1	8.8	5.9
Dielectric loss	0.001	0.004	0.002
TCE (ppm/°C)	6.9	4.5	7.8
Density (g/cm^3)	3.6	3.26	2.45
Bending strength (MPa)	400	350	170
Thermal conductivity (W/m K)	18	170	2
Single layer thickness (μm)	125	125	96
Conductor material	W, Mo	W	Au, Ag

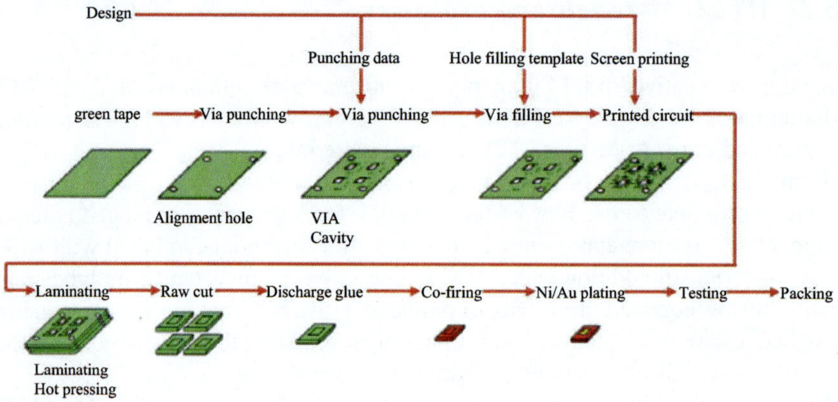

Fig. 7.33 HTCC process flow

7.6.3 *Organic Materials and Processes*

There are many kinds of organic materials. The printed circuit board (PCB) materials used to make antennas mainly include glass fiber epoxy resin, liquid crystal polymer, and ceramic-filled PTFE. The characteristics of organic materials are listed in Table 7.6. Due to the limitations of the characteristics of organic materials, PCB boards cannot be used as hermetic packaging materials for active MMIC chips. In some applications that require high reliability, MMIC chip packaging still needs to use ceramic or metal packaging materials.

Table 7.6 Organic material characteristics

Material model brand	Rogers 5880	Rogers 4350B	Rogers 6035HTC
TCE (ppm/°C)	X: 31 Y: 48 Z: 237	X: 10 Y: 12 Z: 32	X: 19 Y: 19 Z: 39
Density (g/cm^3)	2.2	1.86	2.2
Thermal conductivity (W/m K)	0.2	0.69	1.44
Dielectric constant	2.2	3.48	3.5
Dielectric loss	0.0009	0.0037	0.0013
Single layer thickness (μm)	127	101	254
Material	Glass microfiber reinforced PTFE	Glass fiber reinforced hydrocarbon/ceramic laminate	Ceramic-filled PTFE
Conductor material	Cu	Cu	Cu

Table 7.7 Molding compound dielectric properties

Performance	Material A	Material B		
Testing method	Resonance method	Free space method		
Frequency (GHz)	24–36	40–60	75–110	110–170
Dielectric constant	3.34	3.61	3.62	3.61
Loss tangent	0.015	0.0045	0.0055	0.009

Molding compound is the material used to package MMIC chips in the Fan Out-Wafer Level Chip Scale Packaging (FO-WLCSP) process, and it is also trying to be used in active packaging antennas recently. It can be found from Table 7.7 that the dielectric constant of the molding compound does not change with frequency, while the loss tangent increases with frequency. Due to the limitation of the characteristics of the molding compound material itself, its dielectric properties and mechanical properties will change significantly with the temperature change. In addition, molding compound materials face difficulty in solving the problem of resistance to humid environments and cannot be used as high-reliability active packaging antenna materials.

The use of organic materials to make active packaged antennas usually uses PCB technology. The core layer of the traditional PCB process uses organic dielectric materials. In order to prevent warping of the overall structure, the thickness of the core layer needs to be at least 400 μm, and a balanced layout is implemented on the upper and lower layers of the core layer. Therefore, the line width and spacing (L/S) depend on the thickness of the dielectric layer and the metal layer, and the typical value is L/S = 50/50 μm. A typical PCB process flow is shown in Fig. 7.34.

The FO-WLCSP process differs from LTCC or PCB processes in that it no longer requires laminated substrates but uses molding compounds and reconfiguring metal and dielectric layers. Figure 7.35 shows that the microstrip antenna radiator is realized by the redistribution layer (RDL), and the metal layer on the system board realizes the microstrip antenna ground layer.

7.7 Applications

Figure 7.36 shows a millimeter-wave four-elements active package antenna array. The design idea of the active package antenna is modularization and chip based. With the module as a basic structural unit, a larger sized active packaged antenna array can be constructed through building block expansion. The low profile, light mass, and low power consumption of the large-scale array antenna can be realized by chipping the internal components of the antenna array.

The active antenna-in-package array uses a four-elements active AiP array as the basic module. The four-elements packaged antenna array adopts a distributed architecture and tile-type stacked assembly. It includes four antenna elements, four microwave receiving/transmitting link chips, a power supply, a beam control chip, an

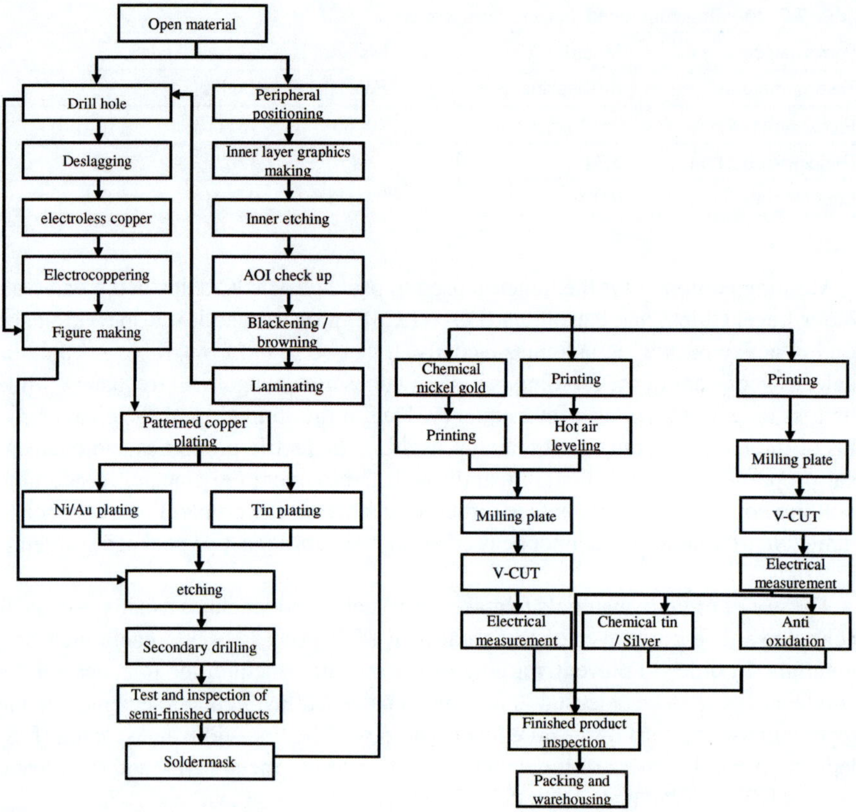

Fig. 7.34 PCB process flow

Fig. 7.35 Schematic diagram of the FO-WLCSP AiP

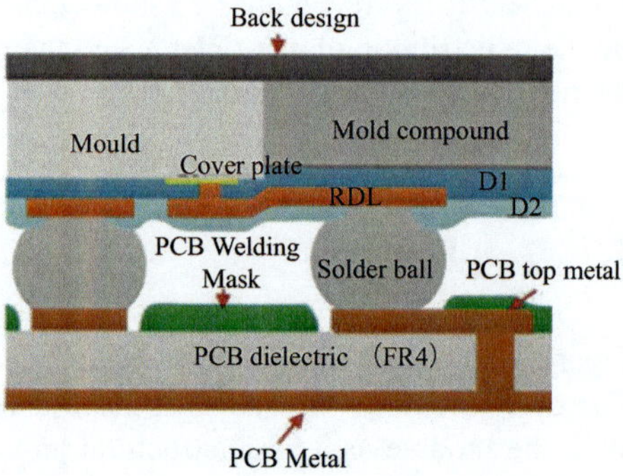

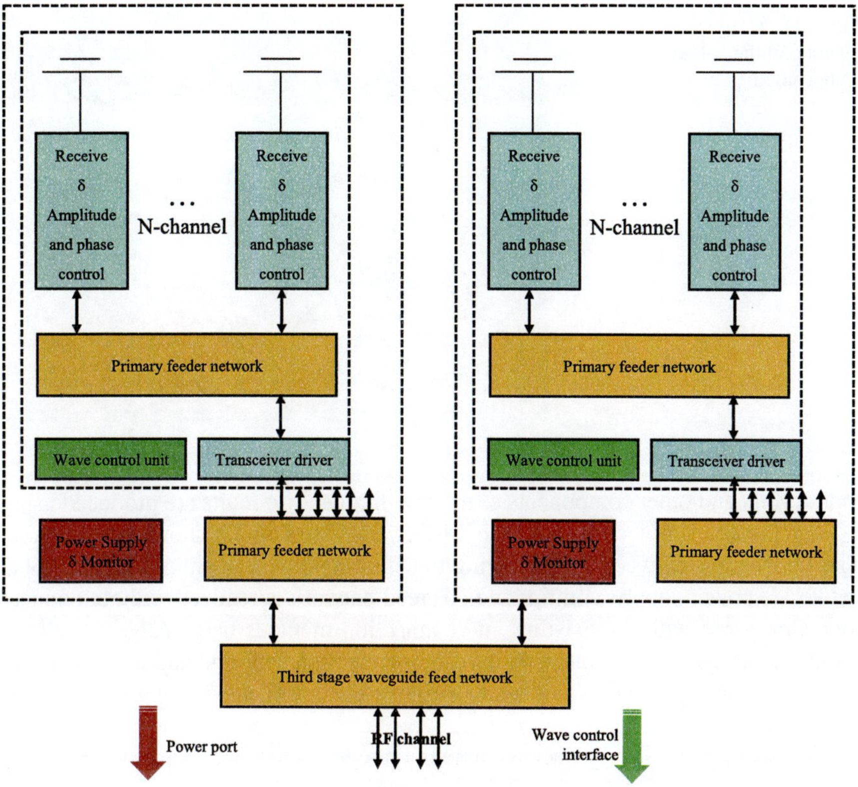

Fig. 7.36 Operation principle block diagram of the active array AiP

energy storage capacitor, a power combination/division network, and other auxiliary structural components, as shown in Fig. 7.37. The basic module of the four-elements packaged antenna array adopts LTCC integrated package, and the upper layer is a microstrip patch antenna occupying eight layers. The antenna adopts the microstrip line coupling feeding method, the transmitting, receiving feeding and the antenna element adopts a vertical interconnection method. The four-elements active channels are fed by two sets of one-by-two Wilkinson power-combination networks. The Wilkinson power division and combination network uses 18-layers LTCC as the substrate. The two power splitters form a four-element packaged antenna array RF input/output port through a stripline. The receiving/transmitting link and the power splitting and combining network are interconnected by gold wire bonding.

The four-elements packaged antenna is used as a module to form a 2 × 8 array. The BGA vertical interconnection with the multi-functional network forms a 64-elements packaged active array antenna. The multi-functional network includes high-density low-frequency circuit and RF signal interconnection, mainly RF circuits, power supply, control circuits, etc. Heterogeneous 3D integration is adopted, the 1/16 RF circuit is made by the LTCC technology. The power supply, control signals,

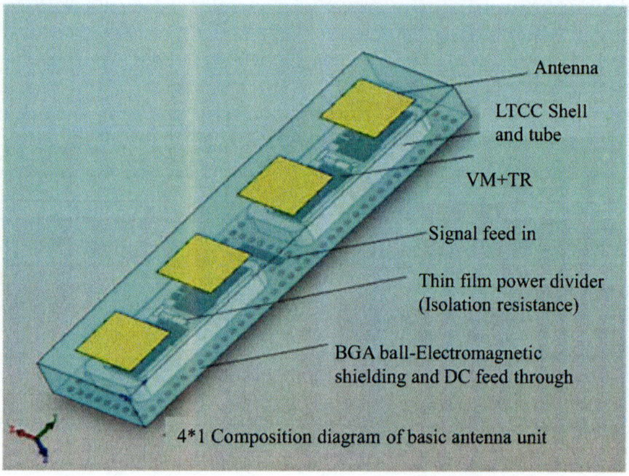

Fig. 7.37 Schematic diagram of the AiP unit composition

monitoring, and other components of the low-frequency circuits are produced using PCB manufacturing technology. The 64-elements structure is shown in Fig. 7.38. The connection between the receiving/transmitting link and the multi-functional network's power supply, control, and RF feed network is realized without connectors. Compared with the existing interconnection methods using cables or elastic blind-mating connectors, intermediate interconnection supports and many connectors are omitted, the section height and weight are reduced, and installation accuracy is improved.

The active phased array package antenna integrates antenna element, RF, power supply, beam control network, RF feed network, and internal vertical interconnection. Compared with the conventional "tile-type" active array antenna, the thickness of the "tile-type" antenna is reduced, and the interconnection support structure between different feeding networks is omitted. The receiving/transmitting link adopts heterogeneous single-chip integration technology, integrating the transmitting power amplifier circuit, receiving low-noise amplifier circuit, phase shifter, attenuator, switch circuit, and other functional chips in the package. As a result,

Fig. 7.38 Internal structure diagram of millimeter-wave active packaged phased array antenna

channel area and weight are reduced by more than 80% compared to discrete component-based modular assemblies. The final wave control uses a single chip to control the amplitude and phase of the receiving/transmitting link. Compared with the traditional beam control method using FPGA plus driver chip, the circuit weight and size are greatly reduced, and the distributed power supply and beam control design are adopted simultaneously. The reliability and stability of the active package array antenna have also been greatly improved.

The millimeter-wave active packaged antenna was measured using far-field measurement method. The prototype of the active packaged 64-elements millimeter-wave array is shown in Fig. 7.39.

The test results are shown in Fig. 7.40. The scanning range of the antenna beam is up to ±60° without grating lobes. At an angle of ±60°, the antenna gain drops less than 3 dB, the antenna gain in the normal direction is greater than 24.2 dB, the beam width is 12.6°, and the transmit power of the packaged active antenna is greater than 3.6 W. The thickness of the four-elements encapsulated antenna is 3 mm, and the weight is only 2.4 g.

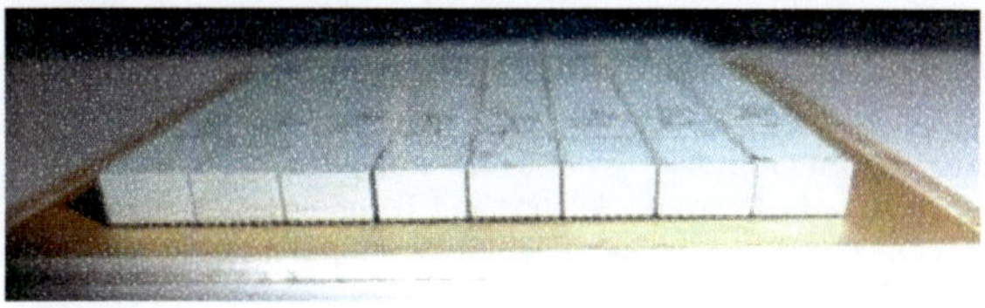

Fig. 7.39 Schematic diagram of the antenna prototype measurement

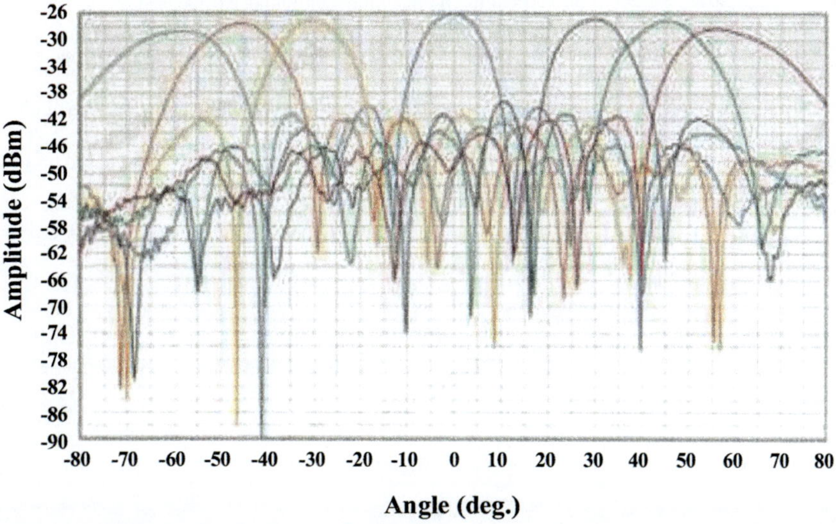

Fig. 7.40 Antenna scanning measurement pattern

References

1. Wang Y, Wang Y (2012) The history and development law on micro-nano electronics. Sci Sinica (Inf) 42(12):1485–1508
2. Zhang YP (2002) Integration of microstrip antenna on ceramic ball grid array package. Electron: Lett 38(1):14–16
3. Sabban A (1983) A new broadband stacked two-layer microstrip antenna. IEEE AP-S International Symposium Digest, Houston, pp 64–66
4. Croq F, Papiernik A (1990) Large bandwidth aperture-coupled microstrip antenna. Electron Lett 26(16):1293
5. Wang W, Cheng S, Zhong S (2004) A broadband slope-strip-fed microstrip patch antenna. Microw Opt Technol Lett 43(2):121–123
6. Bystom A, Berntsen DG (1956) An experimental invertigation of cavity mounted helical antenna. IRE Trans Antennas Propag 4(1):54–58
7. Sun L, Wang W, Guan W (2017) A novel wideband metal cavity antenna. Radar Sci Technol 15(4):439–442
8. Fang X, Wang W, Huang G et al (2019) A wideband low-profile all-metal cavity slot antenna with filtering performance for space-borne SAR application. IEEE Antennas Wirel Propag Lett 18(6):1278–1282
9. Zhang H, Wang W, Zhang Z et al (2011) A broadband single layer edge-fed cavity-backed microstrip antenna. Microw Opt Technol Lett 53(2):2831–2834
10. Xiao-peng LU, Wei WA, Chu GA (2012) Microstrip line fed wideband single-layer patch antenna. Radar Sci Technol 10(1):112–116
11. Jiang Q (2015) Study on millimeter wave wideband high-efficiency back-cavity antenna array technigues. University of Electronic Science and Technology, Chengdu, pp 8–10
12. CassiviI Y, Wu K (2004) Substrate integrated nonradiative dielectric waveguide. IEEE Microwave Wirel Compon Lett 14(3):89–91
13. Wang W, Zheng Z, Zhang H et al (2018) Waveguide slot filtering antenna with metaterial surface. In: 2018 international symposium on antennas and propagation (ISAP2018), Busan, Korea, pp 497–498
14. Wang L (2018) Research on vertical interconnection technologies of phased array frontends. University of Electronic Science and Technology, Chengdu
15. Zhang M (2008) Digital array radar and software radar. Publishing House of Electronics Industry, Beijing, pp 140–192
16. Xie R, Shi X (2004) Precise analysis of the spurious spectrum of DDS in the presence of phase truncation. J Electron Inf Technol 26(3):495–499
17. Lu J, Wu M, Chen S, Fang Z (2000) A calibration method of phased array radar based on FFT. Chin J Radio Sci 15(2):221–224

Chapter 8
Digital Array Antennas

8.1 Introduction

According to the principle of beamforming, antenna beams can be formed in radio frequency (RF), analog, and digital domains, and array antennas in which both receiving and transmitting beams are digitally formed are called digital array antennas. The digital array antenna is an active phased array antenna. Compared with traditional active phased array antennas, it performs amplitude and phase control in the digital domain. Figure 8.1 is a schematic diagram of a typical traditional active phased antenna and a digital array antenna. The digital array antenna digitally processes the signal of each transceiver channel to realize the digitalization of transmitting waveform generation and receiving signal processing. A direct digital frequency synthesizer (DDS) forms a transmitting waveform in the digital domain. Then, an analog/digital converter (A/D) converts the received analog signal into a digital signal. As a result, the amplitude and phase parameters of each transmitting and receiving channel of the digital array antenna are individually controllable, and the amplitude and phase weighting required for the antenna to transmit and receive digital beamforming is completed in the digital domain.

A digital array antenna is generally composed of an antenna radiation front, a calibration/monitoring unit, multiple digital array modules (including multi-channel RF reception, multi-channel RF transmission, digital transceiver, and distributed power supply, etc.), clock and local oscillator division network unit, frequency source unit, and other functional circuits [1], as shown in Fig. 8.2. The function of the receiving channel is to receive, amplify, frequency convert, filter, and digitize the echo signal to form a digital baseband signal. The function of the transmission channel is to generate the transmission signal waveform, frequency convert, filter, amplify, and send the signal to the radiation front of the antenna. The calibration/monitoring unit provides an active transmitting/receiving channel for calibrating, testing, and monitoring the digital array antenna. Finally, the frequency source unit provides the local oscillator signal required for the frequency conversion of the receiving/transmitting channel,

J. Lu et al., *Active Array Antennas for High Resolution Microwave Imaging Radar*,
https://doi.org/10.1007/978-981-99-1475-3_8

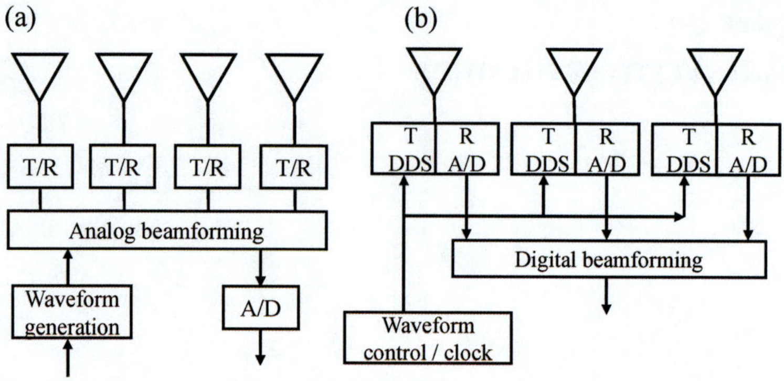

Fig. 8.1 Phased array antenna, **a** active phased array antenna, **b** digital array antenna

the sampling clock required for the digital transceiver circuit, and various reference synchronous clock signals required for signal and data processing, beamforming, and timing control.

The digital array antenna realizes amplitude and phase control in the digital domain, and the RF link is generally composed of a digital receiver (including analog/digital conversion), a digital transmitter (including waveform generation), a frequency source, and a calibration channel, etc., as shown in Fig. 8.3.

In digital array antennas, RF, analog, and digital links are often no longer separate devices but are usually integrated at a certain scale. It is generally composed of a power amplifier, limiting low noise amplifier, up and down converter (except RF direct digitization), analog to digital converter (ADC), DDS, photoelectric conversion, filter, circulator, and so on. Figure 8.4 is a typical RF link topology.

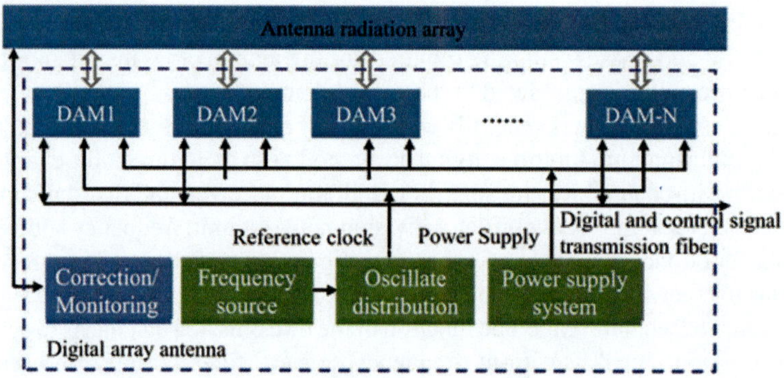

Fig. 8.2 Composition of digital array antenna

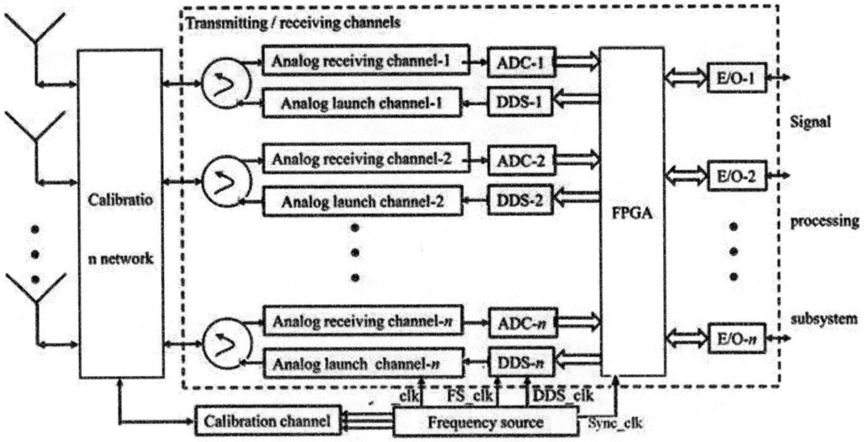

Fig. 8.3 Composition of digital array antenna system

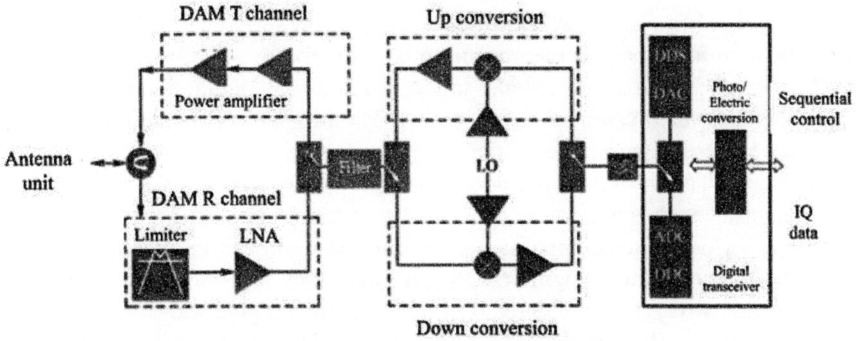

Fig. 8.4 RF link topology diagram

8.2 Digital Signal Generation

The digital transmission chain comprises digital signal generation, matched filter, up-conversion, power amplification chain, and circulator (or isolator). The DDS chip generates the digital signal directly to generate the required waveform signal. Its frequency, bandwidth, modulation form, pulse width, pulse amplitude, initial phase, and other signal characteristics are all controlled by external parameters. The waveform signal generated by DDS generally needs to be purified through matched filtering. Usually, the power amplification chain can be divided into preamplification and final stage amplification, as shown in Fig. 8.5.

The transmission link of the digital array antenna is a multi-channel system. Multiple antenna elements transmit signals separately, and the power combination is realized in space through phase scheduling, which reduces the power requirement

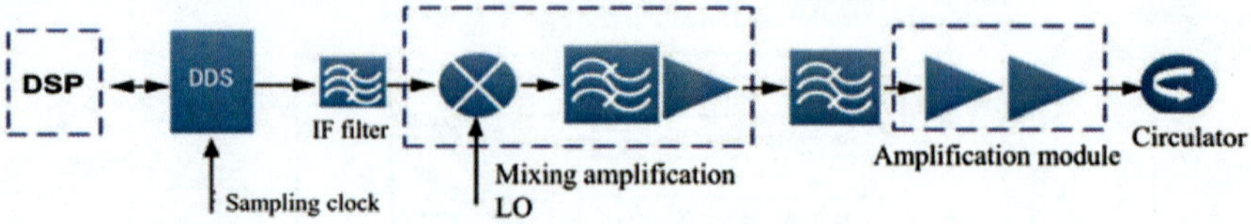

Fig. 8.5 Schematic diagram of the transmission link

for a single transmission link. The up-conversion circuit is the core of the transmission chain. Usually, its key specifications include operating frequency, noise figure, 1 dB compression point, third-order intermodulation, gain, intermodulation frequency interference, and other parameters [2].

Due to the mutual coupling between the antenna elements, when the transmit beam points to a large scanning angle, the active standing wave of the antenna will have a large change. In case the digital array antenna uses a transceiver switch instead of a circulator/isolator, a load-drawing effect on the power amplifier of the transmission link will inevitably appear. This phenomenon will cause the amplitude and phase of its output signal to change with the beam pointing. In severe cases, it will have a relatively harmful impact on the transmitting beam lobe.

The core of the digital array antenna transmission chain is digital signal generation, which includes three essential functions [3]. The first is to generate digital array antenna waveform signals. The second is to provide the phase required for transmit beamforming. The third is to provide the required time delay for the operation of the large instantaneous bandwidth when the beam scanning of the large digital array antenna is performed to compensate for the transition time of the antenna aperture.

Digital direct synthesis signal (DDS) is a frequency synthesis method. Generally, there are two types of DDS realizations, i.e., direct digital waveform synthesis (DDWS) and direct digital frequency synthesis (DDFS). In general cases, DDS refers to the latter.

8.2.1 Phase Accumulators

Assume that the frequency tuning word FTW = 1, the initial value of the phase accumulator is 0, and the phase accumulator will increase by 1 every system clock. For example, if the phase accumulator is 32 bits wide, it takes 2^{32} (over 4 billion) clock cycles before it returns to 0, and the cycle repeats. The truncated output of the phase accumulator is used as the SIN signal (or COS signal) lookup table address. Each lookup table address corresponds to a phase point of the SIN signal from 0° to 360°. The lookup table contains the digital amplitude information corresponding to a complete SIN signal cycle. Thus, a lookup table maps the phase information from the phase accumulator to digital amplitude information to drive the DAC. This process

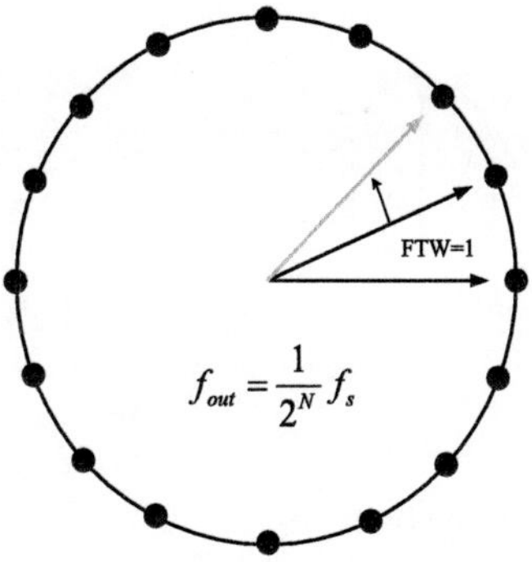

N	Number of Points= 2^N
8	256
12	4096
16	65536
20	1048576
24	16777216
28	268435456
32	4294967296
48	281474976710656

Fig. 8.6 Digital phase wheel (FTW = 1)

is shown graphically as a "phase wheel," as shown in Fig. 8.6. The frequency control word K is a binary number with a word length of N bits, K is the phase increment number, and the value of K can be one of 2^N different numbers, i.e., $0 \leq K \leq 2^N - 1$.

If $N = 32$ and FTW $= 1$, the phase accumulator steps through each of the 2^{32} possible outputs until it overflows and starts over. The corresponding output SIN signal frequency is equal to 2^{32} frequency division of the input clock frequency, i.e., $f_{\text{out}} = \frac{1}{2^N} f_s$. If FTW $= M$, the phase accumulator will "roll and calculate" at the speed of M times, as shown in Fig. 8.7, and the output frequency will also increase by M times, i.e., $f_{\text{out}} = \frac{M}{2^N} f_s$.

It can be seen from the figure that the phase accumulator is updated in each clock cycle, and the output of the phase accumulator is the product of clock cycle and the FTW.

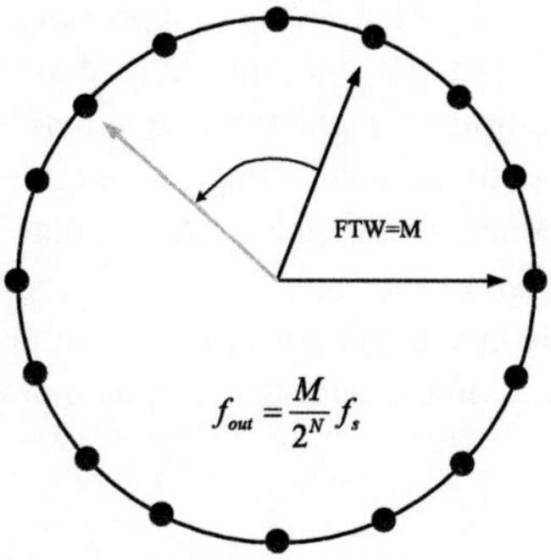

N	Number of Points= 2^N
8	256
12	4096
16	65536
20	1048576
24	16777216
28	268435456
32	4294967296
48	281474976710656

Fig. 8.7 Digital phase wheel (FTW = M)

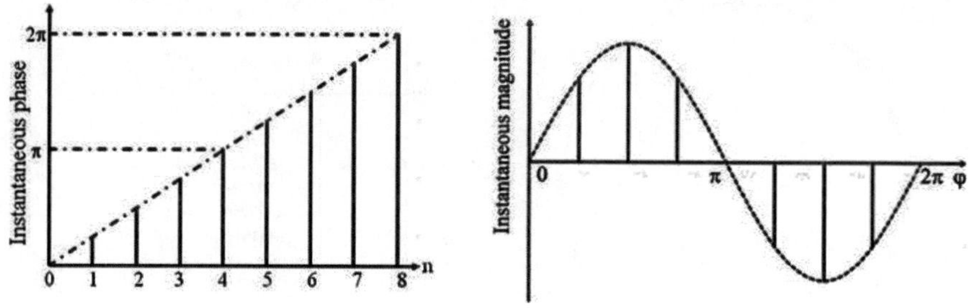

Fig. 8.8 Accumulator and phase/magnitude converter output waveform

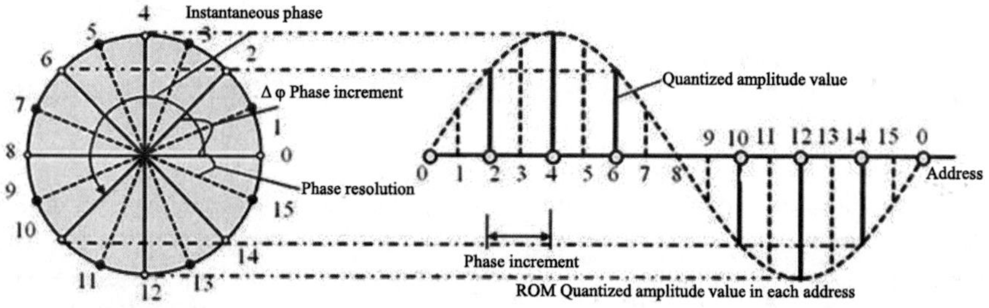

Fig. 8.9 Schematic diagram of phase/amplitude conversion

8.2.2 Phase/Amplitude Converters

The phase/amplitude converter is a waveform memory. In each storage unit, the sine wave amplitude values at different phases are stored with a certain number of binary numbers. In this way, every time a phase word is input, the sine wave amplitude value corresponding to the phase is taken out. The Instantaneous phase value generated by the phase accumulator is used as the input address of the phase/magnitude converter.

When $N = 4$, $K = 2$, the output waveforms of the phase accumulator and phase/amplitude converter in each cycle are shown in Fig. 8.8.

When $N = 4$, $K = 2$, the relationship between the input phase increment of the phase/amplitude converter, the instantaneous phase, and the output quantized amplitude value is shown in Fig. 8.9.

8.2.3 Direct Digital Waveform Synthesis

Direct digital waveform synthesis generally comprises a phase accumulator, an amplitude/phase converter, a digital-to-analog converter, a variable clock generator, and a filter, as shown in Fig. 8.10. According to the predetermined sampling frequency, the time domain characteristics of the signal, the waveform length, and

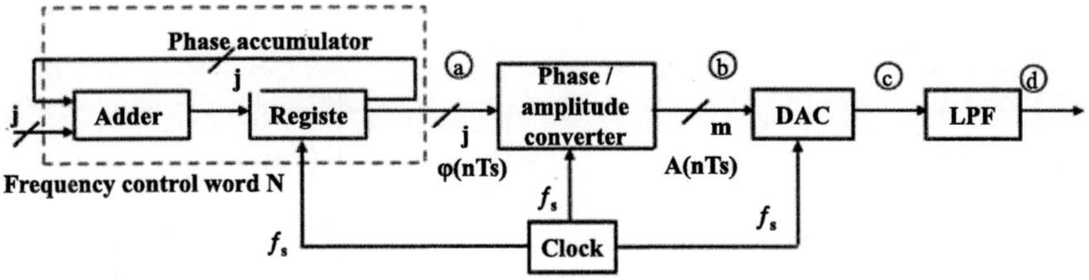

Fig. 8.10 DDWS block diagram

other parameters can be calculated from the mathematical expression of the signal. After quantization, it is mapped to the pre-stored address of the amplitude/phase converter according to the sampling order. The variable clock generator outputs the corresponding clock signal according to the set sampling frequency. On each rising edge of the clock signal, the output address of the address generator is incremented by 1. The output of the address generator addresses the waveform lookup table, reads the waveform data point by point, and generates the corresponding output signal after digital-to-analog conversion. The waveform diagram corresponding to the main nodes in Fig. 8.10 is shown in Fig. 8.11. Assuming the variable clock frequency is f_s, if each period of the periodic waveform consists of n sampling points, then the output frequency of the waveform is

$$f_0 = \frac{1}{T_0} = \frac{1}{\mathrm{nT}_s} = \frac{f_s}{n} \tag{8.1}$$

where T_0 is the output signal waveform period, T_s is the variable clock period.

It can be seen from Eq. (8.1) that the frequency of the output signal depends on the variable clock frequency f_s and the sampling length N of the waveform per cycle. Therefore, the frequency resolution of the output signal is determined by the resolution of the sampling clock.

8.2.4 Direct Digital Synthesis

Digital array antennas generally require high-frequency stability of the transmitted signal, large output dynamic range, good output frequency response, modulation function, high spectral purity, and programmable control of frequency, phase, and amplitude. The waveform generated by the digital method has the characteristics of coherence, repeatability, high stability, and programmable, which can easily realize the agility of waveform parameters and generate arbitrary complex waveforms. Compared with the traditional analog synthesis technology, its frequency switching time is fast, which can reach the nanosecond level. The relative bandwidth of the generated signals is between 0 and $f_s/2$. The frequency resolution is high up tp $\frac{1}{2^N} f_s$.

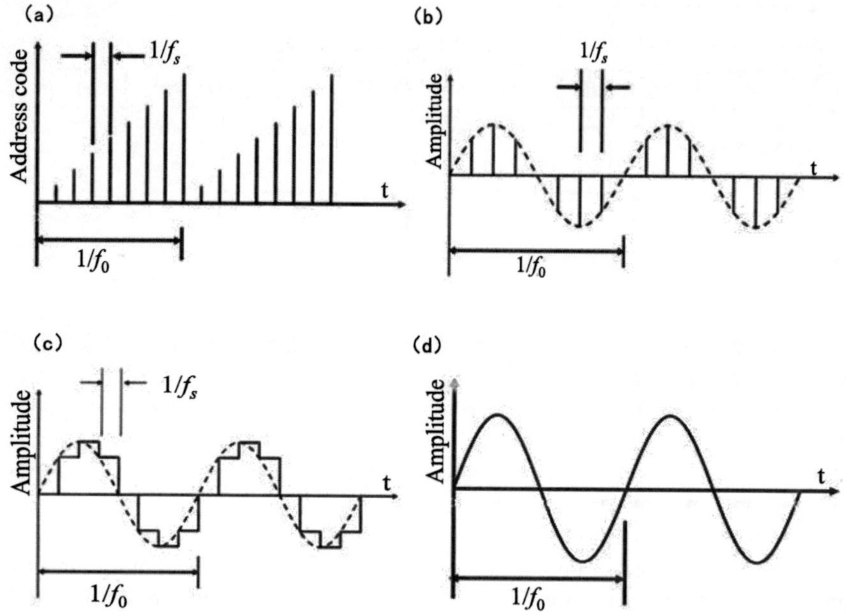

Fig. 8.11 Typical waveform diagram of DDWS link

If $N = 48$, $f_s = 1000$ MHz, the frequency resolution can reach 3.55×10^{-6} Hz. The phase noise is low, comparable to that of a clock signal. Many types of waveforms can be generated in any modulation form. Moreover, it can output ideal two-way signals orthogonal to each other.

Direct digital frequency synthesis generally consists of a fixed clock generator, a phase accumulator, a phase/amplitude converter, a digital-to-analog converter, and a low-pass filter, as shown in Fig. 8.12. Under the control of the sampling clock, the N-bit phase accumulator accumulates with the frequency control word K and outputs the corresponding D-bit amplitude information through the phase/amplitude converter to complete the conversion from waveform phase to amplitude. Next, the output waveform amplitude information is obtained through the digital-to-analog converter to achieve the corresponding analog signal output. Finally, the spurious components are removed by the low-pass filters to ensure the purity of the output waveform.

Theoretically, DDS can generate arbitrary signal waveforms. In other words, DDS technology can directly modulate one, two, or three of the generated signal waveform parameters (such as frequency, phase, and amplitude) at the same time. Taking frequency modulation as an example, the output frequency of a DDS system can be expressed as

$$f_{\text{out}} = K \times \frac{f_s}{2^N} \tag{8.2}$$

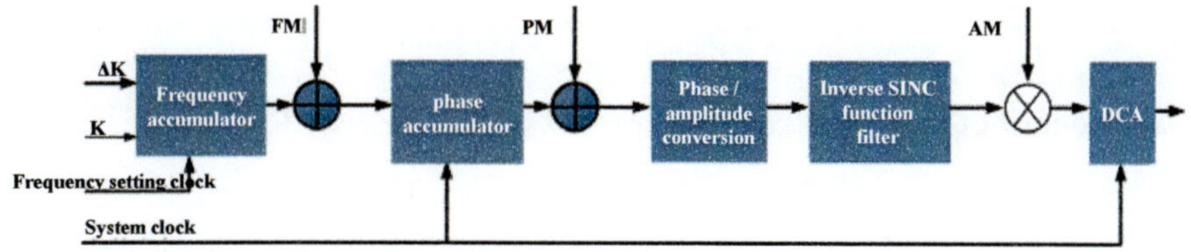

Fig. 8.12 Block diagram of DDFS working principle

where K is the frequency control word, f_s is the DDS input clock frequency, N is the number of bits of the phase accumulator.

For a given DDS, the number of bits of the phase accumulator is a fixed value. When the input clock frequency is set, its output frequency changes with the control word K. Therefore, as long as the frequency control word K is changed according to the regulation of the modulation signal, the required frequency modulation signal can be realized. Precise phase control of the output signal can be achieved through a phase accumulator and a phase/magnitude converter.

Figure 8.13 shows that if the DDS outputs signals with a single tone, the following conditions must be met: frequency change word $\Delta k = 0$, $\mathrm{FTW}(n) = ftw' =$ constant, no frequency, phase, and amplitude modulation.

At the beginning of work, the initial value of the phase accumulator is $\Delta\varphi$, after each system clock, the phase accumulator adds the Frequency Tuning Word (FTW), which is sent to the phase/amplitude conversion. After amplitude scaling, sending to DAC and low-pass filter (LPF), the single tone signal can be output, namely

$$\Delta K = 0 \tag{8.3}$$

The frequency change word $\Delta k = 0$, so the output result of the frequency accumulator is constant, namely

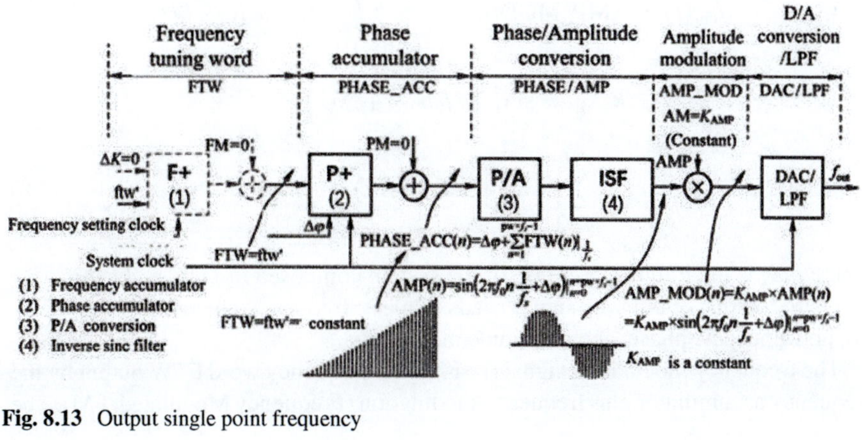

Fig. 8.13 Output single point frequency

$$\mathrm{FTW}(n) = ftw' = \text{constant} \tag{8.4}$$

The output of the phase accumulator per system clock is

$$\mathrm{PHASE}_{\mathrm{ACC}(n)} = \Delta\varphi + \sum_{n=1}^{pw\times f_s-1} \mathrm{FTW}(n)\Big|_{\frac{1}{f_s}} \tag{8.5}$$

where PW is the pulse width of the waveform. After phase/amplitude conversion, the output digital amplitude is

$$\begin{aligned}\mathrm{AMP}(n) &= \sin\left[\Delta\varphi + \sum_{n=1}^{pw\times f_s-1} \mathrm{FTW}(n)\right]\Bigg|_{\frac{1}{f_s}} \\ &= \sin\left(2\pi f_0 n\frac{1}{f_s} + \Delta\varphi\right)\Bigg|_{n=0}^{n=pw\times f_s-1} \\ f_0 &= \frac{ftw'}{2^N} f_s\end{aligned} \tag{8.6}$$

The amplitude scaling is

$$\begin{aligned}\mathrm{AMP\_MOD}(n) &= K_{\mathrm{AMP}} \times \mathrm{AMP}(n) \\ &= K_{\mathrm{AMP}} \times \sin\left(2\pi f_0 n\frac{1}{f_s} + \Delta\varphi\right)\Bigg|_{n=0}^{n=pw\times f_s-1}\end{aligned} \tag{8.7}$$

where K_{AMP} is constant.

After passing through the DAC and LPF, the analog signal is output as:

$$\begin{aligned}f_{\mathrm{out}}(t) &= \mathrm{AMP\_MOD}(nT_s) \times \sum_{n=1}^{pw\times f_s-1} \delta(t - nT_s) * \hbar(t) \\ &= K_{\mathrm{AMP}} \times \sin\left(2\pi f_0 n\frac{1}{f_s} + \Delta\varphi\right)\Bigg|_{n=0}^{n=pw\times f_s-1} \\ &\quad \times \sum_{n=1}^{pw\times f_s-1} \delta(t - nT_s) * \hbar(t)\end{aligned} \tag{8.8}$$

where $\hbar(t)$ is the impulse response of the system composed of DAC and LPF.

DDFS output single tone signal needs to meet: frequency change word $\Delta k = 0$, oo, no frequency, phase, and amplitude modulation.

The frequency modulation signal (FSK) is the frequency word FTW output by the frequency accumulator plus frequency modulation (Frequency Modulated, FM). The phase modulation signal (PSK) is the accumulated phase output by the phase accumulator plus the phase modulation word (Phase Modulation, PM). The amplitude modulation signal (ASK) is the digital amplitude information after phase/amplitude conversion, which is scaled and sent to DAC. The amplitude modulation word is multiplied on this basis. Nonlinear Frequency Modulation (NLFM) signal can be realized in many ways, and it can be regarded as a signal composed of a group of single tone signals output according to certain rules. However, the essences of the single-frequency and NLFM signals are the same. The frequency accumulator can output different frequency words to the phase accumulator according to a certain rule (such as the FTW storage table) at intervals. NLFM can also be regarded as composed of linear frequency modulation (LFM) signals output according to certain rules. Generally, the frequency change word ΔK can output different values to the frequency accumulator according to certain rules (such as ΔK storage table) at intervals.

8.2.5 DDS Spectrum

An ideal DDS should meet the requirement that the output of the phase accumulator is all used to convert to the amplitude, which is an entirely real value, with no quantization error. The resolution of the digital-to-analog converter is infinitely small, i.e., the number of bits of the DAC is infinite, and there is no conversion error, which is completely ideal. DDS is equivalent to an ideal sample/hold circuit. In this case, the structure of DDS can be simplified into the mathematical model shown in Fig. 8.14, where $\delta(t)$ is the impulse function, and $h_r(t)$ is the step reconstruction function, namely

$$h_r(t) = \begin{cases} 1, \; 0 \le t \le T_s (T_s = 1/f_s) \\ 0 \qquad \text{other} \end{cases} \tag{8.9}$$

$$\mu_0(t) = \sum_{-\infty}^{+\infty} \sin(2\pi f_0 t)\delta(t - lT_s) * \hbar(t) \tag{8.10}$$

The discrete Fourier transform of $\mu_0(t)$ can be obtained

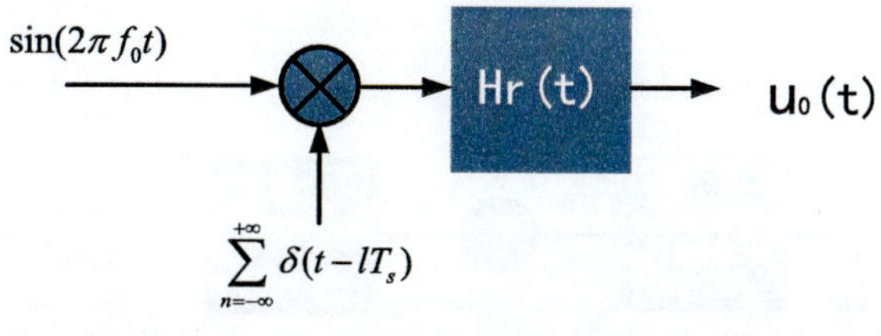

Fig. 8.14 DDS equivalent structure

$$U_0(w) = j\pi Sa\left(\frac{\omega T_s}{2}\right)e^{-\frac{j\omega T_s}{2}}\sum_{-\infty}^{+\infty}[\delta(\omega + \omega_0 - L\omega_S) - \delta(\omega - \omega_0 - L\omega_S)] \tag{8.11}$$

As shown in Fig. 8.15, it can be seen that the spectral line of the ideal DDS output signal is located at $lf_s \pm f_0$ $(l = 0, \pm 1, \pm 2, \ldots)$. This is because they are composed of the sum mixing product of the clock harmonic frequency, the signal frequency, and the various mixing products of the clock harmonic frequency and the signal frequency.

The above discussion is the ideal situation of the DDS system. However, various errors will be introduced due to the limitations and influences of various factors. The error model is shown in Fig. 8.16. There are three main sources of error in the DDS spurious model: phase truncation error $e_p(n)$, amplitude quantization error $e_T(n)$, DAC conversion error $e_{DA}(n)$.

(1) **Phase truncation error**. To obtain a sufficiently high-frequency resolution, a large number of bits N of the phase accumulator is preferred. However, it is impossible to take an infinite number in actual use due to limitations such as volume, speed, and cost. Generally, 32, 48, etc. bits can be used in the early days, DDS used the waveform look-up table method to complete the phase/amplitude

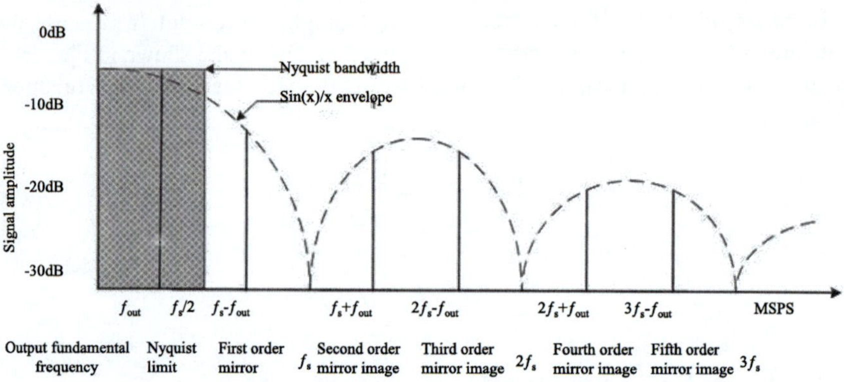

Fig. 8.15 Ideal DDS output spectrum

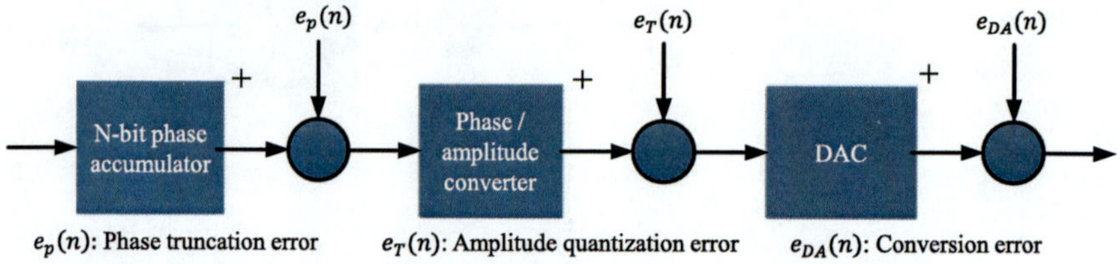

Fig. 8.16 DDS spurious source mathematical model

conversion, and the storage capacity of the ROM table was much smaller than 2^N. Usually, the high M bits of the phase accumulator are used to address the ROM table, and the low $N - M$ bits are discarded. In this way, a phase error is inevitably generated, called a phase truncation error, manifested as a spurious component on the output spectrum. Because the DDS output signal is usually sinusoidal, its phase truncation has obvious periodicity. This is equivalent to periodically introducing a truncation error, the final effect is that the output signal has a certain harmonic component.

It has been proved by scholars [4] that the worst spur/signal ratio due to phase truncation is

$$[C/S] = 2^M \sin\left[\frac{\mathrm{Gcd}\left(K, 2^{N-M}\right)}{2^{N-M}}\right], \ \mathrm{Gcd}\left(K, 2^{N-M}\right) < 2^{N-M} \tag{8.12}$$

where $\mathrm{Gcd}\left(K, 2^{N-M}\right)$ represents the greatest common divisor of K and 2^{N-M}. The phase truncation error occurs only when $\mathrm{Gcd}\left(K, 2^{N-M}\right)$ is smaller than 2^{N-M}. If $\mathrm{Gcd}\left(K, 2^{N-M}\right)$ is equal to or greater than 2^{N-M}, the phase bits below 2^{N-M} are zero, and no phase error occurs.

Phase truncation does not produce spurs at every output frequency point. Its size and distribution depend on three factors: accumulator bit N, addressing bit M, and frequency control word K. Due to the maximum spurious component introduced by the phase truncation error, the relationship between its spurious amplitude and the number of ROM addressing bits M is: spurious (SPUR) = $-6.02 \times M$ (dBc). When $M \geq 15$, the spur introduced by the phase truncation error can be neglected in engineering. Currently, the phase/amplitude converters in DDS are done by algorithms without memory. The phase output from the phase accumulator is directly converted into an amplitude value after pipeline processing. The more pipeline stages, the closer the phase/amplitude conversion result is to the ideal value, and the phase truncation error can be ignored in this case.

(2) **Amplitude quantization error**. DDS used the waveform look-up table in the early days to complete the phase/amplitude conversion. However, because the sinusoidal amplitude values stored in the DDS internal waveform memory are expressed in binary, the sinusoidal amplitude values that exceed the memory word length must be quantized, which introduces quantization errors. Although the phase/amplitude converter in DDS is done by algorithm, it does not need memory. However, the output amplitude value is also quantized, the number of bits of the DAC is limited, and the quantized amplitude needs to be truncated if necessary.

There are two main methods of amplitude quantization, namely, the round-off quantization and truncated quantization. Actual DDSs mostly adopt round-off quantization. In general, the level of spurs introduced by amplitude quantization is lower than that caused by the non-ideal conversion characteristics of the DAC.

(3) **DA conversion error**. The spurs caused by DAC conversion mainly include those caused by DAC nonlinearity and those caused by DAC burrs. DAC nonlinearity includes differential nonlinearity (DNL), integral nonlinearity (INL), settling time, peak energy, slew rate, jitter error, feedthrough error, a limited number of DAC quantization bits, etc. Due to the non-linear existence of DAC, the amplitude sequence obtained from the phase/amplitude conversion will go through a non-linear process from the input to the output of the DAC, so the harmonic component of the output signal f_0 will be generated. Furthermore, because DDS is a sampling system, these harmonics will move with a cycle of f_s, namely

$$f = \text{nf}_s \pm \text{mf}_0 \tag{8.13}$$

where n and m are arbitrary integers. The components that fall into the Nyquist bandwidth will become harmful spurious frequencies. The location of the frequency can be determined, but the magnitude is difficult to determine. The nonlinearity of DAC has become the main source of DDS stray. Especially, when the clock frequency is increased, this problem has become progressively severe.

According to the Nyquist sampling theorem, to restore the ideal waveform, the output frequency cannot exceed $f_s/2$. If it exceeds $f_s/2$, the first-order image frequency will fall within the first Nyquist bandwidth from DC to $f_s/2$. Since the sin x/x envelope brought about by the aperture distortion, the output amplitude of the DDS will drop by several decibels within the first Nyquist bandwidth. Therefore, the DDS chip introduced by some companies contains a pre-distortion filter with the inverse $\sin x/x$ frequency response, which can limit the fluctuation of the output amplitude of the DDS to within ±0.1 dB.

In addition to the spurs caused by the above-mentioned errors, the spurs caused by other errors are generally smaller than the above-mentioned points, generally called background noise or background spurs.

8.2.6 DDS Spurious Suppression

With the development of ultra-high-speed GaAs devices, the limitation of DDS output bandwidth is gradually being addressed. However, the stray is determined by the characteristics of DDS itself, and the stray will become a more and more important factor restricting the application field of DDS technology.

(1) **DDS output spectrum spread**. It can be known from the principle of DDS that the effective range of DDS output effective bandwidth is generally less than $f_s/2$. For a sampling system (including DDS), there will be extended spectrum folding with the sampling frequency f_s as the period. Generally, an analog image

rejection filter is used to filter out the image and select the required signal, but the analog image rejection filter is generally not ideal, so in practice, the effective bandwidth of the DDS output is less than $f_s/2$. Different image frequencies output by DDS have different carrier frequencies, and this feature can be used to expand the output frequency range of DDS. Generally, the baseband, first image, second image, and third image output by DDS are used in projects to expand the output frequency range used by DDS. However, higher-order image signals are rarely used. Higher-order signals have a lower amplitude due to the $\sin x/x$ envelope and the reduced output signal-to-noise ratio, so they cannot meet the system specification requirements. In addition, the analog bandwidth of the DAC itself is limited, and it is impossible to output all image signals flatly. Therefore, the analog method to multiply frequency is also a method of spreading the spectrum. In addition, if the required output frequency range is very wide, use the analog frequency to synthesize multiple frequency standards and then mix and filter with the DDS output signal. In this way, a large frequency range can be covered [5], and the performance index of the signal can also meet the requirements.

(2) **DDS output broadband signal pre-distortion compensation**. Modern radar system requires the radar signal to have a large time-bandwidth product to improve the radar's discovery ability, measurement accuracy, and target resolution ability. Therefore, linear frequency modulation (LFM) has become radar's most important signal form because of its good pulse compression performance and resolution ability. However, since the key factor of broadband radar is the achievable range sidelobe suppression degree, with the improvement of target recognition requirements, the broadband waveform in the actual system may not meet the requirements due to the many links in the signal path. Therefore, it is necessary to perform distortion test calibration, error extraction, and pre-distortion compensation on the broadband signal generation system to suppress the range sidelobe. Thus, much literature has introduced pre-distortion compensation methods for broadband signals [6].

(3) **Typical waveform generator**. GM4940 is a domestic DDS chip. Direct digital frequency synthesizer (DDS) with 4 independent channels, each channel can provide independent phase, frequency, and amplitude control, and its highest operating frequency can reach 1 GHz. In addition, all channels share a common system clock, which is inherently synchronous and supports multi-device synchronization. Figure 8.17 shows the LFM spectrum and intra-pulse spectrum output by the waveform generator based on GM4940.

8.3 Digital Receivers

For the RF receiving link below 2 GHz, the RF signal is usually directly digitized without requiring a frequency mixing link. For the RF link above 2 GHz, the RF frequency is generally moved to the intermediate frequency by primary or secondary

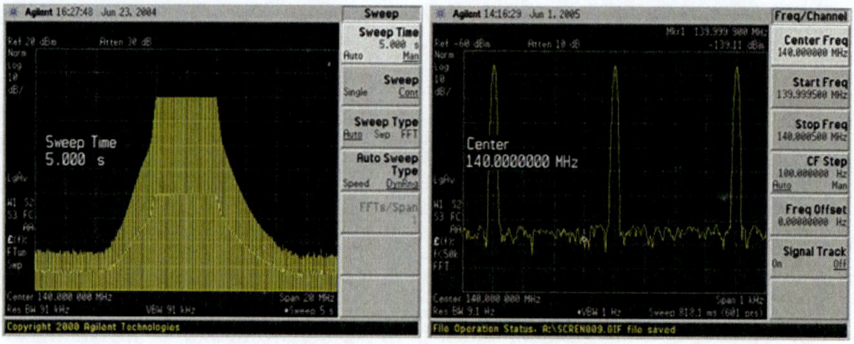

Fig. 8.17 LFM spectrum and intra-pulse spectrum

mixing and then digitized. Then, the sampled digital signal is digitally demodulated in a dedicated Digital Down Conversion (DDC) chip or an FPGA with the help of a Numerical Control Oscillator (NCO). Finally, the digital baseband signal is demodulated. At the same time, to match the data rate of the signal processor, decimation and digital matched filtering is often performed on high-speed digital signals, as shown in Fig. 8.18.

Digital array antennas usually adopt multi-channel digital reception. The multi-channel digital receiver is based on software RF digitization technology and DDC technology to realize the conversion of analog RF signal to digital baseband I/Q signal, as shown in Fig. 8.19.

The implementation structure of digital down-conversion processing includes the Hilbert transform method, interpolation filter method, low-pass filter method, polyphase delay filter method, polyphase frequency mixing post-high-efficiency

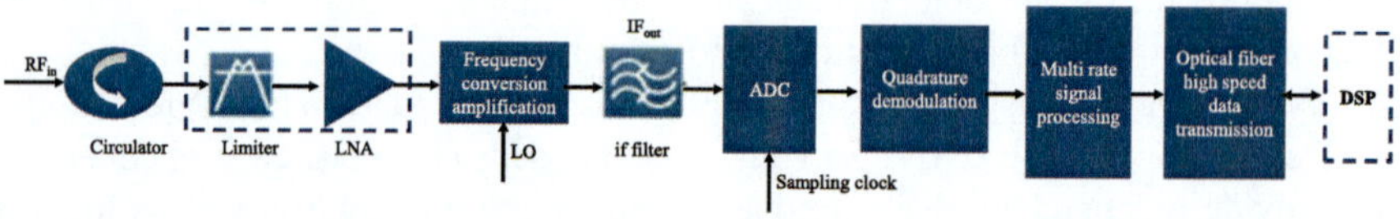

Fig. 8.18 Schematic diagram of receiving chain

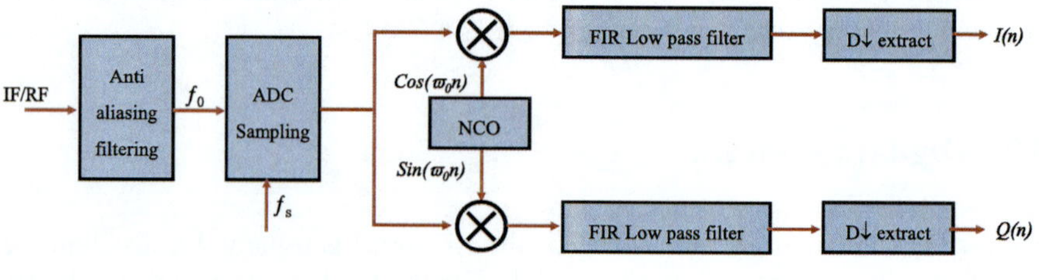

Fig. 8.19 Block diagram of a typical digital receiver

structure, and other implementation methods [7]. The NCO in Fig. 8.19 is a numerically controlled oscillator, essentially a DDS that can output two quadrature signals of sine and cosine. Since digital circuits implement frequency mixing and local quadrature oscillators, they have higher precision and stability. However, its accuracy depends on the suppression of negative spectral components by the low-pass filter FIR.

8.3.1 Digital Sampling

To ensure that the signal sampled by the ADC can recover the original signal without distortion, the ADC sampling needs to satisfy the sampling theorem, and the band-pass sampling theorem needs to be satisfied for the intermediate frequency or RF signal sampling, i.e.

$$\frac{2f_H}{N} \le f_s \le \frac{2f_L}{N-1} \tag{8.14}$$

$$1 \le N \le \text{round}\left(\frac{f_H}{\text{BW}}\right)$$

where f_H and f_L are the signal's upper and lower sideband frequencies, respectively, and BW is the signal bandwidth. In order to simplify the anti-aliasing filter design and DDC digital mixing design, the center frequency f_0 and sampling frequency f_s of the sampling signal are required to meet the requirements of the optimal sampling theorem, namely

$$f_s = \frac{4f_0}{2N-1} \tag{8.15}$$

When the sampling rate satisfies $f_s \ge 2$ BW, $f_0 = \left(M \pm \frac{1}{4}\right)f_s$ (M is an integer ≥ 0, when it is 0, only the "+" sign is taken), when f_s samples the real signal, the positive and negative spectrum will not alias. Sampled

$$x(n) = i(n)\cos\frac{\pi}{2}n - q(n)\sin\frac{\pi}{2}n,\ f_0 = \left(M \pm \frac{1}{4}\right)f_s \tag{8.16}$$

$$= \begin{cases} (-1)^{n/2} i(n), & n \text{ is even} \\ (-1)^{(n+1)/2} q(n), & n \text{ is odd} \end{cases} \tag{8.17}$$

Equation (8.17) shows that the sampling value is an alternating sequence of I and Q, and the I/Q signal can be separated by changing the symbols of the two sequences to positive and aligning the group delay. This is the physical explanation for the low-pass filtering method in the time domain.

8.3.2 Digital Down Conversion

After digital mixing, the I/Q two-way signal is processed by real value low-pass filtering and then decimated to obtain a matching sampling rate. Since the signal complex number indicates that the sampling rate can be reduced to half, the DDC processing extraction ratio is at least 2. When the ADC sampling satisfies the sampling theorem, the optimal sampling is

$$\frac{2f_H}{N} \leq f_s \leq \frac{2f_L}{N-1} \tag{8.18}$$

$$f_s = \frac{4f_0}{2N-1} \tag{8.19}$$

The DDC mixing processing digital local oscillator $e^{-j2\pi f_0 n}$ that satisfies the optimal sampling theorem degenerates into decimation and symbol transformation. In this case, the DDC processing can be realized by the simplified structure of the polyphase delay filtering method [8] shown in Fig. 8.20. The frequency mixing of this structure becomes the direct decimation after the symbol conversion, and the I/Q time delay caused by the decimation is passed through the design of a prototype filter. Finally, the filter is decimated by 1/4 to obtain four polyphase sub-filters. Two phases are selected at intervals to filter the I/Q sub-path, respectively, to realize half-sample point delay calibration and low-pass filter processing.

When the signal's intermediate frequency f_0, sampling frequency f_s, baseband output sampling rate f_{bs}, decimation ratio D, and NCO period M satisfy a specific relationship, it can be realized by using post-mixing polyphase filtering [9] efficient DDC processing structure. Conventional sampling/mixing/filtering/decimation DDC output is

$$\begin{cases} f_0/f_s = m/M, 1 \leq m \leq M-1 \\ f_s/f_{bs} = M = D \end{cases} \tag{8.20}$$

$$y(n) = \left\{\left[s(t)e^{-j\varpi_0 t}\right] * h(t)\right\} \sum_{n=-\infty}^{\infty} \delta(t - nM)$$

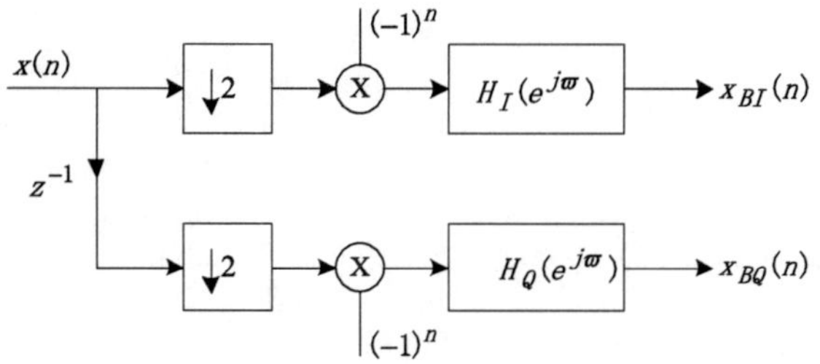

Fig. 8.20 Polyphase delay filtering method DDC processing structure

$$= \sum_{k=0}^{K-1} h(k)s(nM-k)e^{-j\varpi_0(nM-k)} \tag{8.21}$$

where $h(t), t = 0, 1, \ldots, K-1$ is the impulse response of the FIR filter.

By selecting the appropriate intermediate frequency, sampling frequency decimation ratio, and filter coefficients, the digital local oscillator period and decimation ratio are guaranteed to be the same. The local oscillator signal assigned to each polyphase filter-branch is constant. Therefore, frequency mixing can be placed after the polyphase filtering, and therefore the entire DDC implementation structure becomes sampling/decimation/polyphase filtering/digital mixing, as shown in Fig. 8.21.

High-efficiency DDC implementation structure with post-mixing polyphase filtering requires that the center frequency of the intermediate frequency or RF signal be an integer multiple of the sampling rate of the baseband output signal ($f_s = \mathrm{mf}_{\mathrm{bs}}$), and achieve spectrum shift and separation of aliasing through low-pass filtering and subsequent complex phase rotation. When multiple sub-bands are required to output DDC simultaneously, the filter is shared, followed by multiple parallel complex phase rotations. When the number of channels is $M = \log 2(N)$, the filtered phase-rotated multi-channel output can be realized by Inverse Fast Fourier Transform (IFFT)/Inverse Discrete Fourier Transform (IDFT) [10]. The specific implementation structure is shown in Fig. 8.22.

The advantage of the high-efficiency DDC implementation structure of post-mixing polyphase filtering is that the filtering and mixing are processed at a low data rate, which reduces the requirement for processing speed. The filtering process is realized before the frequency mixing and after the ADC. Since it is a real value filtering, the calculation amount and resources are reduced by half. For multi-channel, filter group, or channelized applications, all channels share a filter followed by

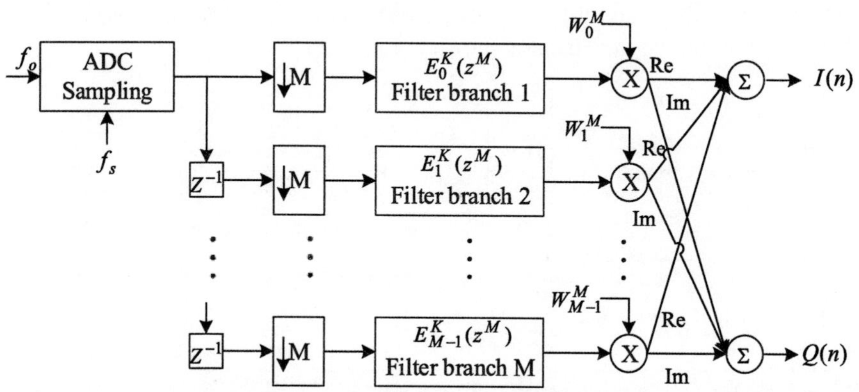

Fig. 8.21 High-efficiency DDC implementation structure with post-mixing polyphase filter

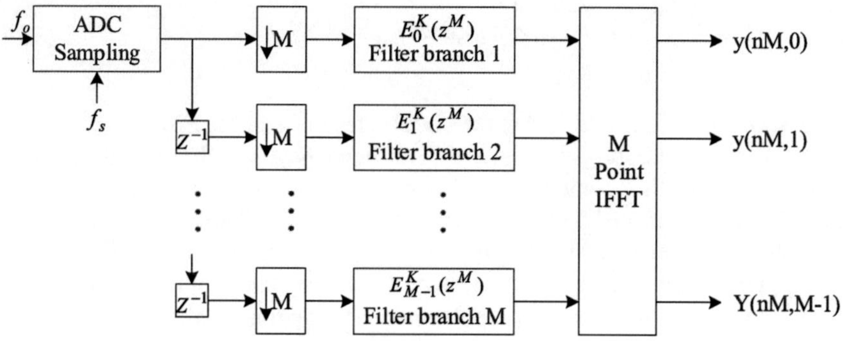

Fig. 8.22 Mixing post-multichannelizer based on IFFT

one, multiple, or IFFT complex phase rotation processing to obtain simultaneous down-converted outputs for different sub-bands.

The high-efficiency structure of the post-mixing polyphase filter can only be used for DDC processing of fixed-position IF signals. For practical applications, there will be receiving blind spots, and the improved post-mixing polyphase filter structure based on the Goertzel algorithm can realize efficient DDC processing of signals at any frequency position [11]. The traditional Goertzel algorithm can realize DFT/IDFT processing at uniformly distributed frequency points. The DFT/IDFT value of any frequency point in the frequency band can be obtained by introducing the concept of Goertzel filtering. The transfer function of the Goertzel filter is

$$H_k(z) = \frac{1 - W_M^k \cdot z^{-1}}{1 - 2 \cdot \cos\left(\frac{2\pi}{M}k\right) \cdot z^{-1} + z^{-2}} \tag{8.22}$$

The direct realization form can be obtained according to the transfer function of the Goertzel filter, and the post-mixing polyphase filter structure based on the Goertzel algorithm improvement is shown in Fig. 8.23.

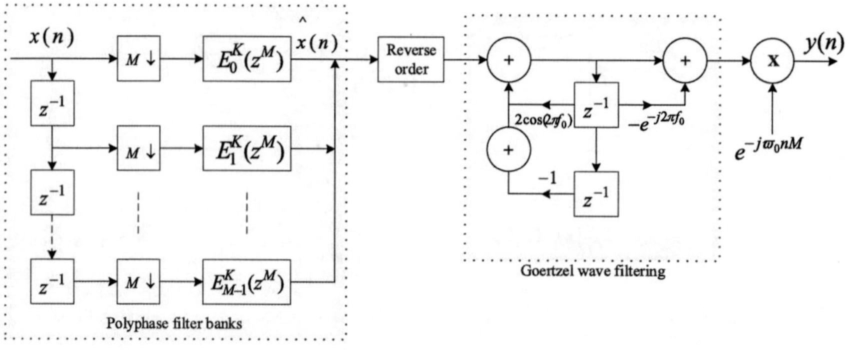

Fig. 8.23 High-efficiency DDC implementation structure based on Goertzel filter

The characteristic of the DDC processing structure is that it inherits all the advantages of the post-mixing polyphase filter structure and realizes no blind zone reception through the precise tuning of the Goertzel filter. In addition, the structure includes polyphase low-pass filtering, Goertzel filtering, and phase-shift processing, and the flexibility of the frequency position of the mixing sequence is achieved at the cost of a certain amount of calculation.

For high-speed broadband DDC processing, the Field-Programmable Gate Arrays (FPGA) cannot directly process at the speed of the sampling frequency, and the sampling rate after decimation also exceeds the processing capability of the FPGA. In this case, the generalized polyphase filter structure [12] can be used to realize it. This structure obtains the generalized polyphase filter structure implemented in parallel by solving the conventional polyphase filter in the frequency domain. It can design broadband parallel DDC processing according to the balance of speed and resources in specific applications.

Conventional DDC handles the whole link decimation ratio as an integer so that it can be realized by single-stage or multi-stage decimation filtering, but for broadband applications such as communication applications and countermeasures/imaging, there are cases where the decimation ratio is a fraction number. In this case, it is necessary to implement DDC processing through a digital resampling filter structure [13], which is realized by interpolating image suppression filtering and decimating anti-aliasing filter multiplexing, selecting an appropriate filter order, and structural transformation.

8.3.3 Noise Figure

The noise figure of the digital receiver system consists of the RF link and the digital receiver noise figure, as shown in Fig. 8.24.

The noise figure of the RF link LNA determines the noise figure of the digital receiver system. Therefore, to reduce the impact of the post circuit, the gain of the low noise amplifier should be increased as much as possible under the condition of satisfying the dynamic range of the system and its stability. Furthermore, to make the low noise amplifier not easy to saturate in the presence of interference, try to increase its output by 1 dB compression point.

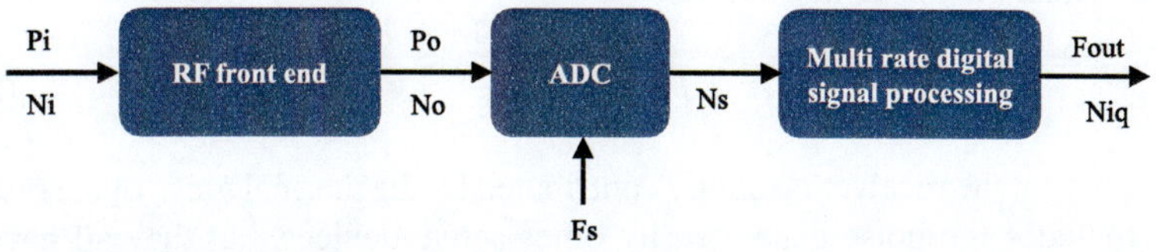

Fig. 8.24 Digital receiver noise block diagram

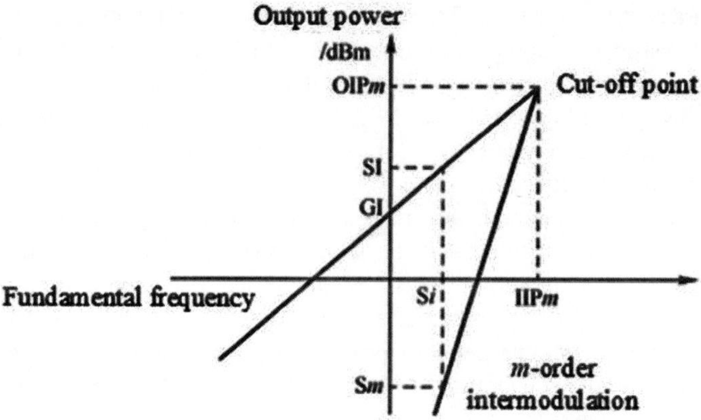

Fig. 8.25 Intermodulation products and *M*-order intercept point diagram

The intermodulation characteristic is an important parameter index to measure the non-linear performance of the receiver. Usually, a two-tone signal is considered as the input, and the magnitude of the third-order intermodulation at the output end is measured to evaluate the intermodulation performance. In the receiving chain, when the power of the input circuit is large enough, the output third-order intermodulation is the same as the power of the amplified fundamental wave signal. In this case, define the input fundamental wave signal as the input third-order intercept point (IIP3), and define the power of the output fundamental wave signal as the output third-order intercept point (OIP3), as shown in Fig. 8.25. The ideal difference between the two concepts is the receive link gain.

Noise is the main factor limiting the sensitivity of the receiver. In general, the noise sources of the receiver mainly include resistor noise, antenna noise, and receiver noise. According to the Nyquist theorem, the resistance element in the circuit will generate an open-circuit thermal noise voltage V_n when the temperature T (unit is K), i.e.

$$V_n = \sqrt{4\ \mathrm{kTRB}_n} \tag{8.23}$$

where k is Boltzmann's constant (1.380658×10^{-23} J/K), R is resistance (Ω), B_n is receiver bandwidth (Hz). When this open circuit voltage is applied to a matched load, the effective noise power is

$$P_n = 4\ \mathrm{kTB}_n \tag{8.24}$$

The noise of the receiver is mainly contributed by the thermal noise of the resistor. It also includes the noise generated by other active devices, but they all have the same spectrum and probability characteristics of thermal noise, so they are generally expressed by noise temperature, i.e.

$$T_n = P_n/4\ \mathrm{kB}_n \tag{8.25}$$

To facilitate the calculation of signal noise, the output noise of the system is often viewed from the perspective of input, and the system noise temperature T_s (unit is K) is defined as

$$\mathrm{kT}_s B_n = P_{\mathrm{no}}/G_0 \tag{8.26}$$

where G_0 is the effective gain of the system, B_n is the system noise bandwidth, P_{no} is the system output noise power, T_s is the system noise temperature converted to the input end of the system, $\mathrm{kT}_s B_n$ is the system output noise power converted to the antenna input end.

If the two-port linear network has a definite input and output, and the source impedance of the input is 290 K, the signal-to-noise ratio of the input to the signal-to-noise ratio of the output is defined as the noise figure of the network. The noise figure of the network is the multiple of the degradation of the signal-to-noise ratio of the network output to the input signal. S_i/N_i is used to represent the receiver input signal-to-noise ratio, S_o/N_o is used to represent the output signal-to-noise ratio, and the noise figure NF is defined as

$$\mathrm{NF} = \frac{S_i/N_i}{S_o/N_o} = \frac{N_0}{\mathrm{GN}_i} = \frac{\mathrm{GN}_i + \mathrm{GN}_r}{\mathrm{GN}_i} = \frac{N_i + N_r}{N_i} = 1 + \frac{T_e}{T_o} \tag{8.27}$$

where G is the receiver's gain, $N_i = \mathrm{kT}_0 B$ is the input noise power of the antenna noise, $N_r = \mathrm{kT}_e B$ is the noise power converted from the output noise power of the receiver to the input terminal. Therefore, the noise figure is irrelevant to the signal power but depends on the input-to-output noise power ratio.

A general receiver is composed of multi-stage amplifiers, mixers, filters, etc., and the noise figure or noise temperature of the cascaded circuit is expressed as (G is the amplifier gain or the reciprocal of frequency conversion/filter loss)

$$\mathrm{NF}_e = \mathrm{NF}_1 + \frac{\mathrm{NF}_2 - 1}{G_1} + \frac{\mathrm{NF}_3 - 1}{G_1 G_2} + \cdots + \frac{\mathrm{NF}_n - 1}{G_1 G_2 \cdots G_n} \tag{8.28}$$

$$T_e = T_1 + \frac{T_2}{G_1} + \frac{T_3}{G_1 G_2} + \cdots + \frac{T_n}{G_1 G_2 \cdots G_n} \tag{8.29}$$

The noise figure is defined when the source temperature is the standard $T_0 =$ 290 K. For example, when the receiver noise figure is 3 dB, the receiver's noise temperature referred to the input end is 290 K. The source and antenna noise temperature T_a is generally not the standard temperature when the radar works. In this case, the actual noise temperature corresponding to the receiver noise figure is converted to the receiver input as

$$T_e = (\mathrm{NF} - 1)T_a \tag{8.30}$$

The ADC and digital signal processing generally cause digitized receiver noise figures. Suppose the ADC equivalent output noise power is N_{ADC}, the noise is band-limited, and the ADC input anti-aliasing filter can ensure that no ADC sampling folding noise enters the ADC, then the ADC output noise N_s is

$$N_s = N_0 + N_{\mathrm{ADC}} \tag{8.31}$$

where N_0 is the input noise power, N_{ADC} is the ADC equivalent output noise power.

The noise figure of the cascaded system is

$$\mathrm{NF}_s = \frac{N_s}{\mathrm{GN}_i} = \frac{N_0 + N_{\mathrm{ADC}}}{\mathrm{GN}_i} = \mathrm{NF} + \frac{N_{\mathrm{ADC}}}{\mathrm{GN}_i} \tag{8.32}$$

Let $M = N_0/N_{\mathrm{ADC}}$, substituting into Eq. (8.32) we get

$$\mathrm{NF}_s = \mathrm{NF}\left(\frac{1+M}{M}\right) \tag{8.33}$$

According to the definition of noise figure, the noise figure of ADC can be expressed as

$$\mathrm{NF}_{\mathrm{ADC}} = P_{\mathrm{FS(dBm)}} - \mathrm{SNR}_{\mathrm{dBFS}} - 10\lg(F_s/2) - \mathrm{KT}_{(\mathrm{dBm/Hz})} \tag{8.34}$$

The noise figure of ADC is not a fixed value, it is related to the sampling clock/sampling rate, input signal amplitude, frequency, digital filter bandwidth, etc. Figure 8.26 shows the relationship between the ADC noise figure and different signal-to-noise ratios and sampling rates. Figure 8.27 shows the ADC output noise spectrum under different input signal amplitudes.

Therefore, the impact of the digital receiver ADC on system noise is directly related to the ratio of front-end noise power to ADC noise power. The reduction of the noise figure of the cascaded system can be solved from two aspects. On the one hand, the equivalent noise power of the ADC should be reduced, and on the other hand, the ADC should be capable of fully quantizing the front-end input noise plus the signal.

The equivalent noise power of the ADC is expressed as

$$N_{\mathrm{ADC}} = N_q + N_t + N_j \tag{8.35}$$

where N_t is the thermal noise, N_q is the ideal quantization noise, N_j is the aperture jitter noise. For an ADC with a resolution of N and an input peak-to-peak value of V_{p-p}, the quantization level is expressed as

$$Q = V_{p-p}/2^N \tag{8.36}$$

In this case, the maximum power of the ADC is

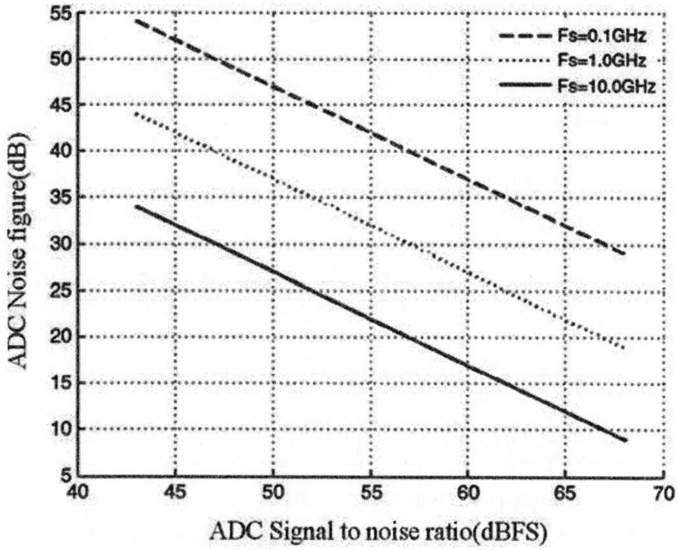

Fig. 8.26 ADC noise figure versus SNR and sampling rate

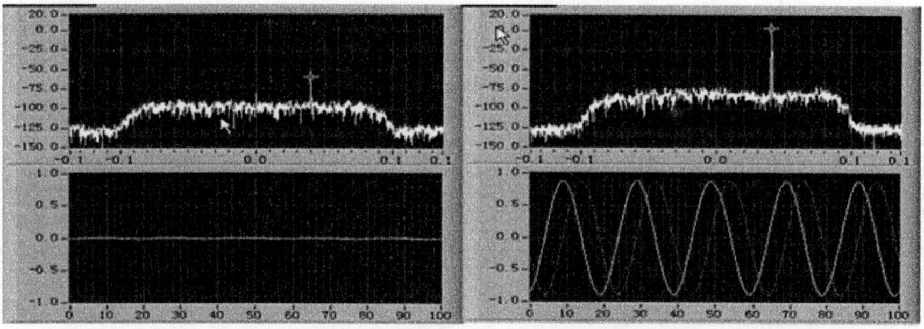

Fig. 8.27 ADC output noise level under different input signal amplitudes

$$P_{\max} = \left(\frac{V_{p-p}}{2\sqrt{2}}\right)^2 = \frac{2^{2N}Q^2}{8} \tag{8.37}$$

When the lowest bit of the ADC is a noise bit, and the noise is uniformly distributed, the quantization noise power is

$$N_q = \frac{Q^2}{12} \tag{8.38}$$

The maximum signal-to-noise ratio of the ideal ADC is

$$\mathrm{SNR_{dBFS}} = 10\lg\frac{P_{\max}}{N_q} = 10\lg\left(\frac{3}{2}\times 2^{2N}\right) = 6N + 1.76(\mathrm{dB}) \tag{8.39}$$

Another key factor affecting the equivalent noise power of the ADC is the noise generated by the clock's aperture uncertainty. Aperture uncertainty includes the conversion of the sampling delay of the ADC's sample-and-hold circuit and the jitter of the upper and lower edges of the sampling clock. Sampling clock jitter and clock phase noise are two different expressions of the same phenomenon, which are closely related to the clock source's thermal noise, phase noise, and spurs. In the case of a single frequency input, the SNR of the ADC limited by the aperture jitter is

$$\mathrm{SNR}_j = -20\lg(2\pi f_a \Delta t_{\mathrm{ms}})(\mathrm{dB}) \tag{8.40}$$

where f_a is the analog input signal frequency, Δt_{ms} is the root mean square value of the aperture jitter. Ideally, the relationship between ADC output SNR and input frequency and clock jitter is shown in Fig. 8.28.

To reduce the noise figure of the digital receiver cascaded system, it is necessary to ensure that the ADC can fully quantize the front-end input noise plus the signal. When a weak signal below the lowest quantization layer level enters the ADC together with a strong signal above the lowest quantization layer and below the highest quantization layer, under the crossover effect of strong signal, the weak signal can pass through the ADC converter together with strong signals. A strong signal can be a real signal or noise, and its strength requires at least 90% of the energy to be quantized by the ADC, which is usually about 6 times the lowest quantization layer level of the ADC. In this case, it is guaranteed that useful weak signals are not lost. Applying

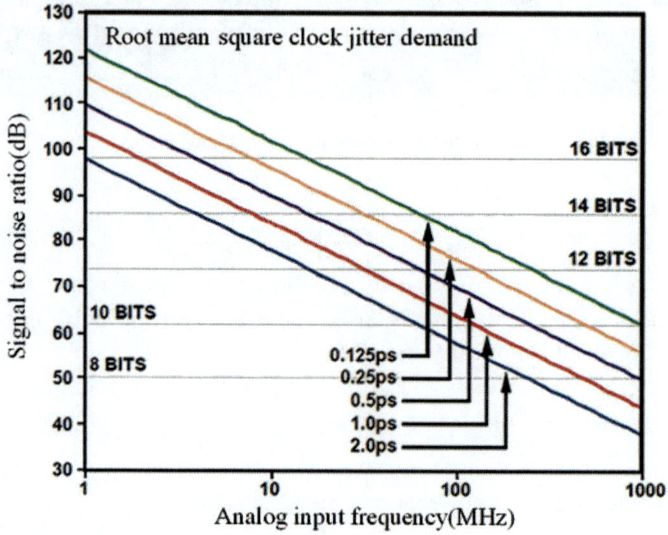

Fig. 8.28 Relationship between ADC ideal SNR and input frequency and clock jitter

the crossover effect of the ADC, relaxing the bandwidth of the RF analog channel can be taken into consideration if conditions permit. To improve the instantaneous dynamic range of the system through digital domain filtering and decimation, the cost will be a higher sampling rate and more digital baseband processing hardware. Figures 8.29, 8.30 and 8.31 are the simulation results of the noise crossover effect of the ADC sampling signal.

Digital receivers generally combine undersampling and oversampling, i.e., the sampling frequency may be lower than the carrier frequency of RF or intermediate frequency signals but much larger than the instantaneous bandwidth of the signal, and the output signal bandwidth after multi-rate signal processing will match the instantaneous bandwidth of the signal. When a multi-rate signal is processed, the filter bandwidths at all levels match the instantaneous bandwidth of the signal, and the

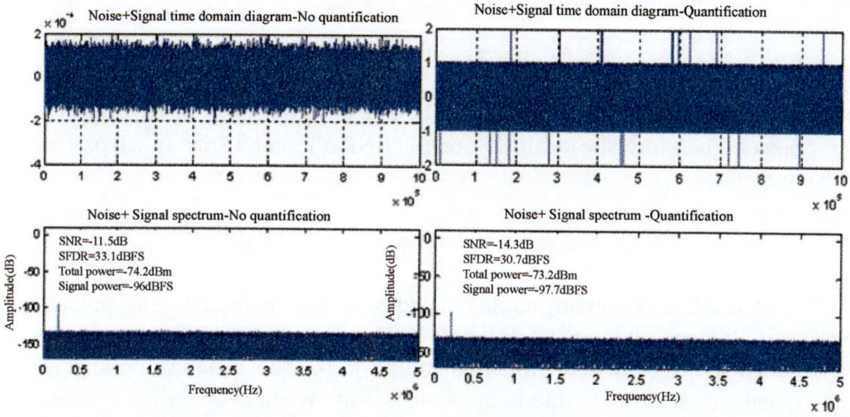

Fig. 8.29 Crossover effect of signal and noise with 1 bit ADC

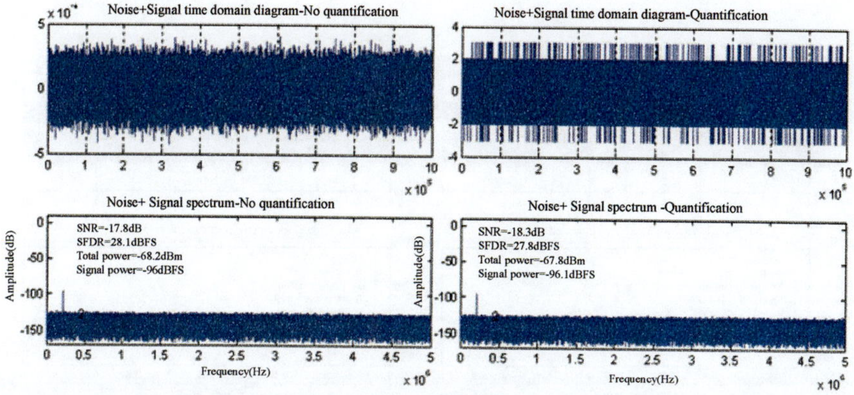

Fig. 8.30 Crossover effect of signal and noise with 2 bit ADC

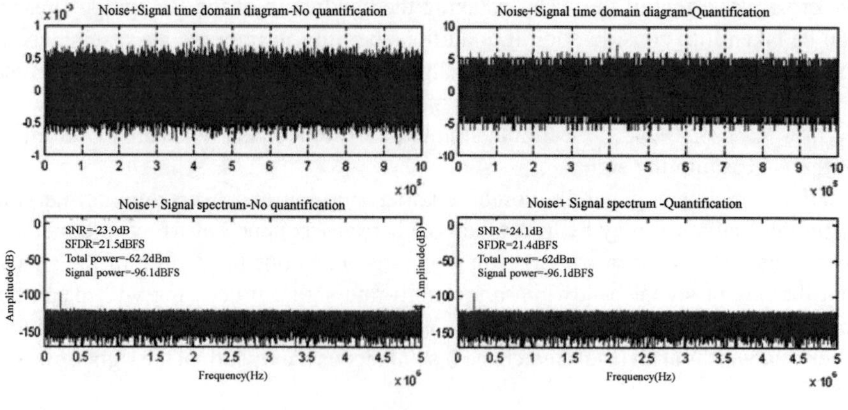

Fig. 8.31 Crossover effect of signal and noise with 3 bit ADC

out-of-band noise is effectively suppressed without noise aliasing due to decimation. In this case, the output noise power will be one part of $F_s/2B$ of the ADC output noise power. Therefore, the ideal ADC output SNR after multi-rate signal processing is expressed as

$$\text{SNR}_{\text{dBFS}} = 6N + 1.76 + 10\lg(F_s/2B)(\text{dB}) \tag{8.41}$$

Multi-rate signal processing mainly includes fixed-point digital signal processing, such as multi-stage filtering and decimation. Therefore, the truncation process must be included in the procedure, which is similar to the full quantization analysis of ADC quantization noise. Furthermore, with the improvement of SNR, it is necessary to leave enough noise bits to ensure the output SNR avoiding the loss of system SNR or the deterioration of the system noise figure. Figures 8.32 and 8.33 show the influence of SNR under different truncation bits.

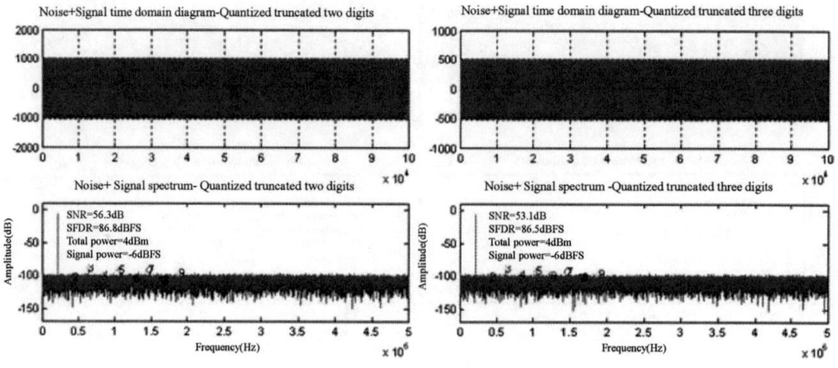

Fig. 8.32 Variation of output SNR when truncation is 2 and 3 bits

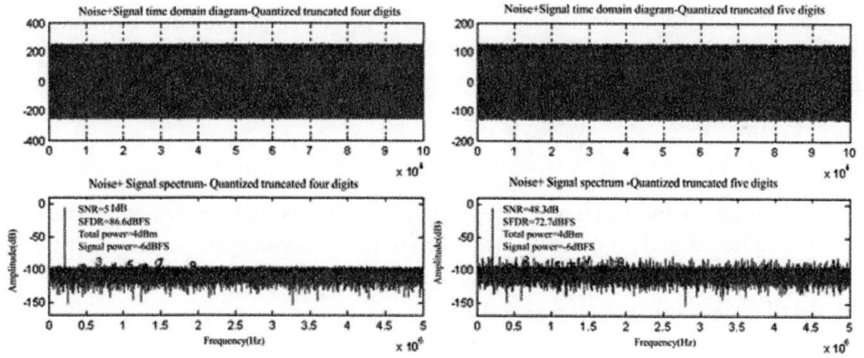

Fig. 8.33 Variation of output SNR when truncation is 4 and 5 bits

8.3.4 Dynamic Range

The dynamic range of the receiving link usually refers to the allowable power range of the input signal at the antenna terminal. The receiver's sensitivity mainly determines its lower limit, and the upper limit is determined by the maximum signal power that the input terminal can handle. It can also be defined as the maximum input power at which the third-order intermodulation product does not exceed the noise floor in a dual-frequency test. The dynamic range obtained under this definition is also called spurious-free dynamic range (SFDR), which is defined as

$$\text{SFDR} = 2 \times (\text{IIP3} - S_{\min})/3 \tag{8.42}$$

where IIP3 is the input third-order intermodulation, $S_{\min}$ is the sensitivity.

Therefore, the spurious-free dynamic range of the receiving chain is related to the third-order intercept point and gain distribution of the selected components. The dynamic range of the receiver is

$$D = D_r + D_\sigma + D_{S/N} + D_f \tag{8.43}$$

where D_r is the variation range of the target echo signal with the distance, D_σ is the variation range of the target RCS, which is determined by the variation range of the cross-sectional area of the radar target/clutter, $D_{S/N}$ is the signal-to-noise ratio required for target detection, which is related to the operating mode and the type of detection target, D_f is the dynamic requirement increase of receiver bandwidth mismatch.

Digital array radar implements active reception, ADC sampling, and digital down-conversion processing at the unit level, digital domain DBF processing, and pulse compression or multi-pulse coherent accumulation processing. Assuming that the DBF processing benefit is D_{DBF}, the digital pulse compression benefit is D_{DPC}, and the multi-pulse accumulation benefit is D_{CI}, the instantaneous dynamic D_{dr} (maximum receiver output SNR) of the single-channel receiver of the phased digital array radar is

$$D_{\mathrm{dr}} = D - D_{\mathrm{DBF}} - D_{\mathrm{DPC}} - D_{\mathrm{CI}} \tag{8.44}$$

The traditional analog phased array antenna performs ADC sampling processing and subsequent signal processing after receiving beamforming in the analog domain. If the signal processing methods are the same, the instantaneous dynamic D_{ar} of the simulated phased array radar receiver is

$$D_{\mathrm{ar}} = D - D_{\mathrm{DPC}} - D_{\mathrm{CI}} \tag{8.45}$$

Therefore, under the same dynamic requirements of the system, the dynamic requirements of the digital phased array antenna receiver are D_{DBF} lower than the dynamic requirements of the analog phased array antenna receiver. Or when the dynamics of the receiver are the same, the total dynamics of the digital array antenna system will be D_{DBF} higher than that of the analog phased array antenna.

There are two commonly used representations of the dynamic range of RF receivers, namely 1 dB compression point dynamic range DR_{-1} (receiver linear dynamic range) and undistorted signal dynamic range $\mathrm{DR}_{\mathrm{SFDR}}$.

(1) The 1 dB compression point dynamic range DR_{-1} is defined as the ratio of the input signal power to the minimum detectable signal or equivalent noise when the receiver output power is large enough to produce 1 dB gain compression, i.e.

$$\mathrm{DR}_{-1} = \frac{P_{i-1}}{P_{i\min}} = \frac{P_{o-1}}{\mathrm{G}P_{i\min}} = \frac{P_{o-1}}{\mathrm{GS}_{\min}} = \frac{P_{o-1}}{\mathrm{GkT_0NFBM}} \tag{8.46}$$

where P_{i-1} is the signal power at the receiver input when 1 dB compression is reached, P_{o-1} is the receiver output signal power when 1 dB compression is reached, G is the receiver gain, NF is the receiver noise figure, B is the receiver bandwidth, $M = 1$ is the identification factor, T_0 is the thermodynamic temperature. Equation (8.46) can also be expressed as

$$\mathrm{DR}_{-1} = P_{o-1(\mathrm{dBm})} + 114 - \mathrm{NF}_{(\mathrm{dB})} - 10\lg B_{(\mathrm{MHz})} - G_{(\mathrm{dB})}(\mathrm{dB}) \tag{8.47}$$

$$\mathrm{DR}_{-1} = P_{i-1(\mathrm{dBSm})} + 114 - \mathrm{NF}_{(\mathrm{dB})} - 10\lg B_{(\mathrm{MHz})}(\mathrm{dB}) \tag{8.48}$$

(2) The dynamic range of the distortion-free signal $\mathrm{DR}_{\mathrm{SFDR}}$ refers to the ratio of the maximum input signal power of the receiver to the third-order intermodulation signal when the third-order intermodulation of the receiver is equal to the minimum detectable signal, i.e.

$$\mathrm{DR_{SFDR}} = \frac{P_i\mathrm{sf}}{P_{i\min}} = \frac{P_o\mathrm{sf}}{\mathrm{G}P_{i\min}} \tag{8.49}$$

Figure 8.34 is the graphic method of the dynamic range of the undistorted signal, where: P_3 is the third-order intermodulation power level, $P_o\mathrm{sf}$ is the maximum signal power output by the receiver when the third-order intermodulation signal of the receiver is equal to the minimum detectable signal. The third-order intermodulation intercept point is the intersection point of the fundamental frequency signal input–output relationship curve and the third-order intermodulation and input signal relationship curve. P_1 is the power of the third-order intercept point of the receiver. Ignoring the phase distortion and amplitude distortion caused by high-order components and nonlinearity, $\mathrm{DR_{SFDR}}$ can be represented as:

$$\mathrm{DR_{SFDR}} = \frac{2}{3}(P_1 - P_{o\min}) = \frac{2}{3}(P_1 - P_{i\min} - G) \tag{8.50}$$

$$P_o\mathrm{sf} = P_{o\min} + \mathrm{DR_{SFDR}} \tag{8.51}$$

$$P_1 = P_{o-1} + 10.65\,(\mathrm{dBm}) \tag{8.52}$$

$$\begin{aligned}\mathrm{DR_{SFDR}} &= \frac{2}{3}(P_{o-1} - P_{i\min} - G + 10.65)\\ &= \frac{2}{3}(P_{o-1} + 114 - \mathrm{NF} - 10\lg B - G + 10.65)\\ &= \frac{2}{3}(\mathrm{DR}_{-1} + 10.65)\end{aligned} \tag{8.53}$$

where $\mathrm{DR_{SFDR}}$ is the dynamic range of the undistorted signal, P_1 is the third-order intercept point power of the receiver, $P_{i\min}$ is the third-order intermodulation signal power of the receiver, P_{isf} is the maximum input signal power of the receiver, G is the receiver gain, P_{osf} is the maximum signal power output by the receiver.

When designing the dynamics of the digital receiver, the dynamics of the receiver is required to match the dynamics of the radar echo signal entering the receiver. That is, the dynamics of the receiver analog RF channel match the dynamics of the receiver input signal, and at the same time, it is required that the dynamics of the RF channel also match the dynamics of the ADC. The RF digital receiver's dynamics and sensitivity are interrelated parameters and restrict each other. They are optimized comprehensively by rationally allocating channel gain, selecting analog device indicators, and selecting ADC indicators.

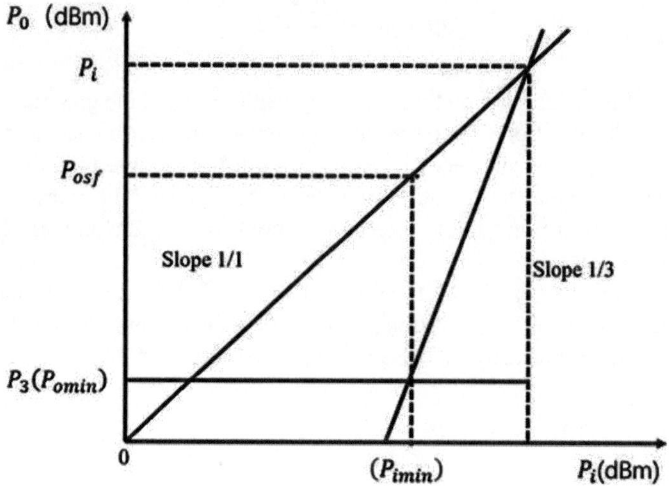

Fig. 8.34 Illustration of the dynamic range of an undistorted signal

The ideal ADC dynamic range is expressed as

$$\mathrm{DR}_{\mathrm{ADC}} = 10\lg\frac{P_{\max}}{P_{\min}} = 10\lg\frac{2^{2N}Q^2/8}{Q^2/8} = 20N\lg 2 = 6N(\mathrm{dB}) \tag{8.54}$$

The dynamic matching of the RF front-end and the ADC dynamics means that when the receiver gain is designed, the maximum input signal will not saturate the ADC. At the same time, the minimum signal input can be fully quantized by the ADC after being amplified by the RF front-end gain without degrading the NF of the receiver. Improving the large linear dynamic range of the receiver can reasonably allocate the gains of the receiver at all levels and select devices with a large dynamic range.

Digital array antennas generally have multiple working modes, corresponding to multiple instantaneous signal bandwidths, and sometimes wide and narrowband compatible working modes (such as target imaging recognition, target tracking, search, etc.). To simplify the receiver design, the receiving channel is often designed according to the widest bandwidth. In the broadband mode, the sensitivity and dynamic requirements of the system are guaranteed, and the sensitivity and instantaneous dynamics of the narrowband working mode can be obtained through subsequent multi-rate signal processing. This requires that the output I/Q signal of digital signal processing must reach a sufficient noise level [14]. This wideband and narrowband integrated receiver design can obtain high sensitivity and large linear dynamic range simultaneously, which is different from traditional analog receivers.

8.3.5 Example

Multi-channel RF digital receivers are generally realized by the integration method of multi-channel parallel RF sampling ADC, high-performance FPGA, and high-speed optical transmission interface. Figure 8.35 is a functional block diagram of a 16-channel digital receiver.

The digital receiver mainly includes bandwidth suppression low-pass filter, IF/RF sampling ADC, digital down-conversion (DDC) processing based on FPGA, optical fiber interface, clock circuit, and power supply circuit. Among them, a surface-mounted low-pass filter based on LTCC technology is used to limit the analog input signal bandwidth of the ADC to suppress out-of-band interference. The ADC chooses eight channels parallel 14-bit 125 MSPS sampling rate, the sampling clock is 120 MHz, and the center frequency of the RF is 270 MHz, which satisfies the optimal band-pass sampling theorem. Therefore, the design requires a 5 MHz bandwidth signal and a single tone signal or five 1 MHz bandwidth signals simultaneously within the instantaneous 20 MHz bandwidth. The main design specifications of the eight-channel parallel ADC are:

(1) Number of parallel ADC channels: 8;
(2) Maximum sampling rate: 125 MSPS;
(3) Resolution: 14bit;
(4) Signal-to-noise ratio SNR $\geq$ 71 dBFS (input frequency 270 MHz, sampling rate 125 MHz);
(5) Spurious free dynamic range SFDR $\geq$ 75 dBFS (input frequency 270 MHz, sampling rate 125 MSPS);
(6) Input analog bandwidth:800 MHz (input signal amplitude 1/2 V_{p-p} optional);
(7) Power supply: 1.8 V/1.12 Wwith built-in single power supply network;

The basic considerations for DDC design are as follows:

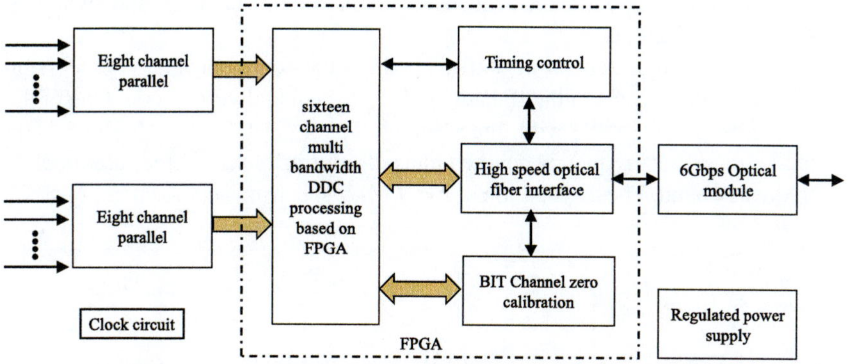

Fig. 8.35 16-channel digital receiver functional block diagram

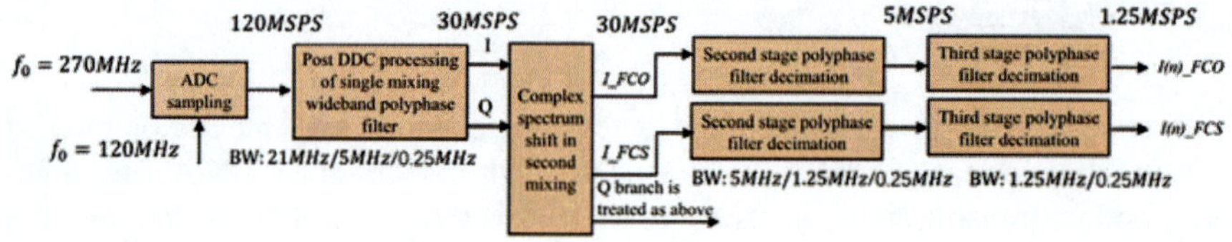

Fig. 8.36 Functional block diagram of multi-sampling rate, multi-bandwidth, multi-channel DDC algorithm

(1) The intermediate frequency is 270 MHz, the sampling rate is 120 MHz, and the intermediate frequency bandwidth is 20 MHz, satisfying the optimal sampling theorem;
(2) There are three instantaneous signal bandwidths: 5/1/0.2 MHz, corresponding to two baseband signal sampling rates: 5/1.25 MHz. At the same time, within the 20 MHz IF bandwidth, frequency division multiplex five sub-band signals with a bandwidth of 1 MHz or only one signal with a bandwidth of 5 MHz;
(3) A single FPGA simultaneously implements 16-channel parallel DDC processing.

The minimization of filter order, multi-sampling rate, multi-bandwidth, and multi-channel DDC is realized by using a second-order digital mixing and three-stage filter decimation structure. The functional block diagram of a single-channel DDC is shown in Fig. 8.36.

The functions of each module are:

(1) The primary mixing broadband polyphase filtering is followed by DDC processing to realize the digital down-conversion processing of the 20 MHz broadband intermediate frequency signal. When the system works in 5 MHz, 1 MHz, and 0.2 MHz bandwidth, the corresponding bandwidth of the filter is 5 MHz, 1.25 MHz, and 0.25 MHz, respectively. The signal filtering processing with different bandwidths is achieved by external control signal and switching the filter coefficients;
(2) Secondary complex spectrum shifting realizes the down-conversion processing of 5 sub-band signals in the IF band under 1 MHz bandwidth mode. One of the subbands is compatible with three bandwidths of 5/1/0.2 MHz. In the 5 MHz mode, its center frequency can be changed arbitrarily within the intermediate frequency band. The distribution of five 1 MHz spectrums is shown in Fig. 8.37;

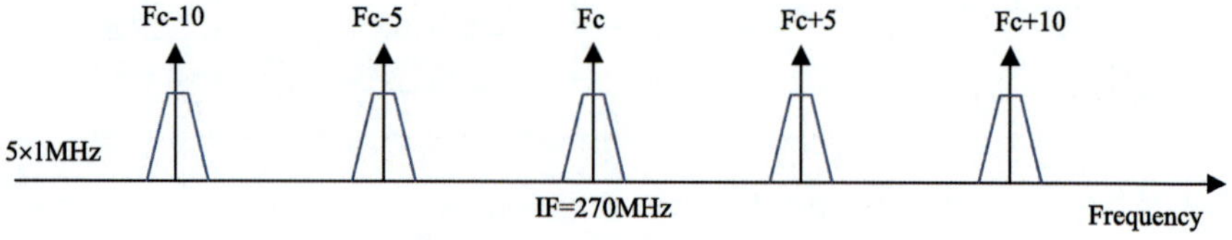

Fig. 8.37 Distribution of signal spectrum in the IF band in 1 MHz bandwidth mode

(3) The second stage of polyphase filter decimation realizes the output of the 5MSPS sampling rate signal. The filter coefficients are transformed according to the working bandwidth to adapt to different bandwidth signal filtering, including three types: 5/1.25/0.25 MHz;
(4) The third polyphase filtering and decimation stage realizes the output of a 1.25 MSPS sampling rate signal. The filter coefficients are transformed according to the working bandwidth to adapt to different bandwidth signal filtering including two kinds of 1/0.25 MHz.

5 MHz, 1.25, and 0.25 MHz bandwidth filter simulation parameters are listed in Tables 8.1, 8.2 and 8.3.

The resources of the DDC algorithm and related modules implemented in FPGA are listed in Table 8.4.

The photograph of the 16-channel RF digital receiver is shown in Fig. 8.38.

The test results of the digital receiver in actual operation are shown in Fig. 8.39.

Table 8.1 Two-stage filter parameter setting of 5 MHz bandwidth

Parameter	FIR1	FIR2
Filter order	23	47
Passband cutoff frequency (MHz)	5.0	5.0
Passband ripple (dB)	0.5	0.2
Stop band suppression (dB)	75	75
Stopband attenuation rate (dB)	55	45

Table 8.2 Two-stage filter parameter setting of 1.25 MHz bandwidth

Parameter	FIR1	FIR2	FIR3
Filter order	23	47	95
Passband cutoff frequency (MHz)	21	1.25	1.25
Passband ripple (dB)	0.5	0.2	0.4
Stop band suppression (dB)	80	85	80
Stopband attenuation rate (dB)	83	50	5

Table 8.3 Two-stage filter parameter setting of 0.25 MHz bandwidth

Parameter	FIR1	FIR2	FIR3
Filter order	23	47	95
Passband cutoff frequency (MHz)	0.25	0.25	0.25
Passband ripple (dB)	0.5	0.2	0.4
Stop band suppression (dB)	80	95	90
Stopband attenuation rate (dB)	102	60	15

Table 8.4 FPGA implementation of 16-channel parallel DDC algorithm module

FPGAmodel	EP4SGX180FF35I4
Multiplier (18 × 18)	800 (87%)
Logical unit (LE)	93,403 (85%)
Maximum operating clock speed (MHz)	148
Memory (Mb)	5.5 (49%)

Fig. 8.38 Photograph of multi-channel RF digital receiver

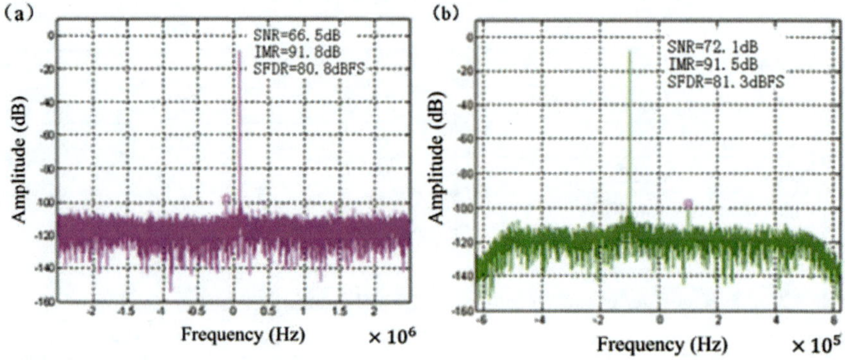

Fig. 8.39 DDC output (frequency offset 0.1 MHz) test results, **a** 5 MHz bandwidth, **b** 1 MHz bandwidth

8.4 Frequency Source

A frequency source is also called a frequency synthesizer. One or more oscillators with high-frequency stability and precision are used as the frequency reference. Through the linear operation in the frequency domain, various clock frequencies required by the digital array antenna are generated, including local oscillator signals, sampling clock signals, waveform generation clock signals, and reference clock signals used by system timers.

Compared with the conventional active phased array antenna, the outstanding advantage of the digital array antenna is the flexible beam steering, including fast beam forming and dynamic multi-beam formation. Multi-channel signal synthesis obtains a high signal-to-noise ratio to achieve high sensitivity and large dynamics

[15]. The signal-to-noise ratio is closely related to the frequency source. In the digital array antenna, it is generally expected that the phase of the local oscillator is synchronous and the clock signal between the antenna elements is coherent, while the noise is non-coherent.

8.4.1 Noise Coherence

Generally, a fully coherent system is used to achieve phase noise cancellation in order to improve the receiving signal-to-noise ratio of the digital array antenna system, especially to improve the radar's ability to detect moving targets in strong clutter phase noise, and reduce the frequency stability requirements of the digital array antenna system. Assuming that the time for the electromagnetic wave to go back and forth to the target is t_d, the equivalent block diagram of the centralized noise coherent frequency source radar is shown in Fig. 8.40.

From the perspective of the frequency domain, the transfer function of the electromagnetic wave round-trip plus coherent detection is

$$H(s) = 1 - e^{-st_d} = 1 - \cos\omega_d t_d + j\sin\omega_d t_d \tag{8.55}$$

where ω_d is the carrier frequency deviation of the signal, i.e., the frequency deviation of the phase noise, t_d is the echo time of the target.

Since phase noise is a power spectral density, there is

$$S_\varphi^*(f) = |H(s)|^2 s_\varphi(f) = 2(1 - \cos\omega_d t_d)s_\varphi(f) \tag{8.56}$$

The phase noise spectral density of the system output signal is changed from $s_\varphi(f)$ to $(1 - \cos\omega_d t_d)s_\varphi(f)$. The phase noise amplitude of the output signal presents a periodic relationship in the Fourier frequency offset of the carrier frequency, which has both cancellation and growth, and the period is the target distance t_d. Therefore, when the clutter is close, the noise coherence can greatly improve the signal-to-noise ratio of the signal, which is why ground radar usually adopts the full coherence system to improve ground clutter suppression. However, for moving targets at different distance points, the radar improvement factor of this system is deteriorated. When the target is far away, such as airborne radar, the cancellation effect of ground clutter

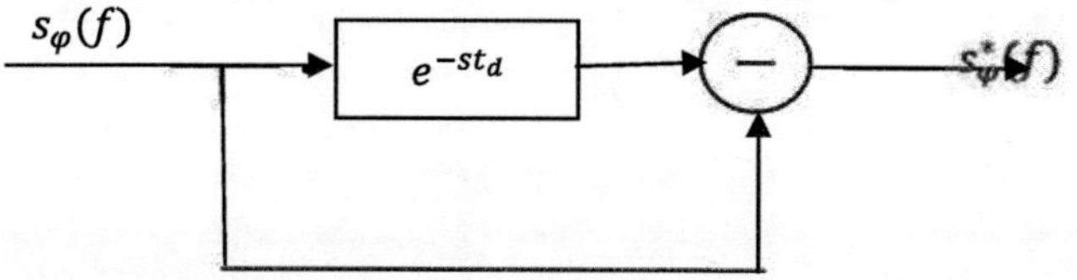

Fig. 8.40 Noise full coherent radar detection principle

will be greatly reduced. For example, at 45 km, τ is 300 μs, i.e., the noise conversion frequency is 330 kHz, in the first cycle, which is the main energy area of radar phase noise, part of the phase noise outside the carrier frequency of 556 Hz is increased instead. Therefore, the benefits of spatial correlation of phase noise are reduced for the airborne radar for target detection in the background of strong ground clutter or sea clutter.

Due to the arbitrary control of the beam of the digital array antenna and the ultra-high sensitivity, it has a strong ability to detect fragile moving targets under strong clutter background and is widely used in modern radars. However, to take advantage of the theoretical combination effect of the digital array antenna to obtain the technical advantages of high SNR, the noise involved must be non-coherent. Otherwise, the expected combination effect will not be achieved. For the amplifier noise of the transceiver channel and the quantization noise of AD and DA conversion, the non-coherent assumption can be satisfied, but for the phase noise of the frequency source system, its characteristics are much more complicated.

Regardless of the characteristics of the target itself, there are two main factors that limit the radar signal-to-noise ratio. The first one is the noise in the receiver, i.e., the receiver channel noise and the frequency source phase noise. The second one is the transmission signal noise, i.e., the transmission amplification link noise and the frequency source phase noise. When the echo signal amplitude is constant, the maximum signal-to-noise ratio of the conventional radar echo is mainly limited by the receiver's noise figure. On the other hand, the thermal noise between each channel of the digital array radar is non-coherent, and noise suppression can be achieved by synthesis. Therefore, the maximum signal-to-noise ratio of the echo is limited by the phase noise of the frequency source and the signal-to-noise ratio of the transmitted signal. For example, digital array antennas use noise-incoherent distributed frequency sources in anticipation of better signal-to-noise ratio benefits.

8.4.2 Frequency Source System

Theoretically, the antenna can improve the signal-to-noise ratio of the transmitted signal and the signal-to-noise ratio and dynamic range of the received signal through multi-channel signal vector synthesis. For an N-element linear array antenna, as shown in Fig. 8.41, the far-field pattern of the antenna can be expressed as

$$S = \sum_{i=0}^{n-1} I_i e^{jikd(\sin\theta - \sin\theta_0)} \tag{8.57}$$

$$I_i = A_{\mathrm{mp}} e^{j\varphi_0} \tag{8.58}$$

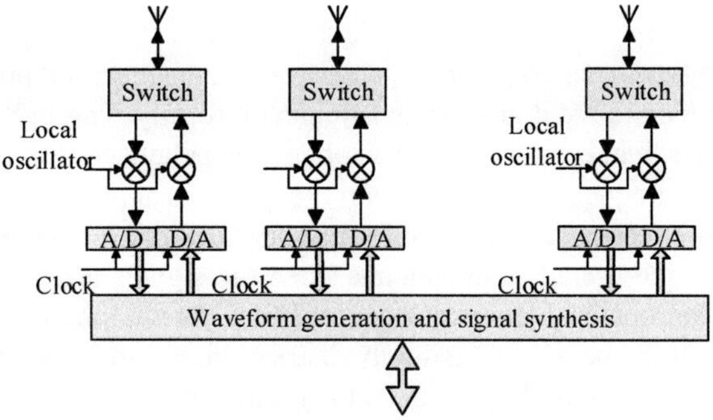

Fig. 8.41 1D linear digital array block diagram

where I_i is the unit excitation weight, d is the antenna element spacing, θ_0 is the antenna beam pointing, A_{mp} is the element excitation amplitude, φ_0 is the initial phase of each excitation element.

Through space synthesis, the digital array antenna signal and the phase noise mathematical model of the signal are

$$S_n = A_n e^{jwt - jPs_n}$$

$$N_n = B_n e^{jwt - jPn_n} \tag{8.59}$$

Neglecting the time-harmonic characteristics of electromagnetic waves, and only analyzing the antenna beam pointing direction, i.e., when $\theta = \theta_0$, the digital array antenna signal and the phase noise of the signal are

$$\begin{cases} G_s = \sum_{i=0}^{n-1} A_i e^{-jPs_i} \\ G_N = \sum_{i=0}^{n-1} B_i e^{-jPn_i} \end{cases} \tag{8.60}$$

where G_s is the signal gain, A_i is the actual excitation amplitude and phase of the i-th unit signal, Ps_i is the actual excitation phase of the i-th unit signal, G_N is the phase noise gain, B_i is the actual excitation amplitude and phase of the i-th unit phase noise, Pn_i is the actual excitation phase of the i-th unit phase noise.

For an N-element linear array antenna, if $A_i = 1$ is uniformly distributed, the initial phases are the same, the signals are coherent, and the gain of space signal synthesis is

$$G_s = N \tag{8.61}$$

If the signals are non-coherent, the gain of spatial signal synthesis is

$$G_s = 0 \tag{8.62}$$

Since the centralized frequency source is usually composed of a reference source and a linear operator [16], it is a coherent signal if it provides the local oscillator and clock to the digital array antenna uniformly. Through spatial signal synthesis, both the signal and the noise can obtain the benefits of Eq. (8.61). If the linear operator of the frequency source is configured on each antenna element or antenna sub-array, and the reference source uniformly provides the reference signal to the linear operator of the digital array antenna, then the signals are coherent. As a result, the signal can obtain the benefit of Eq. (8.61). For the phase noise, the noise caused by the linear operator is randomly distributed within $-180°$ to $180°$, which influences the coherence of the signal, that is to say, the phase noise is non-coherent, and the array space synthesis gain cannot be formed.

The main issues involved in the digital array antenna frequency source are the selection of coherent noise system and noise non-coherent system, noise non-coherent design of local oscillator source, sampling clock synchronization and transmission of high stability reference source, etc. It should be noted that the non-coherent here refers to its phase noise, while the signal must be coherent, and the phase inconsistency of the antenna array elements affects the synthesis effect and beam pointing accuracy. If the sampling clock phase changes seriously, the system's timing will be disordered, so the phase of the distributed source of the digital array antenna must be consistent. Otherwise, the digital array antenna cannot form the expected beam.

8.4.3 Distributed Frequency Source Characteristics

In order to obtain the desired synthesis effect for the digital array antenna, the distributed frequency source needs to meet the stable synchronization of signal amplitude and phase and realize the non-correlation of phase noise. Therefore, there are two main concerns about the distributed frequency source. The first one is the phase consistency between the signals of each unit and the avoidance of multi-phase phenomenon in the state of power-on and frequency agility. The second one is the phase stability between multiple channels, especially the influence of temperature changes on the phase between multiple channels, and phase noise decoherence performance.

Frequency synthesis mainly includes direct and indirect synthesis (with phase-lock as the core). The former is directly synthesized as an open ring structure. Phase consistency depends on the state of the internal counting divider of the linear arithmetic unit. Phase stability depends on the delay stability of the synthesis circuit of the linear operator (mainly composed of active non-linear devices and filters). The time delay changes with temperature and input power (non-linear active circuit), and different circuit units are distributed, so it is difficult to ensure the consistency of the state of the internal counting frequency divider and the consistency of time delay changes. Taking a 440 MHz directly synthesized distributed sampling clock as an example, the experimental results of the phase fluctuation characteristics between the two modules are shown in Fig. 8.42a. The phase fluctuation between the two modules varies greatly by about 10° with time, and the stabilization time is long.

The phase-locked loop is a phase-closed-loop feedback system. For a second-order type 2 system, the phase difference between the output signal and the steady state of the reference source is 0. Therefore, the phase-locked distributed frequency source of the same reference source is used so that the phase between the units will not change due to changes in operating temperature, feedback coefficient, etc., and good phase synchronization can be achieved. Taking the distributed sampling clock of 800 MHz phase-locked synthesis as an example, the phase stability between the two units of the second-order type 2 phase-locked synthesis source is studied. The experimental results of the phase fluctuation characteristics between the two modules are shown in Fig. 8.42b, the phase fluctuation between the two modules is within 1°, and the stabilization time is short.

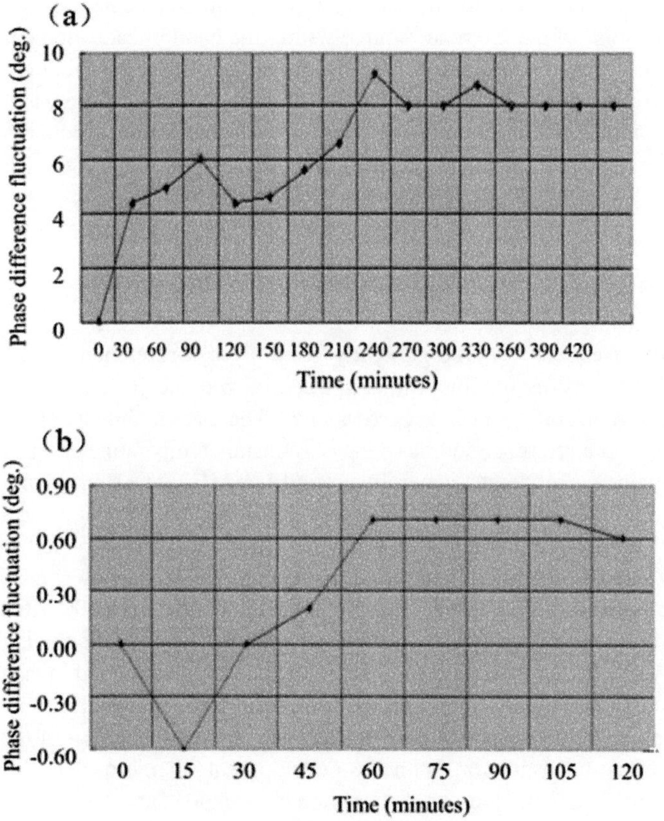

Fig. 8.42 Phase stability between distributed frequency source units, **a** direct synthesis of distributed frequency sources, **b** phase-locked synthesis of distributed frequency sources

The digital array radar of centralized frequency source adopts the form of a split feed, distributing the local oscillator and each clock to each transceiver unit. As a result, the phase noise of each array unit is coherent, and the theoretical synthesis effect of digital array radar cannot be realized. The non-coherent phase noise of each unit is a key issue in the study of distributed frequency sources.

Phase-locked synthesis phase noise mainly comes from the reference source, phase detector, and voltage-controlled oscillator. Since the local oscillator synthesis of each array unit is relatively independent, the phase detector and the noise of the voltage-controlled oscillator are non-coherent, while the phase noise of the reference source is coherent. For a general-scale digital array radar, the additional noise of the sub-synthesis unit of the distributed frequency source (i.e., phase detector noise and voltage-controlled oscillator noise) is the main contributor to the system phase noise. Through the narrow-band filtering of the phase-locked loop, the phase noise between the units of the distributed frequency source is decoherent outside the band.

With the increase in the number of digital array antenna elements, the coherence of the phase noise of the reference source within the band of each element becomes the main constraint for the performance improvement of the distributed frequency source. Therefore, noise decorrelation of the reference source is the key to distributed frequency sources, which mainly includes filter decorrelation, frequency division decorrelation, and $\Sigma - \Delta$ noise decorrelation [17].

8.4.4 Distributed Frequency Source Implementation

A distributed frequency source generally consists of a reference source, a linear operator, and a transmission line similar as a centralized one. Reference sources have a significant impact on synthesizer performance. There are mainly two types of high-performance radar reference sources: One is a constant temperature crystal oscillator, and its frequency is moderate, generally around 100 MHz. The short-term frequency stability is high, the phase noise is relatively small, and the frequency accuracy and long-term stability requirements are high. Low-frequency crystal oscillators (such as 10 MHz) are often applied in this case. Its near-carrier frequency equivalent noise is better than that of higher frequency crystal oscillators (such as 100 MHz), and the long-term frequency stability is better. The second is the atomic clock, its frequency is relatively low (generally around 10 MHz), the phase noise is good in the ultra-near zone, the cost is relatively high, and the long-term frequency stability is high, but the phase noise in the equivalent mid-to-far zone is worse than that of the constant temperature crystal oscillator. With the development of microwave photoelectric technology, more and more research efforts are being paid to photoelectric oscillators. Its frequency is high (generally around 10 GHz), and the phase noise in the middle and far regions is one or two orders of magnitude lower than that of traditional frequency synthesizers.

The linear operator of the distributed frequency source needs to be configured in the digital array antenna element or sub-array, and the miniaturization of the linear operator has attracted more and more attention. On the one hand, with the development of semiconductor technologies such as CMOS, MMIC, and LTCC technologies, chips such as phase-locked loops, frequency dividers, phase detectors, and frequency multipliers in the frequency source are increasingly integrated, and the processes are becoming increasingly mature. All of the above devices can be integrated with a high density. On the other hand, the advancement of multilayer wiring packaging technology, thick film, and co-fired ceramic technology can realize embedded passive devices, such as capacitors and inductors, which can greatly reduce the volume and power consumption of the linear operator module of the entire frequency source.

8.5 Applications

Figure 8.43 is a block diagram of a lightweight broadband digital array antenna, and Fig. 8.44 is a photo of the actual antenna. It is an integrated architecture type of antenna element, light material reflector, digital array module (DAM), photoelectric blind matching, optical fiber transmission, high-speed digital and data processing, and comprehensive design of antenna machine, electrical and thermal multi-physical parameters.

The working parameters of the lightweight broadband digital array antenna are: the working frequency is 995–1400 MHz, The polarization mode is vertical linear polarization, The antenna elements are 4 (elevation) $\times$ 56 (azimuth), The aperture size is 6076 mm (length) $\times$ 800 mm (width), The antenna system weighs more than 220 kg. The antenna system's full-band delay calibration accuracy is better than 5 ns, the amplitude calibration accuracy is better than ± 0.5 dB, and the phase calibration accuracy is better than $\pm 5°$. In addition, the delay calibration accuracy of the antenna system is better than 3 ns in the 10% frequency band, the amplitude calibration accuracy is better than ± 0.3 dB, and the phase calibration accuracy is better than $\pm 2°$.

The typical test patterns of lightweight broadband digital array antennas are shown in Figs. 8.45, 8.46 and 8.47. During the test, the elevation beam of the digital array antenna was pointed at $-6°$. The antenna transmits a beam that adopts equal amplitude weighting, and the antenna receives a difference beam that adopts -30 dB amplitude weighting.

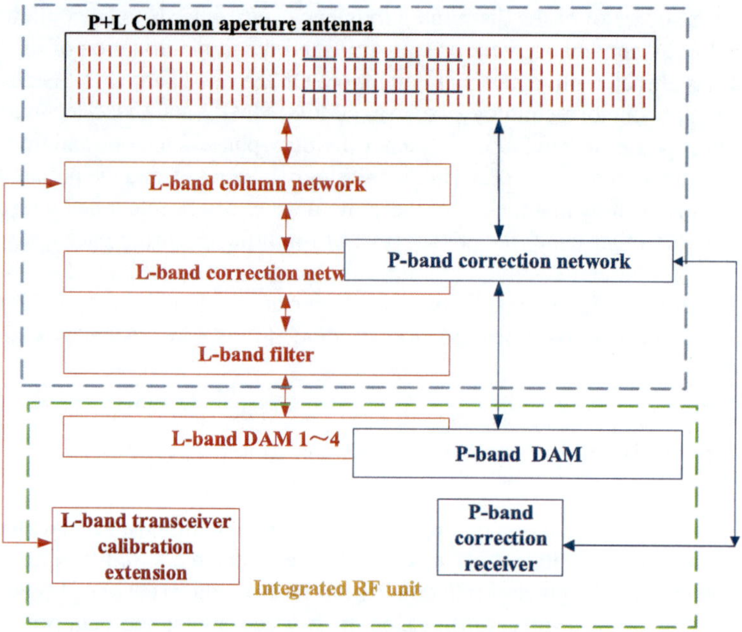

Fig. 8.43 Block diagram of lightweight broadband digital array antenna

Fig. 8.44 Photograph of lightweight broadband digital array antenna

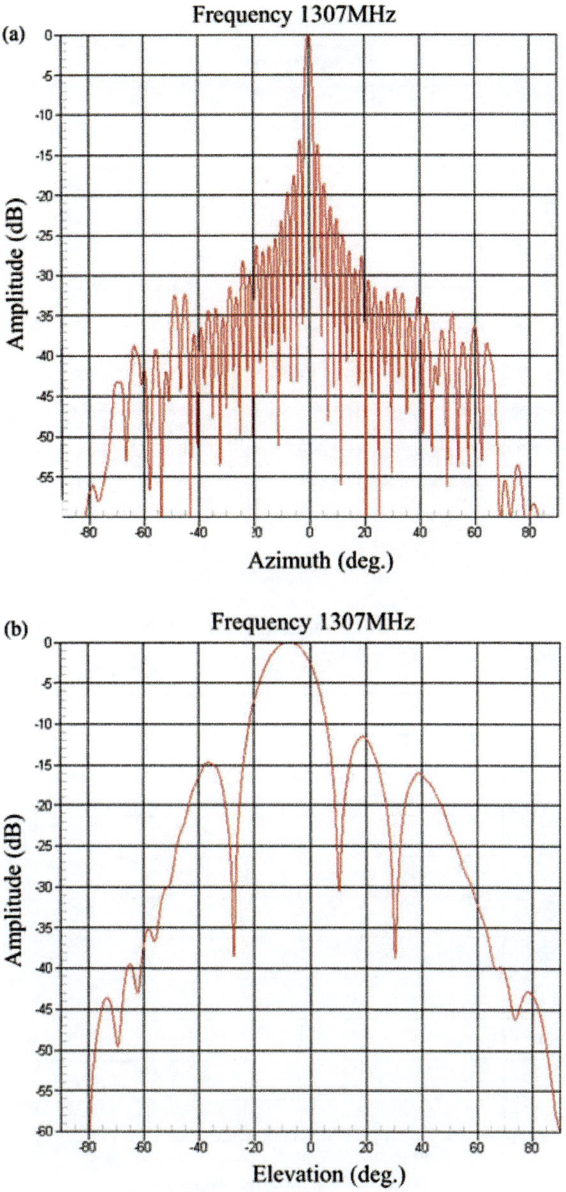

Fig. 8.45 Typical radiation pattern of antenna, **a** azimuth, **b** elevation

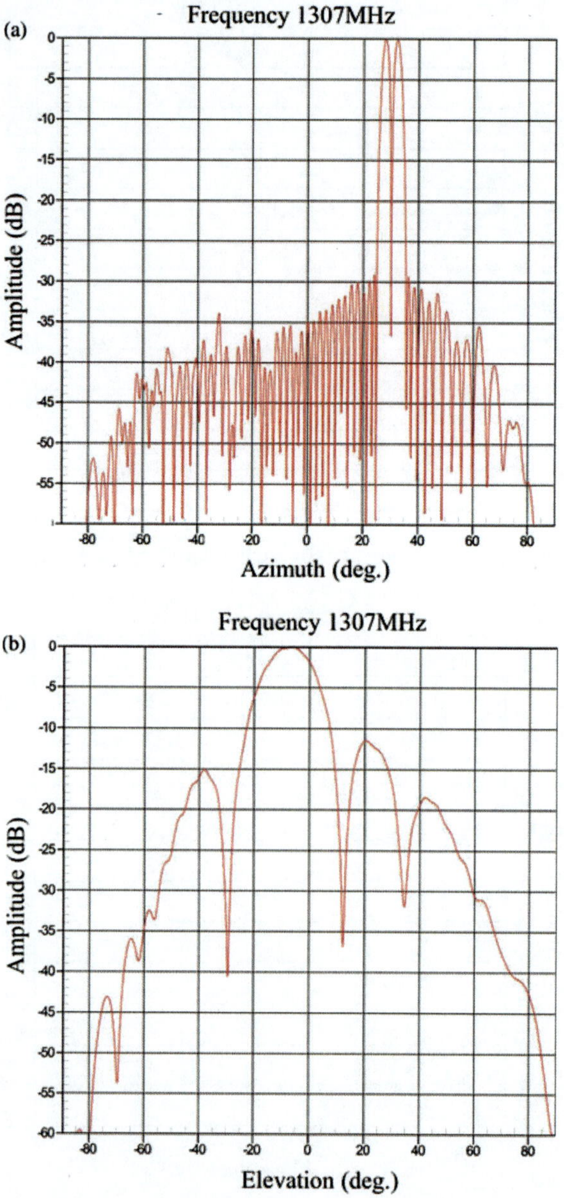

Fig. 8.46 Difference beam pattern of antenna reception, **a** azimuth, **b** elevation

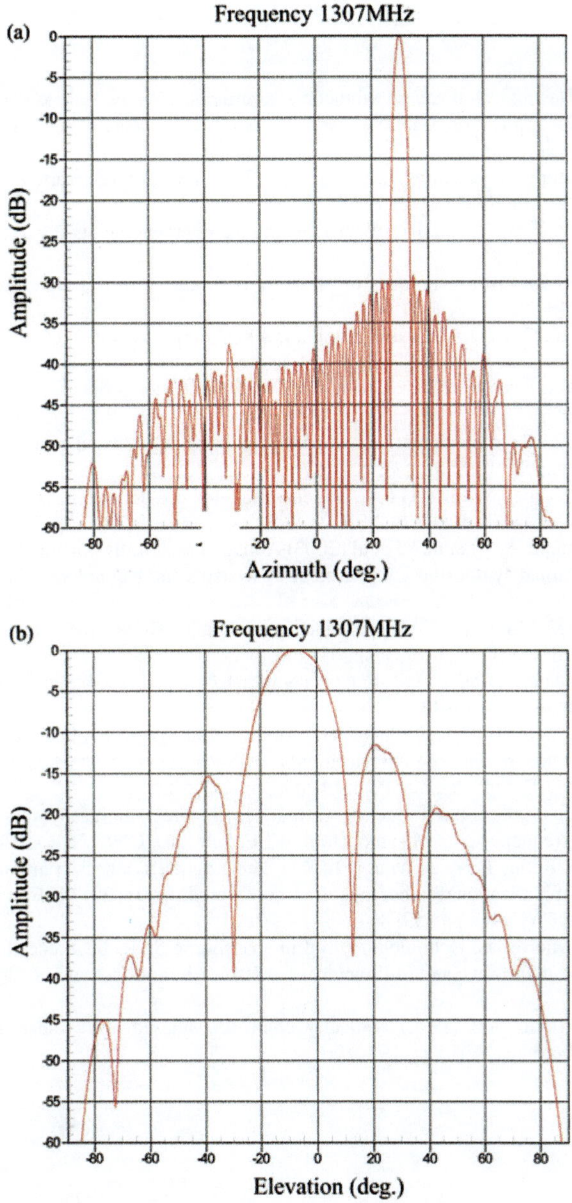

Fig. 8.47 Sum beam pattern of antenna reception, **a** azimuth, **b** elevation

References

1. Lu J (2019) Design techniques of synthetic aperture radar. Wiley, Hoboken, NJ
2. Wen Y (1997) Application of DDS in radar system. In: Symposium of DDS technologies and applications. Chines Institute of Electronics, Hefei, pp 44–47
3. Jiaguo L, Manqin W, Xueming J, Zhengxin F (2003) Active phased-array antenna based on DDS. Acta Electron Sinica 31(2):199–202
4. Chen Z, Zhang Y, Bao Q (2010) Advance in digital array radar and its key technologies. J Natl Univ Def Technol 32(6):1–7
5. Guo C (2009) Receiver techniques of phased array radar. National Defense Industry Press, Beijing, pp 246–269
6. Wu X, Wang B (2015) Matlab simulation and FPGA implementation of wideband DDC and resample processing. Modern Electron Tech (23):6–9
7. Zhang W, Tan J, Chen H (2008) Application of DDS in digital array radar. Radar Sci Technol 6(6):467–471
8. Jin X, Tan J (2004) Implementation and performance of DDS-Based general radar waveform generator. Radar Sci Technol 2(3):183–187
9. Tan J, Sun J, Zhang W, Chen H (2007) Generation and pre-distortion compensation of high-resolution LFM signals. Radar ECM 3(3):29–32
10. Jiaguo L, Manqin W, Xueming J et al (2003) Active phased-array antenna based on DDS. In: IEEE international symposium on phased array systems and technology. IEEE, Boston, MA, pp 511–516
11. Liu M, Wang M, Yang F (2003) Line frequency modulation signal producing and compression. J Proj Rock Missiles Guidance 23(1):205–206
12. Fei L (2004) Shenrui. Radar LFM signal generation based on direct digital frequency synthesis. Electron Technol 31(9):11–13
13. Nicholas HT, Samueli H (1987) An analysis of the output spectrum of direct digital frequency synthesizer in the presence of phase accumulator truncation. In: 41st annual frequency control symposium. IEEE, Philadelphia, Pennsylvania, pp 495–502
14. Reinhardt VS (1993) Spur reduction techniques in direct digital synthesizers. In: IEEE international frequency control symposium. IEEE, Salt Lake City, UT, pp 230–240
15. Thompson D, Kelley R, Yeary M et al (2011) Direct digital synthesizer architecture in multi-channel, dual-polarization weather radar transceiver modules. In: 2011 IEEE radar conference. IEEE, Kansas City, MO, pp 859–864
16. Lu J (2015) Microwave technologies used in spaceborne SAR. In: Proceedings of National conference on microwave and millimeter wave 2015. Chinese Institute of Electronics, Hefei, p 15
17. Forrest SR, Taylor RB (1996) Optically controlled phased array radar. In: Meeting on microwave photonics. IEEE, pp 193–196

Chapter 9
Microwave Photonic Array Antennas

9.1 Introduction

The field of Microwave Photonics emerged as a result of the rapid development and maturity of optical fiber technology, integrated optics, and semiconductor integrated circuit technology. It mainly studies a series of basic technology and application issues derived from the interaction of optical signals and microwave signals. Microwave photonic technology is an emerging technology that combines microwave technology and photonic technology [1].

From the perspective of technical application, although the radar industry has recognized the superiority of the active array antenna, it also has defects such as large system size, heavy weight, narrow operating bandwidth, large transmission loss, susceptibility to interference, and high cost. This greatly limits the potential of active phased array antennas. In view of the excellent characteristics of microwave photonics technology, radar researchers have introduced microwave photonics technology into the active array antenna system. For example, optical fibers are used as radar signal and data transmission lines, optical true delay lines are used to form optical beamformers, and optical processors are used for radar signal and data processing.

According to the functional difference of microwave photonic technology in an antenna array, microwave photonic array antenna is divided into two types: microwave photonic digital array antenna and optically controlled phased array antenna [2]. Researchers have devoted themselves to studying microwave photonic technology based on the above applications in the past few decades and applied this technology to radar, satellite communication, and electronic warfare [3–5].

J. Lu et al., *Active Array Antennas for High Resolution Microwave Imaging Radar*,
https://doi.org/10.1007/978-981-99-1475-3_9

9.1.1 Microwave Photonic Digital Array Antennas

In the microwave and millimeter wave frequency bands, the transmission loss of traditional cables will increase sharply as the frequency increases. For example, the loss of 5.8 GHz microwave signal transmission through coaxial cable is as high as 190 dB/km, and the loss in the 2.4 GHz frequency band is also 145 dB/km. However, compared with traditional transmission media such as coaxial cables or waveguides, the loss of optical fibers is much smaller [6]. For example, the transmission loss of light with a wavelength of 1550 nm in the optical fiber is only 0.2 dB/km, and the loss does not increase with the increase of the transmission signal frequency.

The digital array antenna integrates functional links such as antenna elements, Radio Frequency (RF), analog, and digital. It replaces the T/R module in the traditional phased array antenna with Digital Transmitter and Receiver (DTR) component. The antenna element, RF transceiver, and signal processing unit are integrated for high-density integration, which effectively solves the problems of transmission loss and large physical size [7]. For example, to solve the problem of large transmission loss, the digital array antenna generally uses the antenna radiation front and the Digital Array Module (DAM) to be tightly integrated. With the increase of antenna array scale and higher integration, system complexity, architecture, heat dissipation, power supply, test and maintenance, and other issues have become the bottleneck of analysis and design.

With the development of optoelectronic technology, microwave photonics will significantly impact electronic information systems [8, 9]. The microwave photonic link is used to realize the long-distance transmission of the antenna transmitting and receiving signals so that the digital array antenna is separated from the back-end digital processing and control system. With its unique advantages, microwave photonic links avoid the loss of long-distance transmission from traditional microwave links. For large array antennas, this method retains the inherent characteristics of digital arrays, such as high amplitude and phase control precision, large instantaneous dynamic range, high degree of spatial freedom, and flexible beamforming. And it has the advantages of a wide frequency band, light mass, and anti-electromagnetic interference. Moreover, the structural form of the traditional digital array antenna has been changed, and the antenna system and the feeder transceiver system have been integrated. According to the actual situation of the loading platform or the ground radar position, the antenna system and the back-end digital processing/control system installation device are flexibly set to realize the flexibility and safety of the electronic information system.

When optical links realize long-distance signal transmission, wavelength division multiplexing technology is generally used to simplify the complexity of digital array antennas. Dozens or hundreds of digital array antenna sub-arrays are integrated into a photoelectric transceiver module. Then, hundreds of optical fibers are synthesized into one optical fiber through dense wavelength division multiplexing to complete the signal transmission and long-distance transmission of tens of thousands of transceiver units in the digital array antenna.

The principle of microwave photon digital array radar is shown in Fig. 9.1. The system is composed of a microwave photon array antenna, a back-end photoelectric and a digital signal system, and a long-distance data/signal transmission optical fiber. Microwave photon array antenna comprises the passive antenna array, transceiver switch, digital receiving/transmitting components, photoelectric/electrical-optical converter, an optical multiplexer, etc. The back-end photoelectric and digital processing system consists of photoelectric/electrical-optical converters, digital down-converters, waveform generation, digital beamforming, signal processing, data processing, and system control.

In Fig. 9.1, in order to obtain the desired Signal-to-Noise Ratio (SNR), achieve the larger detection power, and the ability to detect targets under strong clutter of the high-resolution multifunction imaging radar, the digital array antenna generally does

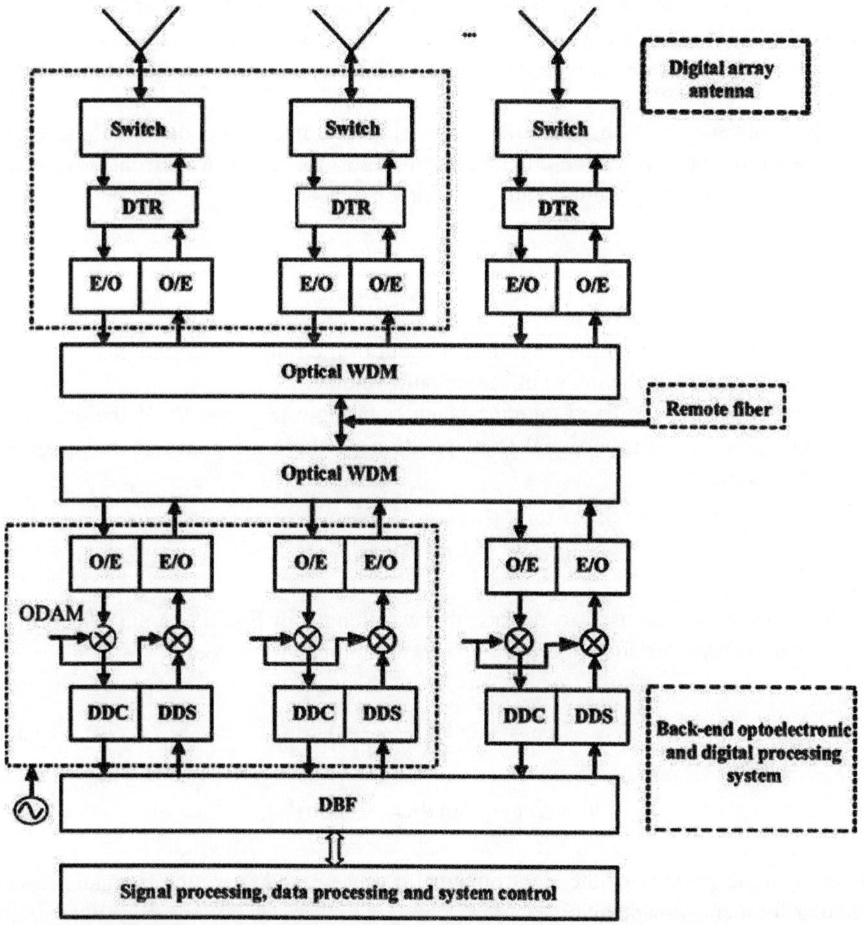

Fig. 9.1 Principle of microwave photon digital array radar

not use a centralized frequency source, but a distributed frequency source with non-correlated noise. At the same time, distributed frequency sources are also suitable for high-density integration of digital array antennas.

9.1.2 Optically Controlled Phased Array Antennas

To solve the problem of high-resolution microwave imaging of targets and improve the ability of radar to distinguish and identify targets, active array antennas must use signals with large instantaneous bandwidths. Since the traditional phased array antenna controls the phase to perform antenna beamforming and scanning, the signals have the same phase at the same frequency, but the delay time is different. This will cause the active array antenna to have a beam-pointing skew phenomenon when the antenna beam is scanning in the case of broadband signals. This is due to the aperture effect caused by the aperture transit time, which limits the instantaneous bandwidth of the signal [10–12].

The beam scanning angle of the active array antenna is to control the phase of each radiation element of the antenna according to the phase wavefront principle, and the angle of the maximum gain of the antenna beam is

$$\theta_0 = \arcsin\left(\frac{\lambda}{y}\frac{\Psi}{2\pi}\right) \tag{9.1}$$

where y is the position of the radiating element of the linear array antenna, λ is the wavelength, and Ψ is the phase of the radiating element.

In order to obtain a fixed antenna beam pointing, the phase Ψ of the antenna radiating element is proportional to its position y, where the phase of the antenna radiating element is

$$\Psi = 2\pi\frac{y}{\lambda}\sin\theta_0 \tag{9.2}$$

If the frequency is used to replace the wavelength in Eqs. (9.1) and (9.2), the narrowband characteristics of the active array antenna will be found, that is

$$\theta_0 = \arcsin\left(\frac{c}{y}\frac{\Psi}{2\pi f}\right) \tag{9.3}$$

$$\Psi = 2\pi\frac{y}{c}f\sin\theta_0 = \frac{\omega y}{c}\sin\theta_0 \tag{9.4}$$

where c is the velocity of the electromagnetic wave, f and ω are the frequency and angular frequency, respectively.

From Eq. (9.3), it can be seen that the change of frequency will lead to the change of antenna beam pointing. Equation (9.4) gives the method to solve the broadband of the antenna array. The phase amount controlled by the antenna element is proportional to the frequency, i.e., in the broadband frequency range, the time delay of the antenna radiating element should be controlled instead of the phase to realize beam scanning.

To realize the wide bandwidth angular scanning of the phased array antenna beam, the True Time Delay (TTD) line is generally used to replace the conventional phase shifter [13]. However, tens of thousands of phase shifters in the phased array antenna bring difficulties to the engineering realization. Therefore, a compromise method is to set a true delay line at the sub-array level of the phased array antenna to perform delay compensation at the sub-array level.

Traditional TTDs are made of waveguides, coaxial cables, or microstrip transmission lines. For a phased array antenna with an aperture length of 20 m, when the beam scanning angle is 60°, the maximum bit delay line length in the true delay line is about 17 m. Such a long waveguide, coaxial cable, or microstrip transmission line is very difficult regarding microwave signal insertion loss and engineering realization. Therefore, if the microwave signal is modulated onto the optical fiber, the optical fiber is used as the true delay line, called Optical True Time Delay (OTTD). Because the optical carrier frequency is extremely high, the signal bandwidth is extremely small relative to the carrier frequency, the optical link has stable transmission characteristics and, at the same time, reduces the volume and weight of the active array antenna.

A common optically controlled phased array antenna modulates the microwave signal onto the optical signal. The optical signal propagates in different lengths of optical fibers to achieve different true delays and then demodulates the microwave signal from the optical signal. Time delay control is performed on the optical link to realize antenna beamforming and scanning [14, 15]. The main feature of the optically controlled phased array antenna is that the optical delay device replaces the phase shifter, which can obtain a large instantaneous bandwidth. Optical transmission and division technology can reduce system weight, reduce volume, improve reliability, and enhance the ability to resist electromagnetic interference.

The basic working principle of the optically controlled phased array antenna is shown in Fig. 9.2. In the transmitting working state, the broadband microwave signal undergoes electro-optic conversion (E/O) and optical isolation, and the optical signal is divided into M channels by a 1: M optical splitter. M channels correspond to M sub-arrays. Each sub-array signal is transmitted to the antenna array through optical fiber after passing through OTTD and then converted into a microwave signal through optical-to-electrical conversion (O/E). After amplified through a 1:N power divider, N array elements in each sub-array are transmitted to N microwave T/R modules and then radiated from N antennas after phase shifting to realize power combination in space. The working state of receiving is similar to that of transmitting, except that the direction of signal transmission is the opposite. Finally, the signals of each antenna sub-array are synthesized into receiving beams.

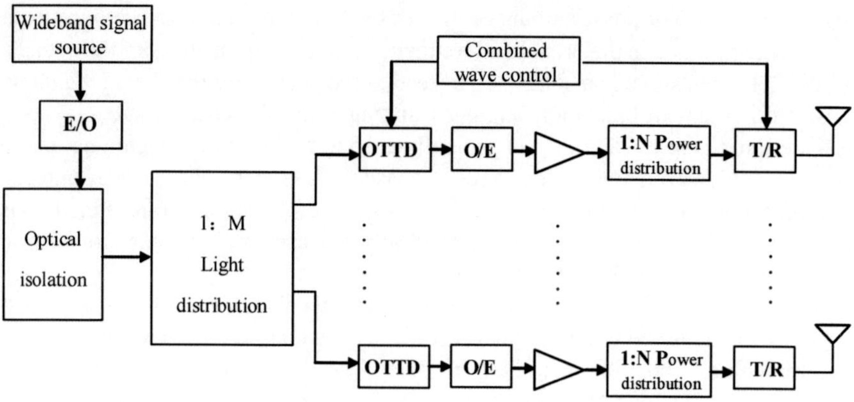

Fig. 9.2 Basic operation principle diagram of optically controlled phased array

9.1.3 *Phase Shifters and Delay Lines*

The broadband digital phase shifter usually requires that the phase shift amount of each phase shift of the phase shifter is constant within the frequency band. In this case, when the RF signal has a relatively large instantaneous bandwidth, and the antenna beam points to a direction other than the normal direction of the antenna, the beam points corresponding to each frequency component are different, which broadens the antenna beam. When the instantaneous bandwidth of the RF signal is narrow, and the operating frequency exceeds the instantaneous bandwidth of the RF signal, the antenna beam pointing will change with the operating frequency. If the direction of the antenna beam is kept constant, the phase of the antenna radiating element should also be changed correspondingly when the operating frequency is changed.

In general, the bit with the largest phase shift in the phase shifter (that is, the first bit) is π, and the *n*th bit is $\pi/2^{n-1}$ accordingly. If the bit with the smallest delay in the time delay line (that is, the first bit) is λ_0, the *m*th bit is $2^{m-1}\lambda_0$. Generally, for phased array antennas with a wide frequency band and large scanning angle, each antenna element is connected to a phase shifter, several antenna elements form a sub-array, and each sub-array is connected to a delay line. The sub-array size depends on the maximum scanning angle and antenna aperture. After selecting the appropriate size of the antenna sub-array, if the size of the antenna sub-array is appropriately reduced, the π bit of the phase shifter is moved to the delay line. That is to say, if $\lambda_0/2$ bits are increased, the performance parameters of the antenna system will remain unchanged, but the total phase shift and delay bits will be reduced.

The boundary between the phase shifter and the delay line is not fixed, and the $\pi/2$ bit of the phase shifter can also be moved to the delay line if necessary. That is to say, by adding $\lambda_0/4$ bits, the digits in the phase shifter and delay line can even be distributed in the power combination/division network of each layer. When both the phase shifter and the delay line have a linear law, the phase shift amount $\Delta\varphi$ of the smallest bit of the phase shifter does not have to be satisfied, i.e., the number of bits n of the phase shifter does not have to be satisfied. It can be determined according to the required minimum beam jump $\theta_{\min}$ and the element spacing d. As for the phase shifter and the remaining bits of the delay line, they are carried out in binary based on $\Delta\varphi$, but the maximum bit of the delay line does not necessarily have to satisfy $2^{m-1}\Delta\varphi$ (m is the sum of the number of bits of the phase shifter and delay line). It can be determined according to the required maximum scan angle $\theta_{\max}$ and the antenna aperture size L.

9.2 True Time Delay Lines

A true time delay (TTD) line is a time delay network, usually placed after the antenna radiation element and controlled according to the delay time (or phase) required for antenna beamforming or scanning. To realize the wide bandwidth angular scanning of the phased array antenna, TTD should generally be used in the phased array antenna. However, in terms of equipment simplification, TTD is usually used on the sub-array level in practice, and the phase shifter is still used in the T/R module. OTTD based on optical transmission has the advantages of low loss, small dispersion, light mass, small size, and no electromagnetic interference. Usually, there are several technical approaches to realizing OTTD [16–20].

(1) **OTTD based on electrical switching and optical fiber ring**

As shown in Fig. 9.3, OTTD is switched by an electric switch to modulate the RF signal to different lasers. Through different lengths of optical fibers, different delays are realized.

The device includes eight fiber delay lines, and the delays are t_0, $t_0 + \Delta t \sim t_0 + 7\Delta t$, thus forming a 3-bit delay structure. There is a bias switch in front of the laser diode on the delay line to control the laser diode on and off. When one of the laser diodes is turned on, the other laser diodes are turned off, the modulation signal obtains a fixed delay, and the eight optical fibers are connected to the coupler. The RF signal output with different delays is obtained through the photodetector.

(2) **OTTD based on the optical switch and optical fiber ring**

As shown in Fig. 9.4, OTTD is based on optical switch switching and optical fiber ring uses optical switches to switch the optical path. The length of the fiber completely determines the delay of this type of OTTD. Theoretically, the transmission time can be arbitrarily long. The switching time, isolation, insertion loss, etc., are mainly determined by the optical switch. The performance of the optical switch is the key to the technical performance of this type of OTTD.

(3) **OTTD based on integrated optics**

OTTD based on integrated optics integrates the optical switching switch and the time delay component (or optical waveguide) on the same substrate. Due to its small

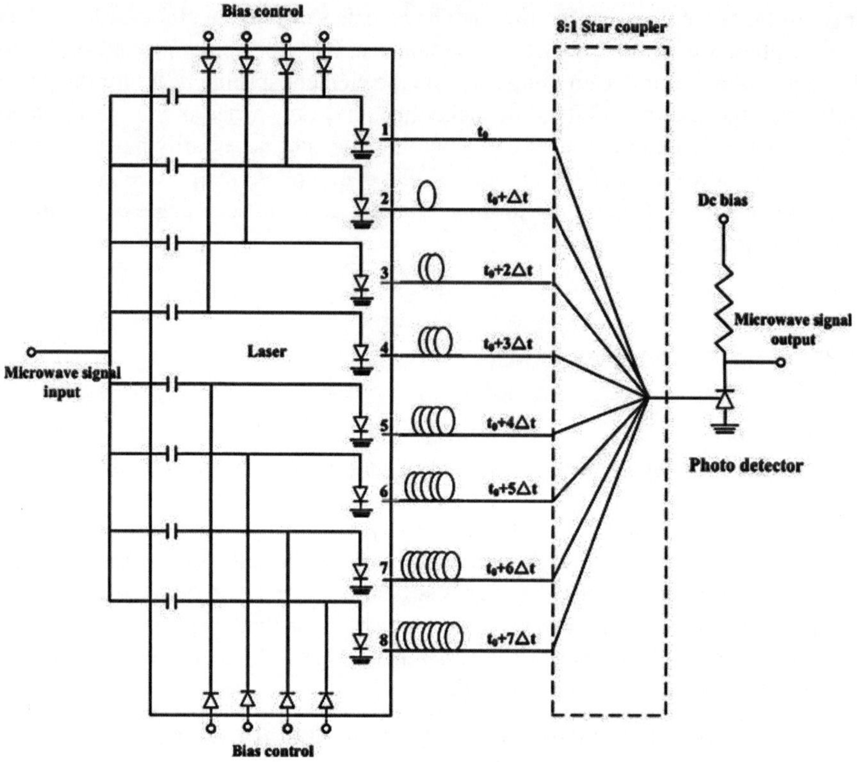

Fig. 9.3 Optical fiber delay line based on the electrical switching control

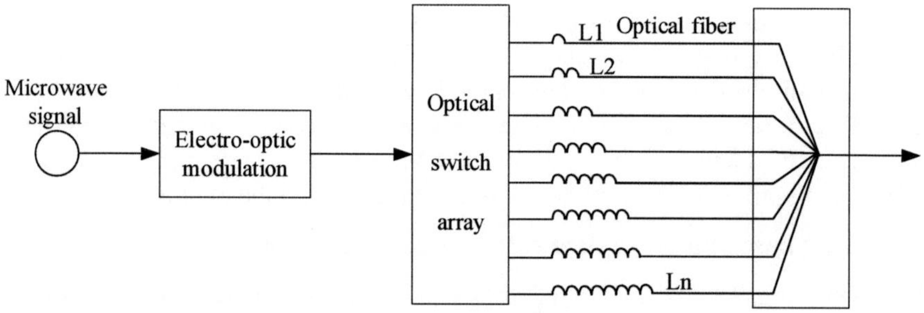

Fig. 9.4 Optical fiber delay line based on the optical switch

size, its maximum delay is usually small (hundreds of picoseconds), so this OTTD is suitable for higher-frequency applications or unit phase shifters.

The principle of the optical delay technology based on the integrated optical waveguide is the same as that of the optical fiber-based delay technology. It uses planar optical waveguide technology to replace the optical fiber with a waveguide. It

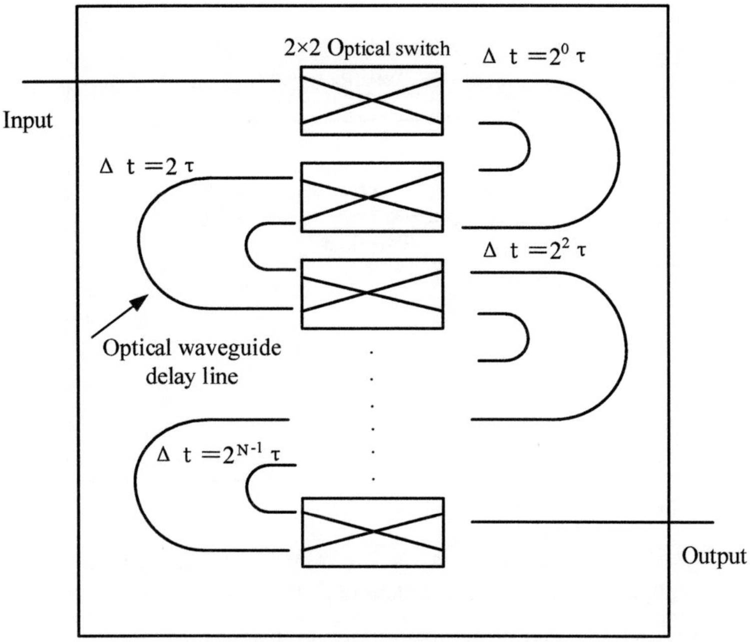

Fig. 9.5 Based on the principle of the optical waveguide delay line

integrates the separated optical switch or modulator on the same substrate to achieve a more compact optical true delay line structure, as shown in Fig. 9.5.

Compared with the delay line using optical fiber, since the integrated optical waveguide preparation process uses photolithography technology, the delay line based on the integrated optical waveguide technology can achieve a delay with an accuracy of picosecond level. Moreover, optical waveguides are also easy to manufacture on a large scale and reduce the cost of using optical delay lines. Still, their disadvantages are that the insertion loss, crosstalk, and polarization sensitivity of the device are relatively large. Therefore, the main research hotspots of optical waveguide delay lines include waveguide structure, loss, material, process, etc. Currently, the optical delay technology based on integrated optical waveguides still has a lot of room for improvement.

(4) **OTTD based on the spatial light path**

The principle of OTTD based on the spatial light path is to control the mirror group to move to different positions so that the light passes through different optical paths to achieve different time delays. The difference between free-space and optical fiber delay lines is that light propagates in space. When the light propagates in the air at speed *c*, the delay time is proportional to the length L of the light in space. Therefore, different optical paths can be realized through the mirror group's movement to achieve a controllable time delay. The principle of the delay line based on the spatial optical path is shown in Fig. 9.6.

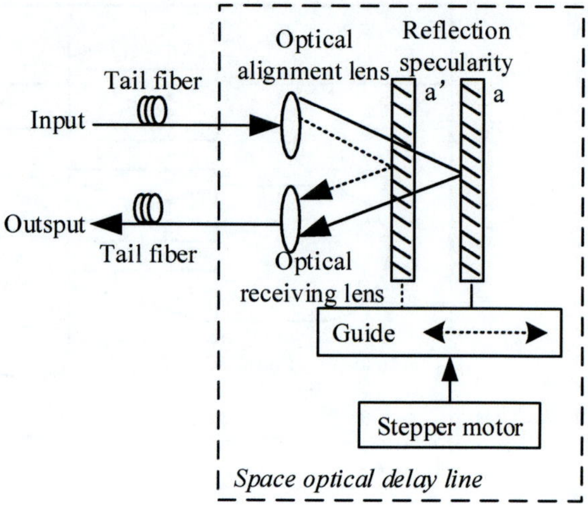

Fig. 9.6 Delay line principle based on the spatial optical path

In Fig. 9.6, the range of light delay is determined by the movable range, and the delay path is approximately twice the moving distance of the reflector.

(5) **OTTD based on tunable lasers and optical dispersion**

The basic working principle of this type of OTTD is to use different wavelengths of lasers to delay the dispersion element or use different wavelengths of lasers to have different reflection characteristics of the grating to achieve different time delays. For example, the delay line based on tunable laser and optical dispersion is shown in Fig. 9.7.

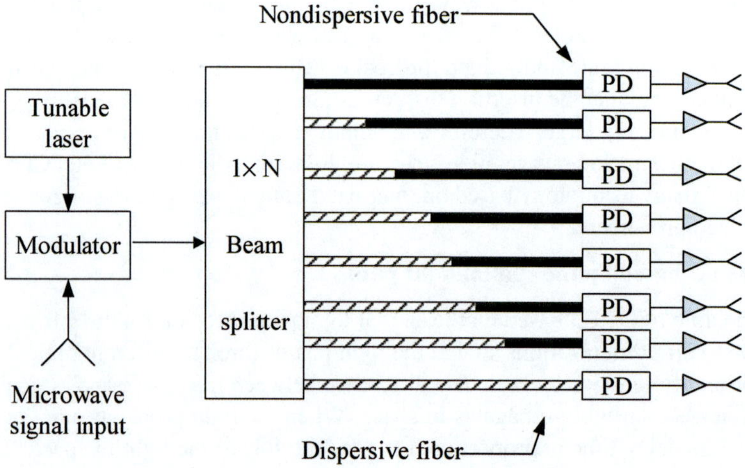

Fig. 9.7 Delay line based on tunable laser and optical dispersion

In Fig. 9.7, the light generated by the laser is modulated by the microwave signal through the electro-optic modulator. It enters different optical fibers after passing through the optical beam splitter. Each optical fiber is composed of a high-dispersion optical fiber and a zero-dispersion optical fiber, but the length ratios of the high-dispersion optical fiber and the zero-dispersion optical fiber are different in each optical fiber. Therefore, when the wavelength of the optical carrier is changed according to a certain rule, the propagation time of each optical signal will change, thereby obtaining different delay differences.

For large-scale phased array radars, due to the large aperture of the antenna array, the delay compensation required is relatively long. For example, for the L-band phased array radar with a 20 m aperture antenna, when the scanning angle is 60°, the maximum delay corresponds to a length of 17 m. Therefore, adopting OTTD based on the optical switch and optical fiber rings is a suitable choice.

The comparison of the characteristics of the five TTD lines is listed in Table 9.1.

9.3 Microwave Photonic Link

The microwave photon transmission link realizes the microwave signal/data transmission between the central station and the remote microwave photon array antenna. It has the characteristics of a wide frequency band and anti-electromagnetic solid interference ability, and the optical fiber channel also has the characteristics of low transmission loss.

9.3.1 Microwave Signal Modulation and Demodulation

The electro-optical modulation methods commonly used in Microwave Photonic Link (MPL) mainly include laser direct modulation (also called direct modulation), external amplitude modulation (abbreviated as external modulation), phase external modulation, and balanced heterodyne detection. Table 9.2 lists the comparison of performance parameters such as dynamic range and environmental stability among the three modulation methods.

In the laser direct modulation method, the microwave signal directly modulates the laser, only one laser is needed for direct modulation, and the amount of equipment is small. After years of development, the technology has matured. There are a large number of communication transmission equipment using this modulation method, but its theoretical dynamic range can only reach 105 $dB/Hz^{2/3}$, and the suitable working frequency band is generally below 18 GHz.

Laser direct modulation is to directly modulate the injection current of the laser with a microwave signal so that the output light intensity of the laser changes with the amplitude of the microwave signal. The advantages of direct modulation are simple structure and low cost, but the instability of the injected light source current will lead

Table 9.1 Comparison of the characteristics of five TTD lines

Types	Working principle	Main medium of delay	Delay characteristics	Size	Environmental adaptability
OTTD based on electrical switching and fiber optic ring	Delay is achieved through electric switching and different length of optical fiber	Optical fiber	Step delay Wide delay range Lower accuracy	Big	Good
OTTD based on optical switch and fiber ring	Delay achieved through optical switching and different length of optical fiber	Optical fiber	Step delay Wide delay range Lower accuracy	Larger	Good
OTTD based on integrated optics	Choose different waveguide paths through integrated optical switches to achieve delay	Optical waveguide	Step delay Small delay range High precision	Small	Poor
OTTD based on spatial light path	Through the movement of the mirror group, different spatial optical path delays are realized	Air	Step delay Small delay range Higher accuracy	Smaller	Sensitive to temperature and vibration
OTTD based on tunable laser and optical dispersion	Different wavelength dispersion characteristics and different lengths of dispersion fibers are used to achieve different delays	Dispersive fiber	Continuous delay Wide delay range Higher accuracy	Smaller	Poor

Table 9.2 Comparison of microwave photonic link modulation methods

Modulation mode	Theoretical dynamic range dB/Hz$^{2/3}$	Optimum operating frequency band (GHz)	Environmental stability
Laser direct adjustment	105	<20	High
External amplitude modulation	115	<40	High
Out-of-phase modulation	120	<40	Moderate

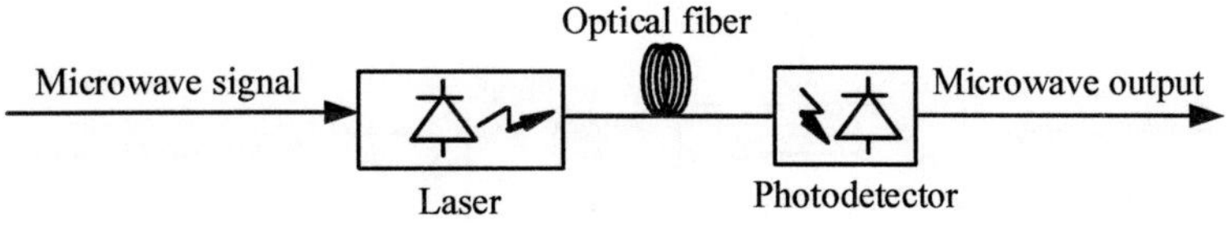

Fig. 9.8 Typical structure of direct modulation microwave photonic link

to frequency shift, and the limited relaxation oscillation frequency of semiconductor lasers (typically 6–9 GHz) limits the modulation bandwidth of microwave photonic links. In addition, the relative intensity noise is larger when the injection current is small. The typical structure of direct modulation microwave photonic link is shown in Fig. 9.8.

Lasers used for direct modulation of microwave photonic links mainly include Fabry-Perot Laser Diode (FP-LD), Vertical Cavity Surface Emitting Laser (VCSEL), and Distributed Feedback (DFB) laser. Among them, the distributed feedback laser is commonly used for direct modulation. When the direct modulation microwave photon link first appeared, the performance was not ideal, the gain was usually about −40 dB, and the noise figure (NF) was as high as 50 dB when transmitting 18 G microwave signals. After years of research and development, the performance of direct modulation links has been greatly improved.

Compared with direct modulation, external modulation can use light sources with constant output light intensity and low relative intensity noise. The externally modulated microwave photonic link is not affected by the relaxation oscillation of the laser, the bandwidth can reach tens of GHz, and the gain of the link can be improved by increasing the optical power. Commonly used external modulations include intensity external modulation and phase external modulation. The most widely used in microwave photonic links is external intensity modulation, composed of a laser and a modulator, which separates the generation of laser from the modulation of the signal, and the laser generates an optical carrier with constant intensity. The modulation of the microwave signal is realized by modulating the continuous light emitted by the laser through the electro-optic modulator. This modulation method can achieve a dynamic range of 115 $dB/Hz^{2/3}$. Furthermore, the laser does not limit its operating bandwidth, its operating frequency can reach up to 40 GHz. Therefore, its environmental stability is good, and it only needs to control the operating point of the modulator to obtain high amplitude and phase stability.

A typical circuit of an externally modulated microwave photonic link based on a Mach-Zehnder modulator is shown in Fig. 9.9, which consists of a laser, a Mach-Zehnder modulator, an optical fiber, and a photodetector. At the transmitting end, the microwave signal is modulated onto the optical carrier through the Mach-Zehnder modulator in the form of intensity modulation, the modulated optical signal is transmitted through the optical fiber, and the receiving end recovers the microwave signal through the photodetector.

Phase external modulation is another way of external modulation, which has a very high dynamic range. Theoretically, it can exceed the dynamic range of 120 $dB/Hz^{2/3}$.

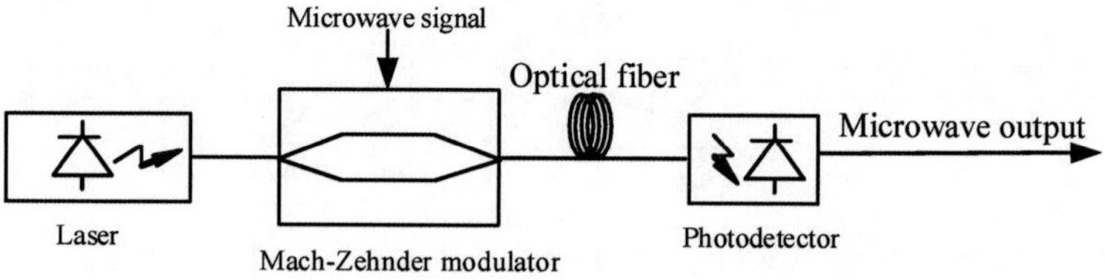

Fig. 9.9 Typical circuit of externally modulated microwave photonic link

However, because the light phase is difficult to keep consistent, it is greatly affected by the environment (such as vibration and temperature). A complex optical phase-locked loop is required, and its environmental stability is relatively poor.

9.3.2 Optical Analog-to-Digital Conversion (ADC)

Analog-to-Digital Conversion (ADC) is the key process to realizing analog signal conversion to a digital signal. With the development of modern digital signal processing technology, more and more analog functions are realized by the digital system. However, due to the influence of sample-and-hold circuit bandwidth, comparator relaxation, and clock jitter, the performance of traditional electronic ADCs has been greatly limited. It can no longer meet the high sampling rate and large bandwidth requirements. The remarkable advantage of using photonic technology to realize ADC is that sampling with picosecond or sub-picosecond ultra-short light pulses can achieve a high sampling rate greater than 100 Gbps. The time jitter of the mode-locked optical pulse can be reduced by two orders of magnitude (up to about 200 fs) compared with the electrical sampling pulse. Therefore, compared with traditional electronic ADC, optical ADC can achieve high quantization accuracy while realizing high-speed, ultra-wideband sampling.

In general, there are two main types of optical ADCs: one is based on time domain stretching, and the other is based on multiplexing technology. The basic principle of optical ADC based on time-domain broadening is to modulate the high-speed input signal on the linearly chirped optical signal and then use fiber dispersion to perform time-domain broadening on the modulated signal to reduce its rate. Then it is converted into an electrical signal by a detector, and a relatively low-speed electrical analog-to-digital converter is used to realize ADC. The basic principle of optical ADC based on multiplexing technology is first to modulate the high-speed input signal on the optical pulse, then use multiplexing technology to divide the modulated optical pulse into several pulse sequences with relatively low repetition frequency. Then it is converted into an electrical signal by a detector, and a relatively low-speed electrical ADC is used to realize ADC. Optical ADC based on multiplexing technology has wavelength division multiplexing optical ADC and time division multiplexing ADC.

The optical ADC based on time domain stretching is shown in Fig. 9.10.

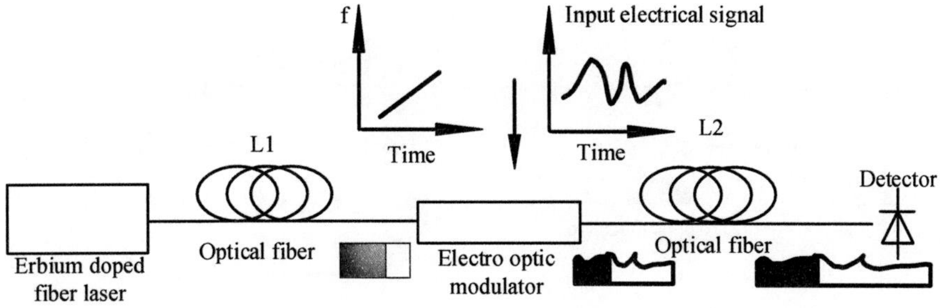

Fig. 9.10 Optical pulse time domain stretching ADC principle

The optical ADC based on time-domain broadening needs to use a linear chirped light source, and the chirp amount of the light source cannot be precisely controlled, so the precise time-domain waveform of the input signal cannot be obtained. At the same time, the optical ADC based on time domain widening cannot realize the continuous true-time processing of the input signal because of the time domain widening of the input signal. Therefore, this method is suitable for processing ultra-high-speed transient signals.

Multiplexing technology includes time division multiplexing technology and wavelength division multiplexing technology. The principle of optical time division multiplexing ADC is shown in Fig. 9.11.

N optical pulses with a repetition rate of f_0 are multiplexed into a sampled optical pulse train with a repetition rate of Nf_0 through time division multiplexing technology, where *N* is the number of channels, and f_0 is the pulse repetition frequency of a single channel. The pulse train modulates the analog signal through the electro-optical intensity modulator and then uses the time division multiplexer to demultiplex the optical pulse train into *N* channels again. The rate of each channel is reduced to f_0, then the photoelectric conversion is performed, and the low-speed electrical ADC is used for acquisition.

The optical ADC based on wavelength division multiplexing is shown in Fig. 9.12.

The basic principles of optical wavelength division multiplexing ADC and optical time division multiplexing ADC are the same. The difference is that the sampling

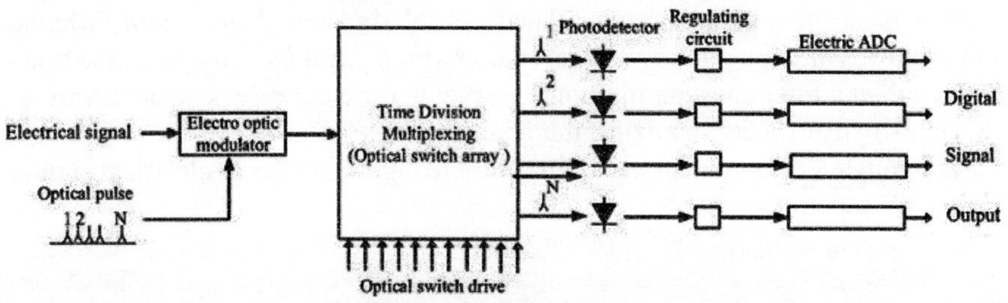

Fig. 9.11 Optical time division multiplexing ADC principle

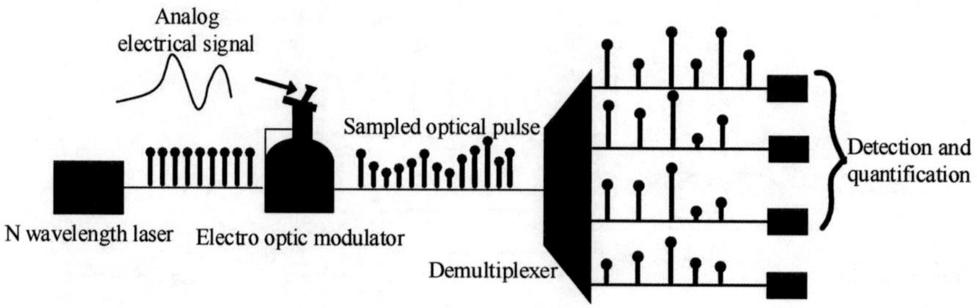

Fig. 9.12 Optical wavelength division multiplexing ADC principle

light pulses of N wavelengths are arranged in different time sequences. The electro-optical modulator is used to sample the analog signal. The modulated optical pulse sequence is divided into parallel N wavelength channels by wave decomposition and multiplexing and then output by photoelectric detection and quantization.

The comparison of three commonly used optical ADC characteristic parameters is listed in Table 9.3.

9.3.3 Microwave Photonic Filtering

The traditional digital filter is an algorithm or device composed of a delay unit, a multiplier, and an adder. Like a digital filter, a microwave photon filter is composed of a delay unit, a multiplier, and an adder. The difference is that the delay, multiplication, and addition of signals are realized in the optical domain. Use lasers and electro-optic modulators to modulate electrical signals into optical signals, realize time delay through different optical fibers, different amplitude weights of differently delayed signals through optical splitters, and realize signal addition through couplers. Compared with traditional electrical filtering technology, microwave photonic filtering technology has the advantages of being broadband, tunable, reconfigurable, and free from electromagnetic interference. The basic structure of the microwave photonic filter is shown in Fig. 9.13.

Figure 9.13 modulates the electrical signal into an optical signal through the electro-optical modulator. The modulated optical signal is divided into different optical powers through a $1 \times N$ optical splitter. It is equivalent to realizing the function of a digital filter constant multiplier, realizing different delays through various lengths of optical fibers, realizing the addition function of signals with different delays through $N \times 1$ couplers, and finally realizing the output of electrical signals through photodetectors.

Table 9.3 Comparison of commonly used optical ADC characteristics

	Optical pulse time-domain broadening ADC	Optical time division multiplexing ADC	Optical wavelength division multiplexing ADC
Operating principle	With the chirp characteristics of the dispersive optical element, high-speed signal is broadened into a relatively low-speed signal, and then a low-speed electrical ADC is used to realize data collection	With characteristics of optical time division multiplexing, the ultra-high-speed sampling optical pulse sequence is constructed to collect high-speed electrical signals directly, and the modulated high-speed optical pulse sequence is converted into low-speed signals through demultiplexing technology, and then low-speed electrical ADC is used to realize data collection	With characteristics of optical wavelength division multiplexing, multiplex optical pulses of different wavelengths into optical pulse sampling sequences, directly optically modulate high-speed electrical signals, and demultiplexing technology is used to convert the dimmed light pulse sequences into low-speed signals, and then low-speed electrical ADC is used to realize data acquisition
Featured components	Dispersive fiber or other dispersive devices	Light pulse with repetition frequency f_0	N wavelength laser
Signal broadening	Yes	No	No
Cost	Low	High	High
Applications	Ultrahigh-speed	Medium and high speed	Medium and high speed

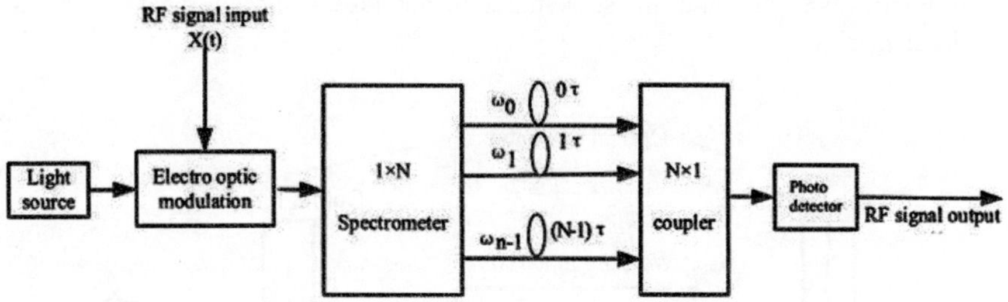

Fig. 9.13 Microwave photonic filter principle

9.4 Microwave Photonic Devices

Microwave photonic devices generally include lasers, detectors, modulators, optical fibers, optical amplifiers, optical splitters, optical wavelength division multiplexers, optical isolators, optical circulators, optical switches, optical phase shifters, etc.

9.4.1 Lasers and Detectors

In a directly modulated microwave photonic link, the role of the laser is to modulate an electrical signal into an optical signal directly. In the externally modulated microwave photonic link, the role of the laser is to provide a stable optical carrier signal and complete the optical modulation of the electrical signal together with the modulator. The laser consists of three parts: working substance, pump source, and optical cavity [21], as shown in Fig. 9.14.

In Fig. 9.14, the working substance is the core of the laser, which provides the energy level structure system for forming the laser, and is the internal cause of the laser. The pump source provides energy excitation to form a laser, which is the external cause of laser formation. The optical resonator provides a feedback amplification mechanism for the laser and has the mode selection function. According to different working substances, lasers can be divided into gas, liquid, solid, and semiconductor. The characteristics of the four types of lasers are compared as listed in Table 9.4.

The most commonly used semiconductor laser in microwave photonic links is Distributed Feedback Laser (DFB). The DFB laser is a side-emitting semiconductor laser with high output power, good single-mode characteristics, a high side-to-mode ratio, and a high modulation rate.

In microwave photonic links, semiconductor lasers should meet the basic requirements:

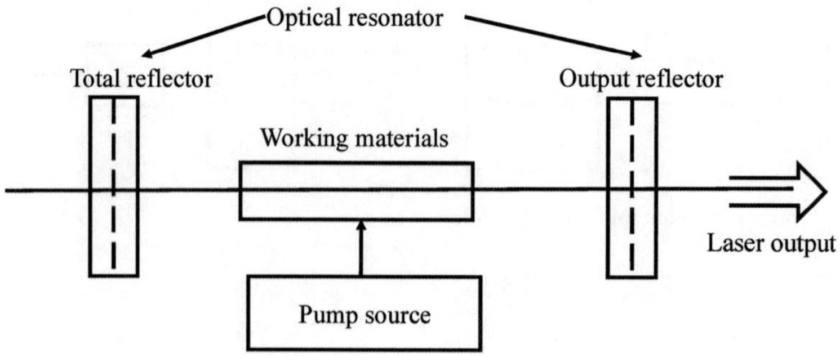

Fig. 9.14 Basic structure of the laser

Table 9.4 Comparison of the characteristics of 4 kinds of lasers

Laser type	Advantage	Insufficient
Gas laser	The simple structure, low cost, and convenient operation, the working medium is uniform, with good beam quality and it can work stably and continuously for a long time	High pump power required, cooling water or heat exchanger required
Liquid laser	Continuously adjustable wavelength (tuning range from ultraviolet to infrared), simple structure, low price	The stability of the dye solution is relatively low
Solid state laser	Small size, easy maintenance, high output power, and suitable for Q-switching to generate high-power pulses, and mode-locking to generate ultra-short pulses	The preparation of the working medium is more complicated and the price is more expensive
Semiconductor laser	Small size, light weight, high energy conversion efficiency, long service life, simple structure, easy monolithic integration, wavelength tuning, and high-speed modulation capabilities	The manufacturing process is complex, with narrow tuning range and the spectral line width is wide

(1) The central wavelength of the output optical signal is around 850, 1310, or 1550 nm;
(2) High electro-optical conversion efficiency and good linearity;
(3) The side mode suppression ratio of the output optical signal is large;
(4) The relative intensity noise of the laser itself is small.

The photodetector is another important device in the microwave photonic link, and its function is to convert the received optical signal into an electrical signal. The basic parameters of photodetectors include frequency bandwidth, responsivity, dark current, and saturated optical power.

In microwave photonic link applications, photodetectors should meet the basic requirements:

(1) The wavelength response range should cover at least 850, 1310, and 1550 nm optical communication windows;
(2) High responsivity, i.e., high photoelectric conversion efficiency;
(3) The dark current is as small as possible, and the noise of photoelectric conversion is small, which is convenient for the reception of weak optical signals;
(4) The saturated optical power is high, i.e., the maximum optical power that can be received is large.

To meet the above requirements, there are two types of photodetectors commonly used in microwave photonic links: Avalanche Photo Diode (APD) and Positive Intrinsic Negative (PIN). Avalanche photodetector refers to the avalanche breakdown of light on a special structure diode with reverse bias so that the concentration of photogenerated carriers increases rapidly, and the junction current increases sharply, thus forming a detectable current. Avalanche photodiodes are characterized by high self-gain, high sensitivity, fast response, and wide operating frequency range. The main application is the rapid detection of extremely weak optical signals. The basic operating principle of the PIN photodetector is that after the modulated light hits the receiving area (p area), the electrons generated in the $p - n$ junction and nearby move under the action of the external voltage, which increases the loop current and forms a detectable current. It has a simple structure, low price, good temperature characteristics, etc., and is suitable for medium, short distance, and medium and low-rate systems. The basic parameters of the photodetector are 3 dB bandwidth, flatness, responsivity, dark current, etc.

9.4.2 Modulators and Demodulators

Directly modulated microwave photonic links use lasers to directly modulate microwave signals onto optical carriers, while externally modulated microwave photonic links require an external electro-optical modulator independent of lasers to modulate electrical signals onto optical carriers.

Directly modulated lasers generally use semiconductor lasers with high emission power, good modulation characteristics, high electro-optical conversion efficiency, and good directivity and coherence of output light [22]. Among semiconductor lasers, DFB lasers with good wavelength stability, narrow line width, and high linearity are more common choices. Direct modulation is realized by controlling the injection current I of the direct modulation laser so that the intensity of its output light is modulated. The relationship between the output optical power P and I is the $P - I$ characteristic of the microwave photonic link, as shown in Fig. 9.15. $I_{\text{sig}}(t)$ is the input microwave signal, Ith is the threshold current, I_{bias} is the DC bias, I_{sat} is the saturation current, and $P_{\text{out\_}D}$ is the output optical power.

It can be seen from Fig. 9.15 that the directly modulated laser is a threshold-type device. The output optical power P varies with I and has several typical stages. When I is small, the spontaneous emission of the laser is dominant, thus emitting fluorescence with a broad spectrum. As I increases, stimulated emission begins to dominate. But when I is still less than the threshold current Ith, the gain of the laser resonator is not enough to overcome its loss. Therefore, a certain mode of oscillation is established so that only strong fluorescence is emitted. Only when I reaches Ith the laser with transparent oscillation mode and the narrow spectral line can be output. When I starts to increase gradually from Ith, the P of the laser maintains a continuous linear relationship with I, but the output optical power P of the laser cannot increase without limit as the operating current I increases. Finally, when I increases to a certain extent (saturation current I_{sat}), the output optical power P reaches saturation.

The Mach-Zehnder modulator based on $LiNbO_3$ crystal is the most widely used external light intensity modulator. It uses the linear electro-optic effect of $LiNbO_3$

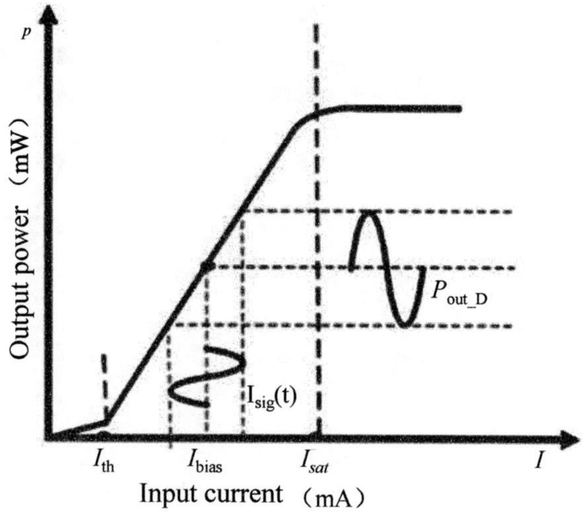

Fig. 9.15 $P - I$ characteristic curve of a directly modulated laser

crystal to realize electro-optic conversion. This type of modulator has the characteristics of a significant electro-optic effect, small insertion loss, small dependence on wavelength, easy production, and mature technology. In addition, the application of Mach-Zehnder modulators in microwave photonic links has other advantages, such as efficient coupling with transmission fibers, modulation chirp can be made small or adjustable, modulation bandwidth is large, etc.

The Mach-Zehnder modulator is usually made of $LiNbO_3$ crystal with an electro-optic effect. The refractive index of the crystal changes with the applied electric field E (voltage U). The structure of the Mach-Zehnder modulator is shown in Fig. 9.16. The modulator includes a Y-shaped splitter, two single-mode optical waveguides with equal optical paths, and a combiner. Applying a variable voltage to one of the two waveguides will cause the light wave to pass through two waveguides with different refractive indices and obtain different phases. In addition, after the combiner recombines the two light waves, constructive or destructive interference will occur, which will cause the intensity of the output optical signal to change with the applied voltage.

The basic parameters of the Mach-Zehnder modulator include optical insertion loss, half-wave voltage, maximum input optical power, modulation bandwidth, and extinction ratio. High-performance microwave photonic links usually require Mach-Zehnder modulators with low optical insertion loss, small half-wave voltage, large input optical power, and wide modulation bandwidth. The typical insertion loss of the modulator is 3–7 dB, and the half-wave voltage is generally 2–5 V. In addition to lithium niobate crystals, new materials such as electro-optic polymers and silicon-based materials have also begun to be considered for producing Mach-Zehnder modulators to achieve the goals of low half-wave voltage, low insertion loss, and small-size devices.

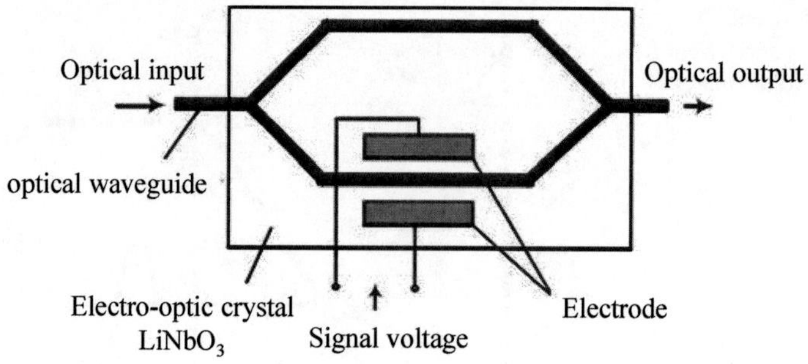

Fig. 9.16 Mach-Zehnder modulator principle

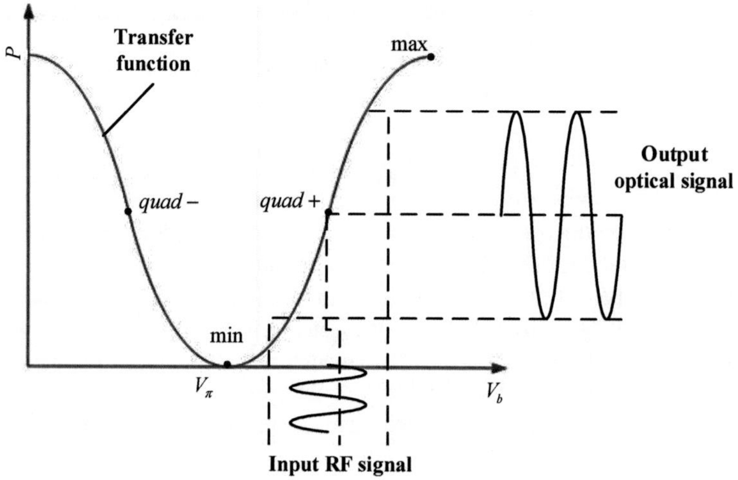

Fig. 9.17 Mach-Zehnder modulator modulation curve

In external modulation, the basic function of the Mach-Zehnder modulator is to realize the electrical-optical conversion of the signal without distortion as much as possible. To ensure the linearity of the modulated signal, the Mach-Zehnder modulator needs to work at the quadrature bias point, as shown in Fig. 9.17.

9.4.3 Optical Fibers and Optical Amplifiers

Figure 9.18 shows the cross-sectional views of three typical optical fibers, and their outer cladding diameters are all 125 μm. The typical core diameter of single-mode fiber is 9 μm, which can only transmit one light mode, and the typical core diameter of multimode fiber is 62.5 μm, which can transmit multiple light modes. Although

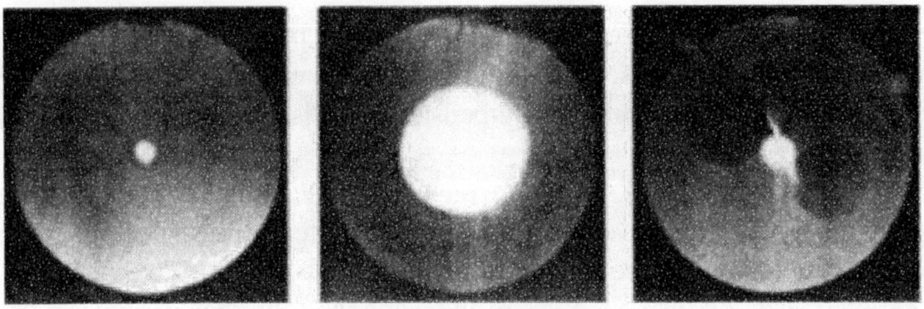

Fig. 9.18 Cross-sectional views of three kinds of optical fibers, **a** single-mode optical fiber, **b** multimode optical fiber, **c** polarization-maintaining optical fiber

multimode fiber has higher fiber-to-device coupling efficiency, it has the disadvantage of intermodal dispersion.

In some applications, such as polarization-sensitive external modulators, the input light with a fixed polarization state is required to maintain maximum modulation efficiency. Generally, a polarization-maintaining fiber is used to connect polarization-sensitive devices, as shown in Fig. 9.18c, which is a cross-sectional view of a panda-type polarization-maintaining fiber.

The loss of the optical fiber is related to the wavelength of the transmitted light. With the development of semiconductor laser and photodetector technology, in the optical fiber communication system, the optical fiber mainly works in the 1310 and 1550 nm bands. The standard single-mode optical fiber has a loss of about 0.2 dB/km in the 1550 nm band, and in the 1310 nm band, the loss is about 0.35 dB/km. In the 1550 nm band, the transmission loss of silica fiber is the smallest, so this band is the best choice for long-distance microwave photonic links. However, the dispersion of the 1310 nm band is zero, so it is also widely used in high-frequency and long-distance microwave photonic links. In an optical fiber, different frequency components or different mode components of the transmitted optical signal propagate at different speeds, and this phenomenon is called the dispersion of the optical fiber. Optical fiber dispersion mainly includes material dispersion, waveguide dispersion, intermodal dispersion, and polarization mode dispersion, among which intermodal dispersion is the unique dispersion of the multimode optical fiber. The phenomenon that different wavelengths of light propagate at different speeds in a single-mode fiber is also called chromatic dispersion, mainly caused by material dispersion and waveguide dispersion. For light in the 1310 nm band, the chromatic dispersion in standard single-mode fiber is zero. The chromatic dispersion of the 1550 nm band is 17 ps/(km nm), the 1 km optical fiber transmits the 30 GHz modulation signal, and the SNR deterioration caused by the dispersion is less than 1 dB. For lower-frequency signals, the effect of dispersion is smaller.

The deterioration of long-distance microwave photonic link performance caused by the dispersion of optical fiber is manifested in the rapid attenuation of gain at a specific frequency point, the introduction of even-order distortion components, and the mutual conversion between phase noise and intensity noise. In recent years, to reduce the deterioration of chromatic dispersion on the performance of long-distance microwave photonic links, some solutions have emerged, e.g., single sideband modulation, adjustable dispersion compensation, pulse shaping, and modulation diversity.

In microwave photonic links, there are many loss mechanisms, including insertion loss, branch loss, and transmission attenuation of optical fibers. The optical amplifier amplifies the signal in the optical domain. It simultaneously reduces the SNR deterioration caused by the shot noise of the signal and the receiver noise.

Optical amplifiers generally consist of the gain media, pump light, and input–output coupling structures. There are three commonly used optical amplifiers: optical fiber amplifiers made of doped optical fibers, among which the Erbium-Doped Fiber Amplifier (EDFA) is the main one. Furthermore, the Semiconductor Optical Amplifier (SOA) is made of a laser diode. Finally, distributed optical amplifier, such as Fiber Raman Amplifier (FRA), is made of conventional optical fiber nonlinear effects. The characteristics of three commonly used optical amplifiers are compared as listed in Table 9.5.

Table 9.5 Comparison of characteristics of the common optical amplifier

Features	Erbium-doped fiber amplifier	Semiconductor optical amplifier	Fiber raman amplifier
Operating principle	The rare earth ions doped in the erbium-doped fiber are used to form the population inversion under the action of pump light source to generate stimulated radiation	The population is reversed to amplify and emit light, and the light-emitting medium is electron-hole pair	Third-order nonlinear effect of strong pump laser in the optical fiber
Gain medium	Erbium-doped fiber	Semiconductor laser	Ordinary fiber or dispersion fiber
Wavelength range	1525–1625 nm	1300–1600 nm	1270–1670 nm full band
Magnification	Single stage approx. 25 dB, cascadable	Greater than 30 dB	Less than 15 Db
Noise figure	4–5 dB	About 5 dB	Less than 1Db
Special application	–	Optical switch, wavelength conversion	Non-linear suppression, dispersion compensation

9.4.4 Optical Splitters and Optical Wavelength Division Multiplexers

The optical splitter divides the optical power transmitted by a single optical fiber into several optical fiber branches and then transmits it to each terminal, thus greatly saving optical fibers and equipment ports. According to the manufacturing process, optical splitters are divided into fused tapered and planar waveguide types. Among them, the fused tapered optical splitter is divided into single, double, and triple windows according to the working window covered by the device performance, and the planar waveguide type is a full-band splitter. In addition, optical splitters can be divided into uniform light splitting and uneven light splitting according to branch power division methods.

Fused taper technology is to bundle two or more optical fibers together, then melts and stretches them on a tapering machine, monitors the change of the split ratio in true time, and ends the melt stretch after the split ratio meets the requirements. One or two optical fibers are reserved at one end as the input end, and the other end is used as the multi-channel output end. Under normal circumstances, the mature taper process can only pull devices below 1×4 at a time, and devices above 1×4 are connected by multiple 1×2 devices and then packaged in a splitter box. However, it is relatively easy to make an optical splitter with uneven light splitting using the fusion tapered technology, and the cost is low.

Planar waveguide technology uses semiconductor technology (photo lithography, etching, development, etc.) to manufacture optical waveguide branch devices. The optical waveguide array is located on the upper surface of the chip, and the shunting function is completed on the chip. For example, more than 1×32 shunts can be realized on one chip, then the input and output multi-channel optical fiber arrays are coupled and packaged at both ends of the chip, and the planar waveguide technology requires high requirements.

The basic characteristics of the two types of splitters are compared as listed in Table 9.6.

A Wavelength Division Multiplexing (WDM) synthesizes optical signals carrying information but with different wavelengths into one bundle and transmits them along a single single-mode optical fiber. Optical devices that separate optical signals of different wavelengths by demultiplexing at the receiving end, as shown in Fig. 9.19. WDM technology has corresponding applications in many optical modules or subsystems.

Commonly used wavelength division multiplexing includes Coarse Wavelength Division Multiplexer (CWDM) and Dense Wavelength Division Multiplexer (DWDM). The difference between the two is that the optical wavelength interval of CWDM is wider, and the standard defined by the industry is 20 nm, while the International Telecommunication Union (ITU) stipulates that the wavelength interval standard of DWDM is 0.8 nm. The wide wavelength spacing characteristics of lasers used in CWDM make the overall number of multiplexing channels relatively low, generally no more than 8 channels, and the upper limit is 18 channels, using 1310 nm

Table 9.6 Comparison of the basic characteristics of splitters

Main features	Fused taper optical splitter	Planar waveguide type optical splitter
Operating wavelength (nm)	1310, 1490 and 1550 ± 40	1260–1650
Power division method	Variable, unequal	Equal division mainly
Spectral uniformity	Big difference	Small difference
Temperature stability	Lower	Better
Wavelength dependent loss	With window, higher	No window, lower
Polarization dependent loss	Higher	Lower
Reliability	Many failure points, low reliability	Few points of failure, high reliability
Size	Larger	Smaller
Fabrication technology	Simpler	More complex
Product price	Less than 1 × 4, cost-effective	More than 1 × 16, cost-effective

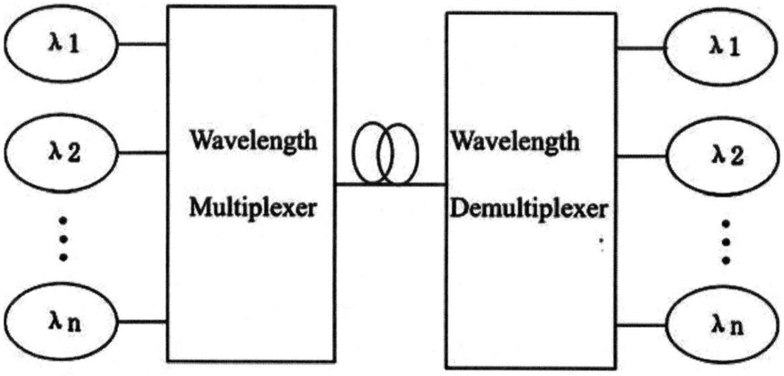

Fig. 9.19 Optical wavelength division multiplexer

transmission, compared with 1550 nm, the transmission loss is larger, and it is not suitable for long-distance transmission. DWDM uses 1550 nm single-mode optical fiber, which usually has 80 channels. The number of channels can be expanded to 150–160 channels, and the transmission rate of each channel can reach more than 10 GB/s. Combined with Erbium- Doped Fiber Amplifier (EDFA) technology, the ultra-long-distance transmission of thousands of kilometers can be achieved.

9.4.5 Optical Isolators and Circulators

The function of the optical isolator is to prevent the reflected light in the optical path due to various reasons from adversely affecting the light source and the optical path system. For example, installing an optical isolator between the semiconductor light source and the optical transmission system can greatly reduce the influence of reflected light on the stability of the output optical power of the light source. Furthermore, in the direct modulation and direct detection optical fiber communication system, reflected light would generate additional noise, deleting the performance of the system. Therefore, an optical isolator is also required to eliminate it.

The principle of the optical isolator is mainly to use the Faraday effect of the magneto-optic crystal. When the linearly polarized light propagating along the direction of the magnetic field passes through the magneto-optic crystal with a thickness L, the rotation angle θ of the linearly polarized light and the magnetic field strength B are proportional to the product of the thickness L of the material, that is

$$\theta = \mathrm{VBL} \tag{9.5}$$

where V is the verdet constant, which is related to the properties of the medium and the frequency of light. The operating principle of the optical isolator is shown in Fig. 9.20. The optical isolator includes two polarizers with an angle of 45° in the light transmission direction and a faraday rotator. When the parallel incident light goes back and forth once, the polarization angle changes by 90° and the reverse light cannot pass through P_1, which can realize the reverse isolation.

The operation principle of the circulator is similar to that of the isolator. The difference is that the circulator generally has three or four ports. For example, in a three-port circulator, the light input at port 1 is output through port 2, and the light input at port 2 is output through port 3, as shown in Fig. 9.21.

The optical circulator can complete the forward and reverse light transmission functions, as shown in Fig. 9.22.

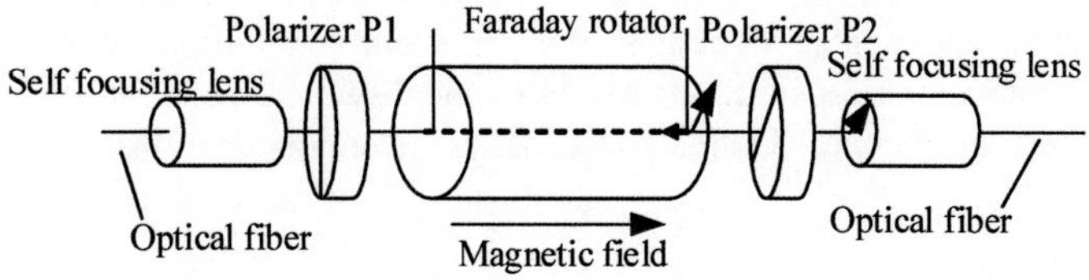

Fig. 9.20 Operation principle of the optical isolator

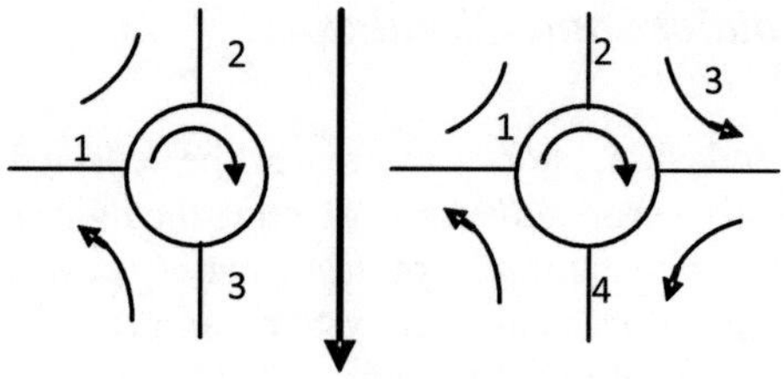

Fig. 9.21 Operation principle of the optical circulator

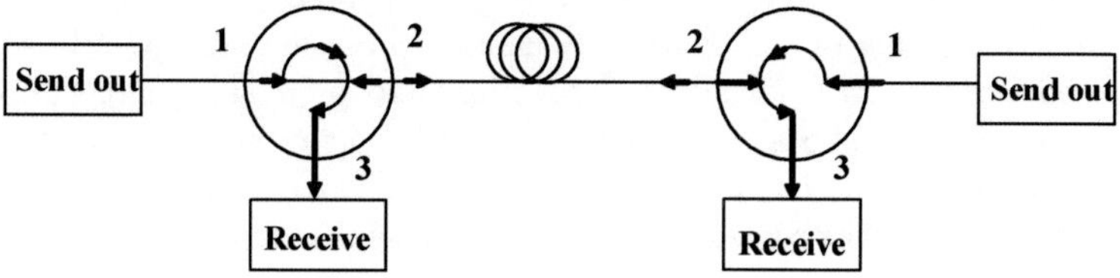

Fig. 9.22 Optical circulator two-way transmission and reception

9.4.6 Optical Phase Shifters and Optical Switches

With the development of silicon-based photonics, silicon-based optical phase shifters emerged. The silicon-based phase shifter is based on a waveguide structure. Its phase is changed by changing the effective refractive index of the waveguide to realize an adjustable phase shift function. The main implementations of silicon-based optical phase shifters are based on the free carrier dispersion effect and the thermo-optic effect. Generally speaking, the phase shifter based on the free carrier dispersion effect can realize high-speed phase shifting, while the phase shifter based on the thermo-optical effect has the characteristics of low loss.

An optical switch is a device with one or more selectable transmission ports which can perform the mutual conversion or logical operation on optical signals in optical transmission lines or integrated optical circuits. Its function is to convert optical circuits and realize optical switching. Commonly used optical switches include mechanical optical switches, MEMS optical switches, and integrated optical waveguide optical switches. Their characteristics are compared as listed in Table 9.7.

9.5 Microwave Photonic Link Analysis

Microwave photonic links are mainly used for point-to-point high-fidelity transmission of microwave signals, but microwave photonic links will introduce loss, noise, and nonlinear distortion to modulated signals. Therefore, some parameters need to be introduced to characterize the performance of microwave photonic links, such as

Table 9.7 Comparison of the characteristics of common optical switches

Description of optical switch	Operating principle	Advantages	Disadvantages
Mechanical optical switch	The optical path is changed by moving the optical fiber or optical element	Low insertion loss (<2 dB), high isolation (>45 dB), independent of polarization and wavelength	The switching time is long (ms), there is bounce jitter, and the repeatability is bad
MEMS optical switch	Miniaturized free-space optical switch made by MEMS technology	Small structure, short switching time, high isolation	The consistency of each channel is poor and the control is difficult
Integrated optical waveguide optical switch	Relying on physical effects to change the refractive index of the waveguide to alter the optical path	Small structure, short switching time, high isolation	The insertion loss is high, with low isolation of around 20 dB

gain, NF, and dynamic range. These parameters are similar to those describing two-port microwave devices. The microwave photonic link can be regarded as a black box with only one microwave input port and one microwave output port.

9.5.1 Noise Source

The noise sources of microwave photonic links mainly include thermal noise, shot noise, and laser relative intensity noise.

Thermal noise mainly comes from the heat-induced random motion of charged ions in each resistive element of the link. Since it exists in all media, it is impossible to cancel the thermal noise, which constitutes the noise floor of communication systems. The thermal noise current is expressed as $I = \sqrt{4\text{KT}\Delta f/R}$, and for the externally modulated microwave photonic link, the thermal noise at the output can be expressed as

$$\text{NFth} = -190 - 20\,\lg(V_\pi)g + 20\,\lg(I_{\text{dc}}) \tag{9.6}$$

where R is the link load impedance, K is the Boltzmann constant with a value of 1.38×10^{-23} J/K, T is the absolute temperature, V_π is the half-wave voltage, I_{dc} is the modulation current, Δf is the absolute bandwidth of the carrier frequency.

Shot noise, also known as flicker noise, is caused by the dispersion of the carriers that form the current. It is one of the main noise sources of semiconductor devices, and the main noise introduced by photodetectors in microwave photonic links is shot noise. Like thermal noise, the spectral density of shot noise also exhibits broad-spectrum characteristics, and its root mean square current spectral density is

$$I_{\mathrm{sh}}^2 = 2eI_{\mathrm{dc}} \tag{9.7}$$

where e is the electronic constant (1.602×10^{-19}), I_{dc} is the output DC value of the detector. If the detector output impedance Z is 50 Ω, then the shot noise power spectral density is

$$\mathrm{NF}_{\mathrm{shot}} = 10\ \lg(2eZI_{\mathrm{dc}}) = -168 + 10\ \lg(I_{\mathrm{dc}}) \tag{9.8}$$

where Z is the output impedance. It can be seen from Eq. (9.8) that the larger the value of the detector output DC, the larger the power spectral density of the shot noise.

Several factors contribute to random fluctuations in laser output power, often called noise. The general measure of the random fluctuation of the output optical power of a laser is defined as relative intensity noise. Laser relative intensity noise is not white noise and has a peak at the relaxation oscillation frequency. Relative Intensity Noise (RIN) is defined as

$$\mathrm{RIN}_{\mathrm{laser}} \equiv \left\langle (\Delta P)^2 \right\rangle / \langle P \rangle^2 \tag{9.9}$$

where $\left\langle (\Delta P)^2 \right\rangle$ is the square mean of optical power fluctuations, $\langle P \rangle$ is the average optical power.

The relative intensity noise is written in logarithmic form and often uses dBc as the unit, with the optical carrier power as the reference value. From the relationship between light intensity and detector current, we can get

$$\left\langle (\Delta P)^2 \right\rangle / \langle P \rangle^2 = \left\langle (\Delta I)^2 \right\rangle / \langle I \rangle^2 \tag{9.10}$$

where P is the light power fluctuation, ΔP is the light average power, ΔI is the light current fluctuation, I is the light average current.

Unlike thermal noise and shot noise, the relative intensity noise spectrum is not flat across the spectrum. At low frequencies, the relative intensity noise spectrum is constant and peaks at the relaxation oscillation frequency, after which the relative intensity noise drops to the level of shot noise. In the relative intensity noise-dominated portion of the spectrogram, the portion below the relaxation peak, the noise power varies as the square of the laser power. For the part above the peak value, that is, the part dominated by shot noise, the noise power changes linearly with the laser power, and the noise power spectral density obtained after the relative intensity noise is output by the detector is

$$\begin{aligned} N_{\mathrm{LaserRIN}} &= 10\ \lg\left[\left\langle (\Delta I)^2 \right\rangle Z\right] - 30 \\ &= 10\ \lg\left(\mathrm{RIN}_{\mathrm{laser}} I_{\mathrm{dc}}^2 Z\right) - 30 \\ &= -13 + 20\ \lg(I_{\mathrm{dc}}) + \mathrm{RIN}_{\mathrm{laser}} \end{aligned} \tag{9.11}$$

From Eq. (9.11), it can be seen that the greater the relative intensity noise of the laser, the greater the power spectral density of the RF noise it produces. And with the increase of the output DC value of the detector, the power spectral density of the RF noise generated by the relative intensity noise of the laser also becomes larger. Assuming that the light intensity input to the detector is 6 ± 1.7 dBm, the power spectral density of the relative intensity noise converted to RF noise output is -168.4 ± 1.45 dBm/Hz when RIN $= -160$ dBc/Hz. At this time, the total noise power spectral density of the microwave photonic link can be expressed as

$$N_{\text{link}} = 10\ \lg\left(10^{N_{\text{shot}}/10} + 10^{N_{\text{RINLaser}}/10}\right) \tag{9.12}$$

where N_{link} is the photon link noise density, N_{shot} is the thermal noise density, N_{RINLaser} is the relative intensity of noise density.

9.5.2 *Noise Figure (NF)*

The noise of microwave photonic links is mainly thermal noise, shot noise, and relative intensity noise. The thermal noise includes the input thermal noise after link amplification or attenuation and the output thermal noise of the link load.

For a typical microwave photonic link as shown in Fig. 9.23, the NF is

$$F_{\text{receive}} = F_1 + \frac{F_{\text{optical path}} - 1}{G_1} + \frac{F_{\text{post-stage}} - 1}{G_{\text{optical path}} G_1} \tag{9.13}$$

where F_{receive} is the NF of the microwave photoelectric receiver, F_1 and G_1 are the NF and gain of the optical path front stage (preselection filter and low noise amplifier), respectively, $F_{\text{optical path}}$ and $G_{\text{optical path}}$ are the NF and gain of the optical path (modulator, optical fiber, and optical demodulator), respectively, $F_{\text{post-stage}}$ is the NF of the post-stage (matched filter, logarithmic detector, and data acquisition)

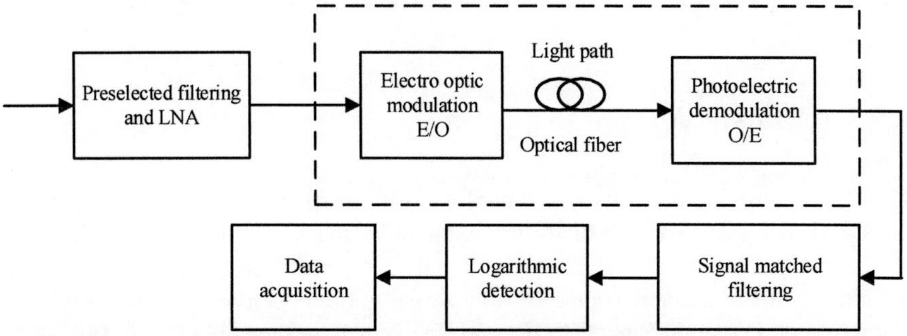

Fig. 9.23 Operation principle of the microwave photoelectric receiver

of the optical path. The NF and gain of the optical path are

$$
\begin{aligned}
N_{\text{optical path}} &= 10 \lg F_{\text{optical path}} \\
&= 10 \lg\left(1 + \frac{I_D^2 10^{\frac{\text{RIN}}{10}} R_{\text{lord}}}{2G_{\text{optical path}}\text{kT}} + \frac{2qI_D R_{\text{lord}}}{G_{\text{optical path}}\text{kT}}\right)
\end{aligned} \tag{9.14}
$$

$$
G_{\text{optical path}} = \frac{4\alpha^2\eta_1^2\eta_d^2 R_0 R_{\text{lord}}}{[sC_L R_L R_0 + R_0 + R_L]^2[sC_D(R_D + R_{\text{lord}})]^2} \tag{9.15}
$$

where I_D is the average value of the output current of the detector, RIN is the relative intensity noise of the laser, R_0 and R_{lord} are the system matching network and load impedance, respectively, C_L and C_D are the equivalent capacitance of the laser and the detector, respectively, R_L is the laser impedance, R_D is detector resistance, η_1 is laser conversion efficiency, α is fiber loss, η_d is detector conversion efficiency, Boltzmann constant $k = 1.38 \times 10^{-23}$ J/K. Absolute temperature $T = 290$ K.

For externally modulated microwave photonic links, assuming that the SNR of the input RF signal is limited by thermal noise, the RF noise figure NF_{rf} can be expressed as

$$
\begin{aligned}
\text{NF}_{\text{rf}} \equiv 10 \lg(\text{snr}_{\text{in}}/\text{snr}_{\text{out}}) &= 10 \lg[(s_i/s_o)\cdot(n_0/n_i)] \\
&= 174 - G_{\text{rf}} + N_t
\end{aligned} \tag{9.16}
$$

where G_{rf} is the total gain of the system, N_t is the total RF noise power spectral density at the system output, snr_{in} is the input SNR, snr_{out} is the output SNR, s_i is the input signal power, s_o is the output signal power, n_i is the input signal noise, n_o is the output signal noise. Equation (9.16) can be further expressed as

$$
\text{NF}_{\text{rf}} = 177 + 20 \lg(V_\pi) + N_{1\,\text{ink}} + 10 \lg(I_{\text{dc}}) \tag{9.17}
$$

According to Eq. (9.17), the RF noise figure is related to the modulator half-wave voltage V_π, the laser RIN noise, and the detector output direct current I_{dc}. The smaller the half-wave voltage of the modulator, the smaller the NF of the system.

9.5.3 Dynamic Range

The dynamic range of a microwave photonic link describes the power range when transmitting signals without distortion. Generally, the noise and nonlinearity of a microwave photonic link determine its dynamic range. Therefore, there are many definitions of the dynamic range of microwave photonic links, such as 1 dB gain Compression Dynamic Range (CDR) and Spurious Free Dynamic Range (SFDR).

1 dB gain CDR, also known as linear dynamic range, is defined as the ratio of the input signal power to the detectable minimum signal or equivalent noise power when

the output power of the receiver is large enough to produce 1 dB gain compression, that is

$$\mathrm{CDR} = \frac{P_{\mathrm{in},-1\mathrm{dB}}}{\mathrm{KTBNF}} = \frac{1.259 P_{\mathrm{out},-1\mathrm{dB}}}{N_{\mathrm{out}}} \tag{9.18}$$

where $P_{\mathrm{in},-1\mathrm{dB}}$ is the power of the input signal when compressing 1 dB, K is the Boltzmann constant, its value is 1.38×10^{-23} J/K, T is the absolute temperature, B is the bandwidth, $P_{\mathrm{out},-1\mathrm{dB}}$ is the output signal power when compressing 1 dB, N_{out} is the noise power at the output end.

SFDR is a parameter closely related to nonlinear characteristics. SFDR is defined as the ratio of the input signal to the equivalent input noise when the nonlinear distortion component is equal to the output noise or the ratio of the output signal to the output noise when the nonlinear distortion component is equal to the output noise. According to the order M of the distortion component that limits the spurious-free dynamic range, the SFDR of the microwave photonic link can be defined as

$$\mathrm{SFDR}_n = \left(\frac{\mathrm{OIP}_n}{N_{\mathrm{out}} B}\right)^{\frac{n}{n-1}} = \left(\frac{\mathrm{IIP}_n}{\mathrm{FKTB}}\right)^{\frac{n}{n-1}} \tag{9.19}$$

where OIP_n is the output intercept point of the n-order nonlinear distortion component, N_{out} is the link output noise floor, B is the test bandwidth, IIP_n is the input intercept point of the n-order nonlinear distortion component, F is the noise factor. If Eq. (9.19) is written in decibel form, and the test bandwidth $B = 1$ Hz, then the SFDR can be expressed as

$$\mathrm{SFDR}_n = \frac{n}{n-1}(\mathrm{OIP}_n - N_{\mathrm{out}}) \tag{9.20}$$

or

$$\mathrm{SFDR}_n = \frac{n}{n-1}(\mathrm{IIP}_n - \mathrm{NF} + 174) \tag{9.21}$$

It is simplified by $\mathrm{KT} = -174$ dBm. In engineering applications, the two-tone method is usually used to test the SFDR of the system. Generally, the third-order intermodulation output of the system is the main nonlinear distortion component that limits the SFDR of the system. Therefore, the definition of SFDR can be simplified as follows: when the third-order intermodulation signal power is equal to the output noise power in the unit bandwidth, the ratio of the fundamental frequency signal power to the noise power in the unit bandwidth.

The nonlinear transmission spectrum line of RF dual-tone signal transmission gives the SFDR and lists each RF component, as shown in Fig. 9.24.

The formula for calculating SFDR is

$$\mathrm{SFDR} = \frac{2}{3}(\mathrm{OIP3} - N_t) \tag{9.22}$$

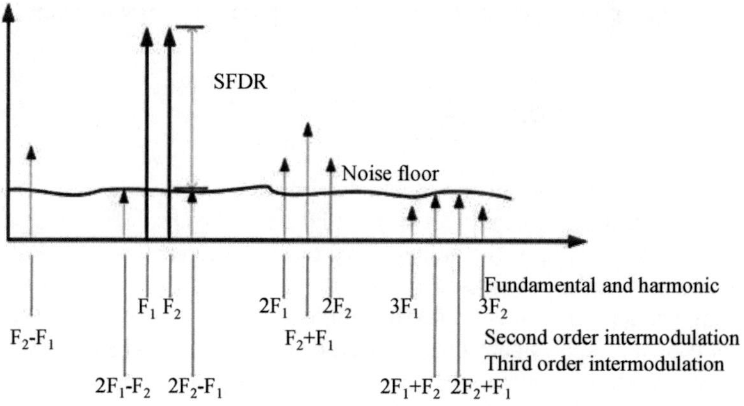

Fig. 9.24 Schematic diagram of the SFDR of a microwave photonic link

where OIP3 is the output third-order intercept point power, N_t is the total output noise power spectral density of the microwave photonic system. By increasing the output third-order intercept point of the link and reducing the link noise, the SFDR of the link can be improved.

For externally modulated microwave photonic links, a Mach-Zehnder modulator is often used, and the third-order intercept point power output by the modulator is $\text{OIP3} = 4I_{\text{dc}}^2 Z$. Assuming that the impedance Z is 50 Ω, the expression for the third-order intercept point power is

$$\text{OIP3} = -7 + 20\ \lg(I_{\text{dc}}) \tag{9.23}$$

When the system output noise is determined by the relative intensity noise of the laser, the formula for calculating SFDR of the microwave photonic system can be written as

$$\text{SFDR}\left[\text{dB} \cdot \text{Hz}^{2/3}\right] = \frac{2}{3}(6 - \text{RIN}_{\text{laser}}[\text{dBc/Hz}]) \tag{9.24}$$

where $\text{RIN}_{\text{laser}}$ is the relative intensity noise of the laser. It can be seen from Eq. (9.24) that when the system output noise is determined by the relative intensity noise of the laser, the SFDR of the system is only related to the relative intensity noise of the laser. The smaller the relative intensity noise, the larger the SFDR of the system.

9.5.4 Isolation

For microwave photonic links, isolation is similar to the isolation between traditional microwave channels, which characterizes the degree of crosstalk between

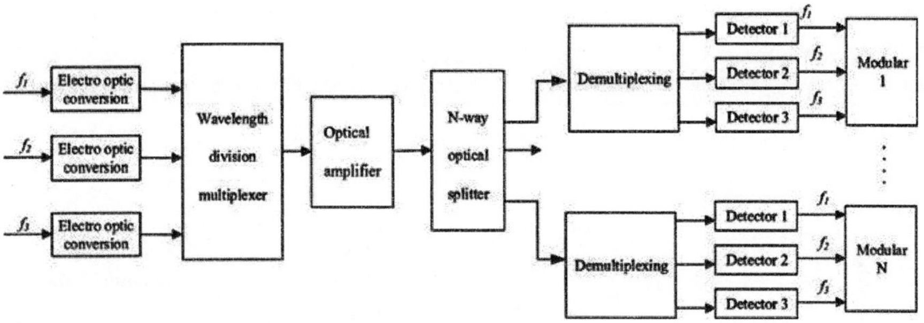

Fig. 9.25 Microwave photonic links using wavelength division multiplexing

two links or signals between multiple channels. Each link uses a separate electro-optical conversion, optical fiber, and optical detector for two completely independent microwave photonic links. If the spatial coupling of the microwave at the port is not considered, the signal leakage caused by the optical path is negligible. For two or more microwave photonic links that share some optical devices, the optical device performance parameters directly affect the isolation.

As shown in Fig. 9.25, f_1, f_2, and f_3 are modulated on optical carriers of different wavelengths, and a wavelength division multiplexer synthesizes one optical signal for transmission. Then, the three light waves are separated at the receiving end by a wave division multiplexer and converted into electrical signals by a photodetector. If the space leakage of the microwave signal is not considered, the isolation between the three links is mainly determined by the wavelength division multiplexing and the isolation between the demultiplexer channels, as shown in Fig. 9.26.

9.5.5 Link Insertion Loss

The loss of the microwave photonic link includes the loss of the light transmitting part, the light link part, and the light receiving part. These losses come from electro-optic, photoelectric conversion, optical path additional loss, impedance adaptation, etc. The link loss is usually compensated at the input or output end of photoelectric conversion.

(1) **Direct modulation link insertion loss**

Direct modulation link insertion loss includes optical-to-electrical conversion gain and electrical gain. In the optical fiber transmission link, the key to the influence of the system characteristics is the photoelectric conversion gain. Under the condition of adopting the impedance transformation technology, this part of the transmission gain has about 30 dB attenuation.

Figure 9.27 shows a typical direct modulation microwave photonic link circuit model. The modulation of the laser constraints the gain of photoelectric conversion

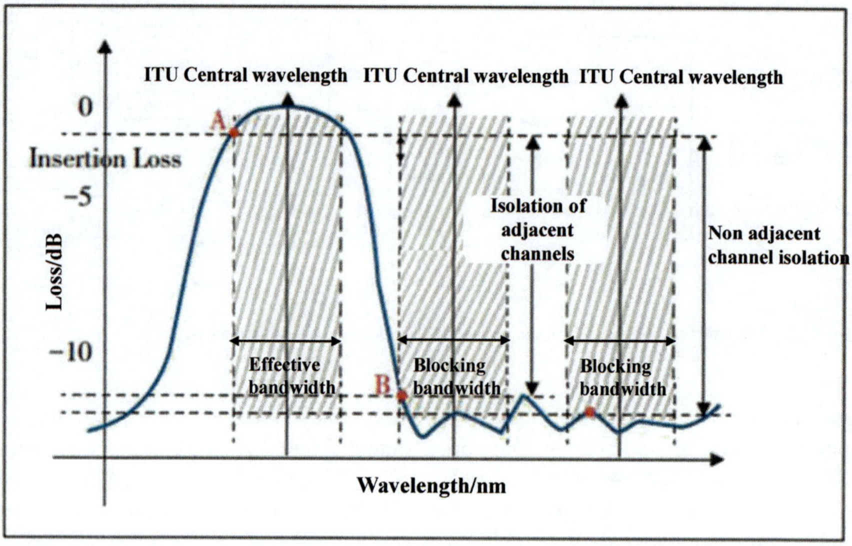

Fig. 9.26 Wavelength division multiplexer/demultiplexer channel isolation

at the transmitting end and the demodulation characteristics of the photodetector at the receiving end, especially their AC characteristics, i.e., the relationship between the modulation degree of the laser and the responsivity of the photodetector. The relationship between the photoelectric conversion gain, the degree of modulation, and responsivity is

$$G_{\text{opt}} = (S_M S_D)^2 \frac{R_D}{R_M} \tag{9.25}$$

where S_M is the laser modulation (W/A), S_D is the photodetector responsivity (A/W), R_M and R_D are the AC characteristic impedances of the laser and the detector, respectively.

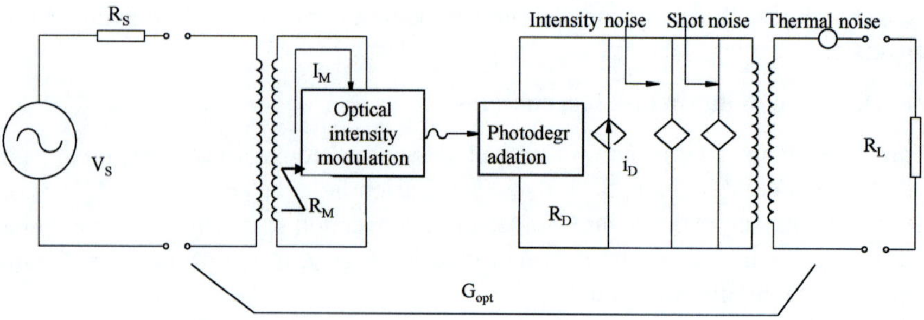

Fig. 9.27 Direct modulation microwave photonic link model

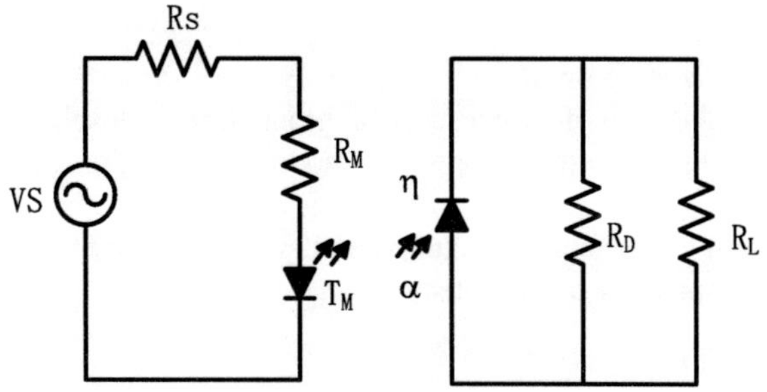

Fig. 9.28 External modulation microwave photonic link model

The modulation degree S_M of the commonly used butterfly laser is 0.15–0.25 W/A, and the responsivity S_D of the photodetector is generally 0.70–0.90 A/W. Therefore, it can be seen from Eq. (9.25) that to optimize the photoelectric conversion gain characteristics, lasers with high modulation degree (i.e., high electro-optical conversion efficiency) can be selected, and detectors with high responsivity (i.e., high photoelectric conversion efficiency) can also be selected. In addition, impedance matching also greatly influences the photoelectric conversion gain.

(2) **External modulation link insertion loss**

A typical model of an externally modulated microwave photonic link based on a Mach-Zehnder modulator is shown in Fig. 9.28.

In Fig. 9.28, α is the optical loss at the fiber joint, η is the detector responsivity, and T_M is the modulator light transmittance. R_S is the equivalent internal resistance of the microwave source, R_M is the equivalent internal resistance of the modulator, R_D is the equivalent internal resistance of the detector, and R_L is the load impedance. Assuming that $V(t)$ is the equivalent microwave voltage loaded on both ends of R_M, according to the circuit theory, the actual output equivalent microwave voltage of the signal source is

$$V_S(t) = \frac{R_M + R_S}{R_M} V_t \tag{9.26}$$

The equivalent microwave signal power actually output by the signal source is

$$P_S = \frac{V_S^2}{2(R_M + R_S)} = \frac{R_M + R_S}{2R_M^2} V_{\text{RF}}^2 \tag{9.27}$$

After the microwave signal is transmitted through the microwave photonic link, the microwave photocurrent is output, and the power shunted to the load is

$$P_L = \frac{V_L'^2}{2R_L} = \frac{R_D^2 R_L}{2(R_D + R_L)^2} I_{\text{PD}}'^2 \tag{9.28}$$

where R_L is the load resistance, V_L' is the effective voltage on the load, I_{PD}' is the detector's current.

The gain of a microwave photonic link is closely related to the optical power, the modulation efficiency of the modulator, and the responsivity of the photodetector. A higher link gain can be achieved by increasing the output power of the light source, reducing the external half-wave voltage of the modulator, and improving the responsivity of the detector.

9.5.6 Gain Flatness

The gain flatness of a microwave photonic link is a parameter used to measure the fluctuation of the gain. It is defined as the value of gain "increase" and "decrease" within a given bandwidth range and is usually expressed in decibels (dB), that is

$$\Delta G = \pm(G_{\max} - G_{\min})/2 \tag{9.29}$$

where ΔG is the flatness of the link gain, $G_{\max}$ is the maximum gain of the link, $G_{\min}$ is the minimum gain of the link.

Factors affecting the gain flatness of microwave photonic links include the gain fluctuation of microwave amplifiers in the link. The amplitude-frequency response characteristics of the electro-optic modulator, i.e., the fluctuation of the signal amplitude with frequency. The amplitude fluctuation caused by the port voltage standing wave ratio in the microwave photonic link includes microwave devices and optical devices.

9.5.7 Amplitude and Phase Error

When microwave photonic links are used for analog optical transmission, lasers, modulators, and detectors will add certain noise to the signal, causing fluctuations in the amplitude and phase of microwave photonic links [23]. In addition, changes in the external ambient temperature will also affect the loss and delay characteristics of optical devices and transmission fibers and ultimately affect the changes in the amplitude and phase of microwave photonic links.

(1) **Amplitude and phase feedback control**

Effective amplitude and phase feedback control technology is the key to high-quality microwave photonic link analog optical transmission technology. The amplitude fluctuation of the microwave photonic link is mainly caused by the optical insertion loss of the optical device, the output power fluctuation of the laser, and the change of the amplitude-frequency characteristic of the detector. The principle of microwave photon link amplitude control is shown in Fig. 9.29.

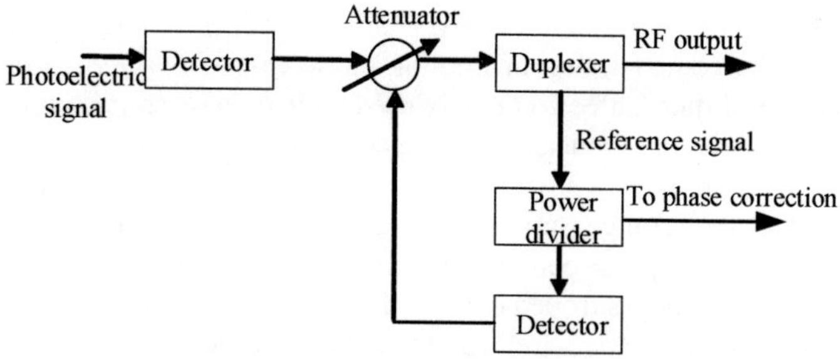

Fig. 9.29 Schematic diagram of the microwave photonic link amplitude control

The RF signal output by the photodetector contains useful and reference signals. A duplexer separates the two signals, and the reference signal is output by a power divider to detect the amplitude and phase of the two signals, respectively. The amplitude adjustment of the microwave signal is realized by a detector [24]. When the detector detects that the link's amplitude changes, the attenuation in the link is adjusted to achieve the consistency of the amplitude between the channels.

The photoelectric phase-locked loop is used to detect and control the phase of the microwave photonic link. As shown in Fig. 9.30, the phase-locked loop has two functions: one is to measure the phase of the microwave photonic link [25]. The second is to control and correct the phase of the microwave photonic link.

To calibrate the system, it is assumed that the amplitude of the reference signal source is A, and the amplitude of the reference signal after passing through the microwave photon link is B. Therefore, it is necessary to move the phase shifter first and calibrate the DC output of the mixer at different phases to form a data table. Then, adjust the phase shifter to the 45° quadrature operating point of the signal source, measure the phase of the microwave photon link, and obtain the phase fluctuation of the microwave photon link according to the DC output of the mixer and the calibrated data sheet.

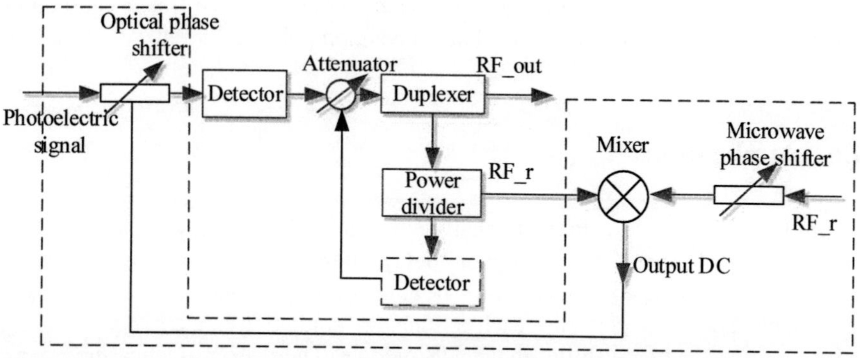

Fig. 9.30 Schematic diagram of the microwave photonic link phase control

The phase fluctuation of the transceiver link is obtained through the DC output of the mixer. The compensation of the link phase can be realized by controlling the optical phase shifter. Three commonly used optical phase shifters with different working modes exist (1) Optical fiber stretcher based on piezoelectric ceramics. The optical phase shifter has small insertion loss, fast speed, and high adjustment accuracy, but the phase compensation amount is small. (2) Fiber delay line and optical phase shifter have much phase compensation, but the insertion loss is generally above 1.5 dB. (3) Temperature-controlled optical fiber ring, the optical phase shifter has a large amount of phase compensation, and the insertion loss can be as low as below 0.1 dB, but the compensation speed of this method is not high. Therefore, you can choose according to the specific requirements of the system.

(2) **Time calibration**

Time calibration consists of time calibration and time system calibration. Based on the microwave photonic receiver, the data acquisition and time measurement are completed at the central station, and the positioning measurement principle of the time difference receiver is to measure the time difference between the arrival of the target signal at different receiving sub-stations, so the system needs to calibrate the processing time of different receivers. The calibration is completed by internal calibration, and the central station sends a time calibration pulse signal, which is sent to the signal input port of each receiving sub-station through the photonic link, and then compared and measured at the central station after passing through the receiver link [1]. In order to ensure consistent paths, wavelength division multiplexing technology is used to make microwave photon signals uplink and downlink through the same optical fiber link. The operation principle of time measurement calibration is shown in Fig. 9.31.

Internal calibration can only calibrate the processing time of multi-substation receivers but cannot accurately calibrate the system time positioning. Positioning accuracy involves many system parameter settings and stability [1], such as baseline design, receiver parameter consistency, threshold detection design, time measurement, positioning equations, etc. This kind of calibration will use system time calibration, i.e., the system will receive and locate the signal to realize system calibration when transmitting a pulse signal at a known position point.

9.6 Applications

Figure 9.32 shows a broadband optically controlled phased array antenna experimental system [26], which uses optical delay lines to provide true time delay to achieve wide instantaneous bandwidth. The antenna array is composed of 24 columns of broadband radiation elements, each column has 4 radiation elements controlled by a 5-bit microwave phase shifter in the T/R module. There are 8 antenna sub-arrays total, and a 5-bit fiber delay line controls each antenna sub-array. The fiber delay line provides a long large delay for the antenna sub-array, while the microwave phase

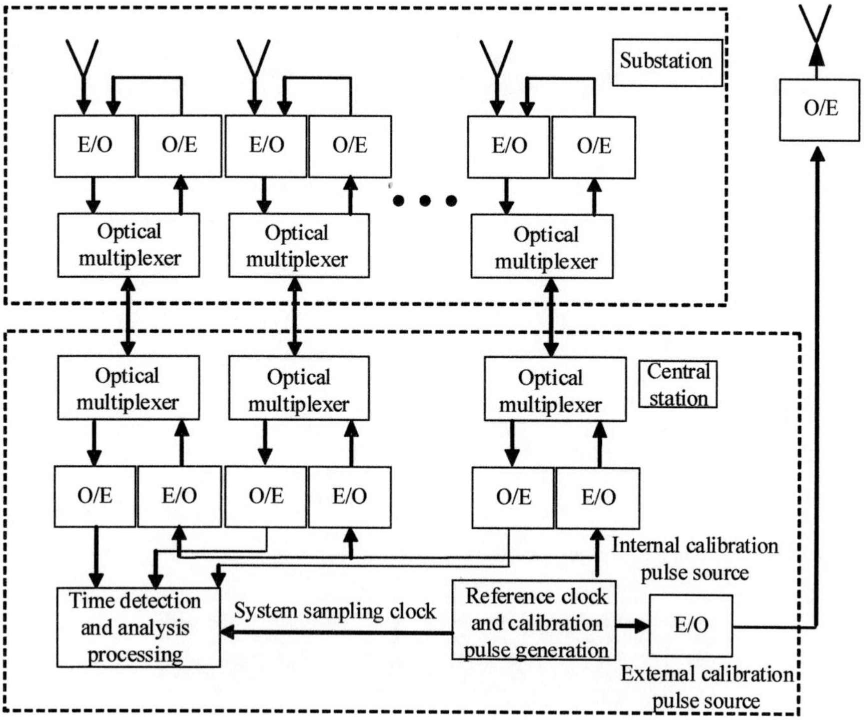

Fig. 9.31 Operation principle diagram of the time measurement calibration

shifter in the T/R module provides phase shifting for the antenna elements, a small fine time delay.

The principle of broadband optically controlled phased array antenna experiment system is shown in Fig. 9.32, and the photo of the real object is shown in Fig. 9.33.

Figure 9.32 shows the time delay network and its formed wavefront, where each element represents a column (4) of array elements, and a T/R module controls each array unit. Each T/R module provides a 5-bit electrical delay, which is used for fine-tuning (0.01–0.5 ns) delay. Each 5-bit optical delay time unit controls 3 T/R modules, and these 3 optical delay time units form a sub-array to control 12 antenna elements. The optical delay unit provides a time delay ranging from 0.25 to 7.75 ns. 32 kinds of optical delays are available for control, which are used to form a large bandwidth and ±45° scan.

Figure 9.34 shows the structure of a 5-bit fiber delay line component based on an RF switch, which provides a large delay with a step size ranging from 0.25 to 7.75 ns for the subarray antenna. The key components inside the fiber delay line assembly are a single-pole 8-throw RF switch, 8 semiconductor lasers, a 4 × 8 fiber coupler, a single-pole 4-throw RF switch, and a post-amplifier.

The microwave signal modulates one of the eight lasers through a single-pole 8-throw RF switch. The laser converts the microwave signal into an optical signal,

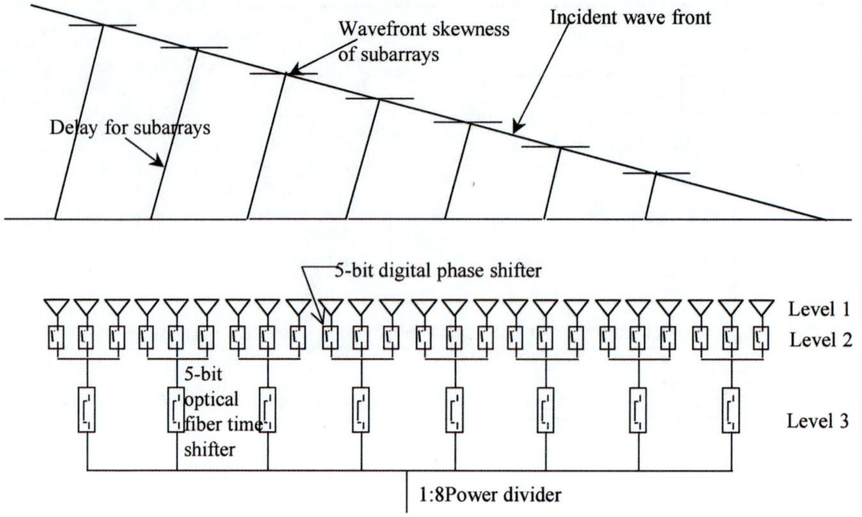

Fig. 9.32 Diagram of the time delay network

Fig. 9.33 Photograph of the broadband optically controlled phased array antenna experiment system

and the optical signal is coupled into a 4 × 8 fiber coupler. After being split by the coupler, the optical signal is incident on the detector. By switching on one of the four detectors through a single-pole four-throw RF switch, an RF signal with one of 32 preset delays can be selected. Then, the RF signal is post-amplified and divided into three paths, respectively fed to the 5-bit microwave phase shifter.

The absolute phase test results of a channel of the fiber delay line are shown in Fig. 9.35. It can be seen from Fig. 9.35 that the phase of the fiber delay line is linear, and the frequency distortion is small. The test results of the delay time of the fiber delay line are shown in Fig. 9.36, and the delay deviation is less than 0.03 ns.

The test results of the fiber delay line are listed in Table 9.8.

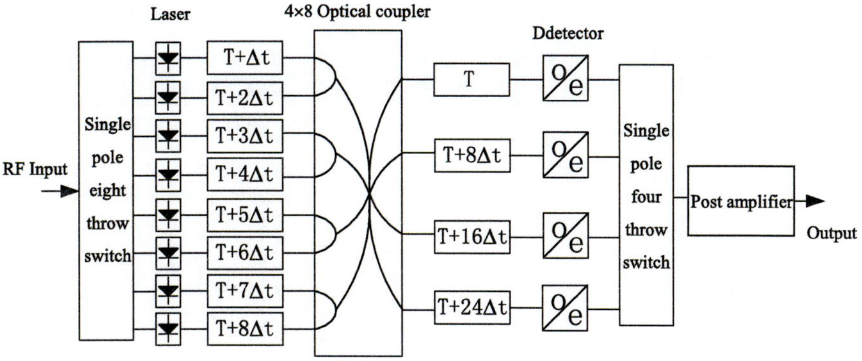

Fig. 9.34 5-bit fiber delay line based on RF switch

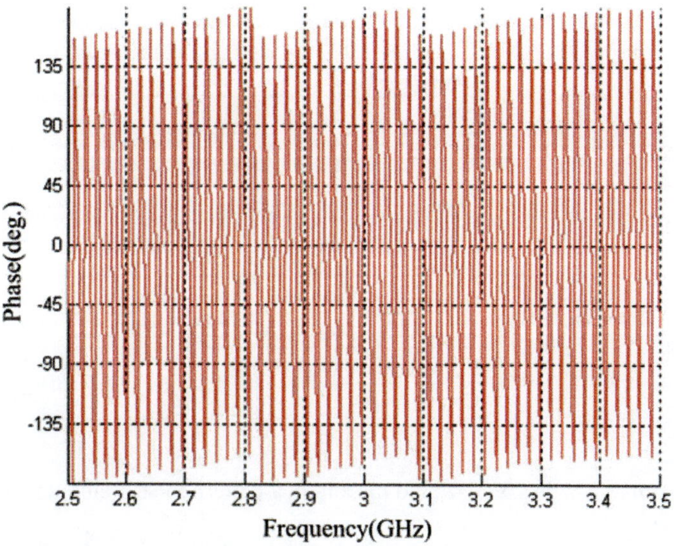

Fig. 9.35 Absolute phase measurement of fiber delay lines

Figures 9.37 and 9.38 show the pattern test results of the broadband optically controlled phased array antenna experimental system. Figure 9.37 shows the test results of the antenna pattern without scanning the antenna beam. Figure 9.37a shows that the delay line compensates for the reference delay, and the phase shifter compensates for the reference phase. Figure 9.37b is the delay line zero state, and the phase shifter compensates for the reference phase. Figure 9.38 shows the measured result of the antenna pattern when the beam is scanned to 45°. Figure 9.38a shows that the delay line compensates for the antenna aperture transit delay, and the phase shifter compensates for the 45° beam scanning phase. Figure 9.38b is the delay line zero state, and the phase shifter compensates for the 45° beam scanning phase. The test

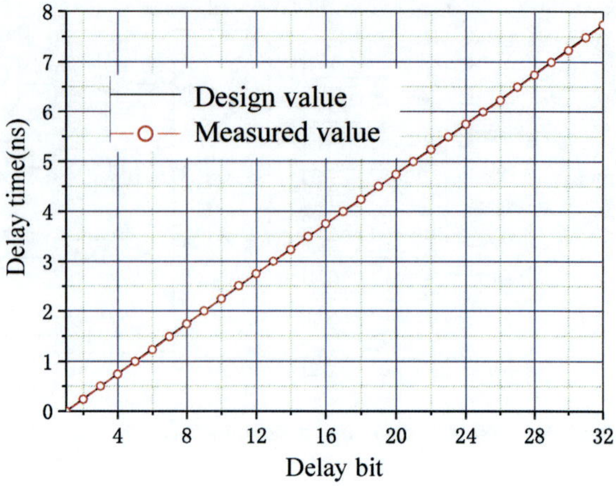

Fig. 9.36 Delay time measurement of the fiber delay line

Table 9.8 Fiber delay line test results

Frequency (GHz)	2.5–3.5
Delay range (ns)	0.25–7.75
Delay step (ns)	0.25
Delay accuracy (ns)	≤0.03
Switching time (μs)	≤10
Input 1 dB compression point (dBm)	≥+10
Amplitude flatness (dB)	≤ ±0.5
Input/output port voltage standing wave ratio	≤2
Input/output connector	SMA
Control level	TTL

frequency range is 2.5–3.5 GHz, and the antenna pattern measurement is performed at 100 MHz.

It can be seen from Figs. 9.37 and 9.38 that in the range of 2.5–3.5 GHz, when the antenna beam is not scanning, there is no true time delay line antenna, and the beam pointing does not shift. When the antenna beam scans, controlling the true time delay line can ensure that the antenna beam points are consistent. If there is no true time delay line, the antenna beam pointing will shift. When the antenna operating frequency bandwidth is wider or the antenna beam scanning angle is larger, the antenna beam pointing angle will be shifted more.

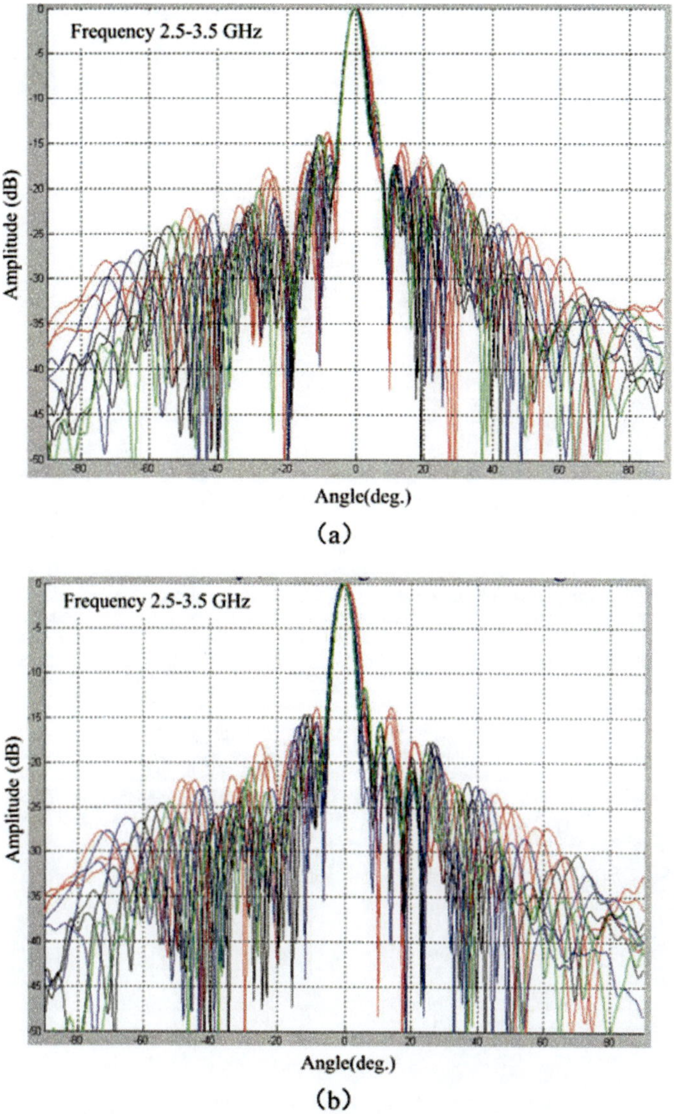

Fig. 9.37 Scanning 0° pattern, **a** delay line and phase shifter, **b** phase shifter

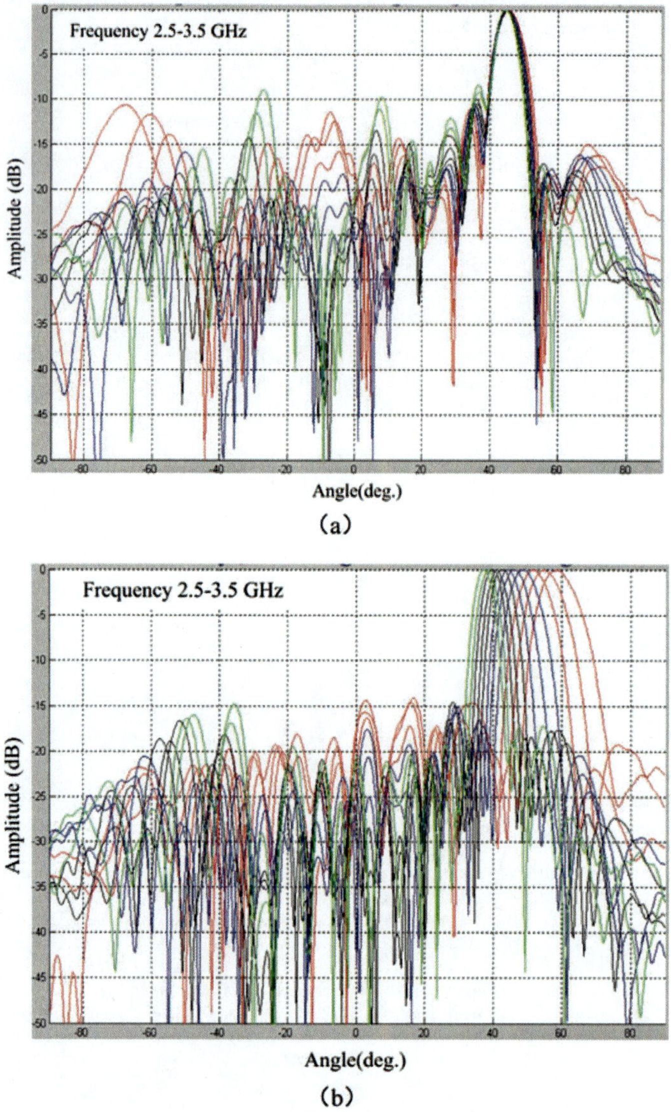

Fig. 9.38 Scanning 45° pattern, **a** delay line and phase shifter, **b** phase shifter

References

1. Lu J (2019) Design techniques of synthetic aperture radar. Wiley, Hoboken, NJ
2. Lee JJ, Loo RY (1995) Photonic wideband array antennas. IEEE Trans Antennas Propag 43(9):966–982
3. Lee JJ, Yen HW, Loo RY et al (1992) System applications of photonics to phased arrays. Proc SPIE Int Soc Opt Eng 1703:491–501

4. Vanblaricum ML (1994) Photonic systems for antenna applications. IEEE Antennas Propag Mag 36(5):30–38
5. Fang L, Li P et al (2013) An idea for future digital array radar based on microwave optoelectronics. Radar Sci Technol 11(6):583–586
6. Frankel MY, Matthews PJ (1997) Fiber-optic true time steering of an ultrawide-band receive array. IEEE Trans Microw Theory Tech 45(8):1522–1526
7. Roman JE, Nichols LT, Wiliams KJ et al (1998) Fiber-optic remoting of an ultrahigh dynamic range radar. IEEE Trans Microw Theory Tech 46(12):2317–2323
8. Li R, Li W, Ding M et al (2017) Demonstration of a microwave photonic synthetic aperture radar based on photonic-assisted signal generation and stretch processing. Opt Exp 25(13):14334–14340
9. Dolfi D, Joffre P, Antoine J et al (1996) Experimental demonstration of a phased array antenna optically controlled with phase and time delays. Appl Opt 35(26):5293–5300
10. Etem Y, Lewis MF (2022) Design and performance of an optically controlled phased array antenna. IN: Meeting on microwave photonics, Mwp 96 Technical Digest. IEEE, pp 209–212
11. Benjamin R, Seeds AJ (1993) Optical beam forming techniques for phased array antennas. In: IEE proceedings H: microwaves, antennas and propagation, vol 139, no 6, pp 526–534
12. Zhang F, Guo Q, Wang Z et al (2017) Photonics based broadband radar for high-resolution and real-time inverse synthetic aperture imaging. Opt Exp 25(14):16274–16281
13. Wu T et al (1993) Optical beam forming techniques for phased array antennas. Int J Infrared Millim Waves 4(6):1191–1199
14. Frankel MY, Esman RD (1995) True time-delay fiber-optic control of an ultrawideband array transmitter/receiver with multibeam capability. IEEE Trans Microw Theory Tech 43(9):2387–2394
15. Matthews PJ, Liu PL, Medberry JB et al (1999) Demonstration of a wide-band fiber-optic nulling system for array antennas. IEEE Trans Microw Theory Tech 47(7):1327–1331
16. Xu L, Taylor R, Forrest SR (1995) The use of optically coherent detection techniques for true-time delay phased array and systems. Erratum 13:1663–1677
17. Monsay EH, Baldwin KC, Caccuitto MJ (1994) Photonic true time delay for high-frequency phased array systems. IEEE Photonics Technol Lett 6(1):118–120
18. Goutzoulis AP, Davies DK, Zomp JM et al (1994) Development and demonstration of the Westinghouse hardware-compressive true time delay (TTD) fiber optic system Oe/lase. Int Soc Opt Photonics 2155:275–286
19. Hu X (2010) Optical devices and their applications. Publishing House of Electronics Industry, Beijing, pp 56–58
20. Ridgway RW, Conway JA, Dohrman CL (2013) Microwave photonics: recent programs at DARPA. In: IEEE international topical meeting on microwave photonics. IEEE, Alexandria, VA, pp 286–289
21. Le C, Yi D, Dongning S, Damin Z, Weilin X, Weisheng H (2012) Influence and suppression of coherent rayleigh noise in fiber-optic-based phase-stabilized microwave-frequency transmission system. Acta Opt Sinica 32(5):59–64
22. Ping Z (2009) A study on errors of multi-channel amplititute and phase calibration for active phased array antenna. Modern Radar 31(10):14–16
23. Li P, Shi R, Chen M et al (2013) Down conversion and linearization of X-and K-band analog photonic links using digital post-compensation. In: Optical fiber communication conference and exposition and the national fiber optic engineers conference. IEEE, Anaheim, CA, pp 1–6
24. Zhang Y, Fang L, Liu Y et al (2017) A time-difference receiver for multipoint location based on microwave optoelectronic technology. Radar Sci Technol 15(3):291–295
25. Sparks RA (1995) The applications of photonics to radar: a system designer's perspective. In: International microwave & optoelectronics conference. IEEE, pp 717–724
26. Jin M, Guan W, Guo J, Qi M, Guo Y (2006) Experimental system of a wideband optically controlled phased array antenna. Acta Electron Sinica 34(6):1127–1129